Probability and Mathematical Statistics (Continued)

RAGHAVARAO · Constructions and Combinatorial Problems in Design of Experiments

RANDLES and WOLFE · Introduction to the Theory of Nonparametric Statistics

RAO · Linear Statistical Inference and Its Applications, *Second Edition*

ROHATGI · An Introduction to Probability Theory and Mathematical Statistics

SCHEFFE · The Analysis of Variance

SEBER · Linear Regression Analysis

SERFLING · Approximation Theorems of Mathematical Statistics

WILKS · Mathematical Statistics

WILLIAMS · Diffusions, Markov Processes, and Martingales, Volume I: Foundations

ZACKS · The Theory of Statistical Inference

Applied Probability and Statistics

ANDERSON, AUQUIER, HAUCK, OAKES, VANDAELE, and WEISBERG · Statistical Methods in Comparative Studies

BAILEY · The Elements of Stochastic Processes with Applications to the Nature Sciences

BAILEY · Mathematics, Statistics and Systems for Health

BARNETT and LEWIS · Outliers in Statistical Data

BARTHOLOMEW · Stochastic Models for Social Processes, *Second Edition*

BARTHOLOMEW and FORBES · Statistical Techniques for Manpower Planning

BECK and ARNOLD · Parameter Estimation in Engineering and Science

BELSLEY, KUH, and WELSCH · Regression Diagnostics: Identifying Influential Data and Sources of Collinearity

BENNETT and FRANKLIN · Statistical Analysis in Chemistry and the Chemical Industry

BHAT · Elements of Applied Stochastic Processes

BLOOMFIELD · Fourier Analysis of Time Series: An Introduction

BOX · R. A. Fisher, The Life of a Scientist

BOX and DRAPER · Evolutionary Operation: A Statistical Method for Process Improvement

BOX, HUNTER, and HUNTER · Statistics for Experimenters: An Introduction to Design, Data Analysis, and Model Building

BROWN and HOLLANDER · Statistics: A Biomedical Introduction

BROWNLEE · Statistical Theory and Methodology in Science and Engineering, *Second Edition*

BURY · Statistical Models in Applied Science

CHAMBERS · Computational Methods for Data Analysis

CHATTERJEE and PRICE · Regression Analysis by Example

CHERNOFF and MOSES · Elementary Decision Theory

£14·25

Helen Milligan
1984

WILEY SERIES IN PROBABILITY
AND MATHEMATICAL STATISTICS

ESTABLISHED BY WALTER A. SHEWHART AND SAMUEL S. WILKS
Editors

Ralph A. Bradley   David G. Kendall
J. Stuart Hunter   Geoffrey S. Watson

Probability and Mathematical Statistics
 ANDERSON · The Statistical Analysis of Times Series
 ANDERSON · An Introduction to Multivariate Statistical
 Analysis
 ARAUJO and GINE · The Central Limit Theorem for Real
 and Banach Valued Random Variables
 BARLOW BARTHOLOMEW, BREMNER, and BRUNK
 · Statistical Inference Under Order Restrictions
 BARNETT · Comparative Statistical Inference
 BHATTACHARYYA and JOHNSON · Statistical Concepts
 and Methods
 BILLINGSLEY · Probability and Measure
 CASSEL, SARNDAL and WRETMAN · Foundations of In-
 ference in Survey Sampling
 DE FINETTI · Theory of Probability, Volume I
 DE FINETTI · Theory of Probability, Volume II
 DOOB · Stochastic Processes
 FELLE · An Introduction to Probability Theory and Its Appli-
 cations, Volume I, *Third Edition*, Revised
 FELLER · An Introduction to Probability Theory and Its
 Applications, Volume II, *Second Edition*
 FISZ · Probability Theory and Mathematical Statistics, *Third
 Edition*
 FULLER · Introduction to Statistical Time Series
 HANNAN · Multiple Time Series
 HANSEN, HURWITZ, and MADOW · Sample Survey
 Methods and Theory, Volumes I and II
 HARDING and KENDALL · Stochastic Geometry
 HOEL · Introduction to Mathematical Statistics, *Fourth Edi-
 tion*
 IOSIFESCU · Finite Markov Processes and Applications
 ISAACSON and MADSEN · Markov Chains
 KAGAN, LINNIK, and RAO · Characterization Problems in
 Mathematical Statistics
 KENDALL and HARDING · Stochastic Analysis
 LAHA and ROHATGI · Probability Theory
 LARSON · Introduction to Probability Theory and Statistical
 Inference, *Second Edition*
 LARSON · Introduction to the Theory of Statistics
 LEHMANN · Testing Statistical Hypotheses
 MATHERON · Random Sets and Integral Geometry
 MATTHES, KERSTAN, and MECKE · Infinitely Divisible
 Point Processes
 PARZEN · Modern Probability Theory and Its Applications
 PURI and SEN · Nonparametric Methods in Multivariate
 Analysis

Practical
Nonparametric
Statistics

Practical Nonparametric Statistics

2ed

W. J. CONOVER
Texas Tech University

John Wiley & Sons
New York   Chichester   Brisbane   Toronto

Library of Congress Cataloging in Publication Data:

Conover, W J
  Practical nonparametric statistics.

  (Wiley series in probability and mathematical statistics)
  Bibliography: p.
  Includes index.
  1. Nonparametric statistics.   I. Title.
QA278.8.C65   1980   519.5'3   80-301

Printed in the United States of America

10 9 8 7 6 5 4 3 2

Preface

I have used the first edition of this book as a text many times. Others who have used it, as students, as teachers, and for reference, have conveyed their thoughts to me. This has helped me to obtain a perspective I didn't have when the first edition appeared. Because the responses were overwhelmingly positive, the essential format and substance of the book are unchanged. Suggestions for changes were reviewed by many people; the ones that survived their critical appraisal were incorporated into this edition.

Several of the changes include a reorganization of the problems into "theoretical" and "applied" sections, the addition of review problems (without answers) at the end of chapters, more discussion of practical matters such as multiple comparison procedures and regression, and discussions of the relationships among the various tests in this book as well as with standard parametric procedures.

In addition, the usual updating was performed. Material on ties, not available when the first edition was published, is now included. More emphasis is placed on using scores based on ranks because of an abundance of recent results showing the value of these methods. References to recent results in the literature are added. Robust methods for regression and experimental designs are introduced, and so forth.

To keep the size of the book within reason, some trimming was required. Since I could find no topic that my advisors agreed could be left out, I used my own judgment regarding deletions. Some topics were removed from prime space and included as problems. Other topics were reduced to a reference in the appropriate section.

I am grateful to the many people who helped me on this revision by offering advice, asking questions, obtaining useful research results that they communicated to me, and encouraging me to keep the book current through revision. However, I am most grateful to the many people who helped make the first edition a success; without them, there would be no second edition.

W. J. Conover

A Note to the Instructor

This book is organized to be as flexible as possible for use as a text. For short courses, such as two semester hours, I suggest using Chapter 3, Chapter 4 through Section 4.5, Chapter 5 through Section 5.8, and the first two sections of Chapter 6, omitting all but a superficial discussion of the theory in those sections.

For a three-semester-hour course, I usually start at the beginning of the book and cover the entire text. This requires that the students read much of the material on their own, without parallel coverage in class lectures. Most students find the text easy to read, so they do not object to this form of instruction. However, coverage of the entire book in one semester is not easy and requires a somewhat shallow treatment of many of the topics. A more in-depth study of some sections may be accomplished by omitting Sections 4.5 to 4.7, 5.9 to 5.12, 6.3, and 6.4.

For longer courses, up to six semester hours, the material in the text may be supplemented by the literature. There are a sufficient number of references supplied throughout the text to enable students to select individual topics to research. Seminars by the students are valuable experiences, as are technical reports on the topics researched. Instructors often forget how difficult that first technical seminar was to present or that first technical report was to write. These research experiences will usually expose unsolved problems and unanswered questions that may be used as topics for theses and dissertations.

The Exercises are similar in format to the Examples in the text. They are designed to give readers an opportunity to use the methodology on some fresh data. The Problems are intended to reinforce the learning of the theory behind the tests. The Review Problems provide students with some mild challenges; the result of this is that students will be able to use these methods properly when the occasion arises.

Several statistical packages are available for use with this course. Two that I have found especially useful are "TUSTAT-II" by Dr. Young Ko at the University of Nevada, Reno, and "MINITAB" by Drs. Thomas Ryan and Barbara Ryan at Penn State and Brian Joiner at the University of Wisconsin. Slight differences in these methods from those in this text may confuse some students but, overall, the use of the computer is beneficial in a course such as this.

W. J. C.

Preface to
First Edition

This book is intended as an introductory textbook and as a reference book for applied research workers. As an introductory text, it requires only algebra as a prerequisite. While it is expected that most courses employing this as a text will require a previous course in elementary statistics, such a requirement may be deleted if one is willing to ignore the occasional references to "usual parametric counterparts." In fact, because of the simple nature and general applicability of nonparametric statistics, it may be more practical to introduce the student first to nonparametric statistics, and then to the usual parametric statistics as a special area. Although this book could serve as a text for such a course, I have used it (in preliminary form) as a text for several years in a graduate-level introductory course for nonstatisticians who have had a previous course in statistics. The entire text may be covered in three semester hours by covering approximately one section, including the problems, in each lecture. This allows ample time for discussion of the more interesting problems, and for examinations. For a short course on probability, Chapter 1 may be studied by itself. Or, for a short course on nonparametric statistics, the methods and examples of Chapters 3 to 7 may be studied without the accompanying theory, and without Chapters 1 and 2.

To use this book as a "book of recipes," the chart seen on the back endpaper may guide in the selection of an appropriate test. Each method is described in a self-contained, clear-cut format. Examples using actual numbers are given to assist in clearing up any ambiguities in the written description. Applications are drawn from the fields of psychology, biology, statistics, engineering business, education, economics, medicine, agriculture and jurisprudence.

I am grateful to the many people who assisted in this work. More than 100 students read the text in various forms and contributed to its clarity. About a dozen professional men read the manuscript and contributed to its validity. Editorial and financial assistance were provided by the publisher. Kansas State University, its Department of Statistics and Computer Science, and its Agricultural Experiment Station supported much of the research that is reported in various places in the book. I was also aided by the National Science Foundation Grant GP-7667 and, in the final stages of the manuscript, by the National Institutes of Health Research Career Development Award, Grant Number 1-KO4-GM42351-01.

Many people contributed to the improvement of the manuscript. I would appreciate personal communications concerning strong points and weak areas, since these may affect the form of any possible later editions.

W. J. Conover

Contents

Practical
Nonparametric
Statistics

Introduction

One of the dictionary definitions of the word "science" is given as "truth ascertained by observation, experiment, and induction." A vast amount of time, money, and energy is being spent by society today in the pursuit of science. This pursuit is quite often frustrating because, as any scientist knows, the processes of observation, experiment, and induction do not always lay bare the "truth." One experiment, with one set of observations, may lead two scientists to two different conclusions.

For example, a scientist places a rat into a pen with two doors, both closed. One door is painted red and the other blue. The rat is then subjected to 20 minutes of music of the type popular with today's teenagers. After this experience, both doors are opened and the rat runs out of the pen. The scientist notes which color door the rat chose. This experiment is repeated 10 times, each time using a different rat.

At the end of the composite experiment, the experimenter notes that the rats chose the red door 7 out of 10 times and concludes the "truth" as being that the treatment used causes rats to prefer the red door to the blue door. However, a colleague overhears this conclusion and jokingly tests the scientist: "If I tossed a coin 10 times getting seven heads and, before each toss, I whistled 'Yankee Doodle,' would you conclude that my whistling caused the coin to prefer heads?" Seeing the analogy between a rat choosing one of two doors and a coin landing on one of its two sides, the scientist realizes the error and decides that the outcome of the experiment could easily have been the result of chance.

Later the scientist conducts a second experiment. He injects a certain drug into the bloodstream of each of 10 rats. Five minutes later he examines the rats and finds that 7 are dead, and the other 3 are apparently healthy. However,

1

since only 7 are dead, he recalls the previous experiment and concludes that such a result could easily have occurred by chance and therefore there is no proof that the drug injections are dangerous.

His colleague again interrupts, saying, "With your first experiment each rat had a 50-50 chance of choosing the red door, without the music, and therefore we can compare that experiment to tossing a coin. In this experiment, the chances of a rat dying within 5 minutes are quite slim indeed, if the drug has no effect. Since your experiment resulted in 7 of these rare events out of only 10 possibilities, it seems safe to conclude that the drug injections caused the deaths."

And so goes research. It soon becomes apparent to most scientists that the ideal way of expressing results of experiments, such as the preceding, is to be able to say something like, "Without the treatment I administered, experimental results as extreme as the ones I obtained would occur only about 3 times in 1000. Therefore I conclude that my treatment has a definite effect." In this way every scientist who reads of this experiment knows just how much subjectivity, or opinion, entered into the stated conclusion.

The purpose of that field of science known as "statistics" is to provide the means for measuring the amount of subjectivity that goes into the scientists' conclusions and thus to separate "science" from "opinion." This is accomplished by setting up a theoretical "model" for the experiment, such as the model called "tossing a coin," which was set up for the first experiment discussed. Laws of probability are applied to this model in order to determine what the "chances" (probabilities) are for the various possible outcomes of the experiment under the assumption that chance alone, and not music or drug injections, determines the outcome of the experiment. Then the experimenter has an objective basis for deciding whether the results were a result of the treatments that were applied, or whether the same results could have easily occurred by chance alone with no treatment.

Although it is sometimes difficult to describe an appropriate theoretical model for the experiment, the real difficulty often comes after the model has been defined, in the form of finding the probabilities associated with the model. Many reasonable models have been invented for which no probability solutions have ever been found. Thus statisticians have often changed the model slightly in order to be able to solve for the desired probabilities in the hope that the change in the model was slight enough so that the changed model was still fairly realistic. Then they are able to obtain exact solutions for these "approximate problems." This body of statistics is sometimes called "parametric statistics," and embodies such well-known tests as the "t test," the "F test," and others.

In the late 1930s a different approach to the problem of finding probabilities began to gather some momentum. This approach involved making few, if any, changes in the model, and using simple and unsophisticated methods to find the desired probabilities, or at least a good approximation to those probabilities. Thus approximate solutions to exact problems were found, as opposed to the exact solution to approximate problems furnished by parametric statistics. This new package of statistical procedures became known as "nonparametric statistics."

Besides the advantage of using a simpler model, nonparametric statistical methods often involve less computational work and therefore are easier and quicker to apply than other statistical methods. A third advantage of nonparametric statistical techniques is that much of the theory behind the nonparametric methods may be developed rigorously, using no mathematics beyond high school algebra. A scientist who understands the theory behind the statistical method is less apt to use that method in a situation where such usage would be incorrect and is better able to develop his or her own statistical methods if the model is one that has not yet been considered by statisticians.

The parts of nonparametric statistics that require the use of more advanced mathematics will be presented without deriving them but, whenever convenient, there will be a reference to a source where the proof may be found.

This formulation of parametric statistics versus nonparametric statistics is merely an attempt to give a rough idea concerning the subject of this book. A more precise distinction between the two branches of statistics will be given in Chapter 2, where the philosophy of scientific experimentation is discussed in greater detail. In order to present examples and illustrations in Chapter 2, a preliminary knowledge of some elementary aspects of probability is needed. This is the concern of Chapter 1.

From Chapter 3 on there is a heavy reliance on the concepts introduced in Chapters 1 and 2. These later chapters present various nonparametric procedures, organized according to the type of model that is being analyzed rather than to the type of experiment being conducted. For convenience to the experimenter who wants to examine the body of techniques that may be used in analysis, a cross-referencing table is presented inside the back cover, listing the techniques given in the book according to the type of problem they are intended to solve.

This book attempts to present nonparametric techniques that are already popular among experimenters in a clearer way than is now available in other books and journals. Also, it presents statistical methods that are not widely known because of their recent development. Some nonparametric methods presented in this book have not yet appeared in the literature but are included because it is felt that they will be useful to experimenters.

A word about the numbering of examples, equations, and figures would not be out of place at this time. Example 4.2.3 refers to the third example in Section 4.2. When referring to an example within the same section, only the last number is used. For instance, within Section 4.2, Example 4.2.3 is referred to simply as Example 3. The same is true for equations, figures, and problems. No such economy is used with regard to section numbers, so that Section 4.2 is always called Section 4.2, even within Chapter 4.

For those who wish to obtain more information about nonparametric procedures, many references are included at the end of each appropriate section.

Most of these references are recent; and earlier, sometimes more important, papers are usually not mentioned. This is because the references given generally refer in turn to the earlier papers on that topic, so there was no need to repeat them here. The bibliography by Savage (1962) is quite useful for obtaining additional references on each topic.

CHAPTER 1

Probability Theory

PRELIMINARY REMARKS

One of the attractive qualities of nonparametric statistical methods is that it is not necessary to be an expert in probability theory to understand the theory behind the methods. With a few easily learned, elementary concepts, the basic fundamentals underlying most nonparametric statistical methods become quite accessible. This chapter introduces those basic concepts. All that is required is patience, confidence, and a good understanding of high school algebra.

This book is arranged so that readers can go directly to the statistical procedure they want to use and follow the step-by-step instructions from beginning to end. However, they will not necessarily understand what they are doing or why they are doing it. Such lack of understanding often leads to mishandled data and misstated conclusions. By spending a little time in Chapters 1 and 2, readers should understand thoroughly the nonparametric procedure being used and may even be able to adapt it slightly so it will apply better to the particular set of data being analyzed.

The recommended procedure for studying each section is to read the text, pencil through the examples, and then work the problems at the end of the section. This will prepare readers for the next section and will develop the patience and confidence first mentioned.

1.1. COUNTING

The process of computing probabilities often depends on being able to count, in the usual sense of counting, "1, 2, 3," and so on. The usual way of counting becomes quite tedious in some complicated situations, so some sophisticated methods of counting are developed in this section to handle those complicated situations.

When we speak of tossing a coin, we will consider only two possible outcomes: either a head (H) appears, or a tail (T) appears. If a coin is tossed once there are two possible outcomes: H or T. If a coin is tossed twice there are $2^2 = 4$ possible outcomes: HH, HT, TH, TT, where HT means a head occurs on the first toss and a tail on the second. Each time we consider one additional toss of the coin, the number of possible outcomes is doubled, since the last toss may result in either of two outcomes. Thus, if a coin is tossed n times there are 2^n possible outcomes.

Generalizing this discussion somewhat, we may refer to the tossing of a coin as one example of an *experiment*. Whether the coin is tossed once, twice or, in general, n times, the procedure may be considered to be an experiment. Since tossing a coin three times may be considered to be an experiment and is a composite of three separate experiments where the coin is tossed only once each time, we may refer to the shorter experiments as *trials* and the collection of trials as "the experiment."

Few scientists seriously consider coin tossing as an experiment worthy of merit by itself. The value of coin tossing is that it serves as a prototype for many different models in many different situations. If an unbiased coin is being considered, one in which each face is equally likely to result, the experiment is not unlike experiments involving rats that have two choices of doors, consumers choosing between two products, educators determining which of two teaching methods is more effective, market analysts deciding whether the market tends to be higher or lower on Mondays, and many other situations.

If we allow the coin to be biased, where one face is more likely to turn up than the other, a much broader class of experiments is included under the same model. Examples include experiments where a drug is injected into the bloodstream of rats to see if the drug is lethal, a new cure is tested on sick patients, a consumer is given several choices of a product and asked to choose one where only one of the products is manufactured by Company X, and other situations. In each case there are two outcomes of interest, such as "life" versus "death," "cure" versus "no cure," "our brand" versus "other brands," and the two outcomes might not be equally likely to occur.

Throughout this chapter and the next, models involving coin tossing, dice rolling, drawing chips from a jar, placing balls into boxes, and so on, will be discussed as if they were experiments worthy of merit, while actually the value of these models lies mainly in the fact that they serve as useful and simple prototypes of many more complicated models arising from experimentation in diverse areas such as electron physics, psychology, sociology, education, biology, economics, chemistry, etc. An excellent study of the diversity of such models is given by Feller (1968). Some justification for the study of these models will be presented in this chapter, but for the most part the justification will be deferred until later chapters where the various nonparametric procedures are introduced.

Thus we may refer to coin tossing as an experiment and each individual toss of the coin as a trial. The possible outcomes of one trial, several trials, or the entire experiment will be called *events*. The coin tossing experiment just described consists of n trials, where each trial may result in either the event H or the event T. A combination of events may itself be an event. Therefore it is permissible to consider each of the 2^n possible outcomes of the experiment as an event. Examples of other events would include the event "at least one head," the event "a tail on the fourth toss," and the event "at least twice as many heads as tails."

Further generalization leads to the following rule:

RULE 1. If an experiment consists of n trials where each trial may result in one of k possible outcomes, there are k^n possible outcomes of the entire experiment.

Example 1. Suppose an experiment is composed of seven trials, where each trial consists of throwing a ball into one of three boxes. The first throw may result in one of three different outcomes. Thus there are $3^2 = 9$ outcomes associated with the first two trials combined. This reasoning extends to the seven throws comprising the experiment, resulting in $3^7 = 2187$ different outcomes of the experiment.

Now consider a box containing n plastic chips numbered 1 to n. One chip is selected from the box and placed on the table so the number is showing. This chip could be any of the n chips that were in the box, so we say there are n ways of selecting the first chip. A second chip is then selected from the chips remaining in the box and placed next to the first chip, so that its number is showing also. Since there were $n-1$ chips remaining in the box, the second chip could be selected in any one of $n-1$ different ways. Since each of the n ways of drawing the first chip has associated with it $n-1$ ways of drawing a second chip, there are all together $n(n-1)$ ways of drawing first one chip and then a second chip. A third chip can be drawn in $n-2$ different ways and placed on the table next to the second chip. Now there are $n(n-1)(n-2)$ ways of drawing three chips in sequence. If the process is continued until the last chip is drawn (there is only 1 way of drawing the last chip, since only 1 chip is left in the box) we can see that there are

$$(1) \qquad n(n-1)(n-2)\cdots(3)(2)(1) = n!$$

(read "n factorial") ways of drawing n numbered chips out of a box, or $n!$ ways of arranging any n distinguishable objects into a row. Note that for convenience we will define $0! = 1$, in accordance with conventional usage.

RULE 2. There are $n!$ ways of arranging n distinguishable objects into a row.

Example 2. Consider the number of ways of arranging the letters A, B, and C in a row. The first letter can be any of the three letters, the second letter can be chosen two different ways once the first letter is selected, and the remaining letter becomes the final letter selected, for a total of $(3)(2)(1) = 6$ different arrangements. The six possible arrangements are *ABC*, *ACB*, *BAC*, *BCA*, *CAB*, and *CBA*.

Example 3. Suppose that in a horse race there are eight horses. If you correctly predict which horse will win the race and which horse will come in second and wager to that effect, you are said to "win the exacta." Suppose you want to be sure to win the exacta. That means you need to purchase $(8)(7) = 56$ betting tickets, one for each of the 56 possible ways the first and second places might result. The complete race results, for all eight positions at the finish line, could occur in any one of $8! = 40{,}320$ different ways.

If the n objects are distinguishable one from another, then each of the $n!$ arrangements is unique. But suppose two of the objects are identical. Then for each arrangement of the n objects, there is a second arrangement that is indistinguishable from the first—the arrangement in which $n - 2$ of the objects are in the same position as in the first arrangement, but the two identical objects are interchanged. Each of the $n!$ arrangements may be paired in this manner with another identical arrangement. The number of different arrangements is thus $n!/2$, or $n!/2!$.

Suppose three of the objects are identical, and $n - 3$ are distinguishable from each other. If we divide the $n!$ arrangements into groups of identical arrangements, we find there are $3!$ arrangements in each group. This is because the three identical objects may be placed $3!$ different but indistinguishable ways into their three positions, using Rule 2. Then the number of different arrangements, equal to the number of groups of identical arrangements, is $n!/3!$. If exactly n_1 objects are identical, the $n!$ arrangements may be divided into groups of identical arrangements, each group being of size $n_1!$. If there are n_1 identical objects of type 1, and n_2 identical objects of a different type 2, then for each arrangement of the objects of type 1 there are $n_2!$ identical arrangements of type 2. So there are, in all, $n_1! \, n_2!$ arrangements in each group of identical arrangements. Therefore, the number of groups is $n!/(n_1! \, n_2!)$. This leads to another counting rule:

RULE 3. If a group of n objects is composed of k identical objects of one kind and the remaining $(n - k)$ objects are identical objects of a second kind, the number of distinguishable arrangements of the n objects into a row, denoted by $\binom{n}{k}$, is given by

$$(2) \qquad \binom{n}{k} = \frac{n!}{k! \, (n - k)!}$$

Furthermore, if a group of n objects is composed of n_1 identical objects of type 1, n_2 identical objects of type 2, ..., n_r identical objects of type r, the number of distinguishable arrangements in a row, denoted by $\begin{bmatrix} n \\ n_i \end{bmatrix}$, is

(3) $$\begin{bmatrix} n \\ n_i \end{bmatrix} = \frac{n!}{n_1!\, n_2! \cdots n_r!}$$

Throughout this book we will use the convention that $\binom{n}{k} = 0$ if k is greater than n. This is natural because there is no way of considering arrangements of n objects where more than n of them are alike.

To justify the use of Rule 3, let us divide the $n!$ arrangements into groups of identical arrangements. Each group then has $n_1!\, n_2! \cdots n_r!$ arrangements in it. Since no arrangement may appear in two different groups, the number of groups is $n!/(n_1!\, n_2! \cdots n_r!)$. We may assume without loss of generality that $n_1 + n_2 + \cdots + n_r = n$, because some of the n_i may equal 1, representing objects that are similar only to themselves. Since $1! = 1$, and since dividing Equation 3 by 1 does not affect the numerical value, Rule 3 remains unaffected by the preceding assumption. It is also apparent now that Rule 2 is a special case of Rule 3, where all of the $n_i = 1$.

Example 4. In Example 2 we listed the six ways of arranging the letters A, B, and C in a row. Suppose now that the letters A and B are identical. We will denote them by the letter X. Then the arrangements ABC and BAC become indistinguishable, denoted by XXC. Also, ACB and BCA become XCX. The original $3! = 6$ arrangements are reduced to

$$\binom{3}{2} = \frac{3!}{2!\,1!} = \frac{(3)(2)(1)}{(2)(1)(1)} = 3$$

distinguishable arrangements, that is XXC, XCX, and CXX.

Example 5. In a coin tossing experiment where a coin is tossed five times, the result is two heads and three tails. The number of different sequences of two heads and three tails equals the number of distinguishable arrangements of two objects of one kind and three objects of another, which is $\binom{5}{2} = 10$.

Note that the 10 arrangements are as follows, where $H = $ "head" and $T = $ "tail."

HHTTT	THHTT	TTHHT
HTHTT	THTHT	TTHTH
HTTHT	THTTH	TTTHH
HTTTH		

How many different groups of k objects may be formed from n objects? We can use Rule 3 to answer this question. Suppose that the n objects are lined up in a row, and we have k identical tags to place on k of the n objects. It is easy to see that the number of ways of placing the k tags on k of the n objects which, in turn, equals the number of distinguishable arrangements of k tagged positions and $n-k$ untagged positions, is $\binom{n}{k}$, as given by **Rule 3**. In this situation $\binom{n}{k}$ is often read "the number of ways of taking n things k at a time."

Example 6. Consider again the three letters A, B, and C. The number of ways of selecting two of these letters is $\binom{3}{2} = 3$, that is, AB, AC, and BC. To see how this relates to the previous discussion, we will "tag" two of the three letters with an asterisk (*) denoting the tag.

$$A^*B^*\, C \text{ gives } AB$$
$$A^*BC^* \text{ gives } AC$$
$$\text{and } AB^*C^* \text{ gives } BC$$

Note the similarity between this example and Example 4.

For still another way of using the term $\binom{n}{k}$, consider the expression $(x+y)(x+y)\cdots(x+y)$. The term x^n occurs only when the x term from the first factor is multiplied by the x term from the second factor, and so on for all n factors. The term $x^{n-1}y$ results from multiplying the x term from $n-1$ of the factors times the y term from one factor. Since the y term may be selected from any one of the n factors, expansion of $(x+y)^n$ results in n terms involving $x^{n-1}y$. Similarly, for each value of k, the term $x^k y^{n-k}$ results from the selection of k xs from k of the factors, and the $n-k$ ys from the remaining $n-k$ factors There are $\binom{n}{k}$ ways of selecting k factors for the xs, with the remaining factors contributing ys. Therefore the term $x^k y^{n-k}$ appears $\binom{n}{k}$ times in the expansion of $(x+y)^n$. Since all terms in the expansion are added together, we may write

$$(4) \qquad (x+y)^n = x^n + \binom{n}{n-1}x^{n-1}y^1 + \binom{n}{n-2}x^{n-2}y^2 + \cdots$$

$$+ \binom{n}{2}x^2 y^{n-2} + \binom{n}{1}x^1 y^{n-1} + y^n$$

Recall that $0! = 1$, so $\binom{n}{0} = 1$ and $\binom{n}{n} = 1$. If we use the notation

$$\sum_{i=a}^{b} C_i = C_a + C_{a+1} + C_{a+2} + \cdots + C_{b-1} + C_b$$

which is read as "the sum of the terms C_i as i goes from a to b," we may write

(5)
$$(x+y)^n = \sum_{i=0}^{n} \binom{n}{i} x^i y^{n-i}$$

which is known as the "binomial expansion" and is found in most high school algebra textbooks. This illustrates why the term "binomial coefficient" is often used to describe the symbol $\binom{n}{k}$. Similarly, it may be noted that the coefficient of $x_1^{n_1} x_2^{n_2} \cdots x_r^{n_r}$ in the expansion of $(x_1 + x_2 + \cdots + x_r)^n$ is given by the "multinomial coefficient" $\begin{bmatrix} n \\ n_i \end{bmatrix}$.

Example 7. We will use the binomial expansion to evaluate $(2+3)^4$. Of course, we know the answer is $5^4 = 625$. From the binomial expansion in Equation 5 we have

$$(2+3)^4 = \sum_{i=0}^{4} \binom{4}{i} 2^i 3^{4-i}$$

$$= \binom{4}{0} 2^0 3^4 + \binom{4}{1} 2^1 3^3 + \binom{4}{2} 2^2 3^2 + \binom{4}{3} 2^3 3^1 + \binom{4}{4} 2^4 3^0$$

$$= (1)(1)(81) + (4)(2)(27) + (6)(4)(9) + (4)(8)(3) + (1)(16)(1)$$

$$= 81 + 216 + 216 + 96 + 16 = 625$$

EXERCISES

1. How many 4-digit numbers (from 0000 to 9999) may be formed using the 10 digits, where each digit may be repeated any number of times?

2. How many different 4-letter arrangements are there, using the 26 letters in the alphabet, where each letter may be used repeatedly?

3. How many ways are there of arranging the letters L, O, V, E into four-letter "words," where each letter is used only once?

4. How many ways are there of seating five people in a row?

5. In how many ways may a committee of three be chosen from a club with 12 members?

6. What is the coefficient of $x^3 y^3$ in the expansion of $(x+y)^6$?

7. What is the coefficient of $x^2 y^4 z$ in the expansion of $(x+y+z)^7$?

8. What is the coefficient of $x^2 y^5$ in the expansion of $(w+x+y+z)^7$? (*Hint.* $x^2 y^5 = w^0 x^2 y^5 z^0$)

9. Evaluate $\sum_{i=1}^{3} \binom{4}{i}$.

10. Evaluate $\sum_{i=0}^{3} \binom{4}{i} \left(\frac{1}{2}\right)^2$.

11. Evaluate $\sum_{i=3}^{5} \binom{6}{i}\left(\frac{1}{3}\right)^{i}\left(\frac{2}{3}\right)^{6-i}$

12. Evaluate $\sum_{i=1}^{4} 5$.

PROBLEMS

1. How many ways are there of choosing n_1 objects of the first kind, n_2 objects of the second kind, and so forth, to n_k objects of the kth kind, where there are altogether N_1 objects of the first kind, N_2 objects of the second kind, and so on? How many ways are there if n_i is greater than N_i for some i?

2. Show that $\sum_{i=0}^{n} \binom{n}{i} p^i (1-p)^{n-i} = 1$.

1.2. PROBABILITY

Now we are ready to apply the three counting rules given in Section 1.1 to find some interesting and useful probabilities. But first we must introduce some standard terminology used in statistics. Correct understanding of the terms defined in this section and elsewhere will make communication of statistical concepts much easier.

We will define the important terms *sample space* and *points in the sample space* in connection with an experiment.

Definition 1. The *sample space* is the collection of all possible different outcomes of an experiment.

Definition 2. A *point in the sample space* is a possible outcome of an experiment.

Each experiment has its own sample space, which consists essentially of a list of the different outcomes of the experiment that are possible. It is tacitly assumed that the sample space is subdivided as finely as reasonably possible with each subdivision being called a point. Also, it is tacitly assumed that each possible outcome is represented by one and only one point.

Example 1. If an experiment consists of tossing a coin twice, the sample space consists of the four points *HH*, *HT*, *TH*, and *TT*.

Example 2. An examination consisting of 10 "true or false" questions is administered to one student as an experiment. There are $2^{10} = 1024$ points in the sample space, where each point consists of the sequence of possible answers to the ten successive questions, such as "*TTFTFFTTTT*."

It is now possible to define *event*, in terms of the points in the sample space.

Definition 3. An *event* is any set of points in the sample space.

In Example 1 we may speak of the event "two heads," which consists of the single point *HH*, the event "one head," which consists of the two points *HT* and *TH*, the event "at least one tail," which consists of the points *TH*, *HT*, and *TT*, as well as the event "four heads," which has no points in it. A set with no points in it is sometimes called *the empty set.* The event consisting of all points in the sample space is sometimes called *the sure event* because it is certain to occur every time the experiment is performed.

Two different events may have points common to both. The events "at least one tail" and "at least one head" have the two points *TH* and *HT* in common. If two events have no points in common, they are called *mutually exclusive* events because the occurrence of one event automatically excludes the possibility of the other event occurring at the same time.

If all of the points in one event are also contained in a second event, we say that the first event *is contained in* the second event, or that the second event *contains* the first event. The event "at least one head" contains the event "two heads." Each event therefore contains itself.

To each point in the sample space there corresponds a number called *the probability of the point* or *the probability of the outcome.* These probabilities may be any number from 0 to 1. If we can conceive of a long series of repetitions of the experiment under fairly uniform conditions, the relative frequency of the occurrence of the point or event in mind represents an approximation to the probability of that point or event.

Definition 4. If *A* is an event associated with an experiment, and if n_A represents the number of times *A* occurs in *n* independent repetitions of the experiment, the *probability of the event A*, denoted by $P(A)$, is given by

(1)
$$P(A) = \lim_{n \to \infty} \frac{n_A}{n}$$

which is read "the limit of the ratio of the number of times *A* occurs to the number of times the experiment is repeated, as the number of repetitions approaches infinity."

A formal definition of *independent* is deferred until later. For now we may think of experiments as independent if the outcome of any one experiment does not influence the outcome of the other experiments.

The definition of the probability of an event includes the definition of the probability of an outcome as a special case, since an event may be considered as consisting of a single outcome. It is apparent from the definition that the probability of an event equals the sum of the probabilities of all outcomes comprising the event, since the number of times the event occurs equals the sum of the numbers of times the mutually exclusive outcomes comprising the event occur.

In practice, the set of probabilities associated with a particular sample space is seldom known, but the probabilities are assigned according to the experimenter's preconceived notions. That is, the experimenter formulates a model as an idealized version of the experiment. Then the sample space of the model experiment is examined, and the probabilities are assigned to the various points of the sample space in some manner which the experimenter feels can be justified.

Example 3. In an experiment consisting of the single toss of an unbiased coin, it is reasonable to assume that the outcome H will occur about half the time. Thus we may assign the probability 1/2 to the outcome H, and the same to the outcome T. We write this as $P(H) = 1/2$, $P(T) = 1/2$.

Example 4. In an experiment consisting of three tosses of an unbiased coin, it is reasonable to assume that each of the $2^3 = 8$ outcomes HHH, HHT, HTH, HTT, THH, THT, TTH, TTT is equally likely. Thus the probability of each outcome is 1/8. Also $P(3 \text{ tails}) = 1/8$, $P(\text{at least one head}) = 7/8$, and $P(\text{more heads than tails}) = P(\text{at least 2 heads}) = 4/8 = 1/2$.

We have been working with *probability functions* in the previous two examples.

Definition 5. A *probability function* is a function that assigns probabilities to the various events in the sample space.

In Example 3 the probability function was given by $P(H) = 1/2$, $P(T) = 1/2$. It is necessary that the probability function assign a probability to each point in the sample space. Then the probabilities of all events in the sample space are automatically specified by the probabilities of the sample points contained in the events.

Several properties of probability functions become apparent. Let S be a sample space and let A be any event in S. Then, if P is a probability function, $P(S) = 1$, because

$$P(S) = \lim_{n \to \infty} \frac{n}{n} = 1$$

$P(A) \geq 0$, because $n_A \geq 0$, and therefore

$$\lim_{n \to \infty} \frac{n_A}{n} \geq 0$$

and $P(\bar{A}) = 1 - P(A)$, where $\bar{A}$ is the event "the event A does not occur," because $n_{\bar{A}} = n - n_A$, and

$$\lim_{n \to \infty} \frac{n_{\bar{A}}}{n} = \lim_{n \to \infty} \frac{n - n_A}{n} = \lim_{n \to \infty} \left(1 - \frac{n_A}{n}\right) = 1 - \lim_{n \to \infty} \frac{n_A}{n} = 1 - P(A)$$

We mentioned earlier that while the various outcomes of an experiment are mutually exclusive, the various events associated with an experiment do not necessarily have that property. In our experiment of tossing a coin three times the events "three heads" and "at least two heads" may both occur at the same time. Now consider the probability of the event "three heads" if we are given that the event "at least two heads" has occurred. If at least two heads have occurred, we know that several points in the sample space, that is, *TTT, TTH, THT*, and *HTT*, may be eliminated. The possible outcomes of the experiment are reduced to four equally likely points. Therefore the probability of each point is now 1/4, and hence the probability of the event "three heads," or *HHH*, is 1/4, if we are given the fact that at least two heads have occurred. The additional information that we are given has the effect of eliminating some of the outcomes from consideration and thus artificially reducing the sample space.

In another experiment, consider rolling a die. Let S be the sample space, let A be the event "a 4, 5, or 6 occurs," and let B be the event "an even number (2, 4, or 6) occurs," as depicted by Figure 1. The probability that the event A has occurred, given that B has occurred, is written $P(A \mid B)$ and is usually read "the probability of A given B." Since we know that B has occurred, we may not only eliminate the points that are in neither A nor B, that is, the points 1 and 3, but we may even eliminate the point in A that is not in B, or 5. Thus all points not in B are eliminated, and the sample space is just the set of points in B. The only points in B that can result in the event A are the points in both A and B, or 4 and 6. These points represent the event "both A and B occur."

Definition 6. If A and B are two events in a sample space S, the event "both A and B occur," representing those points in the sample space that are in both A and B at the same time, is called *the joint event A and B* and is represented by AB. The probability of the joint event is represented by $P(AB)$.

Then the probability of "A given B" is given by the probability of "AB" relative to the reduced sample space "B." Or, symbolically,

$$(2) \qquad P(A \mid B) = \frac{P(AB)}{P(B)}$$

Looking at it another way, suppose that the preceding experiment is repeated n times. However, only those outcomes resulting in the event B are

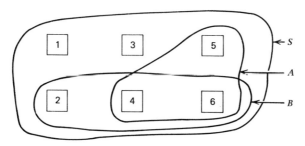

Figure 1

recorded and the outcomes not resulting in B are ignored. Let n_B represent the number of times B occurs, and let n_{AB} represent the number of times A occurs when B occurs. Then

$$(3) \qquad P(A \mid B) = \lim_{n \to \infty} \frac{n_{AB}}{n_B} = \lim_{n \to \infty} \frac{n_{AB}/n}{n_B/n} = \frac{P(AB)}{P(B)}$$

We have intuitively justified the following definition.

Definition 7. The *conditional probability* of A given B is the probability that A occurred given that B occurred and is given by

$$(4) \qquad P(A \mid B) = \frac{P(AB)}{P(B)}$$

where $P(B) > 0$. If $P(B) = 0$, $P(A \mid B)$ is not defined.

Example 5. Consider the rolling of a fair die, so that each of the six possible outcomes has probability 1/6 of occurring. As before, let A be the event "a 4, 5, or 6 occurs" and let B be the event "an even number occurs." Then $P(AB) = P(4 \text{ or } 6) = 2/6 = 1/3$. Also, $P(B) = 3/6 = 1/2$. Then the conditional probability $P(A \mid B)$ is given by

$$P(A \mid B) = \frac{P(AB)}{P(B)} = \frac{1/3}{1/2} = \frac{2}{3}$$

We should note the reasonableness of this answer, since we are given that an even number (i.e., event B) has occurred and the outcome of the experiment is either a 2, 4, or 6. We now want to know the probability that a number greater than 3 (i.e., event A) also occurred. Since two of the three even numbers are greater than 3, our answer is 2/3.

The idea of conditional probability leads quite naturally into the idea of independent events. If the probability of A, given that B occurs, is the same as the probability of A without any information on the occurrence or nonoccurrence of B, we feel that the occurrence or nonoccurrence of A is independent of whether or not B occurs. That is, we feel that A is independent of B if $P(A \mid B) = P(A)$. In fact, this may be used as the definition of independence, but it is not clear from this form of the definition whether B then is also independent of A. So it is better to substitute $P(A)$ for $P(A \mid B)$ in Equation 4, the definition of conditional probability. This leads to the following.

Definition 8. Two events A and B are *independent* if

$$(5) \qquad P(AB) = P(A)(B)$$

Because of the symmetry of Equation 5 it is readily apparent that if A is independent of B, B is also independent of A, and so it is better to say "A and B are independent," where it is meant that they are independent of each other.

Example 6. In an experiment consisting of two tosses of a balanced coin, the four points in the sample space are assumed to have equal probabilities. Let

A be the event "a head occurs on the first toss" and let *B* be the event "a head occurs on the second toss." Then *A* has the points *HH* and *HT*. *B* has the points *HH* and *TH*, and *AB* has the point *HH*. Also $P(A) = 2/4$, $P(B) = 2/4$, and $P(AB) = 1/4$. Therefore Equation 5 is satisfied and *A* and *B* are independent.

The following example illustrates that the independence of two events is not always intuitively obvious and should always be determined directly from the definition and Equation 5.

Example 7. Consider again the experiment consisting of one roll of a balanced die, where the sample space consists of the six equally likely points 1, 2, 3, 4, 5, and 6. Let *A* be the event "an even number occurs." including the points 2, 4, and 6. Let *B* be the event "at least a 4 occurs," including the points 4, 5, and 6. Finally, let *C* be the event "at least a 5 occurs," including the points 5 and 6. Then *A* and *B* are not independent, because $P(A)P(B) = (1/2)(1/2)$, or $1/4$ while $P(AB) = 1/3$. However, *A* and *C* are independent, because $P(A)P(C) = (1/2)(1/3)$, or $1/6$, the same as $P(AC)$.

Sometimes the notions of *independent events* and *mutually exclusive events* are confused with each other, because both notions give the impression that the "the two events do not have anything to do with each other." The property of independence depends not only on the two events being considered but also on the particular probability function defined on the sample space. It is possible for $P(AB)$ and $P(A)P(B)$ to be equal to each other with one set of probabilities and to be unequal with another set of probabilities. But "mutually exclusive" simply means the two events have no points in common, and no matter what probability function is defined on the sample space, *AB* is empty, so $P(AB) = 0$. If *A* and *B* are mutually exclusive, they will be independent only if either $P(A)$ or $P(B)$ equals zero, since Equation 5 must be satisfied.

Now we will define the concept of *independent experiments*.

Definition 9. Two experiments are *independent* if for every event *A* associated with one experiment and every event *B* associated with the second experiment,

$$P(AB) = P(A)P(B)$$

It is equivalent to define two experiments as independent if every event associated with one experiment is independent of every event associated with the other experiment.

It is quite tedious to examine every pair of events associated with two experiments to see if they satisfy Definition 9. However, it is sufficient to verify the definition only for those events consisting of a single point each. Then the definition is automatically verified for all other events.

In practice, the model is usually set up assuming independence, and the assumption of independence is then used to find $P(AB)$ using $P(A)$ and $P(B)$

in Definition 9. This is the main value of the definition of independence. Thus it is reasonable to extend the definition of independent experiments to cover the eventuality of more than two experiments being involved.

Definition 10. n experiments are mutually independent if for every set of n events, formed by considering one event from each of the n experiments, the following equation is true:

(6) $$P(A_1 A_2 \cdots A_n) = P(A_1)P(A_2) \cdots P(A_n)$$

where A_i represents an outcome of the ith experiment, for $i = 1, 2, \ldots, n$.

The word "mutually" may be omitted in the preceding definition if no confusion results.

Example 8. Let an experiment consist of one toss of a biased coin, where the event H has probability p and the event T has probability $q = 1 - p$. Consider three independent repetitions of the experiment, where a subscript will be used to denote the experiment with which the outcome is associated. Thus $H_1 T_2 H_3$ means the first experiment resulted in H, the second in T, and the third in H. Because of our assumption of independence,

$$P(H_1 T_2 H_3) = P(H_1)P(T_2)P(H_3) = pqp$$

If we consider the event "exactly two heads" associated with the combined experiments, this may occur $\binom{3}{2} = 3$ ways, and hence

$$P(\text{exactly two heads}) = 3p^2 q$$

Obviously the preceding might just as well have been described as one experiment with three independent trials. The extension to considering an experiment consisting of n independent tosses may be made. The probability of obtaining "exactly k heads" then equals the term $p^k q^{n-k}$ times the number of times that term can appear. Therefore, in n independent tosses of a coin,

(7) $$P(\text{exactly } k \text{ heads}) = \binom{n}{k} p^k q^{n-k}$$

where $p = P(H)$ on any one toss.

The four preceding definitions, as is true for all definitions, work both ways. Example 6 presents a situation where the satisfaction of Equation 5 implies that two events are independent. Example 8 presents a situation where the assumption of independence implies that Equation 6 is satisfied. It follows then that if Equation 6 is not satisfied, the experiments are not independent, and conversely if the experiments are not independent, Equation 6 is not satisfied for at least one set of events $A_1 A_2 \cdots A_n$.

EXERCISES

1. In an experiment consisting of three tosses of a coin, where the order of the tosses (first to third) is important, list the points in the sample space.

2. Referring to Exercise 1, give:
(a) Two mutually exclusive events.
(b) Two events that are not mutually exclusive.

3. If the probability of rain is .15, what is the probability of no rain?

4. If the probability of arriving at an intersection while the traffic light is green is .35 and the probability of the light being yellow is .10, what is the probability that the light is red?

5. If a football team has an equal probability of winning or losing each game (assuming no ties occur), what is the probability of the team losing at least seven games in an eight-game season?

6. If the probability of getting a torn dollar bill is .05, what is the probability that two out of the three dollar bills obtained are torn (assume independence)?

7. If 60% of all stolen cars are recovered and 2% of all cars are stolen each year, what is the probability of a person having a car stolen and never recovered?

8. The probability of a customer buying a certain bottle of cleaner is .15. Forty percent of the customers that buy that cleaner also buy a dispenser. What is the probability that a customer buys both?

9. In three independent tosses of an unbiased coin, what is the probability of obtaining three heads?

10. In three independent tosses of an unbiased coin, what is the probability of obtaining at least one tail?

11. In three independent tosses of an unbiased coin, what is the probability of obtaining three heads if we know that at least one head has occurred?

12. In three independent tosses of an unbiased coin, what is the probability of obtaining three heads if we know that the first toss resulted in a head? (*Note.* Exercises 11 and 12 have different answers.)

PROBLEMS

1. Show that in a sample space with n points, there are exactly $2^n - 1$ events containing at least one point.

2. Consider three events A, B, and C where $P(A) = 1/3$, $P(B) = 1/3$, $P(C) = 1/3$, $P(AB) = 1/9$, $P(AC) = 1/9$, $P(BC) = 1/9$, and $P(ABC) = 0$.
(a) Are events A and B independent?
(b) Are events A, B, and C mutually independent?
(c) Find the probability that none of the events A, B, or C occurs. (*Hint.* Consider drawing a number from a hat, where the numbers 1 to 9 are in the hat. If the number is 1, 2, or 3, event A occurs. A 1, 4, or 5 drawn produces event B, and a 2, 4, or 6 produces event C.)

1.3. RANDOM VARIABLES

Outcomes associated with an experiment may be numerical in nature, such as the score on an examination, or nonnumerical, such as the choice "red door" by a rat escaping from a pen. In order to analyze the results of an experiment, it is necessary to assign numbers to the points in the sample space. Any rule for assigning such numbers is called a *random variable.*

Definition 1. A *random variable* is a function that assigns real numbers to the points in a sample space.

We will usually denote random variables by the capital letters W, X, Y, or Z, with or without subscripts. The real numbers assigned by the random variables will be denoted by lowercase letters.

Example 1. In an experiment where a consumer is given a choice of three products, soap, detergent, or Brand A, the sample space consists of the three points representing the three possible choices. Let the random variable assign the number 1 to the choice "Brand A" and the number 0 to the other two possible outcomes. Then $P(X = 1)$ equals the probability that the consumer chooses Brand A.

At times it is convenient to define more than one random varaible for a single sample space, such as in the following example.

Example 2. Six girls and eight boys are each asked whether they communicate more easily with their mother or their father. Let X be the number of girls who feel they communicate more easily with their mothers and let Y be the total number of children who feel they communicate more easily with their mothers. If $X = 3$, we know the event "3 girls feel they communicate more easily with their mothers" has occurred. If, at the same time, $Y = 7$, we know that the event "3 girls and $7 - 3 = 4$ boys feel they communicate more easily with their mothers" has occurred.

If X is a random variable, "$X = x$" is a short-cut notation that we use to correspond to some event in the sample space, specifically the event consisting of the set of all points to which the random variable X has assigned the value "x."

Example 3. In an experiment consisting of two tosses of a coin, let X be the number of heads. Then "$X = 1$" corresponds to the event containing only the points *HT* and *TH.*

Thus "$X = x$" is sometimes referred to as "the event $X = x$," when the intended meaning is "the event consisting of all outcomes assigned the number x by the random variable X."

Because of this close correspondence between random variables and events, the definitions of *conditional probability* and *independence* apply equally well to random variables.

Definition 2. The conditional probability of X given Y, written $P(X = x \mid Y = y)$, is the probability that the random variable X assumes the value x, given that the random variable Y has assumed the value y.

The equation for determining conditional probabilities may be obtained from Definition 1.2.7 as

(1) $$P(X = x \mid Y = y) = \frac{P(X = x, \, Y = y)}{P(Y = y)} \quad \text{if} \quad P(Y = y) > 0$$

Example 4. Let X be the number of girls that communicate more easily with their mothers out of six girls, as in Example 2, and let Y be the total number of children who communicate more easily with their mothers. For convenience let $Z = Y - X$, so Z equals the number of boys, out of eight boys, who communicate more easily with their mothers. Assume that the answers given by the children are independent of each other and that each child has the same probability p (unknown) of saying they communicate more easily with their mother. We will find the conditional probability $P(X = 3 \mid Y = 7)$.

First, by the preceding assumptions, $X = 3$ and $Z = 4$ are independent events. Since the event $(X = 3, \, Y = 7)$ is the same as the event $(X = 3, \, Z = 4)$ we have the joint probability

$$P(X = 3, \, Y = 7) = P(X = 3, \, Z = 4)$$
$$= P(X = 3)P(Z = 4) \text{ by independence,}$$

(2) $$= \binom{6}{3} p^3 (1 - p)^3 \binom{8}{4} p^4 (1 - p)^4$$

because of Example 1.2.8. By the same example, we conclude that

(3) $$P(Y = 7) = \binom{14}{7} p^7 (1 - p)^7$$

so the conditional probability $P(X = 3 \mid Y = 7)$ is

(4) $$P(X = 3 \mid Y = 7) = \frac{P(X = 3, \, Y = 7)}{P(Y = 7)} = \frac{\binom{6}{3}\binom{8}{4}}{\binom{14}{7}} = .408$$

because all of the factors involving the unknown p cancel each other.

Just as the points in a sample space are mutually exclusive, the values that a random variable may assume are mutually exclusive. That is, for a single outcome of an experiment, the random variable defined for that experiment furnishes us with only one number. Thus the entire set of values that a random variable may assume has many of the same properties as a sample space. The individual values assumed by the random variable correspond to the points in a sample space, a set of values corresponds to an event, and the probability of the random variable assuming any value within a set of values equals the sum of the probabilities associated with all values within the set. For example,

$$P(a < X < b) = \sum_{a < x < b} P(X = x)$$

where the summation extends over all values of x between, but not including, the numbers a and b, and

$$P(X = \text{even number}) = \sum_{x \text{ even}} P(X = x)$$

where the summation applies to all values of x that are even numbers. Because of this similarity between the set of possible values of X and a sample space, the description of the set of probabilities associated with the various values X may assume is often called the *probability function of the random variable X*, just as a sample space has a probability function. However, the probability function of a random variable is not an arbitrary assignment of probabilities, as is the probability function for a sample space, because once the probabilities are assigned to the points in a sample space and once a random variable X is defined on the sample space, the probabilities associated with the various values of X are known and the probability function of X is thus already determined.

Definition 3. The probability function of the random variable X, usually denoted by $f(x)$, is the function that gives the probability of X assuming the value x, for any real number x. In other words,

(5) $$f(x) = P(X = x)$$

The probability function always equals 0 at values of x that X cannot assume.

Sometimes it is convenient to represent the probability function as a bar graph, with the values of the random variable as the abscissa (along the horizontal axis) and the probabilities as the ordinate (the height of the bar). For instance, if $P(X = 1)$ equals .3, $P(X = 2)$ cquals .4, and $P(X = 4)$ equals .3, the bar graph of the probability function looks like this. The heights of the bars

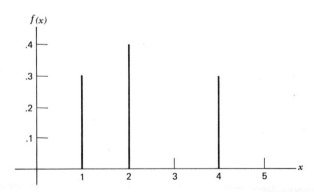

represent the various probabilities associated with the random variable X.

It is not always convenient to use $f(x)$ to denote the probability function of a random variable. Other expressions that may be used include $f_0(x)$, $f_1(x)$, $f_2(x)$, $g(x)$, $h(x)$, and so on. However, the meaning of the various expressions used will always be clear from the context.

We have seen that the distribution of probabilities associated with a random variable may be described by a probability function. Another way of accomplishing the same thing is by means of a *distribution function*, which describes the accumulated probabilities.

Definition 4. The *distribution function of a random variable X*, usually denoted by $F(x)$, is the function that gives the probability of X being less than or equal to any real number x. In other words,

$$(6) \qquad\qquad F(x) = P(X \le x) = \sum_{t \le x} f(t)$$

where the summation extends over all values of t that do not exceed x.

Distribution functions also may be represented graphically, with the x as the abscissa and $F(x)$ as the ordinate. As an illustration, suppose, as before, that $P(X=1)=.3$, $P(X=2)=.4$, and $P(X=4)=.3$. Then the graph of $F(x)$ looks like this.

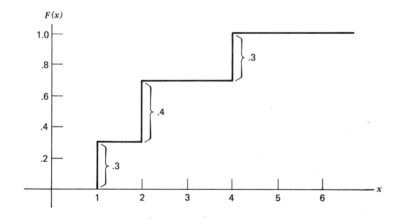

The graph actually consists only of the horizontal lines; the vertical lines are drawn in merely to give the graph a somewhat "connected" appearance, and to assist in the finding of *quantiles* as explained in the next section. The lengths of the vertical lines are the same as the lengths of the bars in the graph of the probability function.

Some probability distributions are well known and consequently have been given names.

Definition 5. Let X be a random variable. The *binomial distribution* is the probability distribution represented by the probability function

(7) $$f(x) = P(X = x) = \binom{n}{x} p^x q^{n-x} \qquad x = 0, 1, \ldots, n$$

where n is a positive integer, $0 \le p \le 1$, and $q = 1 - p$. Note that we are using the usual convention that $0! = 1$.

The distribution function is then

(8) $$F(x) = P(X \le x) = \sum_{i \le x} \binom{n}{i} p^i q^{n-i}$$

where the summation extends over all possible values of i less than or equal to x. Table A3 (see appendix) gives the values of $F(x)$ for some selected values of the parameters n and p.

Example 5. An experiment consists of n independent trials where each trial may result in one of two outcomes, "success" or "failure," with probabilities p and q, respectively, such as with the tossing of a coin. Let X equal the total number of "successes" in the n trials. Then, as was shown by Equation 1.2.7,

$$P(X = x) = \binom{n}{x} p^x q^{n-x}$$

for integer x from 0 to n. Thus X has the binomial distribution.

Another useful probability distribution is the *discrete uniform distribution.*

Definition 6. Let X be a random variable. The *discrete uniform distribution* is the probability distribution represented by the probability function

(9) $$f(x) = \frac{1}{N} \qquad x = 1, 2, \ldots, N$$

Thus X may assume any integer value from 1 to N with equal probability, if X has the discrete uniform probability function.

Example 6. A jar has N plastic chips, numbered 1 to N. An experiment consists of drawing one chip from the jar, where each chip is equally likely to be drawn. The sample space has N points, representing the N chips that may be drawn. Let X equal the number on the drawn chip. Then X has the discrete uniform distribution.

When several random variables are defined on the same sample space or when several experiments, each with one or more random variables defined for them, are considered as a combined experiment, it becomes useful to consider joint distributions, described by *joint probability functions* and *joint distribution functions.*

Definition 7. The *joint probability function* $f(x_1, x_2, \ldots, x_n)$ of the random variables $X_1, X_2, \ldots, X_n$ is the probability of the joint occurrence of $X_1 = x_1, X_2 = x_2, \ldots,$ and $X_n = x_n$. Stated differently,

(10) $$f(x_1, x_2, \ldots, x_n) = P(X_1 = x_1, X_2 = x_2, \ldots, X_n = x_n)$$

Definition 8. The *joint distribution function* $F(x_1, x_2, \ldots, x_n)$ of the random variables $X_1, X_2, \ldots, X_n$ is the probability of the joint occurrence of $X_1 \leq x_1, X_2 \leq x_2, \ldots,$ and $X_n \leq x_n$. Stated differently,

(11) $$F(x_1, x_2, \ldots, x_n) = P(X_1 \leq x_1, X_2 \leq x_2, \ldots, X_n \leq x_n)$$

Example 7. Consider the random variables X and Y as defined in Example 2. Let $f(x, y)$ and $F(x, y)$ be the joint probability function and the joint distribution function, respectively. Then, from Example 4,

(12) $$f(3, 7) = P(X = 3, Y = 7) = \binom{6}{3}\binom{8}{4}p^7(1-p)^7$$

and

(13) $$F(3, 7) = P(X \leq 3, Y \leq 7) = \sum_{\substack{0 \leq x \leq 3 \\ x \leq y \leq 7}} f(x, y)$$

where

$$f(x, y) = \binom{6}{x}p^x(1-p)^{6-x}\binom{8}{y-x}p^{y-x}(1-p)^{8-(y-x)}$$

and where the summation in Equation 13 extends over all values of x and y such that $x \leq 3$ and $y \leq 7$, with the usual restriction that x and $y - x$ be nonnegative integers. Note that Equations 12 and 13 cannot be evaluated without knowing the value of p.

Definition 9. The *conditional probability function* of X given Y, $f(x \mid y)$, is

(14) $$f(x \mid y) = P(X = x \mid Y = y)$$

From Equation 1 we see that

$$f(x \mid y) = P(X = x \mid Y = y) = \frac{P(X = x, Y = y)}{P(Y = y)}$$

(15) $$= \frac{f(x, y)}{f(y)}$$

where $f(x, y)$ is the joint probability function of X and Y and $f(y)$ is the probability function of Y itself.

Example 8. As a continuation of Example 7, let $f(x \mid y)$ denote the conditional probability function of X given $Y = y$. Then

$$f(3 \mid 7) = P(X = 3 \mid Y = 7) = .408$$

from Equation 4. To find a formula for $f(x \mid y)$ in general (i.e., for any values of x and y we may choose), first let $f(x, y)$ denote the joint probability function of X and Y. This is given in Example 7 as

$$f(x, y) = \binom{6}{x} p^x (1-p)^{6-x} \binom{8}{y-x} p^{y-x} (1-p)^{8-(y-x)}$$

which originally was a general form for Equation 2. Also, let $f(y)$ be the probability function of Y. From Example 4 again we can generalize to get

$$f(y) = P(Y = y) = \binom{14}{y} p^y (1-p)^{14-y}$$

By Definition 9 we can now write the conditional probability function of X given $Y = y$.

$$(16) \qquad f(x \mid y) = \frac{f(x, y)}{f(y)} = \frac{\binom{6}{x}\binom{8}{y-x}}{\binom{14}{y}} \qquad \begin{array}{l} 0 \le x \le 6 \\ 0 \le y - x \le 8 \end{array}$$

where all of the terms involving the unknown parameter p conveniently cancel out.

In the previous examples we work with a probability distribution known as the *hypergeometric distribution*. In its more general form we usually refer to having A objects of one kind and B objects of a second kind (the total numbers of girls and boys in the examples). Then the probability of selecting x of the A objects, given that altogether k of the $A + B$ objects total are selected, under the assumption that each object has the same chance of being selected, is given by the hypergeometric probability function.

Definition 10. Let X be a random variable. The *hypergeometric distribution* is the probability distribution represented by the probability function

$$(17) \qquad f(x) = P(X = x) = \frac{\binom{A}{x}\binom{B}{k-x}}{\binom{A+B}{k}} \qquad \begin{array}{l} 0 \le x \le A \\ 0 \le k - x \le B \end{array}$$

where A, B, and k are nonnegative integers and $k \le A + B$.

Mutually independent random variables may be defined in a manner similar to Definitions 1.2.9 and 1.2.10 of independent experiments.

Definition 11. Let $X_1, X_2, \ldots, X_n$ be random variables with the respective probability functions $f_1(x_1), f_2(x_2), \ldots, f_n(x_n)$ and with the joint probability function $f(x_1, x_2, \ldots, x_n)$. Then $X_1, X_2, \ldots, X_n$ are *mutually independent* if

(18) $$f(x_1, x_2, \ldots, x_n) = f_1(x_1)f_2(x_2) \cdots f_n(x_n)$$

for all combinations of values of $x_1, x_2, \ldots, x_n$.

Example 9. Consider the experiment described in Example 8. Then the probability function of X is given by

(19) $$f_1(x) = P(X = x) = \binom{6}{x}p^x(1-p)^{6-x}$$

and the probability function of Y is given by

(20) $$f_2(y) = P(Y = y) = \binom{14}{y}p^y(1-p)^{14-y}$$

Since

$$f(x, y) = P(X = x, Y = y) = P(X = x \mid Y = y)P(Y = y)$$

the use of Equations 16 and 20 results in the joint probability function of X and Y being given by

$$f(x, y) = \frac{\binom{6}{x}\binom{8}{y-x}}{\binom{14}{y}}\binom{14}{y}p^y(1-p)^{14-y}$$

$$= \binom{6}{x}\binom{8}{y-x}p^y(1-p)^{14-y}$$

But, since

$$f_1(x)f_2(y) = \binom{6}{x}\binom{14}{y}p^{x+y}(1-p)^{20-x-y}$$

we see that

$$f(x, y) \neq f_1(x)f_2(y)$$

and, therefore, X and Y are not independent.

EXERCISES

1. If $f(x)$ is the binomial probability function with $n = 6$ and $p = 1/3$, find
(a) $f(6)$.
(b) $f(0)$.
(c) $f(2.5)$.
(d) $F(2.5)$.
(e) $F(-3)$.
(f) $F(7)$.

(g) Draw a bar graph of the probability function.

(h) Draw a graph of the distribution function.

2. Suppose $f(x)$ is the discrete uniform probability function with N equal to 12. Find

(a) $f(2)$.

(b) $f(12)$.

(c) $f(0)$.

(d) $f(1.5)$

(e) $F(0)$.

(f) $F(3.1)$.

(g) $F(1000)$.

(h) $F(-1000)$.

(i) Draw a bar graph of the probability function.

(j) Draw a graph of the distribution function.

3. Let X and Y be independent, binomially distributed random variables, with parameters $n = 3$, $p = 1/2$ for X, and $n = 4$, $p = 1/2$ for Y. Let $f(x, y)$ denote the joint probability function of X and Y. Find

(a) $f(0, 0)$.

(b) $f(0, 1)$.

(c) $f(1, 0)$.

(d) $f(3, 4)$.

(e) $f(4, 4)$.

(f) $F(0, 0)$.

(g) $f(1, 1)$.

(h) $F(3, 4)$.

4. Let $f(x \mid k)$ be the hypergeometric probability function, where $A = 3$ and $B = 4$. Find

(a) $f(0 \mid 0)$.

(b) $f(1 \mid 1)$.

(c) $f(2 \mid 1)$.

(d) $f(1 \mid 5)$.

(e) $f(1 \mid 6)$.

5. A diner selects one sandwich at random out of six possible sandwich varieties.

(a) What is the sample space?

(b) What is the probability function on the sample space?

(c) Define a random variable on the sample space such that the random variable has a discrete uniform distribution.

6. Seven boys and 10 girls take an examination, and each student has probability .2 of failing the examination.

(a) What is the sample space for this experiment?

(b) Given that three students failed the examination, what is the probability that all three are boys?

(c) What is the name of the probability distribution you are using?

(d) If the probability of each failure is .8 instead of .2, what is the answer to part b?

PROBLEMS

1. Which of the following functions are possible probability functions? Justify your answer.

(a) $f(x) = 1/6$ for $x = 1, 2, 3, 4$,

     $= 0$ elsewhere

(b) $f(x) = (1/4)^x$ for $x = 1, 2, 3, 4, \ldots$,

     $= 0$ elsewhere

(c) $f(x) = (1 - p)p^x$ for $x = 0, 1, 2, \ldots$,

     $= 0$ elsewhere, where p is a constant between 0 and 1

2. Assume that every patient with a particular type of disease has probability .1 of being cured within a week, if the patient is given no treatment for the disease. Ten patients with that type of disease are given a new type of drug. After one week 9 out of the 10 patients are cured.
   (a) What is the probability of at least 9 patients being cured if the drug is assumed to have no curative effects?
   (b) In your opinion, would you consider this drug to be beneficial?
   (c) What sample space did you use in this analysis?
   (d) What probability function did you define on the sample space?
   (e) What random variable did you define on the sample space?
   (f) What is the name of the probability distribution of your random variable?

1.4. SOME PROPERTIES OF RANDOM VARIABLES

We have already discussed some of the properties associated with random variables, such as their probability functions and their distribution functions. The probability function describes all of the properties of a random variable that are of interest, because the probability function reveals the possible values the random variable may assume and the probability associated with each value. A similar statement may be made concerning the distribution function. At times, however, it is inconvenient or confusing to present the entire probability function to describe a random variable, and some sort of a "summary description" of the random variable is needed. We will now introduce some other properties of random variables that may be used to present a brief, but incomplete, description of the distribution of the random variable.

The most common method used in this book for summarizing the distribution of a random variable is by giving some selected *quantiles* of the random variable. The term "quantile" is not as well known as the terms "median," "quartile," "decile," and "percentile," yet these latter terms are popular names given to particular quantiles. The median of a random variable, for example, is some number that the random variable will exceed with probability one-half or less and will be smaller than with probability one-half or less. This definition may be extended as follows.

Definition 1. The number x_p, for a given value of p between 0 and 1, is called the pth *quantile of the random variable X*, if $P(X < x_p) \le p$ and $P(X > x_p) \le 1 - p$.

If more than one number satisfies the definition of the pth quantile, we will avoid confusion by adopting the convention that x_p equals the average of the largest and the smallest numbers that satisfy Definition 1.

That is, X is less than x_p with probability p or less, and X exceeds x_p with probability $1 - p$ or less. The *median* is the .5 quantile, the third *decile* is the .3 quantile, the *upper and lower quartiles* are the .75 and .25 quantiles respectively, and the sixty-third *percentile* is the .63 quantile.

Perhaps the easiest method of finding the pth quantile involves using the graph of the distribution function of the random variable. The pth quantile is the abscissa of the point on the graph which has the ordinate value of p, as illustrated in the following example.

Example 1. Let X be a random variable with the following probability distribution.

$$P(X=0)=\tfrac{1}{4}$$
$$P(X=1)=\tfrac{1}{4}$$
$$P(X=2)=\tfrac{1}{3}$$
$$P(X=3)=\tfrac{1}{6}$$

Then the distribution function of X may be represented by the following graph. The .75 quantile $x_{.75}$, called the upper quartile, may be found by drawing a horizontal line through .75 on the vertical axis, as indicated by the dotted line in Figure 2. The value of x where the dotted line intersects the graph is the upper quartile, which equals 2 in this example. Therefore we may say that $x_{.75}=2$, which may be verified directly from the definition, since

$$P(X<2)=\tfrac{1}{2}$$

which is less than .75, and since

$$P(X>2)=\tfrac{1}{6}$$

which is less than $1-.75$.

Similarly, the median is found by drawing a line through .5 on the vertical scale. The median is any value from 1 to 2 inclusive, and it is easy to see that any of these values satisfies the definition of the median. By our convention we select 1.5 as the median.

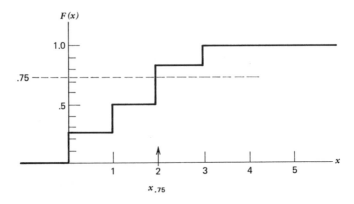

Figure 2

Certain random variables called "test statistics" play an important role in most statistical procedures. These test statistics are useless unless their distribution functions are at least partially known. Most of the tables in the appendix give information concerning the distribution functions of various test statistics used in nonparametric statistics. This information is condensed with the aid of quantiles; otherwise the tables would be inconveniently bulky.

Often we will define a random variable and, instead of working with that random variable, we will work with a function of the random variable. A real valued function of a random variable X is a rule for assigning new real numbers to the sample space instead of the usual numbers assumed by X. For example, if $Y = X + 4$, Y is a real valued function of X; if $X = x$, $Y = x + 4$. If $X = 3$, $Y = 7$. This is usually written as $Y = u(X)$, where $u(X)$ in this case is $X + 4$. Other functions might include $u(X) = X^2$, $u(X) = X$, and $u(X) = (X - a)^2$ for some constant a. Since Y also assigns real numbers to points in the sample space, even though Y uses X in the process, we see that Y is a random variable. It is true in general that a real valued function of a random variable is also a random variable.

Another very useful property of a random variable is its *expected value*. First we will present a general definition of expected value; it will be followed by some particular cases.

Definition 2. Let X be a random variable with the probability function $f(x)$ and let $u(X)$ be a real valued function of X.
  The *expected value of $u(X)$, written $E[u(X)]$,* is

(1) $$E[u(X)] = \sum_x u(x)f(x)$$

where the summation extends over all possible values of X. If the sum on the right side of Equation 1 is infinite, or does not exist, we say that the expected value of $u(X)$ does not exist.

Our interest is confined mainly to two special expected values, the *mean* and the *variance* of X.

Definition 3. Let X be a random variable with the probability function $f(x)$.
  The *mean of X,* usually denoted by μ, is

(2) $$\mu = E(X)$$

From Equation 1 we have

(3) $$\mu = E(X) = \sum_x xf(x)$$

which shows our mean to be the same as the "centroid" in physics. The mean, as does the centroid, marks a central point, a point of balance. If weights, in proportion to the various probabilities, were to be placed on a yardstick at the appropriate values of X, the yardstick would balance right at the mean.

Because of this tendency to "locate" the center of the distribution, the mean is sometimes called a "location parameter." The mean and the median, discussed previously, are the two most commonly used location parameters.

Example 2. Consider a simple experiment that results in "success" with probability p or "failure" with probability q equal to $1-p$. Let X equal 1 if a "success" occurs and 0 if a "failure" occurs. Thus X has the binomial distribution with n equal to 1. From Equation 3 the expected value of X is determined as follows.

(4) $$E(X) = 1(p) + 0(1-p) = p$$

The mean of X is equal to p. If the outcomes have equal probability, p equals 1/2 and the mean of X equals 1/2.

Example 3. A certain business man always eats lunch at a certain restaurant, which has lunches priced at \$1.00, \$1.50, \$2.00, and \$2.50. The businessman knows from past experience that on any given day he will select the \$1.00 lunch with probability .25, \$1.50 lunch with probability .35, and the remaining two lunches with probability .20 each. Let X be the price of the lunch, in dollars. The probability function of X is

$$P(X = 1) = .25$$
$$P(X = 1.5) = .35$$
$$P(X = 2) = .20$$
$$P(X = 2.5) = .20$$

The mean of X is found using Equation 3.

$$E(X) = (1)(.25) + (1.5)(.35) + (2)(.20) + (2.5)(.20) = 1.675$$

Over a long period of time the businessman may expect the average luncheon expense to be somewhere near \$1.67$\frac{1}{2}$, even though no single lunch will cost that amount.

Just as the mean and the median are called location parameters, the name "scale parameter" is given to properties of the random variable that measure the amount of spread, or variability, of the random variable. One scale parameter based on quantiles is the *interquartile range*, the number obtained by subtracting $x_{.25}$ from $x_{.75}$. Another scale parameter, based more directly on the probability function, is the *range*, which equals the largest possible value of the random variable minus its smallest possible value. The most common scale parameter is the *standard deviation*, which equals the square root of the *variance*, defined as follows.

Definition 4. Let X be a random variable with mean μ and the probability function $f(x)$. The *variance of* X, usually denoted by σ^2 or by Var(X), is

(5) $$\sigma^2 = E[(X - \mu)^2]$$

Using Equation 1 the variance of X may be written as

$$\sigma^2 = \sum_x (x - \mu)^2 f(x)$$

$$= \sum_x (x^2 - 2\mu x + \mu^2) f(x)$$

(6)
$$= \sum_x x^2 f(x) - 2\mu \sum_x x f(x) + \mu^2 \sum_x f(x)$$

Because $\sum_x f(x)$ equals 1, and because of Equation 3, Equation 6 becomes

(7)
$$\sigma^2 = E(X^2) - 2\mu^2 + \mu^2 = E(X^2) - \mu^2$$

which is often a more useful form of the variance for computing purposes.

The positive square root of the variance is called the *standard deviation* of X and is usually denoted by σ.

Example 4. If X has the binomial distribution with n equal to 1, then $P(X=1) = p$ and $P(X=0) = 1-p$. In Example 2 the mean of X was found to equal p. Therefore, from Equation 6,

$$\sigma^2 = (1-p)^2(p) + (0-p)^2(1-p)$$

$$= p(1-p)$$

(8)
$$= pq$$

Alternatively, we might use Equation 7 to compute σ^2. Then we would first compute $E(X^2)$ using Equation 1.

$$E(X^2) = (1)^2(p) + (0)^2(1-p)$$

$$= p$$

The variance of X is then found to be

$$\sigma^2 = E(X^2) - \mu^2$$

$$= p - p^2$$

$$= p(1-p)$$

as before.

Example 5. There are six identical chips numbered 1 to 6. A monkey has been trained to select one chip and give it to its trainer, whereupon it receives a number of bananas equal to the number on the chip. The sample space is the chip selected by the monkey. Let X be the number on the chip. If each chip has probability 1/6 of being selected, X has the discrete uniform distribution. For the mean of X we have

$$E(X) = 1(\tfrac{1}{6}) + 2(\tfrac{1}{6}) + 3(\tfrac{1}{6}) + 4(\tfrac{1}{6}) + 5(\tfrac{1}{6}) + 6(\tfrac{1}{6})$$

$$= 3\tfrac{1}{2}$$

The expected value of X^2 is given by

$$E(X^2) = 1(\tfrac{1}{6}) + 4(\tfrac{1}{6}) + 9(\tfrac{1}{6}) + 16(\tfrac{1}{6}) + 25(\tfrac{1}{6}) + 36(\tfrac{1}{6})$$
$$= 15\tfrac{1}{6}$$

The variance of X is computed using Equation 7.

$$\text{Var}\,(X) = E(X^2) - \mu^2$$
$$= 15\tfrac{1}{6} - (3\tfrac{1}{2})^2$$
$$= 2\tfrac{11}{12}$$

The standard deviation is the square root of the variance and, for this example, equals 1.71.

Definition 2 defined the expected value of a function of a single random variable. An extension of the definition may be made to include functions of several random variables considered jointly. This extended definition leads us into consideration of the *covariance* of two random variables and enables us to find the mean and variance of the sum of several random variables.

Definition 5. Let $X_1, X_2, \ldots, X_n$ be random variables with the joint probability function $f(x_1, x_2, \ldots, x_n)$, and let $u(X_1, X_2, \ldots X_n)$ be a real valued function of $X_1, X_2, \ldots, X_n$. Then the *expected value of* $u(X_1, X_2, \ldots, X_n)$ is

(9) $E[u(X_1, X_2, \ldots, X_n)] = \sum u(x_1, x_2, \ldots, x_n) f(x_1, x_2, \ldots, x_n)$

where the summation extends over all possible combinations of values of $x_1, x_2, \ldots, x_n$.

To show that Definition 5 is consistent with Definition 2, consider a function of only one random variable, which we may call $u(X_1)$, although any X_i other than X_1 may be considered in the same way. From Equation 9 we have

$$E[u(X_1)] = \sum u(x_1) f(x_1, x_2, \ldots, x_n)$$

where the summation extends over all combinations of values of $x_1, x_2, \ldots, x_n$. Since

$$f(x_1, x_2, \ldots, x_n) = P(X_1 = x_1, X_2 = x_2, \ldots, X_n = x_n)$$
$$= P(X_2 = x_2, \ldots, X_n = x_n \mid X_1 = x_1) \cdot P(X_1 = x_1)$$

we have

$$E[u(X_1)] = \sum_{x_1} \sum_{x_2} \cdots \sum_{x_n} u(x_1) P(X_2 = x_2, \ldots, X_n = x_n \mid X_1 = x_1) \cdot P(X_1 = x_1)$$
$$= \sum_{x_1} u(x_1) P(X_1 = x_1) \sum_{x_2} \cdots \sum_{x_n} P(X_2 = x_2, \ldots, X_n = x_n \mid X_1 = x_1)$$

However, the last $n - 1$ summations equal unity, because they represent the summing of the conditional probabilities over all of the points in the reduced sample space. We are left with

$$E[u(X_1)] = \sum_{x_1} u(x_1)P(X_1 = x_1)$$

which is the same as in Definition 2. Hence we see that Definition 2 is a special case of Definition 5.

One of the simpler functions of $X_1, X_2, \ldots, X_n$ is

(10) $$Y = X_1 + X_2 + \cdots + X_n$$

That is, each value of the random variable Y associated with the combined experiment involving the X_is is obtained simply by adding the values achieved by all the X_is. Then

$$E(Y) = \sum (x_1 + \cdots + x_n)f(x_1, \ldots, x_n)$$

(11) $$= \sum x_1 f(x_1, \ldots, x_n) + \cdots + \sum x_n f(x_1, \ldots, x_n)$$

where each summation extends over all possible combinations of the values of $x_1, \ldots, x_n$. Using Definition 5, Equation 11 immediately becomes

(12) $$E(Y) = E(X_1) + \cdots + E(X_n)$$

The result of these calculations may be stated as a theorem.

Theorem 1. *Let $X_1, X_2, \ldots, X_n$ be random variables and let*

$$Y = X_1 + X_2 + \cdots + X_n$$

Then $E(Y) = E(X_1) + E(X_2) + \cdots + E(X_n)$.

The statement in Theorem 1 holds true in all cases, whether the random variables are independent or not. Often the apparently difficult problem of finding the mean of the sum of several random variables reduces to a trivial exercise with the use of this theorem.

The results of the next two examples will be used in later chapters.

Example 6. Let Y be the total number of "successes" in n independent trials, where each trial results in either "success" or "failure" with probability p and $q = 1 - p$, respectively. Then Y has the binomial distribution with parameters n and p. However, Y may be regarded as the sum of n independent random variables $X_1, X_2, \ldots, X_n$, where $X_i = 1$ if the ith trial results in "success" and $X_i = 0$ if the ith trial results in failure, for each i from 1 to n. Then

$$Y = X_1 + X_2 + \cdots + X_n$$

and, from Theorem 1,

$$E(Y) = E(X_1) + E(X_2) + \cdots + E(X_n)$$

In Example 2 the mean of X_i was found to equal p. Therefore

(13)
$$E(Y) = np$$

gives the mean for the binomial distribution.

Note that in the binomial distribution the trials are assumed to be independent and, therefore, the X_i are independent. This assumption is not needed in order to find the mean.

The following lemma is needed in Example 7. This lemma presents a convenient equation for expressing the sum of consecutive integers.

Lemma 1.

$$\sum_{i=a}^{N} i = \frac{(N+a)(N-a+1)}{2} \quad and \quad \sum_{i=1}^{N} i = \frac{(N+1)N}{2}$$

Proof. The desired sum may be written two ways. Let $S = \sum_{i=a}^{N} i$. Then

$$S = a + (a+1) + (a+2) + \cdots + (N-1) + N$$
$$S = N + (N-1) + (N-2) + \cdots + (a+1) + a$$

Adding the two equations together gives

$$2S = (N+a) + (N+a) + (N+a) + \cdots + (N+a) + (N+a)$$
$$= (N+a)(N-a+1)$$

Therefore

$$S = \sum_{i=a}^{N} i = \frac{(N+a)(N-a+1)}{2}$$

For $a = 1$, this becomes

$$\sum_{i=1}^{N} i = \frac{(N+1)N}{2}$$

completing the proof.

Example 7. There are N chips in a jar, numbered from 1 to N. One by one, n of those chips, where n is less than N, are drawn from the jar, the number noted, and they are put aside. Let Y be the sum of the numbers on the n drawn chips. Assume the drawings are random; that is, each chip is equally likely to be selected.

The mean of Y would be difficult to find without using Theorem 1. The successive drawings are not independent, because once a number is recorded, no other chip can have that same number. However, we may regard Y as the sum of the random variables $X_1, X_2, \ldots, X_n$, where each X_i is the number on the ith chip drawn.

Now the chip drawn on the ith drawing is just as likely to be any one chip as any other chip. Therefore the probability distribution of X_i, considered by itself, is the discrete uniform distribution, with the probability function

$$P(X_i = k) = \frac{1}{N}, \qquad \text{for} \qquad k = 1, 2, 3, \ldots, N$$

Therefore, with the assistance of Lemma 1, we have the following.

$$
\begin{aligned}
E(X_i) &= \sum_{k=1}^{N} k\left(\frac{1}{N}\right) \\
&= \frac{1}{N} \sum_{k=1}^{N} k \\
&= \frac{1}{N} \frac{N(N+1)}{2} \\
&= \frac{N+1}{2}
\end{aligned}
$$

(14)

Equation 14 furnishes us with the mean of a discrete uniform random variable. Since Y equals $X_1 + X_2 + \cdots + X_n$, we have

$$
\begin{aligned}
E(Y) &= E(X_1) + E(X_2) + \cdots + E(X_n) \\
&= n\frac{N+1}{2}
\end{aligned}
$$

(15)

A particularly useful function of two random variables is $[X_1 - E(X_1)] \times [X_2 - E(X_2)]$, whose expected value is called the *covariance of* X_1 *and* X_2. In particular, a comparison of Definition 4 with the following reveals that the variance of X_1 may be considered as the covariance of X_1 with itself.

Definition 6. Let X_1 and X_2 be two random variables with means μ_1 and μ_2, probability functions $f_1(x_1)$ and $f_2(x_2)$, respectively, and joint probability function $f(x_1, x_2)$. The *covariance of* X_1 *and* X_2 is

(16) $\qquad \text{Cov}(X_1, X_2) = E[(X_1 - \mu_1)(X_2 - \mu_2)]$

The definition of expected value, Definition 5, may be used to give

$$
\begin{aligned}
\text{Cov}(X_1, X_2) &= E[(X_1 - \mu_1)(X_2 - \mu_2)] \\
&= \sum (x_1 - \mu_1)(x_2 - \mu_2)f(x_1, x_2)
\end{aligned}
$$

(17)

where the summation extends over all values of x_1 and x_2. This expands as

$$
\begin{aligned}
\text{Cov}(X_1, X_2) &= \sum (x_1 x_2 - \mu_1 x_2 - \mu_2 x_1 + \mu_1 \mu_2)f(x_1, x_2) \\
&= \sum x_1 x_2 f(x_1, x_2) - \mu_1 \sum x_2 f(x_1, x_2) - \mu_2 \sum x_1 f(x_1, x_2) \\
&\qquad\qquad\qquad\qquad + \mu_1 \mu_2 \sum f(x_1, x_2) \\
&= E(X_1 X_2) - \mu_1 \mu_2 - \mu_2 \mu_1 + \mu_1 \mu_2 \\
&= E(X_1 X_2) - \mu_1 \mu_2
\end{aligned}
$$

(18)

Equation 18 is often easier to use than Equation 17 when calculating a covariance.

Example 8. An insurance company has noticed that the probability of any particular person having an automobile accident within a given year is about .1. However, this probability becomes .3 if it is known that the person had an automobile accident the previous year.

Let X_1 equal 0 or 1, depending on whether a particular person has no accidents or at least one accident, respectively, during the first year of his or her insurance period. Let X_2 be similarly defined for the second year of that same person's insurance period. The probability function of X_1, and therefore X_2 also, is

$$P(X_1 = 0) = .9$$
$$P(X_1 = 1) = .1$$

From example 2 we obtain

$$E(X_1) = .1$$
$$E(X_2) = .1$$

The joint probability function of X_1 and X_2 at $X_1 = 1$ may be found as follows.

$$f(1, 1) = P(X_1 = 1, X_2 = 1)$$
$$= P(X_2 = 1 \mid X_1 = 1)P(X_1 = 1)$$
$$= (.3)(.1)$$
$$= .03$$

The computation of $E(X_1 X_2)$ follows directly from Definition 5.

$$E(X_1 X_2) = (1)(1)f(1, 1) \text{ plus "zero" terms}$$
$$= .03$$

The covariance of X_1 and X_2 is then obtained by using Equation 18.

$$\text{Cov }(X_1, X_2) = E(X_1 X_2) - E(X_1)E(X_2)$$
$$= .03 - (0.1)(0.1)$$
$$= .02$$

We will now define the *correlation coefficient,* which is used as a measure of linear dependence between two random variables. Although we will not prove it here, the correlation coefficient is always between -1 and $+1$. It equals zero when the two random variables are independent, although it may equal zero in other cases also.

Definition 7. The *correlation coefficient* between two random variables is their covariance divided by the product of their standard deviations. That is, the correlation coefficient, usually denoted by ρ, between two random variables X_1 and X_2 is given by

(19)
$$\rho = \frac{\text{Cov}(X_1, X_2)}{\sqrt{\text{Var}(X_1)\,\text{Var}(X_2)}}$$

A lemma that will be used in the next example is now presented. This lemma furnishes us with a convenient formula for expressing the sum of the squares of the first N consecutive integers.

Lemma 2.

$$\sum_{i=1}^{N} i^2 = \frac{N(N+1)(2N+1)}{6}$$

Proof. Let $S = \sum_{i=1}^{N} i^2$. Then

$$
\begin{aligned}
S &= 1^2 + 2^2 + 3^2 + 4^2 + \cdots + N^2 \\
&= 1 + 2 + 3 + 4 + \cdots + N \\
&\quad + 2 + 3 + 4 + \cdots + N \\
&\quad\quad + 3 + 4 + \cdots + N \\
&\quad\quad\quad + 4 + \cdots + N \\
&\quad\quad\quad\quad \cdots \quad \cdots \\
&\quad\quad\quad\quad\quad + N
\end{aligned}
$$

where the sum of the numbers in the ith column is i^2. However, instead of adding down the columns, we will add across the rows. The sum of the numbers in the jth row from the top is found by Lemma 1 to be

$$j + (j+1) + (j+2) + \cdots + N = \frac{(N+j)(N-j+1)}{2}$$

$$= \tfrac{1}{2}(N^2 + N + j - j^2)$$

Adding these row sums together gives

$$S = \sum_{j=1}^{N} \tfrac{1}{2}(N^2 + N + j - j^2)$$

$$= \frac{1}{2}\sum_{j=1}^{N} N^2 + \frac{1}{2}\sum_{j=1}^{N} N + \frac{1}{2}\sum_{j=1}^{N} j - \frac{1}{2}\sum_{j=1}^{N} j^2$$

$$= \tfrac{1}{2}(N \cdot N^2) + \tfrac{1}{2}(N \cdot N) + \tfrac{1}{2} \cdot \frac{(N+1)N}{2} - \tfrac{1}{2}S$$

since the last sum in the middle equation is denoted by S in this proof. Rearranging gives

$$\tfrac{3}{2}S = \tfrac{1}{4}(2N^3 + 3N^2 + N) = \tfrac{1}{4}N(N+1)(2N+1)$$

so that

$$S = \sum_{i=1}^{N} i^2 = \frac{N(N+1)(2N+1)}{6}$$

completing the proof.

Example 9. A jar contains N plastic chips numbered 1 to N, as in Example 7. An experiment consists of drawing n of these chips from the jar, where $n \le N$. We assume that each chip is equally likely to be selected and that the drawing is without replacement. Let $X_1, X_2, \ldots, X_n$ be random variables, where X_i equals the number on the ith chip drawn from the jar, for $i = 1, 2, \ldots, n$. In this example we will find the covariance of X_i and X_j. From Example 7, we have

$$E(X_i) = \frac{N+1}{2}$$

Also, from Equation 7 and Lemma 2 we have

$$\text{Var}(X_i) = E(X_i^2) - [E(X_i)]^2 = \sum_{k=1}^{N} k^2 \frac{1}{N} - \left(\frac{N+1}{2}\right)^2$$

$$= \frac{1}{N} \frac{N(N+1)(2N+1)}{6} - \left(\frac{N+1}{2}\right)^2$$

(20) $$= \frac{(N+1)(N-1)}{12}$$

Now consider jointly two random variables X_i and X_j, where $i \ne j$. Their joint probability function is

$$f(x_i, x_j) = P(X_i = x_i, X_j = x_j) = P(X_i = x_i \mid X_j = x_j) \cdot P(X_j = x_j)$$

(21) $$= \frac{1}{N-1} \cdot \frac{1}{N}, \quad \text{for } x_i, x_j = 1, 2, \ldots, N; \; x_i \ne x_j$$

because once X_j is known to equal x_j, X_i may equal any integer from 1 to N except the integer x_j. Hence there are $N-1$ equally likely values for X_i, each with a probability $1/(N-1)$.

The covariance of X_i and X_j, using Definition 6, is

$$\text{Cov}(X_i, X_j) = E\{[X_i - E(X_i)][X_j - E(X_j)]\}$$

Since the mean of both X_i and X_j is $(N+1)/2$, we have

$$\text{Cov}(X_i, X_j) = \sum_{\substack{k=1 \\ k \ne s}}^{N} \sum_{s=1}^{N} \left(k - \frac{N+1}{2}\right)\left(s - \frac{N+1}{2}\right)\frac{1}{(N-1)N}$$

where the summation extends over all k and s from 1 to N, except that k does not equal s because X_i and X_j cannot equal the same number at the same time. If we, at the same time, add and subtract those terms for $k = s$ the covariance becomes

$$\text{Cov}(X_i, X_j) = \sum_{k=1}^{N} \sum_{s=1}^{N} \left(k - \frac{N+1}{2}\right)\left(s - \frac{N+1}{2}\right)\frac{1}{(N-1)(N)}$$

$$- \sum_{k=1}^{N} \left(k - \frac{N+1}{2}\right)^2 \frac{1}{(N-1)N}$$

$$(22) \qquad = \frac{1}{(N-1)N} \sum_{k=1}^{N} \left(k - \frac{N+1}{2}\right) \sum_{s=1}^{N} \left(s - \frac{N+1}{2}\right)$$

$$- \frac{1}{N-1} \sum_{k=1}^{N} \left(k - \frac{N+1}{2}\right)^2 \frac{1}{N}$$

To simplify Equation 22 we note that

$$(23) \qquad \sum_{i=1}^{N} \left(i - \frac{N+1}{2}\right) = \sum_{i=1}^{N} i - \sum_{i=1}^{N} \frac{N+1}{2} = \frac{N(N+1)}{2} - \frac{N(N+1)}{2} = 0$$

and therefore the first term in Equation 22 equals zero. Also, from the definition of variance and from Equation 20 we have

$$(24) \qquad \text{Var}(X_i) = \sum_{k=1}^{N} \left(k - \frac{N+1}{2}\right)^2 \frac{1}{N} = \frac{(N+1)(N-1)}{12}$$

Substitution of Equations 23 and 24 into Equation 22 yields

$$(25) \qquad \text{Cov}(X_i, X_j) = -\frac{N+1}{12}$$

The fundamental importance of the covariance is based on what happens to the covariance in the case of two independent random variables. Let X_1 and X_2 be two independent random variables with probability functions $f_1(x_1)$ and $f_2(x_2)$, and means μ_1 and μ_2, respectively. Then the covariance of X_1 and X_2 is

$$\text{Cov}(X_1, X_2) = E(X_1, X_2) - \mu_1\mu_2$$
$$= \sum x_1 x_2 f(x_1, x_2) - \mu_1\mu_2$$

where the summation extends over all combinations of values of x_1 and x_2. Since X_1 and X_2 are independent,

$$f(x_1, x_2) = f_1(x_1)f_2(x_2)$$

and

$$\text{Cov}(X_1, X_2) = \sum_{x_1, x_2} x_1 x_2 f_1(x_1)f_2(x_2) - \mu_1\mu_2$$
$$= \left[\sum_{x_1} x_1 f_1(x_1)\right]\left[\sum_{x_2} x_2 f_2(x_2)\right] - \mu_1\mu_2$$
$$= \mu_1\mu_2 - \mu_1\mu_2 = 0$$

Therefore independence of two random variables implies that their covariance is zero, which in turn implies that their correlation coefficient equals zero.

Theorem 2. *If X_1 and X_2 are independent random variables, the covariance of X_1 and X_2 is zero.*

The converse of Theorem 2 is not necessarily true. That is, zero covariance does not necessarily imply that the random variables are independent, even though the implication is an error made often in practice.

Example 10. Define the joint probability function of two random variables as follows.

$$P(X=0, Y=0)=\tfrac{1}{2}$$
$$P(X=1, Y=1)=\tfrac{1}{4}$$
$$P(X=-1, Y=1)=\tfrac{1}{4}$$

The probability function of X is then

$$P(X=0)=\tfrac{1}{2}$$
$$P(X=1)=\tfrac{1}{4}$$
$$P(X=-1)=\tfrac{1}{4}$$

and the probability function of Y is

$$P(Y=0)=\tfrac{1}{2}$$
$$P(Y=1)=\tfrac{1}{2}$$

The expected values of X and Y are

$$E(X)=0$$
$$E(Y)=\tfrac{1}{2}$$

The covariance of X and Y is

$$Cov\,(X, Y) = E(XY) - E(X)E(Y)$$
$$= (1)(\tfrac{1}{4}) + (-1)\tfrac{1}{4} - (0)(\tfrac{1}{2})$$
$$= 0$$

However, X and Y are not independent, because

$$P(X=0, Y=0)=\tfrac{1}{2}$$

which is not equal to

$$P(X=0)P(Y=0) = (\tfrac{1}{2})(\tfrac{1}{2})$$
$$= \tfrac{1}{4}$$

Therefore X and Y have zero covariance, even though they are not independent.

We are now equipped to find the variance of the sum of several random variables. Let Y equal $X_1 + X_2 + \cdots + X_n$, where the X_is may or may not be independent. We wish to find the variance of Y.

$$\begin{aligned}
\mathrm{Var}\,(Y) &= E\{[Y - E(Y)]^2\} \\
&= E\{[X_1 + X_2 + \cdots + X_n - E(X_1) - E(X_2) - \cdots - E(X_n)]^2\} \\
&= E\{[X_1 - E(X_1) + X_2 - E(X_2) + \cdots + X_n - E(X_n)]^2\} \\
&= E\left\{\sum_{i=1}^{n}[X_i - E(X_i)]^2 + \sum_{\substack{i=1\ j=1 \\ i \neq j}}^{n}\sum[X_i - E(X_i)][X_j - E(X_j)]\right\}
\end{aligned}$$

But since the expected value of a sum of random variables equals the sum of the expected values of the random variables,

$$\mathrm{Var}\,(Y) = \sum_{i=1}^{n} E\{[X_i - E(X_i)]^2\} + \sum_{\substack{i=1\ j=1 \\ i \neq j}}^{n}\sum E\{[X_i - E(X_i)][X_j - E(X_j)]\}$$

$$(26) \qquad = \sum_{i=1}^{n} \mathrm{Var}\,(X_i) + \sum_{\substack{i=1\ j=1 \\ i \neq j}}^{n}\sum \mathrm{Cov}\,(X_i, X_j)$$

If $X_1, \ldots, X_n$ are mutually independent then, from Theorem 2, we have $\mathrm{Cov}\,(X_i, X_j) = 0$, and

$$(27) \qquad\qquad \mathrm{Var}\,(Y) = \sum_{i=1}^{n} \mathrm{Var}\,(X_i)$$

We may summarize this as a theorem.

Theorem 3. *Let $X_1, X_2, \ldots, X_n$ be random variables and let*

$$Y = X_1 + X_2 + \cdots + X_n$$

Then

$$\mathrm{Var}\,(Y) = \sum_{i=1}^{n} \mathrm{Var}\,(X_i) + \sum_{\substack{i=1\ j=1 \\ i \neq j}}^{n}\sum \mathrm{Cov}\,(X_i, X_j)$$

Furthermore, if $X_1, X_2, \ldots, X_n$ are mutually independent,

$$\mathrm{Var}\,(Y) = \sum_{i=1}^{n} \mathrm{Var}\,(X_i)$$

Example 11. In continuation of Example 9, let X_i equal the number on the ith chip drawn, as before, and let Y equal the sum of the X_is, as in Example 7. Then Theorem 3 gives us

$$\mathrm{Var}\,(Y) = \sum_{i=1}^{n} \mathrm{Var}\,(X_i) + \sum_{\substack{i=1\ j=1 \\ i \neq j}}^{n}\sum \mathrm{Cov}\,(X_i, X_j)$$

The various terms in this equation are given by Equations 20 and 25. The variance term appears n times, and the covariance term appears $n(n-1)$ times.

$$\text{Var}(Y) = n\frac{(N+1)(N-1)}{12} + n(n-1)\left(-\frac{N+1}{12}\right)$$

(28)
$$= \frac{n(N+1)(N-n)}{12}$$

Note that $\text{Var}(X_i)$ is only a special case of $\text{Var}(Y)$ for $n=1$.

The variance of a random variable that has the binomial distribution is found in Example 12.

Example 12. Consider n independent trials, where each trial may result in "success" with probability p, or "failure" with probability q, where $p+q$ equals 1. As in Examples 4 and 6, let X_i equal 0 or 1, depending on whether the ith trial results in "failure" or "success," respectively, and let Y equal the total number of "successes" in the n trials. Since the X_i are mutually independent, Theorem 3 states that

$$\text{Var}(Y) = \sum_{i=1}^{n} \text{Var}(X_i)$$

From Example 4, $\text{Var}(X_i)$ equals pq, so

$$\text{Var}(Y) = npq$$

furnishes us with the variance of Y, which has the binomial distribution.

The results of some of the preceding examples will be used later in this book, and so they are stated separately as theorems, for convenience.

Theorem 4. *Let X be a random variable with the binomial distribution*

$$P(X=k) = \binom{n}{k}p^k q^{n-k}$$

Then the mean and variance of X are given by

$$E(X) = np$$
$$\text{Var}(X) = npq$$

Theorem 5. *Let X be the sum of n integers selected at random, without replacement, from the first N integers 1 to N. Then the mean and variance of X are given by*

$$E(X) = \frac{n(N+1)}{2}$$

$$\text{Var}(X) = \frac{n(N+1)(N-n)}{12}$$

Example 13. An advertising agency drew 12 sample magazine ads for one of their customers and ranked the ads from 1 to 12 on the basis of the agency's opinion of which ads would be the most effective in selling the product. The "most effective" ad was given the rank 1, and so on. The customer, the manufacturer of the product, selected 4 ads for purchase. They were ranked 4, 6, 7, and 11 by the agency.

Assuming that the customer's choice and the agency's rankings were independent, the sum of the ranks on the selected ads should be distributed the same as the sum of the numbers on 4 chips selected at random out of 12 chips numbered 1 to 12. Let X equal the sum of the ranks of 4 ads if they are selected independently of the ranks. Then Theorem 5 states that the mean of X is

$$E(X) = \frac{(4)(12+1)}{2} = 26$$

and the variance of X is

$$\text{Var}(X) = \frac{(4)(12+1)(12-4)}{12} = 34\frac{2}{3}$$

The standard deviation of X is

$$\sigma = \sqrt{\text{Var}(X)} = 5.9$$

The observed value of X is

$$X = 4+6+7+11 = 28$$

which is close to the mean of X under the preceding assumptions.

EXERCISES

1. If $P(X=0)=1/3$, $P(X=1)=1/3$, $P(X=2)=1/6$, and $P(X=3)=1/6$, find
   (a) $E(X)$ (b) $\text{Var}(X)$.
   (c) $E(X^2+2X)$.   (d) The median.
   (e) $x_{1/3}$.   (f) The fourth decile.

2. If $P(X=0)=0$, $P(X=1)=1/2$, $P(X=2)=1/4$, and $P(X=4)=1/4$, find
   (a) $E(X)$.   (b) $\text{Var}(X)$.
   (c) $E(-X)$.   (d) The median.
   (e) The upper quantile.   (f) The thirty-seventh percentile.

3. If $P(X=0, \ Y=0)=1/4$, $P(X=0, \ Y=1)=1/4$, $P(X=1, \ Y=0)=1/4$, and $P(X=1, \ Y=1)=1/4$, find

   (a) $E(X)$.   (b) $E(Y)$.
   (c) $E(XY)$.   (d) $E(X+Y)$.
   (e) $\text{Cov}(X, Y)$.   (f) $P(X=0)$.
   (g) $P(X=1)$.   (h) Are X and Y independent?

4. If $P(X=0, \ Y=0)=1/8, \ P(X=0, \ Y=1)=3/8, \ P(X=1, \ Y=0)=3/8,$ and $P(X=1, \ Y=1)=1/8,$ find
   (a) $E(X)$.   (b) $E(Y)$.
   (c) $E(XY)$.   (d) $E(X^2Y)$.
   (e) $\text{Cov}(X, Y)$.   (f) $P(X=x)$ for all x.
   (g) Are X and Y independent?

5. What is the sum of the 66 integers from 1 to 66?

6. What is the sum of the 30 integers from 70 to 99?

7. If X equals the number of spots showing on one roll of a balanced die, find
   (a) $E(X)$.
   (b) $\text{Var}(X)$.
   (c) $E(X^2+X)$.

8. If 30 tickets are numbered consecutively from 1 to 30, and 2 tickets are selected at random without replacement, find
   (a) E(sum of the numbers on the two tickets).
   (b) Var(sum of the numbers on the two tickets).

PROBLEMS

1. Prove that a vertical line drawn at the mean of a random variable X divides the distribution function of X in such a way that the area under the distribution function, to the left of the mean, and above 0, equals the area above the distribution function, to the right of the mean, and below 1.

2. In the same way that Lemma 2 was obtained using Lemma 1, prove Lemma 3,

$$\sum_{i=1}^{N} i^3 = \frac{N^2(N+1)^2}{4}$$

using Lemma 2. [A general extension is given by Iman (1970).]

1.5. CONTINUOUS RANDOM VARIABLES

All of the random variables that we have introduced so far in this chapter have one property in common: their possible values can be listed. The list of possible values assumed by the binomial random variable is $0, 1, 2, 3, 4, \ldots, n-1,$ n. No other values may be assumed by the binomial random variable. The list of values that may be assumed by the discrete uniform random variable could be written as $1, 2, 3, \ldots, N$. Similar lists could be made for each random variable introduced in the previous definitions and examples.

These lists may be infinite in length, such as in an experiment where the random variable X equals the number of times a monkey pushes the "wrong" button before finally pushing the "right" button and getting a reward. Then X may equal zero if the right button is pushed the first time, or X may equal 1000 if the monkey has difficulty finding the right button. Theoretically there is no limit to the number of times the monkey may choose the wrong button before pushing the correct one. The possible values of X may be listed, even though the list may be infinitely long. The infinitely long list of possible values is a characteristic of the model and not of the actual experiment in this example and, in most situations, because real factors such as the eventual death of the monkey, the absence of research funds, or the waning enthusiasm of the experimenter will prevent the actual experiment from being prolonged to the point of absurdity. Nevertheless, the model may be reasonable, and the random variable in the model may have an infinity of possible values.

A more precise way of stating that the possible values of a random variable may be listed is to say that there exists a *one-to-one correspondence* between the possible values of the random variable and some or all of the positive integers. This means that to each possible value there corresponds one and only one positive integer, and that positive integer does not correspond to more than one possible value of the random variable. Random variables with this property are called *discrete*. All of the random variables we have considered so far are discrete random variables.

Definition 1. A random variable X is *discrete* if there exists a one to one correspondence between the possible values of X and some or all of the positive integers.

The distribution function of a discrete random variable is always a *step function*, that is, the graph looks like a series of stair steps, although the steps may not be very uniform in appearance, and there may be an infinite number of steps. If any portion of the graph seems to rise gradually instead of rising only in clear-cut steps, the associated random variable is not discrete.

If the graph of a distribution function has no steps but rises only gradually where it rises, the distribution function is called *continuous*, and the random variable with that distribution function is called a *continuous random variable*. Figure 3 is an example of the graph of a continuous distribution function. Saying that a distribution function has no steps is the same as saying that no two horizontal lines will intersect the graph at the same value as measured along the horizontal axis. That is, if there is a step in the graph of a distribution function, at least two horizontal lines may be drawn, say at heights p and p',

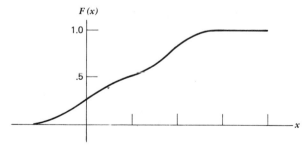

Figure 3

closely enough to each other so that they intersect the graph at the same value, as measured along the horizontal axis. Since this describes the graphical method of finding quantiles, we may say that if there is a step in the distribution function, there are at least two quantiles x_p and $x_{p'}$ that are equal to each other. Conversely, if there are no two quantiles exactly equal to each other, there are no steps in the distribution function, and the function is continuous. This leads to a method of defining *continuous random variable.*

Definition 2. A random variable X *is continuous* if no two quantiles x_p and $x_{p'}$ of X are equal to each other, where p is not equal to p'. Equivalently, a random variable X is continuous if $P(X \leq x)$ equals $P(X < x)$ for all numbers x.

Example 1. The distribution function graphed in Figure 4 is a continuous distribution function, and any random variable with the distribution function $F(x)$ is a continuous random variable. Typical continuous random variables involve measuring time, weight, distance, volume, and so forth.

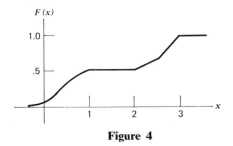

Figure 4

In practice, no actual random variable is continuous, because the observed values of actual random variables are always the result of measurements of some sort, and measurements are made with tools that have only a finite capacity for discriminating between two values. Continuous random variables exist only in theory, such as in a model of an actual experiment. At times it is preferable to assume a random variable to be continuous, even though it is known to be discrete, as in Example 2.

Example 2. The time it takes a racehorse to run a mile race is a continuous quantity, because time is generally a continuous quantity. In practice, however, the time is usually measured to the nearest 1/5 second. It is not unusual for a horse to run two races in identical lengths of time (i.e., measured time). The actual lengths of time will be exactly equal with probability zero; therefore it is reasonable to assume that the time of a race, measured exactly, is a continuous random variable that is approximately equal to the measured time of the race, a discrete random variable. If two horses run in the same race and cross the finish line ahead of all other horses, with identical measured times, the winner of the race is then determined by

examining a photograph taken at the finish line at the moment the horses
crossed. Only rarely does this fail to determine the actual winner of the race.
This is meant to illustrate that even though the random variable (measured
time) is discrete, it is assumed to be continuous because the actual order in
which the horses finished the race still may be determined even though two
or more measured times seem to be identical.

Another reason for considering continuous random variables is that the
distribution function of a discrete random variable sometimes may be approximated by a continuous distribution function, resulting in a convenient method
for computing desired probabilities associated with the discrete random variable. Two continuous distribution functions commonly used for this purpose
are the *normal distribution* and the *chi-square distribution.*

The distribution function in the following definition might frighten those who
are unfamiliar with elementary calculus. There is no cause for alarm, however,
because the distribution function is well tabulated. Such tables may be found in
most statistics texts, as well as in Table A1.

Definition 3. Let X be a random variable. Then X is said to have the *normal
distribution* if the distribution function of X is given by

(1) $$F(x) = P(X \le x) = \int_{-\infty}^{x} \frac{1}{\sqrt{2\pi}\sigma} e^{-\frac{1}{2}[(y-\mu)/\sigma]^2} dy$$

where it can be shown (using calculus) that the parameters μ and σ are the
mean and standard deviation of X. The *standard normal distribution* is the
normal distribution with μ equal to 0 and σ equal to 1.

The normal distribution function cannot be evaluated directly, and so Table
A1 may be used to find approximate probabilities associated with normal
random variables. Table A1 gives selected quantiles of the standard normal
random variable. Quantiles of normal random variables with mean μ and
variance σ^2 may be found from Table A1, with the aid of the equations given
without proof in the following theorem.

*Theorem 1. For a given value of p, let x_p be the pth quantile of a normal random
variable with mean μ and variance σ^2, and let w_p be the pth quantile of a
standard normal random variable. The quantile x_p may be obtained from w_p by
using the relationship*

(2) $$x_p = \mu + \sigma w_p$$

Similarly, w_p may be obtained from x_p with the aid of the relationship

(3) $$w_p = \frac{x_p - \mu}{\sigma}$$

Example 3. Let X be a random variable with the standard normal distribution. To find the probability that X will not exceed 1.42, Table A1 is used.

From Table A1 we see that

$$P(X \le 1.4187) = .922$$

and

$$P(X \le 1.4255) = .923$$

Therefore, as an approximation, we may either interpolate to get an approximate probability

$$P(X \le 1.42) \cong .9222$$

or we may simply use the number closest to 1.42 to obtain

$$P(X \le 1.42) \cong .922$$

Example 4. Let X be the IQ of a person selected at random from a large group of people; assume that X has the normal distribution with mean 100 and standard deviation 15.

Suppose we want to find the probability that X will exceed 125. We have

$$P(X > 125) = 1 - P(X \le 125)$$

so it suffices to find $P(X \le 125)$. The quantile w_p, of the standard normal random variable, corresponding to the quantile $x_p = 125$ is found from Equation 3.

$$w_p = \frac{x_p - \mu}{\sigma}$$

$$= \frac{125 - 100}{15}$$

$$= 1.67$$

From Table A1 we see that if $w_p = 1.67$, $p = .95$. Therefore 125 is the .95 quantile of X.

$$P(X \le 125) = .95$$

$$P(X > 125) = .05$$

The desired probability is .05.

To find the upper 1 percentile, called the 99th percentile, we want to find the number $x_{.99}$, where

$$P(X \le x_{.99}) = .99$$

Since, from Table A1, $w_{.99} = 2.3263$, $x_{.99}$ may be found from Equation 2 to be

$$x_{.99} = \mu + \sigma w_{.99}$$

$$= 100 + 15(2.3263)$$

$$= 134.9$$

Therefore the probability of the randomly selected person having an IQ less than 135 is about .99.

Example 5. A railroad company has observed over a period of time that the number X of people taking a certain train seems to follow a normal distribution with mean 540 and standard deviation 32. How many seats should the company provide on the train if it wants to be 95% certain that everyone will have a seat?

We wish to find the 95th percentile. From Equation 2 we have that

$$x_{.95} = \mu + \sigma w_{.95}$$
$$= 540 + 32(1.6449)$$
$$= 592.6$$

where $w_{.95}$ is obtained from Table A1. The company needs 593 seats on the train so that they can be 95% certain that there will be enough seats for everyone on any one run of that train.

In Example 5 the random variable X is actually a discrete random variable that assumes only the nonnegative integers as values. Therefore X cannot possibly have a normal distribution. The normal approximation to the distribution of X was used partly for convenience and partly out of necessity, because a realistic discrete distribution might be difficult to formulate. In other problems a discrete distribution function that agrees well with the data may be known, yet the normal approximation still might be used for ease in calculations. The validity of using the normal approximation usually depends on the central limit theorem.

The so-called central limit theorem appears in many different forms. All forms have in common the purpose of stating conditions under which the sum of several random variables may be approximated by a normal random variable. The theorem says that the distribution function of the sum of several random variables approaches the normal distribution function, as the number of random variables being added becomes large (i.e., goes to infinity), and when certain other general conditions are met. These "other general conditions" may be stated many different ways, giving rise to the many forms for the central limit theorem. Although a thorough discussion of this theorem is well beyond the scope of this book, frequent reference in this book to the use of the theorem invites a brief attempt to dispel some of the mystery that might otherwise build.

Theorem 2. (*Central Limit Theorem*) *Let Y_n be the sum of the n random variables $X_1, X_2, \ldots, X_n$, let μ_n be the mean of Y_n, and let σ_n^2 be the variance of Y_n. As n, the number of random variables, goes to infinity, the distribution function of the random variable*

$$\frac{Y_n - \mu_n}{\sigma_n}$$

approaches the standard normal distribution function, if one of the following sets of conditions holds.

> *Set A: The X_i are independent and identically distributed, with $\infty > \text{Var}(X_i) > 0$ (Fisz, 1963, p. 197).*
>
> *Set B: The X_i are independent but not necessarily identically distributed, but $E(X_i^3)$ exists for all i and satisfies certain conditions (Fisz, 1963, p. 203).*
>
> *Set C: The X_i are neither independent nor identically distributed, but represent the successive drawings, without replacement, of values from a finite population of size N, where N is greater than 2n. Also a condition stated in Fisz (1963, p. 523) should be satisfied.*

Some of the conditions that must be met for the central limit theorem to apply are not stated here because they are somewhat mathematical and would not add to the intuitive understanding of the theorem. A convenient reference is given for the benefit of the interested reader.

When the central limit theorem is stated in terms of Set A, it is known as the Lindeberg-Lévy theorem (Lindeberg, 1922, and Lévy, 1925). When the theorem is stated with Set B conditions it is usually called the Lapunov theorem (Lapunov, 1901). The theorem with conditions given by Set C was proved by Erdös and Rényi (1959).

In practice, the number of random variables summed never goes to infinity. But the value of the central limit theorem is that in situations where the theorem holds, the normal approximation is usually considered to be "reasonably good" as long as n is "large." The terms "reasonably good" and "large" are subjective terms; therefore much latitude exists in the practice of using the normal approximation.

A useful illustration of a situation where the Set A conditions hold is presented in Example 6.

Example 6. Let Y_n be a random variable with the binomial distribution (Definition 1.3.5) with mean np and variance npq (Theorem 1.4.4). Then Y_n may be regarded as the sum of n independent random variables, each with the binomial distribution where n equals 1 (Example 1.4.12). Since these other random variables have a variance of pq, which is greater than zero for $0 < p < 1$, the Set A of conditions holds. For large n, the random variable

$$\frac{Y_n - np}{\sqrt{npq}}$$

is distributed approximately the same as a standard normal random variable. This is equivalent to saying that for large n, the distribution function of Y_n may be approximated by the normal distribution with mean np and variance npq (Theorem 1).

The conditions given by set C are met in Example 7, which will be useful to us in Chapter 5.

Example 7. consider the sampling scheme where n integers are selected at random, without replacement, from the first N integers, 1 to N. Let X_i be the ith integer selected, and let

$$Y_n = X_1 + X_2 + \cdots + X_n$$

be the sum of the integers selected. For large n and large N and $n < N/2$, the set C conditions hold and the distribution function of

$$\frac{Y_n - \dfrac{n(N+1)}{2}}{\left(\dfrac{n(N+1)(N-n)}{12}\right)^{\frac{1}{2}}}$$

(Theorem 1.4.5) may be approximated by the standard normal distribution function. In other words, the distribution function of Y_n may be approximated by the normal distribution function with mean $n(N+1)/2$ and variance $n(N+1)(N-n)/12$ (Theorem 1).

The widespread applicability of the central limit theorem makes it a very useful theorem. Since it justifies to some extent the use of the normal approximation, the normal distribution is a valuable distribution. Other distributions that are related to the normal distribution also become important, such as the *chi-square distribution.*

In the following definition, the chi-square distribution function is given using the "integral" notation of calculus and the "gamma function" $\Gamma(k/2)$. This notation needs no explanation or even understanding because the tabulated values of the chi-square distribution function, given in Table A2, will be used whenever values of the distribution function are needed. A more extensive table is given by Harter (1964). A convenient nomogram is provided by Boyd (1965).

Definition 4. A random variable X has the *chi-square distribution with k degrees of freedom* if the distribution function of X is given by

$$(4) \qquad F(x) = P(X \le x) = \int_0^x \frac{y^{(k/2)-1} e^{(y/2)}}{2^{k/2} \Gamma(k/2)} \, dy \qquad \text{if } x > 0$$

$$= 0 \qquad\qquad\qquad \text{if } x \le 0$$

The distribution function Equation 4 shows that a chi-square random variable may assume only nonnegative values, since $F(x) = 0$ for negative values of x. The degrees of freedom, k, is merely a parameter. The values of k are usually restricted to the integers 1, 2, 3, and so forth. For different values of the parameter k, the distribution functions are different also. Table A2 gives some selected quantiles of a chi-square random variable, for $k = 1, 2, 3$, up to

30, and for some values of k greater than 30. For k greater than 100 the central limit theorem may be used to obtain approximate quantiles, which will be justified later in this section.

It is shown in most introductory books on mathematical statistics that if X is a random variable with the chi-square distribution with k degrees of freedom, the mean and variance of X are given by

(5) $$E(X) = k$$

(6) $$\text{Var}(X) = 2k$$

The following theorem is proved in Freund (1962, p. 194).

Theorem 3. *Let* $X_1, X_2, \ldots, X_k$ *be* k *independent and identically distributed standard normal random variables. Let* Y *be the sum of the squares of the* X_i.

(7) $$Y = X_1{}^2 + X_2{}^2 + \cdots + X_k{}^2$$

Then Y *has the chi-square distribution with* k *degrees of freedom.*

Example 8. A child psychologist asks each of 100 children to tell which of two trucks they would rather play with. The two trucks are identical in all respects, except that one is red and the other is green. The psychologist is interested in knowing whether children have a color preference.

Forty-two children selected the green truck, and the other 58 chose the red truck. In the model "no preference" is assumed, so the random variable X, equal to the number of children who selected the green truck, should have the binomial distribution with mean $np = 50$ and variance $npq = 25$. The normal approximation to the distribution function of X seems appropriate, so

$$\frac{X - 50}{5}$$

is considered to be approximately the same as a standard normal random variable. However, the psychologist is interested in determining differences in either direction; that is she wants to know whether X is much smaller than 50 as well as whether X is much larger than 50, so she uses the square of the difference essentially, but actually examines the random variable

$$X^* = \left(\frac{X - 50}{5}\right)^2$$

because it may be compared with a chi-square random variable with 1 degree of freedom. In this experiment $X^* = [(42 - 50)/5]^2$, or 2.56. The probability of getting a number smaller than 2.56, corresponding to a value of X closer to 50, is found by interpolation in Table A2, $k = 1$, to be about .88. Therefore the psychologist concludes that there is some indication of a color preference among the children. (More will be said concerning this method of drawing a conclusion in later chapters.)

In Example 8 the distribution function of the random variable $(X-50)/5$ was considered to be approximately equal to the standard normal distribution function. Therefore the chi-square approximation, with 1 degree of freedom, was used for the distribution of X^*. The desired probability could have been found by using both tails of the normal distribution function. That is,

$$P(X^* \le (-1.6)^2) = P\left(-1.6 < \frac{X-50}{5} < +1.6\right)$$

The probability on the left, found from Table A2, should equal the probability on the right, obtained from Table A1. The only difference between the two probabilities results from using interpolation in the two tables. If more than 1 degree of freedom is involved, then Table A1 may not be used as an alternative to Table A2.

Example 9. In continuation of the experiment described in Example 8, the psychologist obtains two toy telephones, identical except that one is white and the other is blue. She asks each of 25 children to choose one to play with. Seventeen children chose the white telephone, and the other 8 preferred the blue telephone. Let Y be the random variable equal to the number of children selecting the white telephone. Since

$$\frac{Y-np}{\sqrt{npq}} = \frac{Y-(1/2)(25)}{5/2}$$

is approximately a standard normal random variable under the assumption of no color preference, the random variable

$$Y^* = \left(\frac{Y-(1/2)(25)}{5/2}\right)^2$$

may be compared with a chi-square random variable with one degree of freedom. Since $Y=17$, $Y^*=3.24$. The probability of a chi-square random variable with 1 degree of freedom being less than 3.24 is found from Table A2 to equal about .92, using interpolation. Therefore, if the assumption that each toy was equally likely to be chosen is in fact true, such a large deviation from the expected value of 12.5 would occur only about 8% of the time.

Since the experiment in Example 8 and this one were designed for the same purpose, it would seem desirable to be able to combine the results in some way. If X^* and Y^* may be considered as independent random variables, a reasonable consideration here, Theorem 3 may be used to combine X^* and Y^* as

$$W - X^* + Y^*$$

and the distribution function of W may be approximated by the chi-square distribution function with 2 degrees of freedom. Then

$$W = 2.56 + 3.24$$
$$= 5.80$$

The probability of a chi-square random variable with two degrees of freedom being greater than 5.80 is only about .06, which was obtained by interpolation in Table A2.

In this example more information concerning the presence of color preference among children was obtained by combining the information gained in the two studies.

It should be noted that if Y had been defined as the number of children preferring the blue telephone, instead of the way Y was defined in this example, Y^* would still have the same value, because the deviation of Y from the mean was squared, eliminating the directional influence of the difference.

In Example 9, two approximate chi-square random variables were added, and their sum was an approximate chi-square random variable with 2 degrees of freedom. This method of combining independent chi-square random variables is valid in general. More discussion of this method is given by Radhadrishna (1965) and Nelson (1966).

The following theorem may be found in Freund (1962, p. 194).

Theorem 4. *Let* $X_1, X_2, \ldots, X_n$ *be independent chi-square random variables with* $k_1, k_2, \ldots, k_n$ *degrees of freedom, respectively. Let Y equal the sum of the* X_i. *Then Y is a chi-square random variable with k degrees of freedom, where*

$$k = k_1 + k_2 + \cdots + k_n$$

Theorem 4 will be used later in this book to approximate the distribution function of the sum of several random variables, where the random variables may be assumed to be independent and to be distributed approximately as chi-square random variables.

Since a chi-square random variable with k degrees of freedom may be considered to be the sum of k independent and identically distributed random variables, each having the chi-square distribution with 1 degree of freedom, the set A conditions on the central limit theorem are met. The mean and variance of a chi-square random variable with k degrees of freedom are given by Equations 5 and 6 to be k and $2k$, respectively. Therefore, if Y is a chi-square random variable with k degrees of freedom, the distribution function of

$$\text{(8)} \qquad Z = \frac{Y - k}{\sqrt{2k}}$$

may be approximated by the standard normal distribution function. From Theorem 1, if x_p is a quantile from Table A1, the quantile y_p for Table A2 may be approximated, for large k, by

$$\text{(9)} \qquad y_p = k + \sqrt{2k}\, x_p$$

This is not as good as the approximations

$$\text{(10)} \qquad y_p = \tfrac{1}{2}(x_p + \sqrt{2k - 1})^2$$

or

(11)
$$y_p = k\left(1 - \frac{2}{9k} + x_p\sqrt{\frac{2}{9k}}\right)^3$$

given at the bottom of Table A2.

EXERCISES

1. Let X be a standard normal random variable. Find
  (a) $P(X \le 0)$.   (b) $P(X \le 1.96)$.
  (c) $P(X > 1)$.   (d) $P(-1 < X < 1)$.
  (e) $P(-4 < X < 0)$.   (f) the upper quartile of X.

2. Let X be a normal random variable with mean 0.5 and standard deviation 3. Find
  (a) $P(X \le 0)$.   (b) $P(X \le 1)$.
  (c) $P(X > -0.5)$.   (d) $P(-1 < X < 1)$.
  (e) The median of X.   (f) The upper quartile of X.

3. Suppose that X is the amount of time (in minutes) it takes a certain high school athlete to run 1 mile. Assume that X has a normal distribution with mean 4.30 and standard deviation 0.05. What is the probability that the athlete will break the school record of 4.15 minutes at the annual track meet?

4. Let X be the number of policyholders who make at least one claim to a large insurance company. Assume that there are 2000 policyholders and that each one has probability .2 of making at least one claim during the year. What is the probability that no more than 500 policyholders will make claims during any given year?

5. If the distribution of weights of a certain class of individuals is approximately normal, with mean 160 and variance 400, how high should a set of bathroom scales be calibrated so that about 99% of the people will be able to weigh themselves?

6. If Y is a binomial random variable with parameters $n = 60$, $p = .5$, find the probability that the random variable

$$\frac{(Y - np)^2}{np(1 - p)}$$

will exceed 5.

7. Let X be a chi-square random variable with k degrees of freedom. Find
  (a) The .95 quantile of X, if $k = 4$.
  (b) The .95 quantile of X, if $k = 8$.
  (c) The .99 quantile of X, if $k = 200$.

8. If X, Y, and Z are independent chi-square random variables with 3, 2, and 3 degrees of freedom, respectively, find the probability that W will exceed 15, where $W = X + Y + Z$.

PROBLEMS

1. Let Z be a chi-square random variable with 100 degrees of freedom. Compare the approximations for the .95 quantile of X obtained using Equations 9, 10, and 11, with the exact value obtained from Table A2.
2. Let X be a binomial random variable with parameters $n = 100$, $p = .3$. Estimate

$$P(20 \le X \le 40)$$

using Tables A1 and A2.

1.6. REVIEW PROBLEMS FOR CHAPTER 1

1. A customer is equally likely to select each of six product brands, labeled "brand one," and so forth. Let X equal the brand number selected. Let Y equal 3 if one of the first three brands is selected and 6 if one of the last three is selected. Let Z equal 1 if the brand number is even and 2 if the brand number is odd.
   (a) List the points in the sample space.
   (b) Describe the probability function on the sample space.
   (c) What is the name of the distribution of X?
   (d) Find the interquartile range of Y.
   (e) Find the variance of Z.
   (f) Find the covariance of X and Z.
   (g) Are Y and Z independent?
2. Twelve diamonds are ranked from 1 to 12 according to quality. Three diamonds are selected at random without replacement from the 12. Let X equal the sum of the ranks of the 3 diamonds.
   (a) How many points are in the sample space?
   (b) Decribe any one point in the sample space.
   (c) Find $P(X = 3)$.
   (d) If $f(x)$ is the probability function of X, find $f(10)$.
   (e) If $F(x)$ is the distribution function of X, find $F(10)$.
   (f) Use the central limit theorem to approximate $F(10)$ and compare this approximation with the exact value found in part e.
3. The top 10 students in a large high school graduation class are ranked from 1 (best) to 10 (tenth best). Assume that each rank is equally likely to be assigned to a male student or a female student. Let X equal the *sum* of the ranks (from 1 to 10) that are assigned to female students; that is if all of the top 10 students are girls, $X = 1 + 2 + \cdots + 10 = 55$.
   (a) How many points are in the sample space?
   (b) Describe one point in the sample space.
   (c) Describe the probability function on the sample space.
   (d) Find $P(X = 0)$.
   (e) Find $P(X = 1)$.
   (f) If $f(x)$ is the probability function of X, find $f(3)$.
   (g) If $F(x)$ is the distribution function of X, find $F(3)$.

4. About 51% of all human births are female and 49% are male. In a family with five children:

(a) What is the expected number of girls?

(b) What is the median number of girls?

(c) What is the probability that there are four boys and one girl?

(d) What is the most likely distribution of boys and girls?

(e) What additional assumptions did you need to make in order to answer these questions?

5. Consider two independent rolls of a balanced die. Let $\bar{X}$ be the average number of spots (i.e., $\bar{X} = (X_1 + X_2)/2$ where X_1 and X_2 are the number of spots on rolls 1 and 2).

(a) Find the probability that $\bar{X} = 2$.

(b) Draw a bar graph of the entire probability distribution of $\bar{X}$.

(c) Draw a graph of the distribution function of $\bar{X}$.

(d) Find the mean and variance of $\bar{X}$.

(e) Find the exact value of $F(3)$.

(f) Find an approximate value of $F(3)$ using the central limit theorem and compare it with part e.

6. Assume that about 10% of the people who make airline reservations on a particular flight do not show up for the flight. In order to accommodate as many people as possible, airlines customarily make more reservations than the airplane will hold because of the people who do not show. If the airplane holds 100 passengers, how many reservations can the airline make and still be 90% sure that everyone who has reservations can be accommodated?

CHAPTER 2

Statistical Inference

PRELIMINARY REMARKS

The concepts of probability theory introduced in the previous chapter do not cover the entire field of probability theory. But this small glimpse into the area of probability theory is all that we need to understand the basic principles behind most of the nonparametric methods that we use. We now bridge the gap between probability theory and its application to data analysis. In this chapter we introduce concepts of the basic science for data analysis called *statistics.*

The field of statistics owes many, if not most, of its significant ideas to people in the applied sciences who had difficult questions concerning their data. These people all had some mathematical ability, some training in mathematics, and a great deal of common sense. Their ideas gradually evolved into a few basic concepts that we present in this chapter.

2.1. POPULATIONS, SAMPLES, AND STATISTICS

Much of our knowledge concerning the world we live in is the result of samples. We eat at a restaurant once and we form an opinion concerning the quality of the food and service at that restaurant. We know 12 people from England and we feel we know the English people. Quite often the opinions we form from the sample are not accurate. However, in most cases, the opinions are more accurate than if no sample had been observed. And usually the larger the sample, the more accurate the opinion.

Our process of forming opinions may be placed within the framework of an investigation. To investigate the quality of a restaurant, we eat there once. To investigate (or study) the English people, we recall our experiences with English people.

We will refer to the collection of all elements under investigation as the *population*. A *sample* is a collection of some elements of a population. Scientific investigations are often concerned with obtaining information about some population. Suppose a psychologist wishes to study the effect of constantly interrupted sleep on the emotional balance of a person. He might consider the population to be all human beings of contemporary times. To conduct his experiment, he uses paid volunteers, obtained through an ad in a college newspaper. He can hardly consider his subjects to be representative of the population because they are all college students, at one university, in a rather restricted age group, and possessing an emotional makeup that prompts them to reply to an ad in a newspaper and volunteer for a somewhat personal study. And yet he is forced to use this type of sample for his experiment for practical reasons such as limited funds and limited time available for research or to abandon his experiment entirely. Thus it is advisable to speak of two populations: the population under investigation and the population actually sampled. The population about which information is wanted is called the *target population*. The population to be sampled is called the *sampled population*. Our example considered all contemporary human beings as the target population and all human beings who responded to the ad as the sampled population. All experimenters must necessarily work with the sampled population, and the validity of their experiment rests on the assumption that the sampled population is similar to the target population, at least with respect to the properties under investigation.

In order to obtain accurate information about a population, it would seem desirable to examine every element in that population. Usually this is impossible or impractical, so only a sample from that population is observed. The sample may consist of those elements that are easily accessible, such as the citizens of England who are known to the observer. The sample may consist of a haphazard selection of elements from the population, such as the names of people obtained from a mailing list. Perhaps only "typical" elements of the population are selected for study; that is, elements that seem to be average or nearly average. Experimentation that involves discomfort or inconvenience often relies on a sample of volunteers. None of these four methods of obtaining a sample permits the use of statistical techniques to aid in making inferences about the population, because they do not result in a random sample. Usually we assume that the sample is random even if it is not, but it is much better actually to have a random sample. A random sample may be obtained by numbering all of the elements of the population from 1 to N and then drawing n numbers in a random manner. The n numbers drawn correspond to the n elements in the population that are to be included in the sample. The statistical methods presented in this book usually assume that the sample is a random sample, so it is important to discuss the idea of a random sample.

If the population has a finite number of elements, the following definition of random sample is appropriate.

Definition 1. A sample from a finite population is a *random sample* if each of the possible samples was equally likely to be obtained.

The definition may seem a little strange in that the term "random" does not really refer to the sample itself but to the method by which the sample was obtained. In fact, we cannot look at a sample to see if it is a random sample or not. Instead, we look at the means by which the sample is obtained. If the finite population has N elements total, then, as seen in Section 1.1, there are $\binom{N}{n}$ possible samples of size n if the sample is obtained without replacement. If the sampling is with replacement, there are N^n possible samples. If each of these possible samples is equally likely to be obtained, the method of sampling is considered to be random, and the resulting sample is a random sample.

The preceding definition of a random sample seems to be satisfactory for most situations where the population is finite. But suppose we are examining the number of dreams a certain individual has in one night. We think of a "random sample" in this case as the number of dreams she has in one night and the number she has another night, and so on for, say, seven nights. Even under ideal conditions the sampling method does not fit into the framework of Definition 1, with a concept of "equally likely." What is equally likely? Not the individual, because presumably we are studying only the individual, not a representative of some population (although this may be the ultimate objective in the back of our minds). Are we to select the nights for study in some equally likely fashion out of the remaining nights that the individual can expect to be alive? Clearly this is impossible. We may conclude that at least one more definition of random sample is needed.

The definition of random sample, which is standard among mathematical statisticians, is the following.

Definition 2. A *random sample of size n* is a sequence of n independent and identically distributed random variables $X_1, X_2, \ldots, X_n$.

This definition requires an explanation. First, what we call a random sample in Definition 2 is called a *simple random sample* by many authors. We will not make a distinction between the two expressions.

Second, each random variable in Definition 2 may actually be a multivariate random variable. That is, X_i may really represent the k-variate random variable $(Y_{i1}, Y_{i2}, \ldots, Y_{ik})$, where the X_is are still independent and identically distributed but where the individual Y_{ij} random variables within each X_i may or may not be independent and/or identically distributed. As an example, consider the "dream" experiment just described. The random variable X_i could be the number of dreams counted during the ith night of observation. Then it may not be too unreasonable to assume that the X_is are independent, as defined by Definition 1.3.11, and identically distributed (meaning each X_i

has the same distribution function). But suppose that each night the experimenter records not only the total number of dreams but also the total amount of sleep, which we will call Y_{i1} and Y_{i2}, respectively. The number of dreams and the length of sleep during any one night may be related variables, so Y_{i1} and Y_{i2} are probably not independent. However, the sleep pattern on one night may be independent of the sleep pattern on another night. Mathematically, this means that the joint probability function of Y_{i1}, Y_{i2}, Y_{j1}, Y_{j2} may be factored as follows

$$(1) \qquad f(y_{i1}, y_{i2}, y_{j1}, y_{j2}) = f_1(y_{i1}, y_{i2}) f_2(y_{j1}, y_{j2})$$

where f_1 and f_2 are the joint probability functions of (Y_{i1}, Y_{i2}) and (Y_{j1}, Y_{j2}), respectively. If the joint probability distribution of the sleep patterns does not change from one night to the next, f_1 is identical with f_2, and we say that (Y_{i1}, Y_{i2}) and (Y_{j1}, Y_{j2}) are identically distributed. A more convenient method of expressing the facts of "between" independence but not necessarily "within" independence and "between" identical distributions but not necessarily "within" identical distributions is to let X_i represent both Y_{i1} and Y_{i2} jointly. X_i is called a *bivariate random variable*, and a value of X_i actually consists of two numbers, one for Y_{i1} and one for Y_{i2}. Then all of the prior statements may be summarized by saying, "The X_is are independent and identically distributed."

Similarly, we may consider k measurements being taken each night, and the resulting k random variables $Y_{i1}, Y_{i2}, \ldots, Y_{ik}$ being represented by X_i, which is called a *k-variate random variable*, or also a *multivariate random variable*. Then independence of the X_is, in the sense of Definition 1.3.11, means that the joint probability distribution of all of the Y_{ij}s may be factored into the product of n joint probability functions, each being the joint probability function of $Y_{i1}, Y_{i2}, \ldots, Y_{ik}$ for some i. Identically distributed X_is means that the joint probability functions just mentioned are identical functions.

The third point of explanation concerning Definition 2 is that even though a random variable is a function that assigns real numbers to the outcomes of the experiment (Definition 1.3.1), complete knowledge of those real numbers is not always necessary in order to use nonparametric statistical methods. This is particularly nice when complete knowledge of those real numbers is not available. An experimental subject who has performed four tasks can usually arrange the tasks in order, from "most difficult" to "least difficult." However, it may be unrealistic for the subject to assign a number to the task where the number represents degree of difficulty. Yet we might speak of the random variable X as measuring degree of difficulty. The random variables that constitute the random sample may be of this type. More will be said concerning various types of measurements later in this section.

We now have two definitions of random sample. The first definition applies only to samples from a finite population and may be directly related to the sample space. If each possible sample (of size n) is represented by one point in the sample space and if each point in the sample space has equal probability of being selected as the sample, the sampling method is random and the resulting sample is a random sample. The concepts of sample space and probability function are used in that definition, but there is no mention, explicit or implicit, of a random variable.

Example 1. A psychologist would like to obtain four subjects for individual training and examination. He advertises and 20 volunteers respond. He has several ways of selecting a sample of 4 from his sampled population of size 20.

He might select the first 4 to volunteer. Thus he may be biasing his selection toward those volunteers who tend to be more prompt or aggressive. This is probably not a random sample.

He might adhere strictly to Definition 1 and consider that there are $\binom{20}{4} = 4845$ ways in which a sample of size 4 may be selected. Then he obtains 4845 pieces of paper that are identical and writes 4 names on each piece of paper, a different combination each time, and puts them in a basket. One slip is randomly drawn, and those 4 people are used. This is a random sample, but such a psychologist would need a psychiatrist.

Another way of obtaining a random sample would be to write each of the names on a slip of paper, 20 slips in all, and one by one draw 4 slips in some random manner, such as from a hat. This method also satisfies the definition of a random sample.

The second definition of random sample is concerned directly with random variables and does not mention the sample space. However, since a random variable is a function defined on a sample space, a sample space is implicitly involved, although it remains in the background. Also, as mentioned in Section 1.3, the set of possible values of a random variable resembles a sample space. At times it will be necessary to list the points in this pseudo sample space in order to solve statistical problems that may arise. In fact, often no confusion will result if the possible measurements themselves (the values assumed by the random variables) are considered to be the points in the sample space. We usually think of these measurements as being numbers, but sometimes the numerical values of the measurements are obscure, as stated earlier in this section. So it would be well to discuss the various types of measurements.

The types of measurements are usually called *measurement scales* and are discussed at some length in various publications, including in an excellent paper by Stevens (1946). We will proceed from the "weakest" scale of measurement, the nominal scale, through the ordinal scale and the interval scale to the "strongest" scale, the ratio scale.

The *nominal scale* of measurement uses numbers merely as a means of separating the properties or elements into different classes or categories. The number assigned to the observation serves only as a "name" for the category to which the observation belongs, hence the title "nominal." We used the nominal scale of measurement when we defined a random variable that equaled 1 if a coin landed as a "head," and 0 if the coin landed as a "tail." We could, just as appropriately, have used the numbers 7.3 and 3.9 to represent head and tail, respectively. Our choice of 1 and 0 was primarily for convenience when we later desired to count the total number of heads in several tosses of the coin. When 12 subjects are arbitrarily numbered 1 to 12, a nominal scale of measurement is being used and the assignment of the numbers is a form of random variable. When classifying objects according to color, the categories may be labeled 1, 2, 3, or blue, yellow, red, or A, B, C. The numbers are merely category names. The numbers may be replaced by other unused numbers, as long as the categories remain intact.

The *ordinal scale* of measurement refers to measurements where only the comparisons "greater," "less," or "equal" between measurements are relevant. The numeric value of the measurement is used only as a means of arranging the elements being measured in order, from the smallest to the largest. It is this need to *order* the elements, on the basis of the relative size of their measurements, that gives the name to the *ordinal scale*. If some of the elements have equal measurements, we say *ties* exist, and the ordering is no longer unique. For many statistical analyses a unique ordering is desired, so it is advisable to exercise sufficient care in measurement so that the number of ties is minimized wherever possible. When a person is asked to assign the number 1 to the most preferred of three brands, the number 3 to the least preferred, and the number 2 to the remaining brand, she is using an ordinal scale of measurement and is using the numbers merely as a convenient way of representing her order of preferences. Instead of the numbers 1, 2, 3, she could have used any three numbers, say 16, 20, 75, as long as the numbers are assigned to the brands in such a way that the relative order of the number represents the relative preference of the brand.

The third scale, the *interval scale* of measurement, considers as pertinent information not only the relative order of the measurements as in the ordinal scale but also the size of the interval between measurements, that is, the size of the difference (in a subtraction sense) between two measurements. The interval scale involves the concept of a unit distance, and the distance between any two measurements may be expressed as some number of units. A good example is the scale by which we usually represent temperature. One unit (degree) increase in temperature is defined by a particular change in volume of mercury in a thermometer; consequently, the difference between any two temperatures may be measured in units, or degrees. The actual numerical value of the temperature is merely a comparison with an arbitrary point called "zero degrees." The interval scale of measurement requires a zero point as well as a unit distance (it is not possible to have the latter without the former), but it is not important which measurement is declared to be zero or which distance is defined to be the unit distance. Temperature has been measured quite adequately for some time by both the Fahrenheit and the Centigrade scales, which have different zero temperatures and different definitions of 1 degree, or unit. The principle of interval measurement is not violated by a change in scale or location or both.

Finally, the *ratio scale* of measurement is used when not only the order and interval size are important, but also the ratio between two measurements is meaningful. If it is reasonable to speak of one quantity being "twice' another quantity,. the ratio scale is appropriate for the measurement, such as when measuring crop yields, distances, weights, heights, income, and so on. Actually, the only distinction between the ratio scale and the interval scale is that the ratio scale has a natural measurement that is called zero, while the zero measurement is defined arbitrarily in the interval scale. As in the interval scale, the unit distance of the ratio scale is arbitrarily defined.

It is not possible to look at the measurements themselves in order to tell which scale of measurement is appropriate. Instead, one looks at the quantities being measured and the method of measurement and then determines the amount of meaning that may be attached to the numeric value of the measurement.

Most of the usual parametric statistical methods require an interval (or stronger) scale of measurement. Most nonparametric methods assume either the nominal scale or the ordinal scale to be appropriate. Of course, each scale of measurement has all of the properties of the weaker measurement scales; therefore statistical methods requiring only a weaker scale may be used with the stronger scales also.

Thus far in this section we have been concerned with populations, samples from populations, and measurement scales for measuring sample properties of interest. Measurement scales relate to random variables, because a system for measuring elements of the sample is in reality a random variable. Therefore measurement scales relate to statistics, because a statistic is a random variable. To a mathematical statistician the term "statistic" is interchangeable with the term "random variable." But popular usage of the word statistic indicates that it is more than just a random variable.

The word statistic originally referred to numbers published by the state, where the numbers were the result of a summarization of data collected by the government. Thus some people think of a statistic as a number that is based on several numbers, such as the average of several numbers in a sample, the proportion of a population that is in a particular category, and so on. In this sense a statistic is just a number. However, if we stop to consider that the numbers being averaged may vary from one sample to the next or that the population may change from one year to the next, we can justify extending our idea of a statistic from being only a number to being a rule for finding the number. Then "the average of the numbers in the sample" is the statistic, and the actual average obtained in one sample is a value of the statistic. As a rule for obtaining a number, a statistic meets the requirements of being a random variable, a function that assigns numbers to the points in the sample space (for an appropriately defined sample space). A statistic also conveys the idea of a summarization of data, so usually a statistic is considered to be a random variable that is a function of several other random variables. Then a value assumed by the statistic is implicitly assumed to be the result of some arithmetic operations performed on other numbers (the data) that, in turn, are the values assumed by several random variables. Since a random variable is a function defined on a sample space, a statistic may be defined as a function defined on a special sample space, a sample space whose points are the possible values of an n-variate random variable. A formal definition and an example may clarify the concept.

Definition 3. A *statistic* is a function which assigns real numbers to the points of a sample space, where the points of the sample space are possible values of some multivariate random variable. In other words, a *statistic* is a function of several random variables.

Each sentence in Definition 3 would suffice as a definition of statistic. Both sentences are included for clarity.

Example 2. Let $X_1, X_2, \ldots, X_n$ represent test scores of n students. Then each X_i is a random variable. Let W equal the average of the test scores.

$$(2) \qquad W = \frac{X_1 + X_2 + \cdots + X_n}{n} = \frac{1}{n} \sum_{i=1}^{n} X_i$$

Then W is a statistic. If $X_1 = 76$, $X_2 = 84$, and $X_3 = 85$ represent the scores of three students, $W = (1/3)(76 + 84 + 85) = 81\frac{2}{3}$. The statistic W satisfies the second sentence in Definition 3 by being a function of the random variables $X_1, X_2, \ldots$, and X_n, the first sentence in Definition 3 is also satisfied because W assigns real numbers to the values of the multivariate random variable $(X_1, X_2, \ldots, X_n)$. In this case, if (X_1, X_2, X_3) assumes the multivariate value $(76, 84, 85)$, W assumes the value $81\frac{2}{3}$, as shown. This particular statistic is used often in statistics (the science). It is called the "sample mean" and will be discussed further in the next section.

We will often have occasion to use a particular class of statistics called *order statistics*, particularly when we are dealing with ordinal-type measurements. Suppose an observation $(x_1, x_2, \ldots, x_n)$ on a multivariate random variable $(X_1, X_2, \ldots, X_n)$ is "ordered"; that is, the elements are arranged from smallest to largest. We will denote the *ordered observation* by $x^{(1)} \le x^{(2)} \le \cdots \le x^{(n)}$.

Definition 4. The *order statistic of rank k*, $X^{(k)}$, is the statistic that takes as its value the kth smallest element $x^{(k)}$ in each observation $(x_1, x_2, \ldots, x_n)$ of $(X_1, X_2, \ldots, X_n)$.

Therefore $X^{(1)}$, the order statistic of rank 1, always takes the smallest element in $(x_1, x_2, \ldots, x_n)$ as its value, and $X^{(n)}$ takes the largest. In Example 2, $X^{(1)} = 76$, $X^{(2)} = 84$, and $X^{(3)} = 85$. If another observation on (X_1, X_2, X_3) yields $(93, 73, 81)$, the values of the order statistics are $X^{(1)} = 73$, $X^{(2)} = 81$, and $X^{(3)} = 93$. If $(X_1, X_2, \ldots, X_n)$ is a *random sample*, sometimes $(X^{(1)} \le X^{(2)} \le \cdots \le X^{(n)})$ is called the *ordered random sample*.

Section 2.2 introduces many other useful statistics. There we will further discuss some uses of statistics in the analysis of experimental results.

EXERCISES

1. A congressional committee wishes to examine the effect of proposed legislation on the nation's high schools. It randomly selects five high schools from the Washington, D.C. area and conducts a study on those five schools.
   (a) What is the target population?
   (b) What is the sampled population?
   (c) If there are 100 high schools in the Washington, D.C. area, how many different samples are possible?
   (d) What is the probability of each sample in part c?

2. A Topeka television station asks the question "Should liquor by the drink be allowed in Kansas" and reports 372 phone calls, in which 164 persons said "no" and the remainder said "yes."
   (a) What was the target population?
   (b) What was the sampled population?
   (c) Was the sample a random sample? Explain.
   (d) Three statistics are indicated in this exercise. What are they, and what numerical values did they assume?
   (e) What measurement scale is used in the voting counts?
   (f) What measurement scale is used in registering each phone call as a "yes" or "no"?

3. A certain track meet awards a trophy to the team that accumulates the most points. A team receives 5, 3, or 1 points each time a member of that team finishes first, second, or third, respectively, in competition.
   (a) What measurement scale is used in awarding points?
   (b) Which statistic is (implicitly) mentioned, and what is it used for?

4. Football players on a team wear numbers on their uniforms. What measurement scale do those numbers represent?

5. What measurement scale is used in the following?
   (a) Postal zip codes.
   (b) Local telephone numbers.
   (c) Telephone area codes.
   (d) Social security numbers.

6. What measurement scale is used in the following?
   (a) Monthly salary.
   (b) Gallons measured on a gasoline pump.
   (c) The price of coffee per pound.
   (d) Intelligence as measured by IQ scores.

7. In order to select a law firm at random, a list of all lawyers of that city was obtained, and a lawyer was selected at random. The law firm to which that lawyer belonged was the selection. Was the law firm selected at random?

8. In order to estimate the number of people watching various TV shows, the following procedure is used. A random sample of 2200 households is obtained. These households must agree to have their TV sets connected to an electronic device that keeps track of the programs watched for more than 8 minutes.
   (a) What was the target population?
   (b) What was the sampled population?
   (c) Comment on the accuracy of the results.

PROBLEMS

1. An experiment consists of n rolls of an unbalanced die. Let X_i be the number of spots showing on the ith roll. Does $X_1, X_2, \ldots, X_n$ constitute a random sample?
2. A random sample of size 4 is to be selected from among the integers 1 to 7, without replacement.
   (a) What is the total number of possible samples?
   (b) What is the probability of each sample?
   (c) What is the probability that the sample has at least one odd number?
   (d) What is the probability that the numbers in the sample sum to 12?
3. A random sample of size n is selected from among the integers 1 to N, without replacement. What is the probability that the sample has at least one odd number?

2.2. ESTIMATION

One of the primary purposes of a statistic is to estimate unknown properties of the population. The unknown properties that may be estimated are necessarily numerical and include items such as unknown proportions, means, probabilities, and so on. Actually, the estimate is based on a sample, a random sample if probability statements are to be made, and the estimate is an educated guess concerning some unknown property of the probability distribution of a random variable, where that random variable represents some quantity of interest in the population. For example, we might use the proportion of defective items in a sample of radio tubes as a statistic to estimate the unknown proportion of defective radio tubes in some population of radio tubes. A statistic that is used to estimate is called, quite naturally, an *estimator*. In this section we will discuss estimators such as the sample mean, the sample variance, and the sample quantiles. But first we will introduce the *empirical distribution function*, an estimator of a somewhat different kind.

The true distribution function of a random variable is almost never known. Sometimes we make an educated guess as to the form of the distribution function and use our guess as an approximation of the true distribution function. One way of making a good guess is by observing several values of the random variable and constructing a graph $S(x)$ that may be used as an estimate of the entire unknown distribution function $F(x)$ of the random variable. The method of constructing the graph is best explained by an example, which follows this definition.

Definition 1. Let $X_1, X_2, \ldots, X_n$ be a random sample. The *empirical distribution function* $S(x)$ is a function of x, which equals the fraction of X_is that are less than or equal to x for each x, $-\infty < x < \infty$.

Example 1. In a physical fitness study five boys were selected at random from the boys in a certain high school. They were asked to run a mile, and the time it took each of them to run the mile was recorded. The times (converted to fractions of a minute) were 6.23, 5.58, 7.06, 6.42, and 5.20,

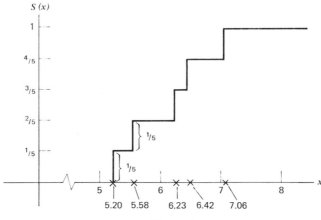

Figure 1

and they are represented on the horizontal axis in Figure 1. The empirical distribution function $S(x)$ is the number of sample values less than or equal to x and, for this particular sample, is represented graphically in Figure 1.

As in Example 1, the empirical distribution function is always a step function, where each step is of height $1/n$ and occurs only at the sample values. The vertical lines in Figure 1 are not part of the empirical distribution function but are included partly for appearance and partly for later convenience in determining sample quantiles. As we look at the graph of the empirical distribution function from left to right, we see that $S(x)$ equals zero until x equals the smallest value in the sample. Then $S(x)$ takes a step of $1/n$ in height. At each of the n sample values, $S(x)$ rises in height another distance of $1/n$. At the largest of the sample values, $S(x)$ reaches a height of 1.0 and remains 1.0 for all larger values of x. $S(x)$ resembles a distribution function in that it is a nondecreasing function that goes from zero to one in height. However, $S(x)$ is empirically (from a sample) determined and therefore its name.

Figure 1 represents merely one observation on $S(x)$. Another sample would have produced another and probably different graph of $S(x)$. This points out the random nature of $S(x)$. In a sense it is a random variable but, since it is a function and its observed values are entire graphs rather than single numbers, $S(x)$ is more properly called a *random function*. It is used as an estimator, since it does a reasonably good job of estimating the distribution function of the random variable, which we will call the population distribution function in order to distinguish it from the empirical (or "sample") distribution function.

In a sense, the observed value of an empirical distribution function may be considered a population distribution function. More precisely, an observed value of $S(x)$, based on the observations $x_1, x_2, \ldots, x_n$ in the sample, is identical to the distribution function of a random variable that may assume any of the numbers $x_1, x_2, \ldots, x_n$, each with probability $1/n$. The distribution function of such a random variable is a step function with jumps of height $1/n$ at each of the n numbers $x_1, x_2, \ldots, x_n$. We could find the mean, variance, and quantiles of the random variable simply by using the definitions of Chapter 1.

Example 2. The random variable, which has a distribution function identical to the function $S(x)$ of Example 1, is the random variable X with the following probability distribution.

$$P(X = 5.20) = .2$$
$$P(X = 5.58) = .2$$
$$P(X = 6.23) = .2$$
$$P(X = 6.42) = .2$$
$$P(X = 7.06) = .2$$

The graph of the distribution function of X is the same as the graph in Figure 1. The median of X is 6.23 by Definition 1.4.1. The mean of X, by Definition 1.4.3, is given by

$$E(X) = \sum_x xf(x)$$
$$= (5.20)(.2) + (5.58)(.2) + (6.23)(.2) + (6.42)(.2)$$
$$+ (7.06)(.2)$$

(1) $$= 6.098$$

Similarly, the variance of X may be found from Definition 1.4.4,

$$\text{Var}(X) = \sum_x (x - E(X))^2 f(x)$$

(2) $$= .424$$

The mean, variance, and quantiles obtained from the sample, as illustrated in Example 2, will be called the *sample mean, sample variance,* and *sample quantiles* to distinguish them from the true "population" mean, variance, and quantiles. In the same way that the empirical distribution function serves as an estimator of the population distribution function, the sample mean, variance, and quantiles may be used as estimators of their population counterparts.

Definition 2. Let $X_1, X_2, \ldots, X_n$ be a random sample. The *pth sample quantile* is that number Q_p that satisfies the two conditions:
1. The fraction of the X_is that are less than Q_p is $\leq p$.
2. The fraction of the X_is that exceed Q_p is $\leq 1 - p$.

The sample quantile may be found from the empirical distribution function in exactly the same way that the population quantile is obtained from the population distribution function. The pth sample quantile is that value of x where $S(x) = p$. If more than one number satisfies the condition that $S(x) = p$, we adopt the convention of using the average of the largest and the smallest numbers that satisfy $S(x) = p$, as we did with the population quantiles. The sample quantile Q_p depends on the random sample for its values; therefore it is a statistic. Note that for simplicity we have defined sample quantiles only for random samples.

Example 3. Six married women were selected at random from among the married women in a ladies' civic club, and the number of children belonging to each was recorded. These numbers were 0, 2, 1, 2, 3, 4. The empirical distribution function is given in Figure 2. The sample median $Q_{.5}$ is 2. The sample quartiles $Q_{.25}$ and $Q_{.75}$ are 1 and 3, respectively. The 1/3 sample quantile $Q_{1/3}$ is the average of 1 and 2 by our convention, which equals 1.5. These numbers are our estimates of the unknown population quantiles.

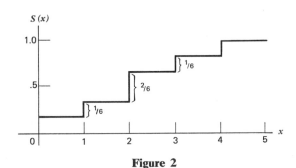

Figure 2

The sample mean and sample variance may be found in a simpler manner than in Example 2 by noting that $f(x) = 1/n$ in Equations 1 and 2, and may be factored out of the summation. This leaves us with simpler computation methods, which are given in the following definition.

Definition 3. Let $X_1, X_2, \ldots, X_n$ be a random sample. The *sample mean* $\bar{X}$ is defined by

(3)
$$\bar{X} = \frac{1}{n} \sum_{i=1}^{n} X_i$$

The *sample variance* S^2 is defined by

(4)
$$S^2 = \frac{1}{n} \sum_{i=1}^{n} (X_i - \bar{X})^2$$

which is equivalent to

(5)
$$S^2 = \frac{1}{n} \sum_{i=1}^{n} X_i^2 - \bar{X}^2$$

The *sample standard deviation* S is the square root of the sample variance.

Example 4. In the random sample 0, 2, 1, 2, 3, 4 of Example 3 the sample mean is

$$\bar{X} = \tfrac{1}{6}(0+2+1+2+3+4)$$

(6) $= 2$

and the sample variance is

$$S^2 = \tfrac{1}{6}(2^2 + 0 + 1^2 + 0 + 1^2 + 2^2)$$

(7) $= 1\tfrac{2}{3}$

Therefore our estimate of the unknown mean is 2 and our estimate of the unknown variance is $1\tfrac{2}{3}$.

The estimators introduced thus far provide a *point estimate* of the unknown population quantity, with the possible exception of the empirical distribution function. That is, our estimate of the unknown mean in the preceding example was provided by the statement, "Our estimate of the mean is 2." The single point "2" is the estimate. It is usually preferred, but more difficult, to state the estimate as follows, "We are 95% confident that the unknown mean lies between 1.3 and 2.7." Such an estimate is called an *interval estimate*. An *interval estimator* consists of two statistics, one for each end of the interval, and the *confidence coefficient*, which is the probability that the interval estimator will contain the unknown population quantity. The confidence coefficient in the preceding statement is .95. The interval estimator and the confidence coefficient together are usually called a *confidence interval* for the unknown quantity.

Point estimation is easy. To make a point estimate we need only to think of a number, any number. However, some point estimators are much better than others. Criteria for comparing point estimators, in order to determine which estimator we prefer, may be found in almost any introductory text in probability or statistics; we will not discuss them here.

In a sense, point estimation is always a nonparametric statistical method, because no knowledge of the form of the unknown distribution function is required in order to make a point estimate. This was shown by the examples in this section, where point estimates were made without knowning anything about the unknown distribution function.

It is more difficult to tell whether the methods of forming confidence intervals are parametric or nonparametric. If no knowledge of the form of the distribution function is required in order to find a confidence interval, that method is clearly nonparametric. If the method requires that the unknown distribution function be continuous, the method is still nonparametric. On the other hand, if the method requires that the unknown distribution function be a normal distribution function (see Definition 1.5.3), or some other specified form, the method is parametric. Several nonparametric methods of forming confidence intervals will be presented later, in Sections 3.1, 3.2, 5.1, 5.5, 5.7, and 6.1.

EXERCISES

1. Ten persons are selected at random from among all persons living in a particular community. The taxable incomes for five of these persons in the previous calendar year were $8600, $15,200, $16,200, $16,400, and $29,600; there was no income for the other five people.
(a) Draw a graph of the empirical distribution function.
(b) Find the sample median income.
(c) Find the sample mean income.
(d) Find the sample variance.
(e) Find the sample standard deviation.

2. In five consecutive games a certain basketball team had scores of 73, 68, 86, 78, and 65.
(a) Draw a graph of the empirical distribution function.
(b) Find the sample upper quartile.
(c) Find the sample interquartile range.
(d) Find the sample mean.
(e) Find the sample standard deviation.

3. Using the same procedure for finding point estimators used in this section, find a point estimator for the probability $P(Y \leq c)$ for a given number c, based on a random sample $X_1, X_2, \ldots, X_n$ with the same distribution function as Y. In other words, if $X_1, X_2, \ldots, X_n$ is a random sample with the distribution function $F(x)$, estimate $F(c)$. Use your estimator to estimate the probability that the score in the next game will exceed 80 in Exercise 2.

4. Using the same procedure for finding point estimators used in this section, find a point estimator for the range of a random variable. Will this sample range ever be larger than the population range? Will it ever be smaller? Is the expected value of the sample range smaller than the population range?

PROBLEMS

1. Since an estimator is a random variable, given enough information we should be able to find the probability distribution of an estimator. Suppose a finite population consists of four elements, with the respective measurements 4, 6, 7, and 10. A random sample of size 2 is drawn without replacement from the population.
(a) How many possible random samples are there?
(b) List the possible samples.
(c) What is the probability of drawing each of the samples listed in part b?
(d) What is the sample median for each of the samples listed in part b?
(e) What is the probability of getting each of the sample medians listed in part d?
(f) Graph the distribution function of the sample median.
(g) Use the same procedure just outlined and obtain the probability function of the sample range.

2. A statistic is said to be an *unbiased estimator* of a population parameter if the mean of the estimator equals the parameter.
(a) From Problem 1, find the mean of the sample median. Does this equal the

population median? Is the sample median an unbiased estimator of the population median?

(b) From Problem 1, find the mean of the sample range. Does this equal the population range? Is the sample range an unbiased estimator of the population range? (Compare with Exercise 4.)

2.3. HYPOTHESIS TESTING

Statistical inference has many forms. The form that has received the most attention by the developers and users of nonparametric methods is called hypothesis testing and is treated in this section and the next.

Hypothesis testing is the process of inferring from a sample whether or not to accept a certain statement about the population. The statement itself is called the hypothesis. Examples of hypotheses include statements such as these.

1. Women are more likely than men to have automobile accidents.
2. Nursery school helps a child achieve better marks in elementary school.
3. The defendant is guilty.
4. Toothpaste A is more effective in preventing cavities than toothpaste B.

In each case the hypothesis is tested on the basis of the evidence contained in the sample. The hypothesis is either *rejected*, meaning the evidence from the sample casts enough doubt on the hypothesis for us to say with some degree of confidence that the hypothesis is false, or *accepted*, meaning that it is not rejected.

A test of a particular hypothesis may be very simple to perform. We may observe a set of data related to the hypothesis, or a set of data not related to the hypothesis, or perhaps no data at all, and arrive at a decision to accept or reject the hypothesis, although that decision may be of doubtful value. However, the type of hypothesis test we will discuss is more properly called a statistical hypothesis test, and the test procedure is well defined. Here is a brief outline of the steps involved in such a test.

1. The hypotheses are stated in terms of the population.
2. A test statistic is selected.
3. A rule is made, in terms of possible values of the test statistic, for deciding whether to accept or reject the hypothesis.
4. On the basis of a random sample from the population, the test statistic is evaluated, and a decision is made to accept or reject the hypothesis.

A more precise description of the testing procedure follows Example 1.

Example 1. A certain machine manufactures parts. The machine is considered to be operating properly if 5% or less of the manufactured parts are defective. If more than 5% of the parts are defective the machine needs remedial attention. The *null hypothesis*

H_0: The machine is operating properly

is the hypothesis to be tested. The *alternative hypothesis* is

H_1: The machine needs attention

H_0 will be tested on the basis of a random sample of 10 parts, from the population of all parts being produced by the machine. The assumption is made that each part has the same probability p of being defective, independently of whether or not the other parts are defective. Therefore, in the assumed model, the original hypotheses H_0 and H_1 are equivalent to

H_0: $p \leq .05$

H_1: $p > .05$

We feel that if too many parts are defective, we should reject H_0. So let the test statistic T be the total number of defective items. Then, according to Example 1.3.5, T has the binomial distribution with parameters p, and 10 for n. From Table A3 we see that if H_0 is true ($p \leq .05$), then

(1) $P(T \leq 2) \geq .9885$

equaling .9885 if $p = .05$, and therefore

(2) $P(T > 2) \leq .0115$

equaling .0115 if $p = .05$. We decide to reject H_0 if T exceeds 2. The set of points in the sample space that correspond to values of T greater than 2 is called the *critical region*. Because the probability of getting a point in the critical region, when H_0 is true, is quite small (less than .0115) the decision rule is this: Reject H_0 if the observed outcome is in the critical region (when T exceeds 2); otherwise, accept H_0.

Suppose a random sample consisting of 10 machined parts is observed and 4 of the parts are found to be defective. Then $T = 4$ and the null hypothesis is rejected. We conclude that the machine needs attention.

The procedure used in Example 1 will now be carefully examined. The hypothesis to be tested is called the *null hypothesis* and is denoted by H_0. The *alternative hypothesis*, denoted by H_1, is the negation of the null hypothesis and usually consists of a statement equivalent to saying "H_0 is not true." As mentioned earlier, the decision to reject H_0 is equivalent to the opinion "H_0 is false," and is equivalent to acceptance of H_1, or the opinion "H_1 is true." The decision to accept H_0 is *not* equivalent to the opinion "H_0 is true" but, instead, represents the opinion "H_0 has not been shown to be false," which could be the result of insufficient evidence. Therefore, if we wish to determine if a statement concerning the population is false, we make it the null hypothesis. If we wish to determine whether a statement is true, we make it the alternative hypothesis. In Example 1 we wanted to determine whether the machine needs attention in the form of inspection and repair, so that statement became the alternative hypothesis.

Then assumptions are made concerning the conditions under which the data are collected and the type of data collected. These assumptions are tantamount to forming a model, or idealized experiment. "Under the model" means "under these assumptions."

Under the model, the original hypotheses may be restated in an equivalent form, usually using statistical terminology. These hypotheses may be classified as either *simple* or *composite.*

Definition 1. The hypothesis is *simple* if the assumption that the hypothesis is true leads to only one probability function defined on the sample space.

Definition 2. The hypothesis is *composite* if the assumption that the hypothesis is true leads to two or more probability functions defined on the sample space.

In the example, the model induces the probability $p^k(1-p)^{10-k}$ on each sample point with k defective items and $10-k$ nondefective items. This represents a whole class of probability functions defined on the sample space, depending on what value p has. (For each point, k is known.) Assume H_0 is true. Still, p may be any value from 0 to .05, so there are several possible probability functions, and H_0 is a composite hypothesis. The same is true for H_1. The hypothesis "$p = .05$" would be a simple hypothesis because, assuming $p = .05$ is true, the probability function assigns the probability $(.05)^k(.95)^{10-k}$ to a point representing k defective parts, and that probability function is well defined (no unknown parameters) and the only one possible.

Definition 3. A *test statistic* is a statistic used to help make the decision in a hypothesis test.

A desirable property of a test statistic is that it should assign real numbers to the points in the sample space in such a way that the points are arranged in some order corresponding to their ability to distinguish between a true H_0 and a false H_0. For example, the points that indicate most strongly that the experimenter should reject H_0 might be given large values by the test statistic, and the points that indicate that the experimenter should accept H_0 might be given small values by the test statistic. Then the larger the value assumed by the test statistic, the more the outcome of the experiment indicates that H_0 should be rejected. In this way, all values of the test statistic greater than a certain number might result in the decision to reject H_0. Furthermore, this enables the experimenter to determine objectively how much smaller or larger the rejection region might have been and still result in the same decision. Such a test, where the rejection region corresponds to the largest values of the test statistic, is called a *one-tailed test.* Similarly, if the ordering is reversed so that the rejection region corresponds to the smallest values of the test statistic, the test is still called a one-tailed test. The test in the example was one tailed. If the test statistic is selected so that the largest values of the test statistic and the smallest values of the test statistic, combined, correspond to the rejection region, the test is called a *two-tailed test*, since the rejection region corresponds to both "tails" of the test statistic's possible values.

Definition 4. The *critical region* is the set of all points in the sample space that result in the decision to reject the null hypothesis.

Sometimes the critical region is called the *rejection region*, and the set of all points in the sample space not in the critical region is called the *acceptance region*, for obvious reasons.

There are two ways of making an incorrect decision in hypothesis testing. If the null hypothesis is true we might make the mistake of rejecting it, thus committing an error known as an *error of the first kind*, or a *type I error*. That is, a type I error occurs when H_0 is true and yet the outcome of our experiment is in the critical region.

Definition 5. A *type I error* is the error of rejecting a true null hypothesis.

The second way of committing an error in hypothesis testing is by accepting the null hypothesis when the null hypothesis is false. This error is known as an *error of the second kind*, or a *type II error*.

Definition 6. A *type II error* is the error of accepting a false null hypothesis.

These two error types have associated with them certain probabilities of the errors being made. Consider first the probability of making a type I error.

Definition 7. The *level of significance*, or α, is the maximum probability of rejecting a true null hypothesis.

The level of significance may be found by first assuming H_0 is true and then ascertaining the probability of getting a point in the critical region. If H_0 is a simple hypothesis, the assumption that H_0 is true leads to only one probability function defined on the sample space, and α may be found by adding the probabilities of all points in the critical region. Usually, however, it is easier to find α by computing the probability that the test statistic will assume one of the values that results in rejection of H_0, under the assumption that H_0 is true.

If H_0 is a composite hypothesis, α is the *maximum* probability of rejecting H_0, where the maximum is obtained by considering all of the probability distributions possible when H_0 is true. In the example H_0 was composite, and the probability of rejecting a true null hypothesis was

$$P(\text{reject a true } H_0) = P(T > 2 \mid H_0 \text{ is true})$$

(3)
$$= \sum_{i=3}^{10} \binom{10}{i} p^i (1-p)^{10-i}; \qquad p \leq .05$$

which differs for each value of p. However the probability in Equation 3 is a maximum when p is a maximum. The maximum value of p, under H_0, is .05, so the level of significance is given by

$$\alpha = \text{maximum } P(T>2 \mid H_0 \text{ is true})$$
$$= P(T>2 \mid p = .05)$$
(4)
$$= .0115$$

from Table A3 or from Equation 2.

The level of significance is sometimes called the *size of the critical region*, for obvious reasons. If H_0 is true the maximum probability of rejecting H_0 is α and, therefore, the minimum probability of accepting H_0, making the correct decision, is $1 - \alpha$.

The probability of committing an error of the second kind is denoted by β. Obviously it is desirable in hypothesis testing for α and β to be close to zero. In practice the sample size helps determine how small α and β may become. Only when the sample includes all of the information contained in the population may the possibility of error be completely eliminated.

If H_0 is false the decision may be to accept H_0, with a probability β, or to reject H_0, with a probability $1 - \beta$. This latter probability represents the power of the test to detect a false null hypothesis.

Definition 8. The *power*, denoted by $1 - \beta$, is the probability of rejecting a false null hypothesis.

Unlike α, the power is not always a unique number. If H_1 is simple the assumption that H_1 is true (equivalent to "H_0 is false") leads to one probability function and hence one probability of rejecting H_0, or getting a point in the critical region. Thus, when H_1 is simple, $1 - \beta$ is unique. If H_1 is composite each probability function, under H_1, has a possibly different value for $1 - \beta$, so the power depends on the various possible probability functions.

		The Decision	
		Accept H_0	Reject H_0
The true situation	H_0 is true	Correct decision probability = $1 - \alpha$	Type I error probability = α (level of significance)
	H_0 is false	Type II error probability = β	Correct decision probability = $1 - \beta$ (power)

Now that the error types have been discussed, we can return to the topic of the critical region. Although the critical region was discussed, no mention was made concerning how it is selected. If the test statistic has been chosen so as to result in a one- or two-tailed test, the selection of a critical region depends only on the experimenter's preference concerning the size of the critical region, the level of significance. Usually a desirable decrease in the level of significance α is accompanied by an undesirable increase in β. Our two objectives in hypothesis testing are to reject H_0 as seldom as possible if H_0 is true and as often as possible if H_0 is false. As a result, the critical region is usually the set of points with the largest value of $1 - \beta$, from among those sets of points of some fixed size α. By convention more than any other reason, α is usually chosen near .05 or .01, and the critical region is then selected in terms of possible values of the test statistic.

The results of a hypothesis test are much more meaningful if the value of the *critical level* is also stated.

Definition 9. The *critical level* $\hat{\alpha}$ is the smallest significance level at which the null hypothesis would be rejected for the given observation.

In example 1, $T = 4$ and $P(T \geq 4 \mid p = .05) = .001$, so the critical level is .001. The critical level is also known as the probability level (*p*-level) and significance level. (See, for instance, Dempster and Schatzoff, 1965.)

Example 2. In order to see if children with nursery school experience perform differently academically than children without nursery school experience, 12 third-grade students are selected for study, 4 of whom attended nursery school. The hypothesis to be tested is

H_0: The academic performance of third-grade children does not depend on whether or not they attended nursery school

The alternative hypothesis is

H_1: There is a dependence between academic performance and attendance at nursery school

The model assumes that the 12 children are a random sample of all third-grade children, and also that the children can be ranked from 1 to 12 (best to worst) academically. The "dependence" in the hypotheses is assumed to mean either the nursery school children tend to do better as a group or they tend to do worse than the nonnursery school children. Under the model the hypotheses may be restated as

H_0: The ranks of the four children with nursery school experience are a random sample of the ranks from 1 to 12
H_1: The ranks of the children with nursery school experience tend to be higher or lower as a group than a random sample of 4 ranks out of 12

We choose as a test statistic T the sum of the ranks of the 4 children who attended nursery school. We decide to let the critical region correspond to values of T that are either very large or very small, so the test is two tailed.

Each possible outcome consists of 4 numbers from 1 to 12, corresponding to the ranks of the 4 children who attended nursery school. Therefore there are $\binom{12}{4} = 495$ points in the sample space. To decide which of these points to include in the critical region, we will assume H_0 is true and keep an eye on α as we decide on the critical region.

If H_0 is true, the ranks of the 4 children should behave as a *random* sample of 4 ranks out of the 12 possible. Therefore each selection of 4 ranks is equally likely, and so each point in the sample space has equal probability, 1/495. Thus H_0 is a simple hypothesis. Since we decided on a two-tailed test, we examine the points that correspond to high and low values of T. The highest and lowest possible values of T are 42 and 10, corresponding to the points (12, 11, 10, 9) and (1, 2, 3, 4), respectively. Other high and low values of T and their corresponding experimental outcomes are given as follows.

T	*Point*	T	*Point*
10	(1, 2, 3, 4)	42	(9, 10, 11, 12)
11	(1, 2, 3, 5)	41	(8, 10, 11, 12)
12	(1, 2, 3, 6)	40	(7, 10, 11, 12)
12	(1, 2, 4, 5)	40	(8, 9, 11, 12)
13	(1, 2, 3, 7)	39	(6, 10, 11, 12)
13	(1, 2, 4, 6)	39	(7, 9, 11, 12)
13	(1, 3, 4, 5)	39	(8, 9, 10, 12)
14	(1, 2, 3, 8)	38	(5, 10, 11, 12)
14	(1, 2, 4, 7)	38	(6, 9, 11, 12)
14	(1, 2, 5, 6)	38	(7, 8, 11, 12)
14	(1, 3, 4, 6)	38	(7, 9, 10, 12)
14	(2, 3, 4, 5)	38	(8, 9, 10, 11)

Note that there are 12 points that correspond to values of $T \le 14$ and 12 points that correspond to values of $T \ge 38$. If the critical region consists of all points that correspond to values of $T \le 14$ or ≥ 38, α is given by

$$\alpha = \frac{\text{number of points in critical region}}{\text{number of points in sample space}}$$

$$= \frac{24}{495}$$

(5) $= .0485$

since all points in the sample space have equal probability under H_0. Our decision rule is: If the observed value of T is ≤ 14 or ≥ 38 we reject H_0; otherwise we accept H_0.

The sample is observed, and the academic ranks of the children who attended nursery school are 2,5,6 and 9, providing a value of

(6) $T = 22$

so we accept H_0. The critical level may be approximated using the normal distribution (see Example 1.5.7). To reject H_0 with $T = 22$ the symmetric two-tailed critical region would include values of $T \le 22$ and ≥ 30, with a critical level

$$\hat{\alpha} = P(T \le 22 \text{ or } T \ge 30)$$
$$= 1 - P(22 < T < 30)$$
$$\cong 1 - P\left(\frac{22-26}{5.9} < Z < \frac{30-26}{5.9}\right)$$

(7) $$= 1 - P(-.68 < Z < +.68)$$

where Z has the standard normal distribution. Table A1 then shows that

$$\hat{\alpha} \cong 1 - (.75 - .25)$$

(8) $$= .50$$

showing that the experimental result is well within the acceptance region.

The test procedure explained in Example 2 is known as the Mann-Whitney test or the Wilcoxon test and will be discussed extensively in Chapter 5 along with its many variations. The data in Example 2 have the ordinal scale of measurement. We did not need to know the numerical value of the academic achievement for each child. In fact, such information usually has little value because each school, even each teacher, has a different interpretation of such numbers, while ranks have a universal interpretation.

Example 1 illustrated the analysis of nominal type data, "defective" or "not defective." The test of Example 1 was based on the binomial distribution. In Chapter 3 this test and other tests based on the binomial distribution will be presented formally.

EXERCISES

1. A new teaching method is being tested to see if it is better than the existing teaching method.
   (a) What are the appropriate H_0 and H_1?
   (b) What does "level of significance" represent in this problem?
   (c) What does "power" represent in this problem?

2. A defendant is being tried by a judge, and it is assumed that the defendant is innocent until proven guilty.
   (a) Who is doing the hypothesis testing?
   (b) What are H_0 and H_1?
   (c) What are the sample and the population?
   (d) What do "level of significance" and "power" mean in this problem?

3. What is the appropriate H_1 for each of the following?
   (a) H_0: Fertilizer B is at least as good as fertilizer A.
   (b) H_0: My opponent is not cheating.
   (c) H_0: The occurrence of sun spots does not affect the economic cycle.

4. What is the appropriate H_0 for each of the following?
   (a) H_1: The subject has extrasensory perception.

(b) H_1: The dowsing rod is effective in finding water.

(c) H_1: Our average yearly temperatures are rising.

5. A coin is tossed five times, and the sequence of heads and tails observed is the outcome. The critical region is the event "at least four heads." If H_0 is true, all outcomes in the sample space are equally likely. What is α? If H_1 is true, "head" has probability .6 of occurring on each toss. What is the power?

6. The sample space contains 10 points, only 1 of which is in the critical region. If H_0 is true, all of the points are equally likely. If H_1 is true, the point in the critical region has probability .91, and the other points each have probability .01. What is α? What is the power?

PROBLEMS

1. There are 12 plastic chips in a jar, and the chips are numbered consecutively from 1 to 12. An experiment consists of drawing 3 chips without replacement. The outcome of the experiment consists of the 3 numbers on the chips, without regard to the order in which they were drawn. Let the test statistic X be the sum of the numbers on the drawn chips and let the critical region correspond to values of X that are less than 11. Suppose that if H_0 is true the drawing of the chips is random. Also suppose that if H_1 is true chips 1, 2, and 3 are each twice as likely to be drawn as each of the other chips.
   (a) List the points in the critical region.
   (b) Find α.
   (c) What is the power?
   (d) Are H_0 and H_1 simple or composite?
   (e) Is the test one tailed or two tailed?

2. Seven chips numbered consecutively from 1 to 7 are placed independently of each other into either of two boxes A and B. The outcome of the experiment consists of numbers of the chips in box A without regard to the order in which they were placed there. Let the test statistic X be the sum of the numbers on the chips in box A and let the critical region correspond to values of X less than 6. Assume that if H_0 is true each chip has probability .5 of being placed in box A and if H_1 is true the corresponding probability is .3.
   (a) List the points in the critical region.
   (b) Find α.
   (c) Find the power.
   (d) Are H_0 and H_1 simple or composite?
   (e) Is the test one tailed or two tailed?

2.4. SOME PROPERTIES OF HYPOTHESIS TESTS

Once the hypotheses are formulated, there are usually several hypothesis tests available for testing the null hypothesis. In order to select one of these tests, we consider carefully several properties of the various tests. One of the most important questions is, "Are the assumptions of this test valid assumptions in my experiment?" If the answer is, "No," that test probably should be discarded. However, before discarding the test, one should be sure that the assumptions behind the test are understood. For example, in most parametric tests one of the stated assumptions is that the random variable being examined has a normal distribution. Further investigation usually reveals that if the random variable has a distribution only slightly resembling a normal distribution, the test is still approximately valid. So the implied assumption is "approximate normality," and the test should not be discarded if the assumptions are "approximately true." However, at least one black mark may be registered against that test. Another result of this criterion is that the test with the fewer assumptions in the model compares favorably with the test which has more assumptions.

The use of a test in a situation where the assumptions of the test are not valid is dangerous for two reasons. First, the data may result in rejection of the null hypothesis, not because the data indicate that the null hypothesis is false, but because the data indicate that one of the assumptions of the test is invalid. Hypothesis tests in general are sensitive detectors not only of false hypotheses but also of false assumptions in the model. The second danger is that sometimes the data indicate strongly that the null hypothesis is false, and a false assumption in the model is also affecting the data, but these two effects neutralize each other in the test, so that the test reveals nothing and the null hypothesis is accepted.

From among the tests that are appropriate, based on the preceding criterion, the best test may be selected on the basis of other properties. These properties, which involve terms that will be defined later in this section, are as follows.

1. The test should be unbiased.
2. The test should be consistent.
3. The test should be more efficient in some sense than the other tests.

Sometimes we are content if one or two of the three criteria are met. Only rarely are all three met. The rest of this section discusses the terms unbiased, consistent, efficiency, and more about the power of a test.

If H_1 is composite, the power may vary as the probability function varies. If H_1 is stated in terms of some unknown parameter, the power usually may be given as a function of that parameter. Such a function is appropriately called a *power function* and may be represented algebraically or graphically. Unlike the power, which is the probability of rejecting H_0 when H_1 is true, the power function is usually defined for all values of the parameter under both H_0 and H_1. In that sense the power function gives more than just the power; it gives the probability of rejecting H_0 whether or not H_0 is true.

Example 1. In Example 2.3.1 the critical region consisted of all points with more than two defectives in the 10 items examined. Under the assumptions of the model, the probability of getting a point in the critical region, the same as the probability of rejecting H_0, is given by

(1) $$P(\text{reject } H_0) = \sum_{i=3}^{10} \binom{10}{i} p^i (1-p)^{10-i} = 1 - \sum_{i=0}^{2} \binom{10}{i} p^i (1-p)^{10-i}$$

where p is the probability of a defective item. The probability of rejecting H_0 is a function of p, and a rough graph of the power function may be drawn with the aid of Table A3.

$P(\text{reject } H_0)$	p	$P(\text{reject } H_0)$	p
.0000	0	.9453	.50
.0115	.05	.9726	.55
.0702	.10	.9877	.60
.1798	.15	.9952	.65
.3222	.20	.9984	.70
.4744	.25	.9996	.75
.6172	.30	.9999	.80
.7384	.35	1.0000	.85
.8327	.40	1.0000	.90
.9004	.45	1.0000	1.00

As indicated in Figure 3, the null hypothesis states that p is between zero and .05. The maximum value of the curve, when H_0 is true, is the level of significance, and is shown by Figure 3 and also by Equation 2.3.4 to equal .0115. The power is seen to range from .0115 for p close to .05 to 1.0000 for p equal to 1.0.

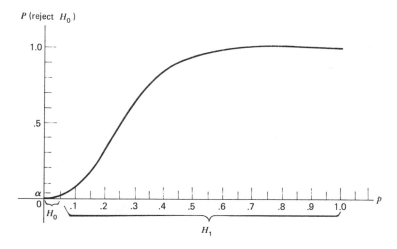

Figure 3.   A power function.

Two tests may be compared on the basis of their power functions. This basis of comparison is discussed again later in this section when relative efficiency is defined.

It is obviously desirable for a test to be more likely to reject H_0 when H_0 is false than when H_0 is true.

Definition 1. An *unbiased test* is a test in which the probability of rejecting H_0 when H_0 is false is always greater than or equal to the probability of rejecting H_0 when H_0 is true.

Thus an unbiased test is one where the power is always at least as large as the level of significance. A test that is not unbiased is called a *biased* test. The test described in Example 2.3.1 and discussed further in Example 1 of this section is an unbiased test, a fact that is readily apparent from Figure 3.

Another desirable property of a test is that of being *consistent.* Although we refer to a test as being "consistent" or "not consistent," the term consistent actually applies to a sequence of tests, because the term applies when the sample size approaches the population size. For convenience we will call the population size "infinity" even though it may be finite. Technically, for each different sample size we have a different test, because the sample space and the critical region depend on the sample size. Thus, as the sample size increases, we consider a sequence of tests, one for each sample size.

Definition 2. A sequence of tests is *consistent against all alternatives in the class H_1* if the power of the tests approaches 1.0 as the sample size approaches infinity, for each fixed alternative possible under H_1. The level of significance of each test in the sequence is assumed to be as close as possible to but not exceeding some constant $\alpha > 0$.

Example 2. We wish to determine whether human births tend to produce more babies of one sex, instead of both sexes being equally likely. We are testing

H_0: A human birth is equally likely to be male or female

against the alternative hypothesis

H_1: Male births are either more likely, or less likely, to occur than female births

The sampled population consists of births registered in a particular country. The sample consists of the last n births registered, for some selected value of n. It is assumed that this method of sampling is equivalent to random sampling as far as the characteristics "male" and "female" are concerned. It is also assumed that the probability p (say) of a male birth remains constant from birth to birth and that the births are mutually independent as far as the events "male" and "female" go. Then the hypotheses are equivalent to the following.

$$H_0: \ p = 1/2$$

$$H_1: \ p \neq 1/2$$

Let the test statistic T be the number of male births. The critical region is chosen to correspond symmetrically to the largest values and the smallest values of T, called the upper and lower tails of T, of the largest size not exceeding .05.

Thus we have described an entire sequence of tests, one for each value of n, the sample size. Each test is two tailed and has a level of significance of .05 or smaller, and T has a binomial distribution. For the various tests the critical regions are given by Dixon (1953) as follows.

n	Values of T Corresponding to the Critical Region			α
5	None			0
6	$T=0$	and	$T=6$	.03125
8	$T=0$	and	$T=8$	.00781
10	$T\leq1$	and	$T\geq9$	.02148
15	$T\leq3$	and	$T\geq12$	.03516
20	$T\leq5$	and	$T\geq15$	.04139
30	$T\leq9$	and	$T\geq21$	.04277
60	$T\leq21$	and	$T\geq39$	.02734
100	$T\leq39$	and	$T\geq61$	.03520

Note that for $n\leq20$ these same values can be obtained from Table A3. For $n>20$ the normal approximation (Example 1.5.6) could have been used, but the exact tables are preferred.

To see if this sequence of tests is consistent, the power functions of the tests are compared. Several of these power functions are plotted on the same graph in Figure 4, from tables given by Dixon (1953). We can see that as the sample size increases, the power at each fixed value of p (except $p=.5$) increases toward 1.0.

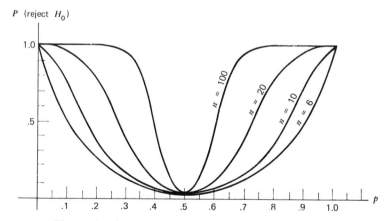

Figure 4. A comparison of several power functions.

This example merely demonstrates the idea behind the term consistent as it applies to a sequence of tests. This demonstration is not a proof that the sequence of tests is consistent. A rigorous proof of consistency usually requires more mathematics than we care to use in this introductory book, and so we will merely state whether or not a sequence of tests (or "a test") is consistent.

Many other properties of statistical tests have been defined and may be found in various books (e.g., Lehmann, 1959). We will limit our discussion to one more property, that of *efficiency*. Efficiency is a relative term and is used to compare the sample size of one test with that of another test under similar conditions. Suppose two tests may be used to test a particular H_0 against a particular H_1. Also suppose that the two tests have the same α and the same β and therefore are "comparable" with respect to level of significance and power. (Note that the condition that β be the same for both tests usually excludes consideration of composite alternative hypotheses, since those usually have more than one value of β.) Then the test requiring the smaller sample size is preferred over the other test, because a smaller sample size means less cost and effort is required in the experiment. The test with the smaller sample size is said to be *more efficient* than the other test, and its *relative efficiency* is greater than one.

Definition 3. Let T_1 and T_2 represent two tests that test the same H_0 against the same H_1, with the critical regions of the same size α and with the same values of β. *The relative efficiency of T_1 to T_2 (or "efficiency of T_1 relative to T_2")* is the ratio n_2/n_1, where n_1 and n_2 are the sample sizes of the tests T_1 and T_2 respectively.

According to Definition 3, if n_1 is smaller than n_2, the efficiency of T_1 relative to T_2 is greater than unity, satisfying our preconceived notions.

If the alternative hypothesis is composite, the relative efficiency may be computed for *each* probability function defined by the alternative hypothesis, resulting in a multitude of values for relative efficiency that may then be represented in a table or, occasionally, graphically.

Example 3. Two tests are available for testing the same H_0 against the same H_1. Both tests have $\alpha = .01$ and $\beta = .14$. The first test requires a sample size of 75. The second test requires a sample size of 50. The first test is therefore less efficient than the second test. The relative efficiency of the first test to the second test is

$$\frac{50}{75} = .67$$

and the efficiency of the second test relative to the first is

$$\frac{75}{50} = 1.5$$

If we know that the efficiency of the first test relative to the second test at $\alpha = .05$, $\beta = .30$, $n_1 = 40$, is .75, the sample size required by the second test may be obtained.

$$\text{relative efficiency} = \frac{n_2}{n_1}$$

$$.75 = \frac{n_2}{40}$$

$$n_2 = 30$$

A sample of size 30 will provide as good an analysis using the second testing method as a sample of size 40 would using the first.

The relative efficiency depends on the choice of α, the choice of β, and the particular alternative being considered if H_1 is composite. In order to provide an overall comparison of one test with another it is clear that relative efficiency leaves much to be desired. We would prefer a comparison that does not depend on our choice of α, or β, or a particular alternative possible under H_1 if H_1 is composite, which it usually is. One way this sometimes may be accomplished is described briefly as follows.

Consider a sequence of tests, all with the same fixed α. If the sequence of tests is consistent, β will become smaller as the sample size n_1 gets larger. Instead of allowing β to become smaller, we could consider a different alternative each time (under the composite alternative hypothesis) for each different value of n_1 where, each time, the alternative considered is one that allows β to remain constant from test to test. Thus, as n_1 becomes larger, α and β remain fixed and the alternative being considered varies. This may be illustrated by considering Figure 2 again. As n_1 becomes larger, the graphs in Figure 2 show that β can remain constant by considering consecutive values of the parameter p that approach closer to $p = .5$. For each value of n_1, a value of n_2 is calculated so the second test has the same α and β under the alternative considered. Then there is a sequence of values of relative efficiency n_2/n_1, one for each test in the original sequence of tests. If n_2/n_1 approaches a constant as n_1 becomes large, and if that constant is the same no matter which values of α and β are being used, then that constant is called the *asymptotic relative efficiency* of the first test to the second test or, more correctly, the first sequence of tests to the second sequence of tests. Sometimes the name *Pitman's efficiency* is used for this definition of asymptotic relative efficiency to distinguish it from other definitions of asymptotic relative efficiency.

Definition 4. Let n_1 and n_2 be the sample sizes required for two tests T_1 and T_2 to have the same power under the same level of significance. If α and β remain fixed the limit of n_2/n_1, as n_1 approaches infinity, is called the *asymptotic relative efficiency* (A.R.E.) of the first test to the second test, if that limit is independent of α and β.

The A.R.E. of two tests is usually difficult to calculate. A comprehensive study of the A.R.E. of various pairs of tests could be the subject of a book by itself. A book by Noether (1967a) contains many of the more important results of studies of A.R.E. See also Stuart (1954) and Ruist (1955) for further discussions.

So the A.R.E. may be given instead of tables of values of relative efficiency, but of what use is A.R.E. if it considers the infinite (and thus impossible) sample size? Studies of the exact relative efficiency for very small sample sizes show that the A.R.E. provides a good approximation to the relative efficiency in many situations of practical interest. Thus the A.R.E. often provides a compact summary of the relative efficiency between two tests.

The term *conservative* is another term we will sometimes use when discussing a test.

Definition 5. A test is *conservative* if the actual level of significance is smaller than the stated level of significance.

At times it is difficult to compute the exact level of significance of a test, and then some methods of approximating α are used. The approximate value is then reported as being the level of significance. If the approximate level of significance is larger than the true (but unknown) level of significance, the test is conservative, and we know the risk of making a type I error is not as great as it is stated to be.

EXERCISES

1. A coin is tossed five times. At each toss the experimenter observes whether it is a head or a tail, and a blindfolded subject being tested for extra sensory perception "states" whether it is a head or a tail. The null hypothesis is that the subject's predictions have probability $p = .5$ of being correct, while the alternative hypothesis is that $p > .5$. The critical region consists of all five correct predictions.
   (a) Find α.
   (b) What is the power function?
   (c) Draw a graph of the power function.
   (d) Is the test unbiased?

2. Two types of shoe leather are being tested to see which is more durable. Eight pairs of shoes are made; the shoes seem to be identical except that one shoe is made from leather A and the other from leather B. These shoes are subjected to normal wear for a period of time and are then judged as to which leather seemed to be more durable for each pair. Let X equal the number of pairs of shoes where leather A is judged to be more durable. The null hypothesis is that $p = .5$, where p is the probability that the shoe made out of leather A was more durable than the other shoe, while H_1 is $p \neq .5$. The critical region corresponds to $X = 0, 1, 7$, and 8.
   (a) Find α.
   (b) Find the power function?
   (c) Draw a graph of the power function.
   (d) Is the test unbiased?

3. Let $T_1, T_2, \ldots,$ represent a sequence of tests, and suppose the power function P_n of T_n is given by $P_n = n/(n + 10)$. Is the sequence of tests consistent?

4. Let $T_1, T_2, \ldots,$ represent a sequence of tests where the power function of test T_n is $n/(2n + 10)$. Is the sequence consistent?

5. The hypothesis $H_0 : p = 1/2$ is being tested against $H_1 : p = 3/4$ using two tests T_1 and T_2 at the same level of significance. If T_1 uses a sample of size 20, T_2 requires a sample of size 35 in order for the power of T_2 to equal the power of T_1.
(a) What is the efficiency of T_2 relative to T_1?
(b) What is the efficiency of T_1 relative to T_2?

6. The hypothesis $H_0 : p = 1/2$ is being tested against $H_1 : p \neq 1/2$ using two tests at the same level of significance. T_2 needs a sample of size 30 when T_1 has a sample of size 15 in order for their power functions to be equal at the particular alternative $p = 1/3$.
(a) What is the efficiency of T_2 relative to T_1?
(b) Is the efficiency necessarily the same at the alternative $p = 2/3$?

PROBLEMS

1. Suppose that the asymptotic efficiency of T_2 relative to T_1 is 0.75, and suppose that the relative efficiency for finite sample sizes is always greater than the asymptotic relative efficiency. If an experimenter prefers to use test T_2 but wishes to have at least as much power as if test T_1 were being used with a sample size 24, what should the minimum sample size be?

2. Suppose the A.R.E. of test T_1 relative to test T_2 is $3/\pi$ and the A.R.E. of test T_3 relative to T_2 is $2/\pi$. What is the A.R.E. of test T_1 relative to T_3?

2.5. SOME COMMENTS ON NONPARAMETRIC STATISTICS

This book purports to be one on nonparametric statistical methods, and yet we have not yet defined the adjective nonparametric, except for an unsatisfactory hand waving distinction between parametric and nonparametric given in the Introduction. There is no agreement among statisticians as to the meaning of the word nonparametric. In fact, there is not even agreement among statisticians concerning whether certain tests should be classified as parametric or nonparametric. For instance, the test presented in Example 2.3.1 is claimed by both camps.

On one extreme, Walsh (1962) states the following:

The viewpoint adopted in this handbook is that a statistical procedure is of a nonparametric type if it has certain properties which are satisfied to a reasonable approximation when some assumptions that are at least of a moderately general nature hold. That is, generality of application is the criterion for deciding on the nonparametric character of a statistical procedure. As an example, the binomial distribution is considered to be of nonparametric interest even though it is of an elementary parametric form.

At the other extreme consider the following statement by Kendall and Sundrum (1953):

It is important to confine the adjectives "parametric" and "nonparametric" to statistical *hypotheses*. They should not be applied to statistics, tests, or types of inference. This may sound rather austere, but we have found a great deal of confusion arising from the use of phrases like "nonparametric tests" and "nonparametric inference." We ourselves are far from guiltless in this respect in previous writings but hope to atone for past errors by correct usage in the future.

Furthermore, Kendall and Sundrum distinguish between the terms "nonparametric" and "distribution-free"; the latter term they refer to as "a convenient though not a perfect, term introduced by American writers." Then they go on to permit the adjective "distribution-free" to apply to test statistics, the distribution of the test statistic, the critical region determined by the test statistic for the null hypothesis, and so on, if the element "does not depend on the parent population association with the (null) hypothesis."

There are other definitions of the terms "nonparametric" and "distribution-free" by other well-qualified statisticians. See, for instance, Bell (1964). We will adopt the low-brow convention of using the two terms interchangeably. We will also further confuse the already confused situation by offering our own definition of the term "nonparametric." Since this definition is appearing here for the first time, we can claim no acceptance, universal or otherwise, of this definition in the world of statisticians.

Definition 1. A statistical method is nonparametric if it satisfies at least one of the following criteria.
   1. The method may be used on data with a nominal scale of measurement.
   2. The method may be used on data with an ordinal scale of measurement.
   3. The method may be used on data with an interval or ratio scale of measurement, where the distribution function of the random variable producing the data is either unspecified or specified except for an infinite number of unknown parameters.

The test in Example 2.3.1 analyzed data with a nominal scale of measurement (defective or nondefective) and therefore the test is nonparametric by the first criterion. The test in Example 2.3.2 analyzed data with an ordinal scale of measurement and therefore, by the second criterion, it too is nonparametric. Nearly all nonparametric hypothesis tests satisfy one of these two criteria. The point estimates of Section 2.2 satisfy the third criterion, and so do the procedures that assume symmetric distributions in Sections 5.1 and 5.11. Therefore we consider them to be nonparametric. There are undoubtedly situations where Definition 1 is inadequate, but we will advocate its use, at least temporarily, as a convenient yardstick.

This book is primarily concerned with hypothesis testing and the forming of confidence intervals. Unfortunately, this emphasis often gives experimenters the false impression that if they do not test some hypothesis or form some confidence interval, they are not using a statistical analysis. Other forms of statistical inference are just as important, such as description of the population, interpretation of the data, prediction of unknown events, and point estimation. These other forms of inference depend to a great extent on the experimenter's maturity and good judgment instead of on complicated probabilistic arguments; therefore we consider them too difficult to present in a book. We are attempting to assist the experimenter who already possesses maturity and good judgment by spelling out the complicated probabilistic arguments associated with hypothesis testing and confidence intervals.

Nonparametric statistical methods have been developed for several types of problems that are not covered in this book. These areas (and some references for the interested reader) include bioassay (Miller, 1973, Chmiel, 1976), survival curves (Susarla and Van Ryzin, 1976, Tarone and Ware, 1977), and longitudinal studies (Ghosh, Grizzle, and Sen, 1973). Multivariate methods are discussed by Bhapkar and Patterson (1977) and are the topic of a book by Puri and Sen (1971). For discrimination analysis see Gessaman and Gessaman (1972), Broffitt, Randles, and Hogg (1976), Randles, Broffitt, Ramberg, and Hogg (1978), and Conover and Iman (1980).

Robust methods are methods that depend to some extent on the population distribution function but are not very sensitive to departures from the assumed distributional form. Robust methods are discussed briefly in Section 5.1. A more complete discussion of robust methods may be found in Govindarajulu and Leslie (1972), Hogg (1974), Pearson and Please (1975), Policello and Hettmansperger (1976), and other references cited in Section 5.1. General overviews of the field of nonparametric statistics may be obtained by reading articles by Blum and Fattu (1954), Savage (1969), and Govindarajulu (1976).

2.6. REVIEW PROBLEMS FOR CHAPTER 2

1. A box contains seven tickets. Five tickets belong to students and the other two belong to faculty. Two tickets are drawn from the box, without replacement, to determine the two winners. The null hypothesis is that the drawing is random. The alternative hypothesis is that the drawing is rigged so that the first ticket drawn belongs to a faculty member and the second ticket is then randomly selected from the remaining six tickets.
   (a) Suppose the decision rule is to reject the null hypothesis if both tickets drawn belong to faculty members. Find α. Find the power.
   (b) Suppose, instead, the decision rule is to reject the null hypothesis if the first ticket drawn belongs to a faculty member. Find α. Find the power.
   (c) Some people might prefer the test in part a because it has a smaller level of significance. Others might prefer the test in part b because it has greater power. Discuss some of the social consequences connected with using each test. Which test would you use?

2. What is the scale of measurement for the following random variables?
    (a) The number of pounds gained (or lost) while on a particular diet.
    (b) The standings of the Kansas City Royals within their league.
    (c) Your student identification number.
    (d) The scoring average for a particular basketball player.
    (e) The score a figure skater receives in an olympic contest.

3. Two students are playing chess to see if they are equal in ability. The rules are that seven games will be played. If either person wins at least six of the games, they agree that they are not of equal chess-playing ability.
    (a) What is H_0?
    (b) What is H_1?
    (c) Write down any one point in the sample space.
    (d) List the points in the sample space that consititue the critical region.
    (e) Find the level of significance.
    (f) Is H_0 simple or composite?
    (g) Is H_1 simple or composite?
    (h) What is the equation of the power function?
    (i) Is the test unbiased?
    (j) What assumptions did you make here?

CHAPTER 3

Some Tests Based on the Binomial Distribution*

PRELIMINARY REMARKS

The binomial probability distribution was introduced in Chapter 1 to describe the probabilities associated with the number of heads when a coin is tossed n times. In its more general form each of n independent trials results either in "success," with probability p, or "failure," with probability $q = 1 - p$. The binomial distribution describes the probability of obtaining exactly k successes. Table A3 presents some of the binomial distribution functions.

Many experimental situations in the applied sciences may be modeled this way. Several customers enter a store and independently decide to buy or not to buy a particular product. Several animals are given a certain medicine and either they are cured or not cured. Examples can be found in almost any field. Data obtained in these situations may be analyzed using some of the simplest statistical methods known, those based on the binomial distribution. In this chapter we present a few of the available methods. The literature abounds with other procedures based on the binomial distribution. After studying the variety of tests presented in this chapter, the reader should be able to invent variations to match a given experimental situation.

3.1. THE BINOMIAL TEST AND ESTIMATION OF p

One example of the binomial test has already been presented. In Example 2.3.1 the binomial test was applied to a quality control problem. This entire

---

* Review Problems for Chapter 3 are included in the Review Problems for Chapter 4.

chapter (Chapter 3) is little more than an elaboration of Example 2.3.1, showing the many uses and amazing versatility of that simple little binomial test. With a little ingenuity the binomial test may be adapted to test almost any hypothesis, with almost any type of data amenable to statistical analysis. In some situations the binomial test is the most powerful test; in those situations the test is claimed by both parametric and nonparametric statistics. In other situations more powerful tests are available, and the binomial test is claimed only by nonparametric statistics. However, even in situations where more powerful tests are available, the binomial test is sometimes preferred because it is usually simple to perform, simple to explain, and sometimes powerful enough to reject the null hypothesis when it should be rejected.

We will now formally present the binomial test and, at the same time, introduce the format for presenting tests. We feel that there is a need for some format in presenting tests both for the convenience of the reader and for ready review by the users of nonparametric techniques.

The Binomial Test

DATA. The sample consists of the outcomes of n independent trials. Each outcome is in either "class 1" or "class 2," but not both. The number of observations in class 1 is O_1 and the number of observations in class 2 is $O_2 = n - O_1$.

ASSUMPTIONS

1. The n trials are mutually independent.
2. Each trial has probability p of resulting in the outcome "class 1," where p is the same for all n trials.

HYPOTHESES. Let p^* be some specified constant, $0 \leq p^* \leq 1$. The hypotheses may take one of the following three forms.

A. (two-tailed test)

$$H_0: p = p^*$$
$$H_1: p \neq p^*$$

B. (one-tailed test)

$$H_0: p \leq p^*$$
$$H_1: p > p^*$$

C. (one-tailed test)

$$H_0: p \geq p^*$$
$$H_1: p < p^*$$

TEST STATISTIC. Since we are concerned with the probability of the outcome "class 1," we will let the test statistic T be the number of times the outcome is "class 1." That is,

$$(1) \qquad\qquad T = O_1$$

DECISION RULE. Depending on which hypothesis is being tested, A, B, or C, the different decision rules are as follows. (Because the test statistic T is discrete, α will seldom be a nice round number.)

A. (two-tailed test) The critical region of size α corresponds to the two tails of the binomial distribution with parameters p^* and n, where the size of the upper tail is α_2, the size of the lower tail is α_1, and $\alpha_1 + \alpha_2$ equals α. That is, from Table A3 for the particular values of p^* and n, find the number t_1 such that

$$(2) \qquad\qquad P(Y \le t_1) = \alpha_1$$

and find the number t_2 such that

$$(3) \qquad\qquad P(Y > t_2) = \alpha_2$$

or, equivalently,

$$(4) \qquad\qquad P(Y \le t_2) = 1 - \alpha_2$$

where Y is a binomial random variable with parameters p^* and n.

The values of α_1 and α_2 should be approximately equal to each other. Then reject H_0 if T exceeds t_2 of if T is less than or equal to t_1. Otherwise accept H_0.

B. (one-tailed test) Since large values of T indicate that H_0 is false, the critical region of size α consists of all values of T greater than t, where t is the number obtained from Table A3, using p^* and n, such that

$$(5) \qquad\qquad P(Y > t) = \alpha$$

or, equivalently,

$$(6) \qquad\qquad P(Y \le t) = 1 - \alpha$$

where Y has the binomial distribution with parameters p^* and n. Reject H_0 if T is greater than t. Accept H_0 if T is less than or equal to t.

C. (one-tailed test) Since small values of T indicate that H_0 is false, the critical region of size α consists of all values of T less than or equal to t, where t is obtained from Table A3, using p^* and n, so that

$$(7) \qquad\qquad P(Y \le t) = \alpha$$

where Y has the binomial distribution with parameters p^* and n. Reject H_0 if T is less than or equal to t. Otherwise accept H_0.

Example 1. Under simple Mendelian inheritance a cross between plants of two particular genotypes may be expected to produce progeny one-fourth of which are "dwarf" and three-fourths of which are "tall." In an experiment to determine if the assumption of simple Mendelian inheritance is reasonable in a certain situation, a cross results in progeny having 243 dwarf plants and 682 tall plants. If "class 1" denotes "tall," $p^* = 3/4$ and T equals the number of tall plants. The null hypothesis of simple Mendelian inheritance is equivalent under the model to the hypothesis

$$H_0: p = \tfrac{3}{4}$$

The alternative of interest is

$$H_1: p \neq \tfrac{3}{4}$$

Since $n = 925$ $(243 + 682)$, the critical region of approximate size $\alpha = .05$ may be obtained using the large sample approximation given at the end of Table A3. Thus the critical region corresponds to all values of T less than or equal to t_1, where

$$t_1 = np^* + w_{.025}\sqrt{np^*(1-p^*)}$$

$$= (925)(\tfrac{3}{4}) + (-1.960)\sqrt{(925)(\tfrac{3}{4})(\tfrac{1}{4})}$$

(8) $$= 667.94$$

and all values of T greater than t_2, where

$$t_2 = np^* + w_{.975}\sqrt{np^*(1-p^*)}$$

$$= (925)(\tfrac{3}{4}) + (1.960)\sqrt{(925)(\tfrac{3}{4})(\tfrac{1}{4})}$$

(9) $$= 719.56$$

The value of T obtained is 682 in this experiment. Therefore the null hypothesis is accepted.

The critical level $\hat{\alpha}$ may be found by considering that the acceptance region in our test is some region on both sides of $np^* = 693.75$. Since the observed value 682 is smaller than np^*, half of $\hat{\alpha}$ is found by calculating $P(T \leq 682)$, assuming the null hypothesis is true. As in Example 1.5.6,

$$P(T \leq 682) = P\left(\frac{T - np^*}{\sqrt{np^*(1-p^*)}} \leq \frac{682 - np^*}{\sqrt{np^*(1-p^*)}}\right)$$

(10) $$\cong P\left(Z \leq \frac{682 - 693.75}{13.17}\right)$$

where Z has the standard normal distribution as given in Table A1.

$$\frac{\hat{\alpha}}{2} = P(T \leq 682) \cong P(Z \leq -0.8922)$$

(11) $$\cong .186$$

Therefore $\hat{\alpha} = 2(.186) = .372$. A level of significance of at least .372 would be required to reject H_0. Thus the data are in good agreement with the null hypothesis.

The previous example illustrates the two-tailed form of the binomial test. The one-tailed binomial test is illustrated in Example 2.3.1.

☐ *Theory.* That the test statistic in the binomial test has a binomial distribution is easily seen by comparing the assumptions in the binomial test with the assumptions in Examples 1.3.5 and 1.2.8. That is, if T equals the number of trials that result in the outcome "class 1," where the trials are mutually independent and where each trial has probability p of resulting in that outcome (as stated by the assumptions), then T has the binomial distribution with parameters p and n. The size of the critical region is a maximum when p equals p^*, under the null hypothesis, and so Table A3 is entered ☐ with n and p^* to determine the exact value of α.

As mentioned earlier, hypothesis testing is only one branch of statistical inference. We will now discuss another branch, *interval estimation.* If we are attempting to make some inferences regarding an unknown parameter associated with some population, it is reasonable to examine a random sample from that population and, on the basis of that sample, to make some statement regarding the population parameter. Such a statement might be "the population parameter lies between a and b," where a and b are two real numbers obtained from the sample. The numbers a and b are computed from the sample and are therefore realizations of two statistics. The two statistics that furnish us with the lower and upper boundary points for the interval will be denoted by L and U, respectively, for "lower" and "upper." The interval from L to U is called the *interval estimator.* The probability that the unknown population parameter lies within its interval estimate is called the *confidence coefficient.* The interval estimator together with the confidence coefficient provide us with the *confidence interval.*

A method for finding a confidence interval for p, the unknown probability of any particular event occurring, is closely related to the binomial test.

Confidence Interval for a Probability

DATA. A sample consisting of observations on n independent trials is examined, and the number Y of times the specified event occurs is noted.

ASSUMPTIONS

1. The n trials are mutually independent.
2. The probability p of the specified event occurring remains constant from one trial to the next.

METHOD A. For confidence coefficients of .95 or .99 use the charts in Table A4. Read from the lower left corner if Y/n is less than .50, and the upper right corner if Y/n is greater than .50. Read horizontally across the chart to the value obtained for Y/n and then read vertically from there to the two lines labeled with the correct sample size n, interpolating if necessary. The ordinates of these two intersections provide the values for L and U, obtained from the left side if Y/n is less than .50 and the right side otherwise. (Note that the values for L and U depend on the values of the random variable Y which, in turn, is a function of the random sample, which shows that L and U are statistics.)

METHOD B. For confidence coefficients other than .95 or .99, and for n less than or equal to 20, Table A3 may be used. Let the desired confidence coefficient be denoted by $1-\alpha$. Compute $P_1 = 1 - \alpha/2$. Enter Table A3 with the sample size n and read across the row for $y = Y - 1$ until the entry in the table equals P_1 (approximately). The value of p found at the top of that column containing P_1 is the value of the lower limit L. Interpolate if necessary.

Read across the next row ($y = Y$) until the entry $P_2 = \alpha/2$ is reached (approximately). The value of p at the top of the column containing P_2 is the value for the upper limit U. Interpolate if necessary.

METHOD C. For n greater than 20 the normal approximation may be used. Use

(12) $$L = \frac{Y}{n} - x_{1-\alpha/2}\sqrt{Y(n-Y)/n^3}$$

and

(13) $$U = \frac{Y}{n} + x_{1-\alpha/2}\sqrt{Y(n-Y)/n^3}$$

where $x_{1-\alpha/2}$ is the quantile of a normally distributed random variable, obtained from Table A1.

For the sake of illustration, all three methods of computing confidence intervals are used in the following example.

Example 2. In a certain state 20 high schools were selected at random to see if they met the standards of excellence proposed by a national committee on education. It was found that 7 schools did qualify and accordingly were designated "excellent." What is a 95% confidence interval for p, the proportion of all high schools in the state that would qualify for the designation "excellent?"

First, we assume that the number of high schools in the state is large enough so the high schools are classified "excellent" or "not excellent" independently of one another. (Actually, for any finite number of schools, the fact that one school is classified one way tends to increase the chances of the next school being classified the other way, since there would then be a slightly higher proportion of schools in the other category among those not yet selected.) Because we assumed the selection was random, p is the same for all schools and represents the probability of a randomly selected school being designated "excellent."

Since n equals 20 and Y equals 7, Y/n equals .35. For Method A we consult Table A4, read across the bottom of the chart to .35, and up to the curves with "20" (for n) written on them. At the lower curve we read $p = .15 = L$ on the left, and at the upper curve the point of intersection is $p = .59 = U$. Therefore the 95% confidence interval for p is $(L, U) = (.15, .59)$, or

$$(14) \qquad\qquad P(.15 < p < .59) = .95$$

Method B involves the use of Table A3. Reading across the row $y = Y - 1 = 6$ in the table for $n = 20$, we are looking for the probability $1 - (1/2)(.05) = .975$. Interpolating between $p = .15$ for the entry .9781, and $p = .20$ for the entry .9133, we obtain $p = .15$ as a value for the lower bound L. To find the upper bound U we read across the next line, $y = 7$, to the entry $.05/2 = .025$. This involves interpolating between .0580 $(p = .55)$ and .0210 $(p = .60)$. The result is $p = .59 = U$. The result is the same as before.

$$(15) \qquad\qquad P(.15 < p < .59) = 95$$

Method C, the use of the normal approximation based on the central limit theorem, gives

$$L = \frac{Y}{n} - x_{.975}\sqrt{Y(n - Y)/n^3}$$

$$= .35 - (1.960)\sqrt{(7)(13)/(20)^3}$$

$$= .35 - .21$$

$$(16) \qquad\qquad = .14$$

and

$$U = .35 + .21$$

$$(17) \qquad\qquad = .56$$

The confidence interval furnished by method C is

$$(18) \qquad\qquad P(.14 < p < .56) = .95$$

Both Methods A and B are exact methods, insofar as graph readings in Table A4 or interpolation in Table A3 may be considered to be exact. However, Method C is an approximation that becomes better as n gets larger and is better for values of p near .5 than it is for values of p near 0 or 1. For n as small as 20, the example shows that the approximation is still pretty close.

We should remark at this point that objections are often made to statements in the form of Equation 14, that is,

$$P(.15 < p < .59) = .95$$

The objection is that either $.15 < p < .59$ is true or else it is not true, and so the probability is either 1.0 or .0, according to the situation. It is fine to say

$$P(L < p < U) = .95$$

because L and U are random variables. We feel that the objection is too subtle to concern us, and we will consider (incorrectly, perhaps) our definition of probability to be stretched sufficiently to include the above statements.

☐ *Theory.* For the exact Methods A and B just described, the confidence interval consists of all values of p^* such that the data obtained in the sample would result in acceptance of

$$H_0: p = p^*$$

if one were using the two-tailed binomial test. More precisely, if we want to form a $(1 - \alpha)$ confidence interval, we observe the sample and determine Y. Then we ask, "For the given value of Y, which values may we use for p^* in the hypothesis

$$H_0: p = p^*$$

such that a two-tailed binomial test (at level α) would result in *acceptance* of H_0?" Those values of p^* would be in our confidence interval. The values of p^* that would result in *rejection* of H_0 would not be in the confidence interval. Since each tail of the binomial test has probability $\alpha/2$, the value of L is selected as the value of p^* that would barely result in rejection of H_0, for the given value of Y, say y, or a larger value. Thus $p_1{}^*$ is selected so that

(19) $$P(Y \geq y \mid p = p_1{}^*) = \frac{\alpha}{2}$$

and then $L = p_1{}^*$. Next, another value of p^* is selected so the same value y is barely in the lower tail. That is, $p_2{}^*$ is selected so

(20) $$P(Y \leq y \mid p = p_2{}^*) = \frac{\alpha}{2}$$

and we set $U = p_2{}^*$. Because Table A3 gives $P(Y \leq y)$, Equation 20 may be solved by finding the value of p that gives

$$P(Y \leq y) = \frac{\alpha}{2}$$

as described by Method B. Equation 19 needs a slight rearrangement into the form

(21) $$P(Y \leq y - 1 \mid p = p_1{}^*) = 1 - \frac{\alpha}{2}$$

because of the structure of Table A3. More information on confidence intervals for the binomial parameter p may be found in Clopper and Pearson (1934).

The large sample approximations for L and U may be obtained by considering Example 1.5.6, which states that if Y is a binomially distributed random variable with parameters p and large n, then

$$(22) \qquad Z = \frac{Y - np}{\sqrt{npq}}$$

is a random variable whose distribution may be approximated by the standard normal distribution. Then, if $x_{1-\alpha/2}$ is the $(1 - \alpha/2)$ quantile from Table A1, and because $x_{\alpha/2} = -x_{1-\alpha/2}$, we have

$$1 - \alpha = P\left(-x_{1-\alpha/2} < \frac{Y - np}{\sqrt{npq}} < x_{1-\alpha/2}\right)$$

$$= P(-x_{1-\alpha/2}\sqrt{npq} < Y - np < x_{1-\alpha/2}\sqrt{npq})$$

Multiplication by (-1) reverses the sense of the inequalities

$$1 - \alpha = P(x_{1-\alpha/2}\sqrt{npq} > np - Y > -x_{1-\alpha/2}\sqrt{npq})$$

and reversal of the reading order gives

$$1 - \alpha = P(-x_{1-\alpha/2}\sqrt{npq} < np - Y < x_{1-\alpha/2}\sqrt{npq})$$

$$= P(Y - x_{1-\alpha/2}\sqrt{npq} < np < Y + x_{1-\alpha/2}\sqrt{npq})$$

Now we divide through by n

$$(23) \qquad 1 - \alpha = P\left(\frac{Y}{n} - x_{1-\alpha/2}\sqrt{\frac{pq}{n}} < p < \frac{Y}{n} + x_{1-\alpha/2}\sqrt{\frac{pq}{n}}\right)$$

Using a further approximation, the estimator Y/n for p under the radical in Equation 23 gives

$$1 - \alpha \cong P\left(\frac{Y}{n} - x_{1-\alpha/2}\sqrt{\frac{Y}{n}\left(1 - \frac{Y}{n}\right)\Big/n}\right.$$

$$\left. < p < \frac{Y}{n} + x_{1-\alpha/2}\sqrt{\frac{Y}{n}\left(1 - \frac{Y}{n}\right)\Big/n}\right)$$

$$(24) \qquad \cong P(L < p < U)$$

where L and U are the same as in Equations 12 and 13. This latter approximation of Y/n for p results in a slight difference between the confidence interval and the hypothesis test, when the large sample approximations are used for both.

Multiplication by the sample size n in the preceding procedure gives nL and nU as the lower and upper bounds of the confidence interval for np, the mean of a binomial random variable. Also, the binomial test may be used to test hypotheses involving the mean of a binomial random variable, because

$$H_0: p = p^*$$

☐ is equivalent to

$$H_0: np = np^*$$

Other methods of obtaining binomial confidence limits are given by Anderson and Burstein (1967 and 1968). Methods dealing with simultaneous confidence intervals for multinomial proportions are given by Quesenberry and Hurst (1964) and Goodman (1965).

EXERCISES

In each of the following exercises clearly state H_0, H_1, T, α, $\hat{\alpha}$, the decision, and the name of the test used, where such information is appropriate.

1. It is known that 20% of a certain species of insect exhibit a particular characteristic A. Eighteen insects of that species are obtained from an unusual environment, and none of these have characteristic A. Is it reasonable to assume that insects from that environment have the same probability of .20 that the species in general has?

2. Of 16 cars inspected during a safety campaign, 6 were found to be unsafe. Test the hypothesis that no more than 10% of the cars in the population are unsafe. (Which assumption is most likely to be false in the application?)

3. In a dice game a pair of dice were thrown 180 times. The event "seven" occurred on 38 of those times.
   (a) Is the probability of "seven" what it should be if the dice were fair?
   (b) Find a 95% confidence interval for P(seven) using Table A4.
   (c) Find a 95% confidence interval for P(seven) using the large sample approximation.

4. In Exercise 2, what is a 90% confidence interval for the true proportion of unsafe cars in the population?

5. Twenty independent observations on a random variable X with the unknown distribution function $F(x)$ resulted in the following numbers.

142	134	98	119	131
103	154	122	93	137
86	119	161	144	158
165	81	117	128	103

   Find a 95% confidence interval for $F(100)$.

6. A civic group reported to the town council that at least 60% of the town residents were in favor of a particular bond issue. The town council then asked a random sample of 100 residents if they were in favor of the bond issue. Forty-eight said yes. Is the report of the civic group reasonable?

PROBLEMS

1. *The continuity correction.* It is obvious that if Y has a binomial distribution,

$$P(Y \le 4) = P(Y \le 4.1) = \cdots = P(Y \le 4.999)$$

because Y takes on only integer values, such as 4 or 5, but no values between integers. Therefore, which number should be used in the normal approximation to the binomial distribution: 4, or 4.1, or what? The *continuity correction* (because we are trying to use a continuous distribution such as the normal to approximate a discrete distribution such as the binomial) says to use the number midway between two adjacent values in the discrete distribution. That is, in the binomial distribution estimate $P(Y \le 4)$, with

$$P(Y \le 4) \cong P\left(Z \le \frac{4 + .5 - np}{\sqrt{npq}}\right)$$

where Z has a normal distribution, because 4.5 is halfway between 4 and 5.

Usually the continuity correction works well when using the normal distribution to approximate binomial probabilities.

(a) For $n = 20$, $p = .1$, find the exact value of $P(Y \le 1)$ from Table A3. Use the normal approximation to estimate $P(Y \le 1)$, first without the continuity correction and then with the continuity correction. Which estimate is closer?

(b) Repeat part a, but change from $p = .1$ to $p = .3$. Now which estimate is closer?

2. Let Y_1 and Y_2 be independent binomial random variables with parameters n and p_1 and n and p_2, respectively.

(a) Show that $Y_1 - Y_2$ has mean $n(p_1 - p_2)$.

(b) Show that $Y_1 - Y_2$ has variance $np_1(1 - p_1) + np_2(1 - p_2)$.

(c) Justify the use of $[Y_1(n - Y_1) + Y_2(n - Y_2)]/n$ as an estimate of the variance of $Y_1 - Y_2$.

(d) If $Y_1 - Y_2$ is approximately normal, show how an approximate $1 - \alpha$ confidence interval for $(p_1 - p_2)$ is given by

$$\frac{Y_1 - Y_2}{n} - x_{1-\alpha/2}\frac{s}{n} < p_1 - p_2 < \frac{Y_1 - Y_2}{n} - x_{\alpha/2}\frac{s}{n}$$

where

$$s = \sqrt{(Y_1(n - Y_1) + Y_2(n - Y_2))/n}$$

and where x_p is obtained from Table A1.

3.2. THE QUANTILE TEST AND ESTIMATION OF x_p

The binomial test may be used to test hypotheses concerning the quantiles of a random variable, in which case we call it the quantile test. For example, we may examine a random sample of values of some random variable X to see if the median of X is greater than zero, or equal to 17 (say). The measurement scale is usually at least ordinal for the quantile test, although the binomial test only required the weaker nominal scale for its measurements. This is because quantiles have little or no real meaning with nominal scale measurements. If the random variable being examined is a continuous random variable, the hypothesis being tested,

$$H_0: \text{The } p^* \text{th quantile of } X \text{ is } x^* \text{ (specified)}$$

is the same as

$$H_0: P(X \le x^*) = p^*$$

from the definition of the word *quantile*. If we represent the unknown probability $P(X \le x^*)$ by p, H_0 becomes

$$H_0: p = p^*$$

which is the same null hypothesis tested with the binomial test. The test statistic equals the number of sample values that are less than or equal to x^*, and the two-tailed binomial test may be used.

The situation is not as simple if the random variable is not assumed to be continuous. Then the null hypothesis

$$H_0: \text{The } p^* \text{th quantile of } X \text{ is } x^*$$

is the same as

$$H_0: P(X \le x^*) \ge p^* \quad \text{and} \quad P(X < x^*) \le p^*$$

Now the binomial test may be used, but the adaptation of the test to this hypothesis is a little tricky, so we will present the procedure as a separate test.

The Quantile Test

DATA. Let $X_1, X_2, \ldots, X_n$ be a random sample. The data consist of observations on the X_i.

ASSUMPTIONS

1. The X_is are a random sample (i.e., they are independent and identically distributed random variables).
2. The measurement scale of the X_is is at least ordinal.

HYPOTHESES. Let x^* and p^* represent some specified numbers, $0 < p^* < 1$. The hypotheses may take one of the following three forms.

A. (two-tailed test)

$$H_0: \text{The } p^* \text{th population quantile is } x^*$$

[This is equivalent to $H_0: P(X \le x^*) \ge p^*$, and $P(X < x^*) \le p^*$, where X has the same distribution as the X_is in the random sample.]

$$H_1: x^* \text{ is not the } p^* \text{th population quantile}$$

[This is equivalent to H_1: Either $P(X \leq x^*) < p^*$ or $P(X < x^*) > p^*$.]

B. (one-tailed test)

H_0: The p^*th population quantile is at least as great as x^*

[This is equivalent to H_0: $P(X < x^*) \leq p^*$.]

H_1: The p^*th population quantile is less than x^*

[This is the same as H_1: $P(X < x^*) > p^*$.]

C. (one-tailed test)

H_0: The p^*th population quantile is no greater than x^*

[Or H_0: $P(X \leq x^*) \geq p^*$.]

H_1: The p^*th population quantile is greater than x^*

[Or H_1: $P(X \leq x^*) < p^*$.]

TEST STATISTIC. We will use two test statistics in this test. Let T_2 equal the number of observations less than x^*, and let T_1 equal the number of observations less than *or equal to* x^*. Then $T_1 = T_2$ if none of the numbers in the data exactly equals x^*. Otherwise, T_1 is greater than T_2.

DECISION RULE. As in the binomial test, the test statistics have a discrete distribution, so α will seldom be a nice round number. The different decision rules, corresponding to the hypotheses A, B, or C, are given next.

A. (two-tailed test) The critical region corresponds to values of T_2 that are too large [indicating that perhaps $P(X < x^*)$ is greater than p^*] and to values of T_1 that are too small [indicating that perhaps $P(X \leq x^*)$ is less than p^*]. The critical region is found by entering Table A3 with the sample size n and the hypothesized probability p^*, as in the two-tailed binomial test. Find the number t_1 such that

(1) $$P(Y \leq t_1) = \alpha_1$$

where Y has the binomial distribution with parameters n and p^* and where α_1 is about half of the desired level of significance. Then find the number t_2 such that

(2) $$P(Y > t_2) = \alpha_2$$

or, equivalently,

(3) $$P(Y \leq t_2) = 1 - \alpha_2$$

where α_2 is chosen so that $\alpha_1 + \alpha_2$ is about equal to the desired level of significance. Reject H_0 if T_1 is less than or equal to t_1 or if T_2 is greater than t_2. Otherwise accept H_0. The level of significance equals $\alpha_1 + \alpha_2$.

B. (one-tailed test) Since large values of T_2 indicate that H_0 is false, enter Table A3 with the sample size n and the hypothesized p^* as p. Find the number t_2 such that

$$\text{(4)} \qquad\qquad P(Y > t_2) = \alpha$$

which is the same as

$$\text{(5)} \qquad\qquad P(Y \leq t_2) = 1 - \alpha$$

for some acceptable level of significance α. Then reject H_0 if T_2 exceeds t_2. Accept H_0 if T_2 is less than or equal to t_2. (This is the same as decision rule B in the binomial test.)

C. (one-tailed test) Small values of T_1 indicate H_0 is false, so enter Table A3 with the sample size n and the specified probability p^* to find t_1 such that

$$\text{(6)} \qquad\qquad P(Y \leq t_1) = \alpha$$

for an acceptable level α, where Y has a binomial distribution with parameters n and p^*. Reject H_0 if T_1 is less than or equal to t_1. Accept H_0 if T_1 exceeds t_1. (This is the same as decision rule C in the binomial test.)

Example 1. Entering college freshmen have taken a particular high school achievement examination for many years, and the upper quartile is well established at a score of 193. A particular high school sends fifteen of its graduates to college, where they take the exam and get the following scores

189	233	195	160	212
176	231	185	199	213
202	193	174	166	248

It is assumed that these fifteen students represent a random sample of all students from that high school who go on to college. One way of comparing college students from that high school with other college students is by testing the hypothesis that the above scores come from a population whose upper quartile is 193. That is,

H_0: The upper quartile is 193

is tested against the alternative

H_1: The upper quartile is not 193

where we are referring to the upper quartile of the test scores of all college students from that high school, past, present, or future.

The two-tailed quantile test is applied. A critical region of approximate size .05 is obtained by entering Table A3 with $n = 15$ and $p = .75$. There it is seen that, for the binomial random variable Y,

$$\text{(7)} \qquad\qquad P(Y \leq 7) = .0173$$

and

$$P(Y \le 14) = .9866$$

(8) $$= 1 - .0134$$

The critical region of size

$$\alpha = .0173 + .0134$$

(9) $$= .0307$$

corresponds to values of T_1 less than or equal to $t_1 = 7$, and values of T_2 greater than $t_2 = 14$.

In this example $T_1 = 7$, the number of observations less than or equal to 193, and $T_2 = 6$, since one observation exactly equals 193. Therefore T_1 is too small, and H_0 is rejected. The upper quartile for students from that high school does not seem to be 193.

Because the observed value of the test statistic T_1 was barely in the rejection region, the level of significance is as small as it could be to still result in rejection of H_0. Therefore the *critical level* in this example equals .0307, the same as the level of significance.

The one-tailed quantile test, with the large sample approximation, is illustrated in the following example.

Example 2. The time interval between eruptions of Old Faithful geyser is recorded 112 times to see whether the median interval is less than or equal to 60 minutes (null hypothesis) or whether the median interval is greater than 60 minutes (alternative hypothesis). If the median interval is 60, 60 is $x_{.50}$, or the median. If the median interval is less than 60, 60 is a p quantile for some $p \ge .50$. Thus $H_0 = P(X \le 60) \ge .50$, and $H_1 = P(X \le 60) < .50$, where X is the time interval between eruptions. Assuming that the various intervals are independent and identically distributed, the one-tailed quantile test may be used, with decision rule C. The test statistic T_1 equals the number of intervals that are less than or equal to 60 minutes, and the critical region of size .05 corresponds to values of T_1 less than

$$t_1 = np^* + w_{.05}\sqrt{np^*(1 - p^*)}$$

$$= (112)(.50) - (1.645)\sqrt{(112)(.50)(.50)}$$

(10) $$= 47.3.$$

Of the 112 time intervals, 8 are 60 minutes or less, so $T_1 = 8$, and H_0 is soundly rejected in favor of the alternative "the median time interval between eruptions is greater than 60 minutes." The critical level is

$$\hat{\alpha} = P(T_1 \le 8)$$

$$= P\left(\frac{T_1 - np}{\sqrt{npq}} \le \frac{8 - np}{\sqrt{npq}}\right)$$

(11) $$\cong P\left[Z \le \frac{8 - (112)(.50)}{\sqrt{(112)(.50)(.50)}}\right]$$

where Z is a standard normal random variable. Then, from Table A1,

$$\hat{\alpha} = P\left(Z \leq \frac{-48}{5.3}\right)$$

$$= P(Z \leq -9.05)$$

(12) $\ll .0001$

which is read "much less than .0001."

☐ *Theory.* First we will explain why the hypotheses within the parentheses in A, B, and C are equivalent to the hypothesis not in parentheses. Perhaps this is most easily seen by referring to the graph of an arbitrary distribution function, as in Figure 1.

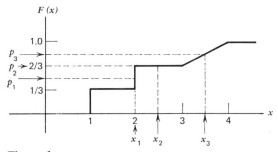

Figure 1

The distribution function at x^* may be in one of three phases: it may be rising vertically, as at x_1; it may be in a horizontal segment, as at x_2; or it may be rising gradually, as at x_3. H_0 in the third set of hypotheses (set C) states that the p^*th population quantile (x_{p*}) is no greater than x^*, or $x_{p*} \leq x^*$. Because every value of x^* is some sort of a quantile, we can say that x^* is the pth quantile for some p, say p_0. (We are temporarily ignoring our convention of choosing only the midpoint of the horizontal segments as the quantile and adhering directly to the definition of quantile.) Because the graph of the distribution function never descends as x gets larger, $x_{p*} \leq x^*$ implies that $p^* \leq p_0$, which may be seen by imagining x^* as being in each of the three phases typified by x_1, x_2, and x_3 in Figure 1. Any value of x_{p*} to the left of x^* implies that the ordinate p^* at x_{p*} is no greater than the ordinate p_0 of x^*. From the definition of quantile, Definition 1.4.1,

(13) $P(X > x^*) \leq 1 - p_0$

which is the same as

(14) $p_0 \leq 1 - P(X > x^*) = P(X \leq x^*)$

Since $p^* \leq p_0$, this implies

(15) $p^* < P(X \leq x^*)$

which is the equivalent form of H_0 in set C of hypotheses. The negation of H_0 is H_1, and the negation of Equation 15 is

$$(16) \qquad\qquad p^* > P(X \le x^*)$$

as stated in the alternative hypothesis. The same reasoning is used to show the other hypothesis to be equivalent.

Briefly, Figure 1 is used to visualize that $x_{p_0} \le x_{p*}$ (H_0 in B) implies that $p_0 \le p^*$. If $x^* = x_{p_0}$ then, by Definition 1.4.1,

$$(17) \qquad\qquad P(X < x^*) \le p_0 \le p^*$$

is true, which furnishes the equivalent form of H_0.

The binomial test is applied directly to test the hypotheses in parentheses. H_0 in B is tested by defining the "class 1" of the binomial test as those observations less than x^*. H_0 in C is tested by considering "class 1" to represent those observations less than or equal to x^*. The two tests in B and C are combined to give the two-tailed test in A. The assumptions of independence and constant probability p in the binomial test are satisfied because the X_i are independent and identically distributed (respectively).

In the previous section we showed how to find a confidence interval for a probability p. The same method is used to find a confidence interval for $F(x_0)$, the distribution function at some specified number x_0. That is, given a number x_0, we can find a "vertical" confidence interval (referring to a graph) for the unknown probability $F(x_0)$. Suppose, however, we are given a probability, say p^*, and asked to find a "horizontal" confidence interval for the unknown quantile x_{p_0}. This second type of confidence interval, a confidence interval for a quantile, is found if we wish to make a statement concerning a specified quantile such as the median, the upper quartile, or any p^* quantile where p^* is a specified constant, $0 \le p^* \le 1$. The statement then takes the form

$$(18) \qquad\qquad P(X^{(r)} \le x_{p*} \le X^{(s)}) = 1 - \alpha$$

where $1 - \alpha$ is a known *confidence coefficient* and where $X^{(r)}$ and $X^{(s)}$ are order statistics (see Definition 2.1.4) with r and s specified. The values for r and s may be determined prior to drawing the sample in the manner described next, with knowledge of only the sample size n and the desired confidence coefficient. The sample $X_1, \ldots, X_n$ needs only to be random. No restrictions are made on the distribution function of the X_i. Thus this statistical method may be applied freely to any random sample from any population.

Confidence Interval for a Quantile

DATA. The data consist of observations on $X_1, X_2, \ldots, X_n$, which are independent and identically distributed random variables. Let $X^{(1)} \le X^{(2)} \le \cdots \le X^{(r)} \le \cdots \le X^{(s)} \le \cdots \le X^{(n)}$ represent the ordered sample, where $1 \le r <$

$s \leq n$. We wish to find a confidence interval for the (unknown) p^*th quantile, where p^* is some specified number between zero and one.

ASSUMPTIONS

1. The sample $X_1, X_2, \ldots, X_n$ is a random sample.
2. The measurement scale of the X_i's is at least ordinal.

METHOD A (*small samples*). For $n \leq 20$ Table A3 may be used to find r and s. Enter Table A3 with the sample size n and the probability $p = p^*$. Read down the column for $p = p^*$ until reaching an entry approximately equal to $\alpha/2$, where $1 - \alpha$ is the approximate confidence coefficient desired. Call that entry α_1, and the corresponding value for y (to the far left of α_1) is $r - 1$. Add 1 to get r. Then continue down the column for $p = p^*$ until reaching an entry approximately equal to $1 - (\alpha/2)$, which we will call $1 - \alpha_2$. The value of y corresponding to the entry $1 - \alpha_2$ is called $s - 1$, and 1 is added to obtain s. Thus we have determined α_1, α_2, r, and s. The exact confidence coefficient is $1 - \alpha_1 - \alpha_2$. The interval estimator is the interval between $X^{(r)}$ and $X^{(s)}$, whose values may be obtained from the data. Then

(19) $$P(X^{(r)} \leq x_{p^*} \leq X^{(s)}) \geq 1 - \alpha_1 - \alpha_2$$

provides the confidence interval. If we assume that the unknown distribution function is continuous, then

(20) $$P(X^{(r)} \leq x_{p^*} \leq X^{(s)}) = 1 - \alpha_1 - \alpha_2$$

as stated in Equation 18 also.

METHOD B (*large sample approximation*). For n greater than 20 the approximation based on the central limit theorem may be used. (See the end of Table A3, or Example 1.5.6.) Compute

(21) $$r^* = np^* + w_{\alpha/2}\sqrt{np^*(1 - p^*)}$$

and

(22) $$s^* = np^* + w_{1-\alpha/2}\sqrt{np^*(1 - p^*)}$$

where the quantiles w_p are obtained from Table A1 and where $1 - \alpha$ is the desired confidence coefficient. In general, r^* and s^* will not be integers. Let r and s be the integers obtained by rounding r^* and s^* upward to the next higher integers. Then the approximate confidence interval is given by Equation 19, or Equation 20 if the unknown distribution function is continuous.

A one-sided confidence interval may be formed by finding only r or s as described. One-sided confidence intervals are of the form

(23) $$P(X^{(r)} \leq x_{p^*}) = 1 - \alpha_1$$

and

(24) $$P(x_{p^*} \leq X^{(s)}) = 1 - \alpha_2$$

if the distribution function is continuous, or

(25) $$P(X^{(r)} \leq x_{p^*}) \geq 1 - \alpha_1$$

and

(26) $$P(x_{p^*} \leq X^{(s)}) \geq 1 - \alpha_2$$

otherwise.

Example 3. Sixteen radio tubes are selected at random from a large batch of radio tubes and are tested. The number of hours until failure is recorded for each one. We wish to find a confidence interval for the upper quartile, with a confidence coefficient close to 90%. Table A3 is entered with $n = 16$ and $p = .75$. Reading down the column for $p = .75$ the probability .0271 is selected as being close to .05. The value of y associated with $\alpha_1 = .0271$ is $y = 8$; therefore r equals 9. The probability closest to .95 is $.9365 = 1 - \alpha_2$, which has a corresponding y of 14. Therefore $s = 15$. The confidence interval is

(27) $$P(X^{(9)} \leq x_{.75} \leq X^{(15)}) = .9094$$

(It is reasonable to assume the time to failure is a continuous random variable, so we can use Equation 20.)

The results of the testing, arranged in increasing order, are as follows.

$X^{(1)} = 46.9$	$X^{(5)} = 56.8$	$X^{(9)} = 63.3$	$X^{(13)} = 67.1$
$X^{(2)} = 47.2$	$X^{(6)} = 59.2$	$X^{(10)} = 63.4$	$X^{(14)} = 67.7$
$X^{(3)} = 49.1$	$X^{(7)} = 59.9$	$X^{(11)} = 63.7$	$X^{(15)} = 73.3$
$X^{(4)} = 56.5$	$X^{(8)} = 63.2$	$X^{(12)} = 64.1$	$X^{(16)} = 78.5$

Because $X^{(9)} = 63.3$ and $X^{(15)} = 73.3$, we may say "the interval from 63.3 hours to 73.3 hours, inclusive, is a 90.94% confidence interval for the upper quartile."

The large sample approximation, furnished by Equations 21 and 22, yields

$$r^* = (16)(.75) + (-1.645)\sqrt{(16)(.75)(.25)}$$

$$= 12 - 2.86$$

(28) $$= 9.14$$

and

$$s^* = 12 + 2.86$$

(29) $$= 14.86$$

Therefore $r = 10$ and $s = 15$, so the 90% confidence interval becomes (63.4, 73.3), slightly smaller than the more precise method used before.

☐ *Theory.* Consider first the simpler case where the distribution function is continuous. If x_{p*} is the $p*$th quantile, we have the exact relationship

(30)
$$P(X \geq x_{p*}) = P(X > x_{p*}) = 1 - p*$$

where the distribution function of X is the same as that of the random sample.

The order statistic of rank 1, $X^{(1)}$, will assume a value larger than some specified constant only if the smallest value in the sample is larger than the constant. Therefore $X^{(1)}$ is greater than the constant only if all n values in the sample are greater than the constant. Choosing x_{p*} as the constant, we may conclude

$$P(x_{p*} < X^{(1)}) = P(\text{all sample values exceed } x_{p*})$$
$$= P(x_{p*} < X_1, x_{p*} < X_2, \ldots, x_{p*} < X_p)$$
$$= P(x_{p*} < X_1) \cdot P(x_{p*} < X_2) \cdot \cdots \cdot P(x_{p*} < X_p)$$

(31)
$$= (1 - p*)^n$$

because the X_is are independent, and they all have the same $p*$th quantile x_{p*}.

If x_{p*} is less than $X^{(2)}$, then either exactly $n - 1$ observations are greater than x_{p*}, in which case $X^{(1)} \leq x_{p*} < X^{(2)}$, or else exactly n observations are greater than x_{p*}, in which case $x_{p*} < X^{(1)} < X^{(2)}$. Therefore

$$P(x_{p*} < X^{(2)}) = P(x_{p*} < X^{(1)}) + P(X^{(1)} \leq x_{p*} < X^{(2)})$$
$$= P(\text{at least } n - 1 \text{ of the } X_i \text{ exceed } x_{p*})$$

(32)
$$= P(1 \text{ or fewer of the } X_i \text{ are } \leq x_{p*})$$

Now the probability in Equation 32 is given by the binomial distribution function, because each X_i has probability $p*$ of being less than or equal to x_{p*}, and the X_i are independent. Therefore Equation 32 leads to

(33)
$$P(x_{p*} < X^{(2)}) = \sum_{i=0}^{1} \binom{n}{i} (p*)^i (1 - p*)^{n-i}$$

With the aid of the binomial distribution function given by Equation 1.3.8, the preceding argument may be extended as follows.

$$P(x_{p*} < X^{(r)}) = P(\text{at least } n - r + 1 \text{ of the } X_i \text{ exceed } x_{p*})$$
$$= P(r - 1 \text{ or fewer of the } X_i \text{ are } \leq x_{p*})$$

(34)
$$= \sum_{i=0}^{r-1} \binom{n}{i} (p*)^i (1 - p*)^{n-i}$$

The confidence coefficient is given by

$$1 - \alpha \cong P(X^{(r)} \leq x_{p*} \leq X^{(s)})$$

(35)
$$= P(x_{p*} \leq X^{(s)}) - P(x_{p*} < X^{(r)})$$

Therefore r and s may be selected, with the aid of Equation 34 and Table A3, so that

$$(36) \qquad 1 - \alpha_2 = P(x_{p*} \leq X^{(s)}) = 1 - \frac{\alpha}{2}$$

and

$$(37) \qquad \alpha_1 = p(\chi_{p*} < X^{(r)}) = \frac{\alpha}{2}$$

Then the confidence coefficient will be $1 - \alpha_2 - \alpha_1 \cong 1 - \alpha$. Note that because the distribution function is assumed to be continuous, we have

$$(38) \qquad P(x_{p*} \leq X^{(s)}) = P(x_{p*} < X^{(s)})$$

so that Table A3 may be used to find s.

If the distribution function of X, and therefore of the X_is, is not necessarily continuous, Equation 30 is not necessarily true. Instead, by Definition 1.5.1 we have

$$(39) \qquad P(X > x_{p*}) \leq 1 - p^*$$

and

$$(40) \qquad P(X \geq x_{p*}) \geq 1 - p^*$$

First we will consider how Equation 39 affects Equation 34 and, therefore, our method for finding r in Equation 37. Because Equation 39 is true, the probability of each observation exceeding x_{p*} may be smaller than it was when X was continuous. Therefore there may be less of a tendency for each of the order statistics to exceed x_{p*} than was formerly the case. That is, the probability $P(x_{p*} < X^{(r)})$ may be smaller than it was in the continuous case, which was then given by Equation 34. So, in general, the following holds true instead of Equation 34.

$$(41) \qquad P(x_{p*} < X^{(r)}) \leq \sum_{i=0}^{r-1} \binom{n}{i}(p^*)^i(1 - p^*)^{n-i}$$

If Table A3 is used to find r in the manner just described, then

$$(42) \qquad P(x_{p*} < X^{(r)}) \leq \alpha_1$$

Now we will consider how Equation 40 affects the probability $1 - \alpha_2$ resulting from our method of selecting s. Because Equation 40 is true, there may be a larger probability for each observation to be greater than *or equal to* x_{p*} than in the continuous case. Therefore the number of observations exceeding or equaling x_{p*} may tend to be larger and the probability of $X^{(s)} \geq x_{p*}$ may be larger than in the continuous case. As a result, Equation 34 is modified in the general case to read

$$(43) \qquad P(x_{p*} \leq X^{(s)}) \geq \sum_{i=0}^{s-1} \binom{n}{i}(p^*)^i(1 - p^*)^{n-i}$$

Therefore, if Table A3 is used to find s in the manner described, we have

$$(44) \qquad P(x_{p*} \le X^{(s)}) \ge 1 - \alpha_2$$

Equations 42 and 44, which are true for all distributions, may be used as follows.

$$P(X^{(r)} \le x_{p*} \le X^{(s)}) = P(x_{p*} \le X^{(s)}) - P(x_{p*} < X^{(r)})$$
$$\ge P(x_{p*} \le X^{(s)}) - \alpha_1$$
$$(45) \qquad \ge 1 - \alpha_2 - \alpha_1$$

Thus the method of finding a confidence interval for a quantile has been justified for the case where exact tables of the binomial distribution function are available.

The large sample method of obtaining r and s is based on the use of the standard normal distribution to approximate the binomial distribution. Different arguments may be advanced for the different possible ways of converting r^* and s^* into the integers r and s, but the method given here of simply rounding upward to the next higher integer seems to provide a sufficiently close approximation.

The usage of one-sided confidence intervals for quantiles in life testing situations is discussed by Barlow and Gupta (1966). Tables of distribution-free confidence limits are given by Van der Parren (1970) for the median and by Van der Parren (1973) for quantiles. Confidence intervals for intervals between quantiles are discussed by Krewski (1976) and Reiss and Rüshendorf (1976).

EXERCISES

1. A random sample of tenth-grade boys resulted in the following 20 observed weights.

142	134	98	119	131
103	154	122	93	137
86	119	161	144	158
165	81	117	128	103

Test the hypothesis that the median weight is 103.

2. In Exercise 1 test the hypothesis that the upper quartile is at least 150.

3. In Exercise 1 test the hypothesis that the third decile is no greater than 100.

4. In Exercise 1 find an approximate 90% confidence interval for the median. What is the exact confidence coefficient? Also compare the results using the exact method with the results obtained using the large sample approximation.

5. It is desired to design a given automobile to allow enough headroom to accommodate comfortably all but the tallest 5% of the people who drive. Former studies indicate that the 95th percentile was 70.3 inches. In order to see if the former studies are still valid, a random sample of size 100 is selected. It is found that the 12 tallest persons in the sample have the following heights.

72.6	70.0	71.3	70.5
70.8	76.0	70.1	72.5
71.1	70.6	71.9	72.8

Is it reasonable to use 70.3 as the 95th percentile?

6. In Exercise 5, what is a 95% confidence interval for the 95th percentile of drivers from which the sample was selected.

PROBLEMS

1. One parametric method for finding a $1 - \alpha$ confidence interval for the median is to assume that the population is normal and use

$$\bar{X} - t_{\alpha/2} S/\sqrt{n-1} < x_{.5} < \bar{X} + t_{\alpha/2} S/\sqrt{n-1}$$

where $\bar{X}$ is the sample mean, S is the sample standard deviation (Definition 2.2.3), n is the sample size, and t_p is the pth quantile from Table A25, $n - 1$ degrees of freedom. Compute the preceding confidence interval on the data in Exercise 1 and compare it with the nonparametric confidence intervals of Exercise 4, where $\alpha = .10$. Which confidence interval is the easiest to justify? Which confidence interval is "best" (in terms of being shortest)?

3.3. TOLERANCE LIMITS

The confidence intervals of Sections 3.1 and 3.2 provide interval estimates for unknown population parameters, such as the unknown probability p or the unknown quantile x_p, and a certain probability $1 - \alpha$ (confidence coefficient) that the unknown parameter is within the interval. Tolerance limits differ from confidence intervals in that tolerance limits provide an interval within which at least a proportion q of the population lies, with probability $1 - \alpha$ or more that the stated interval does indeed "contain" the proportion q of the population. A typical application would be in a situation where we are about to draw a random sample size n, $X_1, X_2, \ldots, X_n$, and we want to know how large n should be so that we can be 95% sure that at least 90% of the population lies between $X^{(1)}$ and $X^{(n)}$, the largest and smallest observations in our sample. We may generalize somewhat and consider the question, "How large must the sample size n be so that at least a proportion q of the population is between $X^{(r)}$ and $X^{(n-1-m)}$ with probability $1 - \alpha$ or more?". The numbers q, r, m, and $1 - \alpha$ are known (or selected) beforehand, and only n needs to be determined.

The preceding tolerance limit would be a two-sided tolerance limit. One-sided tolerance limits are of the form, "At least a proportion q of the population is greater than $X^{(r)}$, with probability $1 - \alpha$," or "At least a proportion q of the population is less than $X^{(n-1-m)}$, with probability $1 - \alpha$."

One-sided tolerance limits are identical with one-sided confidence intervals for quantiles, as will be shown in this section.

The population referred to here is either infinite or, if the population is finite, the sample is drawn with replacement so the X_is are independent. For finite populations where the sampling is without replacement and where the sample size n is small compared to the population size N, these methods are fairly accurate. More precise methods for finite populations may be found in Wilks (1962).

Tolerance Limits

DATA. Choose a confidence coefficient $1-\alpha$, a pair of positive integers r and m, and a fraction q between zero and one. We wish to determine the size n of a random sample $X_1, X_2, \ldots, X_n$, for which we can make the statement, "The probability is $1-\alpha$ that the random interval from $X^{(r)}$ to $X^{(n+1-m)}$ inclusive contains a proportion q or more of the population." Note that we are using the convention $X^{(0)} = -\infty$ and $X^{(n+1)} = +\infty$, so that one-sided tolerance limits may be obtained by setting either r or m equal to zero.

ASSUMPTIONS

1. The $X_1, X_2, \ldots, X_n$ constitute a random sample.
2. The measurement scale is at least ordinal.

METHOD. If $r+m$ equals 1, that is, if either r or m equals zero as in a one-sided tolerance limit, read n directly from Table A5 for the appropriate values of α and q. If $r+m$ equals 2, read n directly from Table A6 for the appropriate values of α and q. If Tables A5 and A6 are not appropriate, use the approximation

(1) $$n \cong \tfrac{1}{4} x_{1-\alpha} \frac{1+q}{1-q} + \tfrac{1}{2}(r+m-1)$$

where $x_{1-\alpha}$ is the $(1-\alpha)$ quantile of a chi-square random variable with $2(r+m)$ degrees of freedom, obtained from Table A2.

Then, with a sample of size n, there is probability at least $1-\alpha$ that at least q [or $(100)(q)\%$] of the population is between $X^{(r)}$ and $X^{(n+1-m)}$ inclusive. That is,

(2) $$P(X^{(r)} \leq \text{at least a fraction } q \text{ of the population} \leq X^{(n+1-m)}) \geq 1-\alpha$$

For one-sided tolerance regions let either r or m equal zero, where $X^{(0)}$ and $X^{(n+1)}$ are considered to be $-\infty$ and $+\infty$ respectively, and proceed as described above.

Example 1. Probably the most widely used two-sided tolerance limits are those where $r = 1$ and $m = 1$. A certain manufactured product varies in length from one item to the next. In order to obtain the proper length box for shipping the item it is desirable that we obtain an upper and a lower limit within which we are 90% certain, 80% or more of all of the manufactured items lies. What must n be so that $X^{(n)}$ and $X^{(1)}$ furnish our upper and lower limits?

Table A6 is entered with $q = .80$ and $1 - \alpha = .90$. The obtained value for n is 18. The approximation furnished by Equation 1 is

$$n \cong \tfrac{1}{4} x_{1-\alpha} \frac{1+q}{1-q} + \tfrac{1}{2}(r + m - 1)$$

$$= \tfrac{1}{4}(7.779) \frac{1.80}{.20} + \tfrac{1}{2}$$

(3) $$= 18.003$$

A sample of size 18 is drawn. The largest value in the sample is

$$X^{(18)} = 7.57 \text{ inches}$$

and the smallest value is

$$X^{(1)} = 7.21 \text{ inches}$$

Therefore there is probability .90 that at least 80% of the manufactured items are equal to or between 7.21 and 7.57 inches in length.

The following is an example of a one-sided tolerance limit.

Example 2. Along with each lot of steel bars, a certain manufacturer guarantees that at least 90% of the bars will have a breaking point above a number specified for each lot. Because of variable manufacturing conditions the guaranteed breaking point is established separately for each lot by breaking a random sample of bars from each lot and setting the guaranteed breaking point equal to the minimum breaking point in the sample. How large should the sample be so that the manufacturer can be 95% sure the guarantee statement is correct?

Table A5 is entered with $q = .90$ and $1 - \alpha = .95$, with the result $n = 29$. In each lot a sample of size 29 is selected at random, and the smallest breaking point of these bars in the sample is stated as the guaranteed breaking point, at which at least 90% of the bars in the lot will still be intact, with probability .95.

☐ *Theory.* A careful examination of the statement furnished by the one-sided tolerance limit reveals the similarity it has with the one-sided confidence interval for quantiles. That is, the one-sided tolerance limit says

(4) $$P(\text{at least } q \text{ of the population is} \le X^{(n+1-m)}) \ge 1 - \alpha$$

However, "at least q of the population is $\leq X^{(n+1-m)}$" is the same as saying "the q quantile is $\leq X^{(n+1-m)}$"; the two statements are merely different ways of stating the same idea. So we have

$$P(\text{at least } q \text{ of the population is} \leq X^{(n+1-m)})$$

$$= P(\text{the } q \text{ quantile is} \leq X^{(n+1-m)})$$

(5) $$= P(x_q \leq X^{(n+1-m)})$$

The probability in Equation 5 was given in Equation 3.2.43 as

(6) $$P(x_q \leq X^{(n+1-m)}) \geq \sum_{i=0}^{n-m} \binom{n}{i} q^i (1-q)^{n-i}$$

The right side of Equation 6 is examined to find the smallest value for n such that the right side of Equation 6 exceeds $1-\alpha$. This may be accomplished by entering Table A3 with $y = n - m$, the parameter p equal to q, and then searching for the lowest value of n for which the entry is greater than or equal to $1-\alpha$. Because the value for y changes as n changes, it is more convenient to rewrite the right side of Equation 6 as

(7) $$\sum_{i=0}^{n-m} \binom{n}{i} q^i (1-q)^{n-i} = 1 - \sum_{i=n-m+1}^{n} \binom{n}{i} q^i (1-q)^{n-i}$$

which is possible because the sum of all the binomial probabilities equals unity. A change of index, $j = n - i$, on the right side of Equation 7 results in

(8) $$\sum_{i=0}^{n-m} \binom{n}{i} q^i (1-q)^{n-i} = 1 - \sum_{j=0}^{m-1} \binom{n}{j} (1-q)^j q^{n-j}$$

Equation 8 could have been obtained immediately by saying, "The probability of $n - m$ or fewer successes equals the probability of m or more failures, which equals 1 minus the probability of $m - 1$ or fewer failures." The combination of Equations 8 and 6 shows that we could find n by solving for the smallest value of n that satisfies

(9) $$\sum_{j=0}^{m-1} \binom{n}{j} (1-q)^j q^{n-j} \leq \alpha$$

which is obtained from the inequality

(10) $$1 - \sum_{j=0}^{m-1} \binom{n}{j} (1-q)^j q^{n-j} \geq 1 - \alpha$$

Then Table A3 may be entered with $y = m - 1$ and $p = 1 - q$ and the pages turned until the entry in the table is less than or equal to α. That value of n is the sample size selected.

The other one-sided tolerance limit is

(11) $$P(X^{(r)} \leq \text{at least } q \text{ of the population}) \geq 1 - \alpha$$

which is equivalent to the statement

(12) $$P(X^{(r)} \leq x_{1-q}) \geq 1 - \alpha$$

because at least $1 - q$ of the population is greater than or equal to x_{1-q}. Equation 12 becomes

(13) $$\alpha \geq 1 - P(X^{(r)} \leq x_{1-q}) = P(x_{1-q} < X^{(r)})$$

From Equation 3.2.41 we see that the solution to Equation 13 is the smallest value of n such that

(14) $$\alpha \geq \sum_{i=0}^{r-1} \binom{n}{i}(1-q)^i q^{n-i}$$

just as in Equation 9.

In fact, it can be shown, with the aid of calculus (see Noether, 1967a), that for the two-sided tolerance limits and for both types of one-sided tolerance limits, the sample size n depends on the solution to

(15) $$\alpha \geq \sum_{i=0}^{r+m-1} \binom{n}{i}(1-q)^i q^{n-i}$$

which is somewhat surprising in that Equation 15 depends on the sum $r + m$ but does not depend on whether we wish to choose as our interval all values to the right of $X^{(r+m)}$, or all values to the left of $X^{(n+1-r-m)}$, or all values between $X^{(r)}$ and $X^{(n+1-m)}$, or any combination of two order statistics whose ranks have a difference of $n + 1 - m - r$.

The use of Table A3 to solve Equation 15 is, at best, frustrating. Therefore Tables A5 and A6 are given for the most popular values $r + m = 1$ and $r + m = 2$. The approximation in Equation 1 is furnished without proof by Scheffé and Tukey (1944). Graphs to aid in finding n are given by Murphy (1948) and Birnbaum and Zuckerman (1949). More extensive tables are given by Owen (1962).

Tolerance limits may also be used with two samples (Danziger and Davis, 1964), with a single censored sample (Bohrer, 1968), or for deciding from which of two possibly multivariate populations a sample was obtained (Quesenberry and Gessaman, 1968). Usage of tolerance limits on discrete random variables is examined by Hanson and Owen (1963). An application of tolerance intervals to the regression problem is discussed by Bowden (1968). Other articles dealing with tolerance limits are given by Mack (1969) and Goodman and Madansky (1962).

EXERCISES

1. What must the sample size be to be 90% sure that at least 95% of the population lies within the sample range?
   (a) Use the exact table.
   (b) Use the approximation.

2. What must the sample size be to be 95% certain that at least 90% of the population equals or exceeds $X^{(1)}$?
(a) Use the exact table.
(b) Use the approximation.
3. What must the sample size be in order for there to be a probability .90 that at least 85% of the population is $\leq X^{(n)}$?
(a) Use the exact table.
(b) Use the approximation.
4. What must the sample size be in order for there to be a 95% chance that 99% of the population is $\geq X^{(2)}$?
(a) Use the exact table.
(b) Use the approximation.
5. What must the sample size be so there is probability .90 that at least 50% of the population is between $X^{(5)}$ and $X^{(n-4)}$ inclusive?
6. What must the sample size be in Exercise 5 if the probability is .95 instead of .90? How about .99?

PROBLEM

1. Use table A3 to solve Exercise 3. Find the exact value of α.

3.4. THE SIGN TEST

After straying from hypothesis testing somewhat, at least in the previous section, we now return to discuss the oldest of all nonparametric tests, the sign test. Actually, the sign test is just the binomial test, with $p^* = 1/2$. But the sign test deserves special consideration because of its versatility, its age (dating back to 1710), and because $p^* = 1/2 = 1 - p^*$ makes it even simpler than the binomial test. The sign test is useful for testing whether one random variable in a pair (X, Y) tends to be larger than the other random variable in the pair. Also, as we will see in Section 3.5, it may be used to test for trend in a series of ordinal measurements or as a test for correlation. In many situations where the sign test may be used, more powerful nonparametric tests are available for the same model. However, the sign test is usually simpler and easier to use, and special tables to find the critical region are sometimes not needed.

The Sign Test

DATA. The data consist of observations on a bivariate random sample $(X_1, Y_1), (X_2, Y_2), \ldots, (X_{n'}, Y_{n'})$, where there are n' pairs of observations. There should be some natural basis for pairing of the observations; otherwise the Xs and Ys are independent, and the more powerful Mann–Whitney test of Chapter 5 is more appropriate.

Within each pair (X_i, Y_i) a comparison is made, and the pair is classified as "+" or "plus" if $X_i < Y_i$, as "−" or "minus" if $X_i > Y_i$, or as "0" or "tie" if $X_i = Y_i$. Thus the measurement scale needs only to be ordinal.

ASSUMPTIONS

1. The bivariate random variables (X_i, Y_i), $i = 1, 2, \ldots, n'$, are mutually independent.
2. The measurement scale is at least ordinal within each pair. That is, each pair (X_i, Y_i) may be determined to be a "plus," "minus," or "tie."
3. The pairs (X_i, Y_i) are internally consistent, in that if $P(+) > P(-)$ for one pair (X_i, Y_i), then $P(+) > P(-)$ for all pairs. The same is true for $P(+) < P(-)$, and $P(+) = P(-)$.

HYPOTHESES

A. (Two-Tailed Test)

$$H_0: P(+) = P(-)$$
$$H_1: P(+) \neq P(-)$$

B. (One-Tailed Test)

$$H_0: P(+) \leq P(-)$$
$$H_1: P(+) > P(-)$$

C. (One-Tailed Test)

$$H_0: P(+) \geq P(-)$$
$$H_1: P(+) < P(-)$$

It should be noted that the sign test is unbiased and consistent when testing these hypotheses. The sign test is also used for testing the following counterparts of the hypotheses, in which case it is neither unbiased nor consistent unless additional assumptions concerning the distributions of (X_i, Y_i) are made.

A. (Two-Tailed Test) The null hypothesis is interpreted as "X_i and Y_i have the same location parameter" and, therefore,

$$H_0: E(X_i) = E(Y_i) \qquad \text{for all } i$$

is tested against the alternative

$$H_1: E(X_i) \neq E(Y_i) \qquad \text{for all } i$$

to see if X_i and Y_i have different means. Similarly, the test may be a test of medians.

$$H_0: \text{The median of } X_i \text{ equals the median of } Y_i \text{ for all } i$$
$$H_1: X_i \text{ and } Y_i \text{ have different medians for all } i$$

B. (One-Tailed Test) The null hypothesis may be considered to indicate that the values of X_i tend to be larger than the values of Y_i. because H_0 states that X_i may be more likely to exceed Y_i than to be less than Y_i. Therefore this one-tailed sign test is sometimes used to test

$$H_0: E(X_i) \geq E(Y_i) \qquad \text{for all } i$$

against the alternative

$$H_1: E(X_i) < E(Y_i) \qquad \text{for all } i$$

A similar set of hypotheses may be stated in terms of the median.

C. (One-Tailed Test) The null hypothesis in the preceding category C may be considered to indicate that X_i has a tendency to assume smaller values than does Y_i; hence this one-tailed sign test may be used to test

$$H_0: E(X_i) \leq E(Y_i) \qquad \text{for all } i$$

against the alternative

$$H_1: E(X_i) > E(Y_i) \qquad \text{for all } i$$

A similar statement may be made concerning the medians.

TEST STATISTIC. Let the test statistic T equal the number of "plus" pairs; that is, T equals the number of pairs (X_i, Y_i) in which X_i is less than Y_i.

$$T = \text{total number of } +\text{'s}$$

DECISION RULE. First, disregard all tied pairs and let n equal the number of pairs that are not ties.

$$n = \text{total number of } +\text{'s and } -\text{'s}$$

Let α represent the approximate level of significance desired.

Use decision rules A, B, or C, depending on whether the hypothesis being tested is classified under the preceding categories A, B, or C.

A. (Two-Tailed Test) For $n \leq 20$, use Table A3 with the proper value of n and $p = 1/2$. Select a table value of about $\alpha/2$ and call it α_1. The value of y corresponding to α_1 is called t. The critical region of size $2\alpha_1$ corresponds to values of T less than or equal to t, or greater than or equal to $n - t$. Reject H_0 if $T \leq t$ or if $T \geq n - t$, at a level of significance of $2\alpha_1$. Otherwise accept H_0.

For n larger than 20 the approximation at the end of Table A3 is used to obtain

(1)
$$t = \tfrac{1}{2}(n + w_{\alpha/2}\sqrt{n})$$

where $w_{\alpha/2}$ is obtained from Table A1. If $\alpha = .05$, $w_{\alpha/2} = (-1.96)$, and Equation 1 becomes approximately

(2)
$$t = \frac{n}{2} - \sqrt{n}$$

which may be easily remembered.

B. (One-Tailed Test) Large values of T indicate that a plus is more probable than a minus, as stated by H_1. Therefore the critical region corresponds to values of T greater than or equal to $n - t$, where t is found by entering Table A3 with $p = 1/2$ and n and finding the table entry that approximately equals α, say α_1. The value of y corresponding to α_1 is t. For n greater than 20, t may be found from the approximation

(3)
$$t = \tfrac{1}{2}(n + w_\alpha \sqrt{n})$$

where w_α is obtained from Table A1. H_0 is rejected at the level of significance α_1 (or α) if T is greater than or equal to $n - t$.

It is equivalent, and may be easier, to consider the test statistic T' equal to the number of "minus" pairs. Then H_0 is rejected if $T' \leq t$, where t is the same as before.

As in Equation 2, for $\alpha = .025$, t may be quickly computed using the equation

(4)
$$t \cong \frac{n}{2} - \sqrt{n}$$

C. (One-Tailed Test) Small values of T indicate that a minus is more probable than a plus, in agreement with H_1. Therefore t is found exactly as in category B. The critical region of size α_1 (or α) corresponds to values of T less than or equal to t. Reject H_0 if $T \leq t$, at a level of significance of α_1 (or α in the case of $n > 20$). Otherwise accept H_0.

Example 1. An item A is manufactured using a certain process. Item B serves the same function as A but is manufactured using a new process. The manufacturer wishes to determine whether B is preferred to A by the consumer, so she selects a random sample consisting of 10 consumers, gives each of them one A and one B, and asks them to use the items for some period of time. The sign test (one tailed) will be used to test

$$H_0: P(+) \leq P(-)$$

against

$$H_1: P(+) > P(-)$$

where "+" represents the event "item B is preferred over item A," and "−" represents the event "item A is preferred over item B." In words, H_0 says, "Item B does not tend to be preferred to item A," while H_1 says, "Item B tends to be preferred to item A." The test statistic T is the number of + signs, the number of consumers who prefer B over A. The critical region corresponds to values of T greater than or equal to $n - t$. However, we need to know how many ties there are before we can find n and, hence, t.

At the end of the allotted period of time the consumers report their preferences to the manufacturer. Eight consumers preferred B to A, 1 preferred A to B, and 1 reported "no preference." Therefore,

$$8 = \text{number of } +\text{'s}$$
$$1 = \text{number of } -\text{'s}$$
$$1 = \text{number of ties}$$
$$n = \text{number of } +\text{'s and } -\text{'s}$$
$$= 8 + 1 = 9$$
$$T = \text{number of } +\text{'s}$$
$$= 8$$

Table A3 is entered with $n = 9$ and $p = 1/2$ and for an entry close to .05. The critical region of size $\alpha_1 = .0195$ corresponds to values of T greater than or equal to

$$n - t = 9 - 1 = 8$$

Since $T = 8$, H_0 is rejected. The critical level $\hat{\alpha} = .0195$, because the observed value of T was barely in the rejection region.

The manufacturer decides that that the consumer population prefers B to A.

A two-tailed sign test illustrating the use of the large sample approximation is presented in the next example.

Example 2. In what was perhaps the first published report of a nonparametric test, Arbuthnott (1710) examined the available London birth records of 82 years and for each year compared the number of males born with the number of females born. If for each year we denote the event "more males than females were born" by "+" and the opposite event by "−," (there were no ties), we may consider the hypotheses to be

$$H_0: P(+) = P(-)$$
$$H_1: P(+) \neq P(-)$$

The test statistic T equals the number of + signs, and the critical region of size $\alpha = .05$ corresponds to values of T less than

$$t = .5(82 - (1.960)\sqrt{82})$$
$$= 32.1$$

and values of T greater than

$$n - t = 82 - 32.1$$
$$= 49.9$$

where t is calculated using Equation 3.

From the records, Arbuthnott obtained 82 plus signs, no minus signs, and no ties, as mentioned earlier. So $T = 82$ and the null hypothesis is rejected.

In fact, H_0 could have been rejected at an α as small as

$$\hat{\alpha} = P(T = 0) + P(T = 82)$$
$$= (\tfrac{1}{2})^{82} + (\tfrac{1}{2})^{82}$$
$$= (\tfrac{1}{2})^{81}$$

To see the versatility of the sign test, consider the following example, which was suggested to Batschelet (1965) by K. Schmidt-Koenig.

Example 3. Ten homing pigeons were taken to a point 25 kilometers west of their loft and released singly to see whether they dispersed at random in all directions (the null hypothesis) or whether they tended to proceed eastward toward their loft. Field glasses were used to observe the birds until they disappeared from view, at which time the angle of the vanishing point was noted. These 10 angles are: 20, 35, 350, 120, 85, 345, 80, 320, 280, and 85 degrees. Let "+" denote directions more eastward than westward (angles from 0 to 90 degrees or from 270 to 360 degrees) and let "−" denote directions away from the loft (between 90 and 270 degrees). The hypotheses

$$H_0: P(+) \le P(-)$$
$$H_1: P(+) > P(-)$$

match set B, so the critical region consists of large values of T, the number of "+" signs. From Table A3, for $n = 10$ and $p = 1/2$, the critical region of size $\alpha = .0547$ corresponds to values of T greater than or equal to $10 - 2 = 8$.

For these data $T = 9$, so the null hypothesis is rejected. The conclusion is that these pigeons tend to fly homeward instead of in random directions. The critical level $\hat{\alpha}$ is .0107.

☐ *Theory.* The event "+" represents the event "$Y_i > X_i$," or "$Y_i - X_i > 0$," which says that the difference $Y_i - X_i$ is positive. Similarly, "−" and "0" represent the events $Y_i - X_i$ is negative, or zero, respectively. Therefore the sign test is a test for comparing the probability of a positive difference with the probability of a negative difference. In the binomial test these were called "class 1" and "class 2" probabilities, respectively. By omitting ties we have

(5) $$P(+) + P(-) = 1$$

and so the hypothesis

$$H_0: P(+) = P(-)$$

is the same as saying

$$H_0: P(+) = \tfrac{1}{2}$$

which is in the same form as that of the binomial test with $p^* = 1/2$. So the same binomial test procedure is used, although a slight simplification results from the symmetry of

$$p^* = \tfrac{1}{2} = 1 - p^*$$

When the sign test is used with the original sets A, B, and C of hypotheses, the sign test is unbiased and consistent (Hemelrijk, 1952). Example 2.4.2 illustrated the binomial test with $p = 1/2$, which is the same as the sign test if there are no ties. Therefore the power functions graphed in Figure 2.4.2, in that example are power functions for the sign test. It is evident from those graphs that the sign test is unbiased and consistent, although such evidence is not conclusive proof.

□

If, in addition to the assumptions in the sign test, we can also assume legitimately that the differences $Y_i - X_i$ are continuous random variables with a symmetric distribution function [the distribution function of a random variable Z is symmetric about some point c if $P(Z \leq c - x) = P(Z \geq c + x)$ for all x], the Wilcoxon test for matched pairs is more appropriate (see Chapter 5). Furthermore, if the differences $Y_i - X_i$ are independent and identically distributed normal random variables, the appropriate parametric test is called the paired t-test. The A.R.E. compared to the paired t test under these conditions is only $2/\pi = .637$. Also, under these conditions the A.R.E. compared to the Wilcoxon test is 2/3. Both small and large sample relative efficiencies have been examined by Walsh (1951), Dixon (1953), Hodges and Lehmann (1956), and Gibbons (1964), among others. Special tables for sample sizes to 1000 are given by MacKinnon (1964). Hemelrijk (1952) discusses ties.

Data that occur naturally in pairs, as in the sign test, are usually analyzed by reducing the sequence of pairs to a sequence of single values, and then the data are analyzed as if only one sample were involved. That is, bivariate samples are usually analyzed using univariate techniques. In the sign test the differences $Y_i - X_i$ were analyzed in the same manner that one would analyze a series of values to see if positive values are more likely than negative values. This principle of reducing bivariate (or even multivariate) data to a simple univariate sample is a useful one to remember.

EXERCISES

1. Six students went on a diet in an attempt to lose weight, with the following results:

Name	Abdul	Ed	Jim	Max	Phil	Ray
Weight Before	174	191	188	182	201	188
Weight After	165	186	183	178	203	181

Is the diet an effective means of losing weight?

2. The reaction time before lunch was compared with the reaction time after lunch for a group of 28 office workers. Twenty-two workers found their reaction time before lunch was shorter, and 2 could detect no difference. Is the reaction time after lunch significantly longer than the reaction time before lunch?

3. Two different additives were compared to see which one is better for improving the durability of concrete. One hundred small batches of concrete were mixed under various conditions and, during the mixing, each batch was divided into two parts. One part received additive A and the other part received additive B. After the concrete hardened, the two parts in each batch were crushed against each other, and an observer determined which part appeared to be the most durable. In 77 cases the concrete with additive A was rated more durable; in 23 cases the concrete with additive B was rated more durable. Is there a significant difference between the effects of the two additives?

4. Twenty-two customers in a grocery store were asked to taste each of two types of cheese and declare their preference. Seven customers preferred one kind, 12 preferred the other kind, and 3 had no preference. Does this indicate a significant difference in preference?

5. An obstetrician claimed that more babies are born at night (6 P.M. to 6 A.M.) than during the day, while his friend the statistician said it only seemed that way. For the next year they kept track of the time of birth of all spontaneous births in that doctor's care to see who was correct. The result was:

Midnight to 3 A.M.—16 births   Noon to 3 P.M.—10 births
3 A.M. to 6 A.M.—17 births   3 P.M. to 6 P.M.—11 births
6 A.M. to 9 A.M.—12 births   6 P.M. to 9 P.M.—12 births
9 A.M. to noon—9 births   9 P.M. to midnight—15 births

Is the statistician correct?

6. In a laboratory, insects of a certain type are released in the middle of a circle drawn on a plain, flat table. A scent, intended to attract that type of insect, is located at one end of the table. Each insect is released singly and observed until it crosses the boundary of the circle. At that time it is recorded whether the insect crossed the half of the boundary "toward" the scent or the half "away" from the scent. At the conclusion of the experiment, 33 insects went "toward" the scent, 16 went "away," and 12 did not cross the boundary within a reasonable time. Does the scent attract those insects?

PROBLEMS

1. If the normal approximation is used in a two-tailed test at $\alpha = .05$, the value of t may be computed using the approximation

$$t_1 = \tfrac{1}{2}(n + 1.9600\sqrt{n})$$

or the approximation

$$t_2 = \tfrac{1}{2}n - \sqrt{n}$$

as given by Equations 1 and 2. For example, if $n = 21$, $t_1 = 6.009$ and $t_2 = 5.917$, so the first critical region includes the integer 6 while the second does not, and the two equations yield different critical regions. For which values of n between 20 and 30 do the two equations result in identical tests? Are the two results equivalent at $n = 16$?

3.5. SOME VARIATIONS OF THE SIGN TEST

Suppose now that the data are not ordinal as in the sign test but nominal, with two categories that we will call "0" and "1." That is, each X_i is either 0 or 1, and similarly for each Y_i. Then a question sometimes asked is, "Can we detect a difference between the probability of $(0, 1)$ and the probability of $(1, 0)$?" Such a question arises when the X_i in the pair (X_i, Y_i) represents the condition (or state) of the subject before the experiment and Y_i represents the condition of the same subject after the experiment. The same procedure as used in the sign test may be used here also, but the test is well known by a different name.

The McNemar Test for Significance of Changes

DATA. The data consist of observations on n' independent bivariate random variables (X_i, Y_i), $i = 1, 2, \ldots, n'$. The measurement scale for the X_i and the Y_i is nominal with two categories, which we may call "0" and "1;" that is, the possible values of (X_i, Y_i) are $(0, 0)$, $(0, 1)$, $(1, 0)$, and $(1, 1)$. In the McNemar test the data are usually summarized in a 2×2 *contingency table*, as follows.

		Classification of the Y_i	
		$Y_i = 0$	$Y_i = 1$
Classification of the X_i	$X_i = 0$	a (the number of pairs where $X_i = 0$ and $Y_i = 0$)	b (the number of pairs where $X_i = 0$ and $Y_i = 1$)
	$X_i = 1$	c (the number of pairs where $X_i = 1$ and $Y_i = 0$)	d (the number of pairs where $X_i = 1$ and $Y_i = 1$)

ASSUMPTIONS

1. The pairs (X_i, Y_i) are mutually independent.
2. The measurement scale is nominal with two categories for all X_i and Y_i.
3. The difference $P(X_i = 0, Y_i = 1) - P(X_i = 1, Y_i = 0)$ is negative for all i, or zero for all i, or positive for all i.

HYPOTHESES

$$H_0 : P(X_i = 0, Y_i = 1) = P(X_i = 1, Y_i = 0) \qquad \text{for all } i$$

$$H_1 : P(X_i = 0, Y_i = 1) \neq P(X_i = 1, Y_i = 0) \qquad \text{for all } i$$

These hypotheses may take a slightly different form if we add $P(X_i = 0, Y_i = 0)$ to both sides of the equation in H_0 to get

$$H_0 : P(X_i = 0, Y_i = 1) + P(X_i = 0, Y_i = 0) = P(X_i = 1, Y_i = 0) + P(X_i = 0, Y_i = 0)$$

The left side of H_0 includes all possibilities for Y_i and hence equals $P(X_i = 0)$. Similarly, the right side includes all possibilities for X_i and so equals $P(Y_i = 0)$. Therefore we have a new set of hypotheses in the form

$$H_0: P(X_i = 0) = P(Y_i = 0) \qquad \text{for all } i$$
$$H_1: P(X_i = 0) \neq P(Y_i = 0) \qquad \text{for all } i$$

Of course, these are also equivalent to

$$H_0: P(X_i = 1) = P(Y_i = 1) \qquad \text{for all } i$$
$$H_1: P(X_i = 1) \neq P(Y_i = 1) \qquad \text{for all } i \quad ,$$

These latter sets of hypotheses are usually easier to interpret in terms of the experiment.

TEST STATISTIC. The test statistic for the McNemar test is usually written as

(1)
$$T_1 = \frac{(b - c)^2}{b + c}$$

However, for $b + c \leq 20$, the following test statistic is preferred.

(2)
$$T_2 = b$$

Note that neither T_1 nor T_2 depends on a or d. This is because a and d represent the number of "ties," and ties are discarded in this analysis.

DECISION RULE. Let n equal $b + c$. If $n \leq 20$, use Table A3. If α is the desired level of significance, enter Table A3 with $n = b + c$ and $p = 1/2$ to find the table entry approximately equal to $\alpha/2$. Call this entry α_1, and the corresponding value of y is called t. Reject H_0 if $T_2 \leq t$, or if $T_2 \geq n - t$, at a level of significance of $2\alpha_1$. Otherwise accept H_0.

If n exceeds 20, use T_1 and Table A2. Reject H_0 at a level of significance α if T_1 exceeds the $(1 - \alpha)$ quantile of a chi-square random variable with 1 degree of freedom. Otherwise accept H_0.

Example 1. Prior to a nationally televised debate between the two presidential candidates, a random sample of 100 persons stated their choice of candidates as follows. Eighty-four persons favored the Democratic candidate, and the remaining sixteen favored the Republican. After the debate the same hundred people expressed their preference again. Of the persons who formerly favored the Democrat, exactly one-fourth of them changed their minds, and also one-fourth of the people formerly favoring the Republican switched to the Democratic side. The results are summarized in the following

2×2 contingency table.

| | | After | | Total |
		Democrat	Republican	Before
Before	Democrat	63	21	84
	Republican	4	12	16
				100

The McNemar test may be used to test H_0: The population voting alignment was not altered by the debate, against H_1: There has been a change in the proportion of all voters who favor the Democrat. Consider the X_i in (X_i, Y_i) to be 0 if the ith person favored the Democrat before, or 1 if the Republican was favored before. Similarly, Y_i represents the choice of the ith person after the debate. (Our choice of whether to represent the Democrat by 0 or 1 does not affect the results, as long as the X_i and the Y_i use the same representation.) The test statistic t_1 in the McNemar test becomes

$$T_1 = \frac{(b-c)^2}{b+c}$$

$$= \frac{(21-4)^2}{21+4}$$

$$= \frac{289}{25}$$

(3) $= 11.56$

The critical region of size $\alpha = .05$ corresponds to all values of T_1 greater than 3.841, the .95 quantile of a chi-square random variable with 1 degree of freedom, obtained from Table A2. Because 11.56 exceeds 3.841, the null hypothesis is rejected, and the conclusion is that the voter alignment has been altered. The critical level is less than .001.

☐ *Theory.* This test is a variation of the sign test, where the event $(0, 1)$ was called "+," the event $(1, 0)$ was called "−," and the events $(1, 1)$ and $(0, 0)$ were called ties. The hypothesis of the McNemar test then takes the form

$$H_0: P(+) = P(-)$$

which is the same as H_0 in the two-tailed sign test. The critical region for T_2 is found just as in the sign test for $n \leq 20$.

For n greater than 20 the sign test suggests using the normal approximation, based on the idea that

(4) $$Z = \frac{T_2 - n(\tfrac{1}{2})}{\sqrt{n(\tfrac{1}{2})(\tfrac{1}{2})}} = \frac{b - n(\tfrac{1}{2})}{(\tfrac{1}{2})\sqrt{n}}$$

has approximately the standard normal distribution when H_0 is true (see Example 1.5.6). Because $n = b + c$, Equation 4 reduces to

$$Z = \frac{b - [(b+c)/2]}{(\frac{1}{2})\sqrt{b+c}}$$

(5)
$$= \frac{b-c}{\sqrt{b+c}}$$

Therefore

$$T_1 = Z^2$$

has approximately a chi-square distribution with one degree of freedom (see Theorem 1.5.3). A two-tailed test involving T_2 or Z is comparable to using the upper tail of $T_1 = Z^2$ for a critical region.

As the sign test was presented in both the two-tailed and the one-tailed forms, so could the McNemar test take both forms. The easiest way of performing a one-tailed McNemar test is just to use the one-tailed sign test. The McNemar test and its variations are discussed by Bennett and Underwood (1970), Ury (1975), Mantel and Fleiss (1975), and McKinlay (1975).

Another modification of the sign test is one introduced by Cox and Stuart (1955), and it is used to test for the presence of *trend*. A sequence of numbers is said to have trend if the later numbers in the sequence tend to be greater than the earlier numbers (upward trend) or less than the earlier numbers (downward trend). This test involves pairing the later numbers with the earlier numbers and then performing a sign test on the pairs thus formed. If there is a trend, one member of each pair will have a tendency to be higher or lower than the other member. On the other hand, if there is no trend and the sequence of numbers actually represents observations on independent and identically distributed random variables, there will be no tendency for one particular member of each pair to exceed the other one.

Cox and Stuart Test for Trend

DATA. The data consist of observations on a sequence of random variables $X_1, X_2, \ldots, X_{n'}$, arranged in a particular order, such as the order in which the random variables are observed. It is desired to see if a trend exists in the sequence. Group the random variables into pairs $(X_1, X_{1+c}), (X_2, X_{2+c}), \ldots, (X_{n'-c}, X_{n'})$, where $c = n'/2$ if n' is even, and $c = (n'+1)/2$ if n' is odd. (Note that the middle random variable is eliminated using this scheme if n' is odd.) Replace each pair (X_i, X_{i+c}) with a "+" if $X_i < X_{i+c}$, or a "−" if $X_i > X_{i+c}$, eliminating ties. The number of untied pairs is called n.

This test may be used to detect any specified type of nonrandom pattern, such as a sine wave or other periodic pattern. The sequence of random variables is merely rearranged so that the smallest numbers, as predicted, will be near the beginning of the sequence and the larger numbers near the end. Then the presence of an upward trend in the rearranged sequence is evidence that the predicted pattern is present in the original arrangement of the sequence.

ASSUMPTIONS

1. The random variables $X_1, X_2, \ldots, X_{n'}$ are mutually independent.
2. The measurement scale of the X_is is at least ordinal.
3. Either the X_is are identically distributed or there is a trend; that is, the later random variables are more likely to be greater than instead of less than the earlier random variables (or vice versa).

HYPOTHESES. The following hypotheses are comparable to their counterparts in the sign test.

A. (Two-Tailed Test)

$$H_0: P(+) = P(-)$$

$$H_1: P(+) \neq P(-)$$

B. (One-Tailed Test)

$$H_0: P(+) \leq P(-)$$

$$H_1: P(+) > P(-)$$

C. (One-Tailed Test)

$$H_0: P(+) \geq P(-)$$

$$H_1: P(+) < P(-)$$

The usual interpretation given to these hypotheses is the following.

A. H_0: No trend exists
   H_1: There is either an upward trend or a downward trend

B. H_0: There is no upward trend
   H_1: There is an upward trend

C. H_0: There is no downward trend
   H_1: There is a downward trend

TEST STATISTIC. The test statistic T, as in the sign test, equals the number of + pairs (the pairs where X_{i+c} exceeds X_i).

DECISION RULE. The decision rule for the Cox and Stuart test for trend is exactly the same as the decision rule for the sign test and therefore is not repeated here.

The following is an example in which the two-tailed Cox and Stuart test for trend is applied.

Example 2. The total annual precipitation is recorded each year for 19 years, and this record is examined to see if the amount of precipitation is tending to increase or decrease. The precipitation in inches was 45.25, 45.83, 41.77, 36.26, 45.37, 52.25, 35.37, 57.16, 35.37, 58.32, 41.05, 33.72, 45.73, 37.90, 41.72, 36.07, 49.83, 36.24, and 39.90. Because $n' = 19$ is odd, the middle number 58.32 is omitted. The remaining numbers are paired.

$$
\begin{array}{ll}
(45.25, 41.05) & (45.37, 41.72) \\
(45.83, 33.72) & (52.25, 36.07) \\
(41.77, 45.73) & (35.37, 49.83) \\
(36.26, 37.90) & (57.16, 36.24) \\
& (35.37, 39.90)
\end{array}
$$

There are no ties, so $n = 9$. The test statistic T equals the number of pairs in which the second number exceeds the first number. The critical region of size .0390 corresponds to values of T less than or equal to 1 and values of T greater than or equal to $9 - 1 = 8$.

For the data obtained, $T = 4$, well within the region of acceptance. The critical level is 1.0. Therefore the null hypothesis "no trend exists" is accepted.

In Example 2 the assumptions of the model on which the test is valid are reasonable assumptions. Thus the test is reasonably valid. However, the assumptions listed are not all necessary. We need only to assume enough to satisfy the model for the sign test. That is, we need only assume:

1. The bivariate random variables (X_i, X_{i+c}) are mutually independent.
2. The probabilities $P(X_i < X_{i+c})$ and $P(X_i > X_{i+c})$ have the same relative size for all pairs.
3. Each pair (X_i, X_{i+c}) may be judged to be $a +$, $a -$, or a tie.

These assumptions are not as readily understood as the set of assumptions given in the test, but they may prove more useful in some applications such as the following.

Example 3. On a certain stream the average rate of water discharge is recorded each month (in cubic feet per second) for a period of 24 months. The hypothesis to be tested is

H_0: The rate of discharge is not decreasing

against the alternative

H_1: The rate of discharge is decreasing

The rate of discharge is known to follow a yearly cycle, so that nothing is learned by pairing stream discharges for two different months. However, by pairing the same months in two successive years the existence of a trend can be investigated. The following data were collected.

Month	First Year	Second Year	Month	First Year	Second Year
Jan	14.6	14.2	Jul	92.8	88.1
Feb	12.2	10.5	Aug	74.4	80.0
Mar	104	123	Sep	75.4	75.6
Apr	220	190	Oct	51.7	48.8
May	110	138	Nov	29.3	27.1
Jun	86.0	98.1	Dec	16.0	15.7

The test statistic T equals the number of pairs where the second year had a higher discharge than the first year, which is 5 in this example. Because the test is to detect a downward trend, the critical region of size .0730 corresponds to all values of T less than or equal to 3 (from Table A3, $n = 12$, $p = 1/2$). Therefore H_0 is accepted. The critical level $\hat{\alpha}$ is given by

$$\hat{\alpha} = P(T \leq 5 \mid H_0 \text{ is true})$$
$$= .3872$$

which is too large to be an acceptable α.

The examples presented in this section represent only a few of the many ways the sign test may be applied to test different types of hypotheses. Two more applications conclude this section. In the first the sign test is used as a simple method of detecting correlation, that is, detecting whether high values of one random variable tend to be paired with high values of a second random variable and low values with low values (positive correlation), or whether high values of one random variable tend to be paired with low values of the second random variable and low values with high values (negative correlation). The test involves arranging the pairs (the pairs remain intact) so that one member of the pair (either the first member or the second) is arranged in increasing order. If there is correlation the other member of the pair will exhibit a trend, upward if the correlation is positive, and downward if the correlation is negative. The Cox and Stuart test for trend may be used on the sequence formed by the other member of the pair.

Example 4. Cochran (1937) compares the reactions of several patients with each of two drugs, to see if there is a positive correlation between the two reactions for each patient.

Patient	Drug 1	Drug 2	Patient	Drug 1	Drug 2
1	+ .7	+1.9	6	+3.4	+4.4
2	−1.6	+ .8	7	+3.7	+5.5
3	− .2	+1.1	8	+ .8	+1.6
4	−1.2	+ .1	9	.0	+4.6
5	− .1	− .1	10	+2.0	+3.4

Ordering the pairs according to the reaction from drug 1 gives

Patient	Drug 1	Drug 2	Patient	Drug 1	Drug 2
2	−1.6	+ .8	1	+ .7	+1.9
4	−1.2	+ .1	8	+ .8	+1.6
3	− .2	+1.1	10	+2.0	+3.4
5	− .1	− .1	6	+3.4	+4.4
9	.0	+4.6	7	+3.7	+5.5

The one-tailed Cox and Stuart test for trend is applied to the newly arranged sequence of observations on drug 2. The five resulting pairs are $(+.8, +1.9)$, $(+.1, +1.6)$, $(+1.1, +3.4)$, $(−.1, +4.4)$, and $(+4.6, +5.5)$. Because we are testing

H_0: There is no positive correlation

against the alternative

H_1: There is positive correlation

we are, in essence, testing for the presence of an upward trend (H_1). The test statistic $T = 5$, because in all five pairs the second observation on drug 2 exceeds the first observation on drug 2. The critical region of size .0312 (obtained from Table A3 for $n = 5$, $p = 1/2$, and hence $t = 0$) corresponds to the single value $T = 5$. Therefore the null hypothesis is rejected, and we may conclude that there is a positive correlation between reactions to the two drugs. The critical level in this example is also .0312.

The final example illustrates how the sign test, or rather the Cox and Stuart test for trend, may be used to test for the presence of a predicted pattern.

Example 5. The number of eggs laid by a group of insects in a laboratory is counted on an hourly basis during a 24-hour experiment, to test

H_0: The 24 egg counts constitute observations on 24 identically distributed random variables

against the alternative

H_1: The number of eggs laid tends to be a minimum at 2:15 P.M., increasing to a maximum at 2:15 A.M., and decreasing again until 2:15 P.M.

The hourly counts are as follows.

Time	Number of Eggs	Time	Number of Eggs	Time	Number of Eggs
9 A.M.	151	5 P.M.	83	1 A.M.	286
10 A.M.	119	6 P.M.	166	2 A.M.	235
11 A.M.	146	7 P.M.	143	3 A.M.	223
Noon	111	8 P.M.	116	4 A.M.	176
1 P.M.	63	9 P.M.	163	5 A.M.	176
2 P.M.	84	10 P.M.	208	6 A.M.	174
3 P.M.	60	11 P.M.	283	7 A.M.	139
4 P.M.	109	Midnight	296	8 A.M.	137

If the alternative hypothesis is true, the egg counts nearest 2:15 P.M. should tend to be the smallest and those nearest 2:15 A.M. should tend to be the largest. Therefore the numbers of eggs are rearranged according to the times, from the times nearest 2:15 P.M. to the times nearest 2:15 A.M.

Time	Number of Eggs	Time	Number of Eggs
2 P.M.	84	8 A.M.	137
3 P.M.	60	9 P.M.	163
1 P.M.	63	7 A.M.	139
4 P.M.	109	10 P.M.	208
Noon	111	6 A.M.	174
5 P.M.	83	11 P.M.	283
11 A.M.	146	5 A.M.	176
6 P.M.	166	Midnight	296
10 A.M.	119	4 A.M.	176
7 P.M.	143	1 A.M.	286
9 A.M.	151	3 A.M.	223
8 P.M.	116	2 A.M.	235

If H_1 is true these numbers should exhibit an upward trend. The Cox and Stuart one-tailed test for trend is used. The first half of the sequence (first column) is paired with the last half of the sequence (second column), with the result that the two egg counts on each line form a pair. In all 12 pairs the number in the second column exceeds the number in the first column, so $T = 12$. For $n = 12$, $p = 1/2$, Table A3 shows that the critical region of size $\alpha = .0193$ corresponds to values of T greater than or equal to $12 - 2 = 10$. Therefore H_0 is rejected, and we conclude that the predicted pattern does seem to be present. The critical level is given as

$$\hat{\alpha} = P(T \geq 12) = .0002$$

Therefore H_0 would have been rejected at any reasonable level of significance.

☐ *Theory.* The Cox and Stuart test for trend is an obvious modification of the sign test and, therefore, the distribution of the test statistic when H_0 is true is obviously binomial. Also, the test is unbiased and consistent when the first sets A, B, and C of hypotheses are being used, but not necessarily so when the later sets are used. Stuart (1956) shows that the test, when applied to random variables known to be normally distributed, has an A.R.E. of .78 with respect to the best parametric test, a test based on the regression coefficient. Under the same conditions it has an A.R.E. of .79 compared to Spearman's or Kendall's rank correlation tests used as tests of randomness, which will be presented in Chapter 5.

If the test is altered so that the middle one-third of the observations are eliminated and only the first one-third of the observations are paired with the last one-third of the observations, the A.R.E. increases to .83 when compared to the best parametric test, under ideal conditions for the parametric test. Apparently the loss of data is small as compared with the gain in larger differences. This suggests another variation, that of pairing from the ends of the sequence. That is, by forming the pairs (X_1, X_n), (X_2, X_{n-1}), and so forth, using all the data, perhaps the larger differences may be preserved, along with no loss in data. The test may still be performed as just described, because the distribution of the test statistic under the null hypothesis remains unchanged.

The test for correlation, described in Example 4, has not been investigated to see what its properties are. One of the difficulties in applying the test for correlation is that if many observations equal each other, there is more than one way of arranging the observations so that the test for trend can be applied. Therefore it is recommended that the original data pairs be arranged using the pair member that has the smallest number of ties. Of the arrangements still possible due to ties, the conservative approach is to choose the arrangement that will be least likely to result in rejection of H_0.

A bivariate sign test for location is discussed by Chatterjee (1966). Other modifications of the sign test may be used to test for trends in dispersion (Ury, 1966) or to compare several treatments with a control (Rhyne and Steel, 1965). Rao (1968) uses the Cox and Stuart test for testing trend in dispersion. The power of the test for trend is discussed further by Mansfield (1962). Olshen (1967) presents tests for testing quadratic trend versus linear trend. Other variations of the sign test appear in Woodbury and Manton (1977) and Altham (1971). A paper by Schaafsma (1973) examines the consequences of order dependence on the sign test; that is, a customer preferring one brand over another may be influenced by which brand he or she was exposed to first.

EXERCISES

1. One hundred thirty-five citizens were selected at random and were asked to state their opinion regarding U.S. foreign policy. Forty-three were opposed to the U.S. foreign policy. After several weeks, during which they received an informative newsletter, they were again asked their opinion; 37 were opposed, and 30 of the 37 were persons who originally were not opposed to the U.S. foreign policy. Is the change in numbers of people opposed to the U.S. foreign policy significant?

2. In Exercise 1, suppose all 37 of the persons opposed to the foreign policy after the experiment were also among those opposed to the U.S. foreign policy before the experiment. Is the change in the number of people opposed to the U.S. foreign policy significant?

3. In a certain city the mortality rate per 100,000 citizens due to automobile accidents for each of the last 15 years was 17.3, 17.9, 18.4, 18.1, 18.3, 19.6, 18.6, 19.2, 17.7, 20.0, 19.0, 18.8, 19.3, 20.2, and 19.9. Is there any basis for the statement that the mortality rate is increasing?

4. For each of the last 34 years a small Midwestern college recorded the average heights of male freshmen. The averages were 68.3, 68.6, 68.4, 68.1, 68.4, 68.2, 68.7, 68.9, 69.0, 68.8, 69.0, 68.6, 69.2, 69.2, 68.9, 68.6, 68.6, 68.8, 69.2, 68.8, 68.7, 69.5, 68.7, 68.8, 69.4, 69.3, 69.3, 69.5, 69.5, 69.0, 69.2, 69.2, 69.1, and 69.9. Do these averages indicate an increasing trend in height?

5. A manufacturer computes the average cost in dollars of producing a certain item for each of 44 months with the resulting averages 13.65, 13.41, 13.53, 13.23, 13.58, 13.43, 13.73, 13.40, 13.70, 13.58, 13.80, 13.40, 13.63, 13.69, 13.92, 13.68, 13.72, 13.42, 13.66, 13.98, 13.81, 13.60, 13.32, 13.45, 13.27, 13.26, 13.28, 13.29, 13.10, 13.09, 13.36, 13.40, 13.35, 13.53, 13.66, 13.10, 13.28, 13.33, 13.02, 13.09, 13.12, 13.16, 12.96, and 12.95. Is there a statistically significant trend in these averages?

6. In an experiment to determine the influence of suggestion, 20 straight lines of varying lengths were shown one at a time to subjects A and B, and the subjects estimated aloud the length of each line. Subject A stated her preference first and, unknown to subject B, was under instructions to overestimate the first 10 lines and underestimate the last 10 lines. After hearing subject A's estimate, subject B stated his estimate. The errors of the estimates, measured by subtracting the true lengths of the lines from the estimated lengths of the lines, were as follows.

	Line									
	1	*2*	*3*	*4*	*5*	*6*	*7*	*8*	*9*	*10*
Error by A	+0.3	+1.1	+0.9	+0.6	+1.0	+1.3	+0.8	+1.6	+1.2	+0.8
Error by B	−0.1	+0.6	+1.0	+0.7	+0.2	+0.9	−0.1	+0.2	0.0	+0.5

	Line									
	11	*12*	*13*	*14*	*15*	*16*	*17*	*18*	*19*	*20*
Error by A	−1.3	−1.1	−1.3	−0.7	−1.4	−1.1	−0.8	−0.5	−1.2	−1.0
Error by B	−0.6	−1.2	−1.0	−0.7	−1.0	−0.1	−0.5	0.0	−0.4	−0.3

Is there a significant positive correlation between subject A's errors and subject B's errors?

7. A certain major league baseball player had compiled the following record over 12 years.

	1968	*1969*	*1970*	*1971*	*1972*	*1973*
Number of Home Runs	7	14	17	15	9	19
Batting Averages	.212	.232	.234	.210	.201	.256

	1974	*1975*	*1976*	*1977*	*1978*	*1979*
Number of Home Runs	16	17	22	17	13	10
Batting Averages	.261	.247	.255	.241	.238	.235

Is there significant correlation between the number of home runs he hit and his batting average for that year?

8. Test the following data to see if there is a significant correlation between the yearly income of a family and the number of children in that family.

Income	Number of Children	Income	Number of Children	Income	Number of Children
$8,720	3	11,660	3	14,470	3
8,832	2	11,787	4	14,650	1
8,861	4	11,975	2	14,687	3
8,941	3	12,012	3	14,776	1
9,000	4	12,118	5	14,831	1
9,166	2	12,272	2	14,902	2
9,328	0	12,465	5	15,067	2
9,392	3	12,786	4	15,317	3
9,568	6	12,812	4	15,617	1
9,747	5	12,937	2	15,899	3
9,916	2	13,005	1	15,945	4
10,050	1	13,070	3	16,182	1
10,111	6	13,267	2	16,474	3
10,218	3	13,330	4	16,799	2
10,486	5	13,564	5	17,055	2
10,691	2	13,734	0	17,221	3
10,979	0	13,851	1	17,853	1
11,096	8	13,957	4	19,672	1
11,106	1	14,122	2	20,270	1
11,317	4	14,349	4	27,843	2

PROBLEMS

1. A barber shop is considering raising the price of haircuts 25 cents and then giving the customers a coupon worth one free refreshing drink at a nearby pub. A survey was conducted, and 200 people, selected at random from the population of real and potential customers were given an explanation of this proposal. Ten percent of the customers in the sample said they would go elsewhere for their haircuts. Five percent of the noncustomers in the sample said they would become customers at that barber shop. Test the null hypothesis that the proposed change will not increase the total number of customers who receive haircuts in that shop, if only 20 people in the sample are presently customers. How does the answer change if there are 60% customers in the sample instead of 20?

2. Data for the McNemar test may be written as bivariate observations X_i, Y_i, where each observation is 0 or 1 "before" and 0 or 1 "after." A parametric test called the "paired t test" is often applied to data of this type. The paired t test uses the differences $D_i = X_i - Y_i$, $i = 1, 2, \ldots, n$. The sample mean $\bar{D}$ and sample standard deviation S are used in the statistic

$$t = \bar{D}\sqrt{n' - 1}/S$$

which is compared with the quantiles in Table A25, $n' - 1$ degrees of freedom. This test is only approximate if the D_is are not normally distributed.

Show that the following relationship holds between t and T_1

$$t^2 = \frac{(n'-1)T_1}{n'-T_1}, \qquad \text{or} \qquad T_1 = \frac{n't^2}{n'-1+t^2}$$

where T_1 is given by Equation 1. That is, as T_1 gets larger, t^2 also gets larger, so the two tests that reject H_0 for large T_1 or large t^2 are equivalent if their critical regions correspond to each other's.

CHAPTER 4

Contingency Tables

PRELIMINARY REMARKS

A contingency table is an array of natural numbers in matrix form where those natural numbers represent counts, or frequencies. For example, an entomologist observing insects may say he observed 37 insects, or he may say he observed:

Moths	Grasshoppers	Others	Total
12	22	3	37

using a 1×3 (one by three) contingency table. He may wish to be more specific and use a 2×3 contingency table.

	Moths	Grasshoppers	Others	Total
Alive	3	21	3	27
Dead	9	1	0	10
Total	12	22	3	37

The totals, consisting of two *row totals*, three *column totals*, and one *grand total*, are optional and are usually included only for the reader's convenience.

4.1. THE 2×2 CONTINGENCY TABLE

In general an $r \times c$ contingency table is an array of natural numbers arranged into r rows and c columns and thus has rc *cells* or places for the numbers. This section is concerned only with the case where $r = 2$ and $c = 2$, the 2×2 contingency table. Because there are four cells, the 2×2 contingency table is also called the *fourfold* contingency table.

One application of the 2×2 contingency table arises when N objects (or persons), possibly selected at random from some population, are classified into one of two categories before a treatment is applied or an event takes place. After the treatment is applied, the N objects are again examined and classified into the two categories. The question to be answered is, "Does the treatment significantly alter the proportion of objects in each of the two categories?" This use of the contingency table was introduced in Section 3.5, and the appropriate statistical procedure was seen to be a variation of the sign test known as the McNemar test. The McNemar test is often able to detect subtle differences, primarily because the same sample is used in the two situations (such as "before" and "after"). Another way of testing the same hypothesis tested with the McNemar test is by drawing a random sample from the population before the treatment and then comparing it with another random sample drawn from the population after the treatment. The additional variability introduced by using two different random samples is undesirable because it tends to obscure the changes in the population caused by the treatment. However, there are times when it is not practical, or even possible, to use the same sample twice. Then the procedure to be described in this section may be used.

In addition to this situation, the procedure may be, and usually is, used to analyze two samples drawn from two different populations to see if both populations have the same or different proportions of elements in a certain category. More specifically, two random samples are drawn, one from each population, to test the null hypothesis that the probability of event A (some specified event) is the same for both populations.

The Chi-Square Test for Differences in Probabilities, 2×2

DATA. A random sample of n_1 observations is drawn from one population (or before a treatment is applied) and each observation classified into either class 1 or class 2, the total numbers in the two classes being O_{11} and O_{12}, respectively, where $O_{11} + O_{12} = n_1$. A second random sample of n_2 observations is drawn from a second population (or the first population after some treatment is applied) and the number of observations in class 1 or class 2 is O_{21} or O_{22}, respectively, where $O_{21} + O_{22} = n_2$. The data are arranged in the following 2×2 contingency table.

	Class 1	Class 2	Total
Population 1	O_{11}	O_{12}	n_1
Population 2	O_{21}	O_{22}	n_2
Total	C_1	C_2	$N = n_1 + n_2$

The total number of observations is denoted by N.

ASSUMPTIONS

1. Each sample is a random sample.
2. The two samples are mutually independent.
3. Each observation may be categorized either into class 1 or class 2.

HYPOTHESES. Let the probability that a randomly selected element will be in class 1 be denoted by p_1 in population 1 and p_2 in population 2. Note that it is not necessary for p_1 and p_2 to be known. The hypotheses merely specify a relationship between them.

A. (Two-Sided Test)

$$H_0: p_1 = p_2$$
$$H_1: p_1 \neq p_2$$

B. (One-Sided Test)

$$H_0: p_1 \leq p_2$$
$$H_1: p_1 > p_2$$

C. (One-Sided Test)

$$H_0: p_1 \geq p_2$$
$$H_1: p_1 < p_2$$

TEST STATISTIC. For the two-sided test the test statistic T is given by

(1)
$$T = \frac{N(O_{11}O_{22} - O_{12}O_{21})^2}{n_1 n_2 C_1 C_2}$$

For the one-sided tests it is simpler to use the signed square root T_1 of T, that is,

$$T_1 = \frac{\sqrt{N}(O_{11}O_{22} - O_{12}O_{21})}{\sqrt{n_1 n_2 C_1 C_2}}$$

DECISION RULE. The exact distribution of T is difficult to tabulate because of all the different combinations of values possible for O_{11}, O_{12}, O_{21} and O_{22}. Therefore the large sample approximation is used, which is the chi-square distribution with 1 degree of freedom. The normal distribution is used for T_1 (see Theorem 1.5.3). Use decision rules A, B, or C, depending on whether the hypothesis being tested is under preceding categories A, B, or C.

A. (Two-Sided Test) Reject H_0 at the approximate level α if T exceeds $x_{1-\alpha}$, the $(1-\alpha)$ quantile of the chi-square random variable with 1 degree of freedom, found in Table A2 and also given in the following material.

B. (One-Sided Test) Reject H_0 at the approximate level α if T_1 exceeds $x_{1-\alpha}$, the $(1-\alpha)$ quantile of the standard normal distribution given in Table A1.

C. (One-Sided Test) Reject H_0 at the approximate level α if T_1 is less than x_α, the α quantile of the standard normal distribution given in Table A1.

For convenience the $(1-\alpha)$ quantiles of a chi-square random variable with 1 degree of freedom are given here as they appear on the first line of Table A2.

$$x_{.750} = 1.323 \qquad x_{.990} = 6.635$$
$$x_{.900} = 2.706 \qquad x_{.995} = 7.879$$
$$x_{.950} = 3.841 \qquad x_{.999} = 10.83$$
$$x_{.975} = 5.024$$

Example 1. Two carloads of manufactured items are sampled randomly to determine if the proportion of defective items is different for the two carloads. From the first carload 13 of the 86 items were defective. From the second carload 17 of the 74 items were considered defective.

	Defective	Nondefective	Totals
Carload 1	13	73	86
Carload 2	17	57	74
Totals	30	130	160

The assumptions are met, and so the two-tailed test is used to test

H_0: the proportion of defectives is equal in the two carloads

using the test statistic

$$T = \frac{N(O_{11}O_{22} - O_{12}O_{21})^2}{n_1 n_2 C_1 C_2}$$
$$= \frac{160((13)(57) - (73)(17))^2}{(86)(74)(30)(130)}$$
$$= 1.61$$

The .95 quantile of a chi-square random variable with 1 degree of freedom is 3.841. Therefore the critical region of approximate size .05 corresponds to values of T greater than 3.841. In this example T is less than 3.841, so H_0 is accepted. By interpolation the critical level is found to be about $1 - .78$, or

$$\hat{\alpha} \cong 1 - .78 = .22$$

Therefore the decision to accept H_0 seems to be a fairly safe one.

The following example illustrates the use of the one-tailed test.

Example 2. At the U.S. Naval Academy a new lighting system was installed throughout the midshipmen's living quarters. It was claimed that the new lighting system resulted in poor eyesight due to a continual strain on the eyes of the midshipmen. Consider a (fictitious) study to test the null hypothesis,

H_0: The probability of a graduating midshipman having 20–20 (good) vision is the same or greater under the new lighting system than it was under the old lighting system

against the one-sided alternative

H_1: The probability of good vision is less now than it was

Let p_1 be the probability that a randomly selected graduating midshipmen had good vision under the old lighting system and let p_2 be the corresponding probability with the new lights. Then the preceding hypotheses may be restated as

$$H_0: p_1 \leq p_2$$
$$H_1: p_1 > p_2$$

which matches the set B of hypotheses. The random samples are taken to be the entire graduation class just prior to the installation of the new lights for population 1, and the first graduation class to spend 4 years using the new lighting system for population 2. It is hoped that these samples will behave the same as would random samples from the entire population of graduating seniors, real and potential.

Suppose the results were as follows.

	Good Vision	Poor Vision	
Old Lights	$O_{11} = 714$	$O_{12} = 111$	$n_1 = 825$
New Lights	$O_{21} = 662$	$O_{22} = 154$	$n_2 = 816$
Totals	1376	265	$N = 1641$

Decision rule B defines the critical region for $\alpha = .05$ to be all values of T_1 greater than 1.6449 from Table A1. Computation of T_1 gives

$$T_1 = \frac{\sqrt{N}(O_{11}O_{22} - O_{12}O_{21})}{\sqrt{n_1 n_2 C_1 C_2}}$$

$$= \frac{\sqrt{1641}[(714)(154) - (111)(662)]}{\sqrt{(825)(816)(1376)(265)}}$$

$$= 2.982$$

so the null hypothesis is clearly rejected. From Table A1 we see that the null hypothesis could have been rejected at a level of significance as small as about .002, by interpolation, so $\hat{\alpha} = .002$.

We may therefore conclude that the populations represented by the two graduation classes do differ with respect to the proportions having poor eyesight, and in the direction predicted. That is, population 2 (with the new lights) has poorer eyesight than population 1 (with the old lights). Whether the poorer eyesight is a *result* of the new lights has not been shown. However, an association of poor eyes with the new lights has been shown in this hypothetical example.

☐ *Theory.* The 2×2 contingency table just presented is actually a special case of the $r \times c$ contingency table presented in the next section, and so the theory involved is a special case of the theory behind the $r \times c$ case. However, the exact distribution of the test statistic is difficult to find unless r and c are very small, so the exact distribution of T is presented now. It should be mentioned, before we proceed, that the form of T given in Equation 1 seems to be different from the general form of T given in the next section. However, with a little algebra the two expressions for $r = 2$ and $c = 2$ may be seen to be equivalent.

The exact probability distribution of T, when H_0: $p_1 = p_2 = p$ (say) is true, may be calculated as illustrated in the following. For the sample from population 1, the probability of exactly x_1 items in class 1 and $n_1 - x_1$ items in class 2 is given by the binomial probability distribution.

$$(2) \quad P\left(\text{Population 1} \begin{array}{|c|c|} \hline \text{Class 1} & \text{Class 2} \\ \hline x_1 & n_1 - x_1 \\ \hline \end{array} \right) = \binom{n_1}{x_1} p^{x_1} (1-p)^{n_1 - x_1}$$

Similarly the probability of the sample from population 2 having exactly x_2 items in class 1 and $n_2 - x_2$ items in class 2 is given by

$$(3) \quad P\left(\text{Population 2} \begin{array}{|c|c|} \hline \text{Class 1} & \text{Class 2} \\ \hline x_2 & n_2 - x_2 \\ \hline \end{array} \right) = \binom{n_2}{x_2} p^{x_2} (1-p)^{n_2 - x_2}$$

Because the two samples are independent the probability of the joint event may be obtained by multiplying the right sides of Equations 2 and 3. Thus

$$(4) \quad P\left(\begin{array}{l|c|c|} & \text{Class 1} & \text{Class 2} \\ \hline \text{Population 1} & x_1 & n_1 - x_1 \\ \hline \text{Population 2} & x_2 & n_2 - x_2 \\ \hline \end{array} \right) = \binom{n_1}{x_1}\binom{n_2}{x_2} p^{x_1 + x_2} (1-p)^{N - x_1 - x_2}$$

In the simple case where $n_1 = 2$ and $n_2 = 2$ there are nine different points in the sample space, corresponding to the nine possible tables that appear on the following page.

The undefined values for T arise from the indeterminate form $0/0$. However, since the two outcomes that result in undefined values for T are strongly indicative that H_0 is true, just as the fifth outcome is strongly indicative that H_0 is true, we may arbitrarily define T to be 0 for the first and last outcomes in agreement with the fifth outcome. Then T has the following probability distribution.

$$p = \tfrac{1}{2} \qquad\qquad\qquad p = 1$$
$$P(T = 0) = \tfrac{3}{8} \qquad\qquad P(T = 0) = 1$$
$$P(T = 4/3) = \tfrac{1}{2}$$
$$P(T = 4) = \tfrac{1}{8}$$

Similarly for any sample sizes n_1 and n_2 the exact probability distributions may be found after the appropriate defining of the undefined values of T. However, the probability function is not unique even when H_0 is assumed to be true, as is seen in the previous example, but, instead, it depends on p. Hence the null hypothesis in the preceding test is a composite hypothesis. It is not easy to show, but the size of the critical region is a maximum when $p = 1/2$. Therefore x may be found in the prior small sample case by setting p equal to $1/2$. If the critical region corresponds to the largest value of T (i.e., $T = 4$), then $x = .125$.

It is not easy to show that the asymptotic distribution of T is the chi-square distribution with 1 degree of freedom, so we will not attempt it here. The interested and well-qualified reader may find the asymptotic distribution derived in Cramér (1946). □

A "correction for continuity" was introduced by Yates (1934) to compensate partially for the inaccuracy introduced by the use of a continuous distribution function (the chi-square) to approximate the discrete distribution function of T. Yates' modification involves using

(5)
$$T = \frac{N[|O_{11}O_{22} - O_{12}O_{21}| - (N/2)]^2}{n_1 n_2 C_1 C_2}$$

instead of Equation 1. The correction consists of reducing the absolute value of $O_{11}O_{22} - O_{12}O_{21}$ by an amount $N/2$, before squaring. However, I (Conover, 1974) tend to agree with Pearson (1947), Plackett (1964), and Grizzle (1967) in recommending against the use of Yates' correction. The use of Equation 5 tends to be overly conservative, and Equation 1 seems to be more in agreement with a chi-square random variable with 1 degree of freedom.

Another use for the 2×2 contingency table appears when each observation in a single sample of size N is classified according to two properties, where each property may take one of two forms. Then there are $(2)(2) = 4$ different combinations of the two properties, and the 2×2 contingency table is a

Tables		Probabilities if H_0 is true	$(p = 1/2)$	$(p = 1)$	T
2	0				
2	0	p^4	1/16	1	Undefined
2	0				
1	1	$2p^3(1-p)$	1/8	0	4/3
2	0				
0	2	$p^2(1-p)^2$	1/16	0	4
1	1				
2	0	$2p^3(1-p)$	1/8	0	4/3
1	1				
1	1	$4p^2(1-p)^2$	1/4	0	0
1	1				
0	2	$2p(1-p)^3$	1/8	0	4/3
0	2				
2	0	$p^2(1-p)^2$	1/16	0	4
0	2				
1	1	$2p(1-p)^3$	1/8	0	4/3
0	2				
0	2	$(1-p)^4$	1/16	0	Undefined

convenient means of tabulating the number of observations in each category. However, this use of the 2×2 contingency table is a special case of the $r \times c$ contingency table and does not have any special variation (such as the one-sided test of this section) that would warrant a separate presentation. Therefore it is presented in the next section.

Confidence intervals may be formed for any unknown probabilities associated with the 2×2 contingency table or any contingency table, for that matter, by applying the procedure described in Section 3.1. Similarly, the test in Section 3.1 may be used on contingency tables, whenever the hypotheses are pertinent and the assumptions of the test are met.

A shortcut rule for the one-sided test is given by Ott and Free (1969). Further discussion of the continuity correction may be found in Mantel and Greenhouse (1968), Pirie and Hamdam (1972), and Maxwell (1976). The power of the test is discussed by Harkness and Katz (1964). The exact test is considered by Gail and Gart (1973), Garside and Mack (1976), and McDonald, Davis, and Milliken (1977). For methods of combining the test statistics in several 2×2 contingency tables, see Radhakrishna (1965), Nelson (1966), Meeker (1978), and Zelen (1971). Possible errors in the marginal totals because of misclassification is the subject of many papers, including recent ones by Chiacchierini and Arnold (1977) and Plackett (1977). Other related papers are by Fienberg and Gilbert (1970), Upton and Lee (1976), and Ray (1976). An excellent book by Fleiss (1973) is concerned primarily with a discussion of 2×2 contingency tables.

EXERCISES

1. A random sample of 135 people was drawn from each of two populations to gauge reaction to pending legislation. In the first sample there were 43 responses of "opposed;" in the second sample there were 37 "opposed." Is there a difference in the proportion of people opposed in the two populations? Does a comparison of this problem with Exercises 1 and 2 in Section 3.5 suggest an advantage in using the same persons in both samples whenever possible, such as in a "before" and "after" situation?

2. Sixty students were divided into two classes of 30 each and taught how to write a program for a computer. One class used the conventional method of learning, and the other class used a new, experimental method. At the end of the courses, each student was given a test that consisted of writing a computer program. The program was either correct or incorrect, and the results were tabulated as follows.

	Correct Program	Incorrect Program
Conventional Class	23	7
Experimental Class	27	3

Is there reason to believe the experimental method is superior? Or could the preceding differences be due to chance fluctuations?

3. One hundred men and 100 women were asked to try a new toothpaste and to state whether they liked or did not like the new taste. Thirty-two men and 26 women said they did not like the new taste. Does this indicate a difference in preferences between men and women in general?

4. Contingency tables may be used to present data representing scales of measurement higher than the nominal scale. For example, a random sample of size 20 was selected from the graduate students who are U.S. citizens, and their grade point averages were recorded.

3.42	3.54	3.21	3.63	3.22
3.80	3.70	3.20	3.75	3.31
3.86	4.00	2.86	2.92	3.59
2.91	3.77	2.70	3.06	3.30

Also, a random sample of 20 students was selected from the non-U.S. citizen group of graduate students at the same university. Their grade point averages were as follows.

3.50	4.00	3.43	3.85	3.84
3.21	3.58	3.94	3.48	3.76
3.87	2.93	4.00	3.37	3.72
4.00	3.06	3.92	3.72	3.91

Test the null hypothesis that the proportion of graduate students with averages of 3.50 or higher is the same for both the U.S. citizens and the non-U.S. citizens.

PROBLEMS

1. In the test for differences in probabilities, find the exact probability distribution of the test statistic when $n_1 = 2$, $n_2 = 3$. Also, let the largest value of T correspond to the critical region and find α.

2. The data in this section may be considered as two independent samples, $X_1, X_2, \ldots, X_{n_1}$ from population 1, and $Y_1, Y_2, \ldots, Y_{n_2}$ from population 2. Each X_i or Y_i equals 0 if the observation is in class 1 and equals 1 if it is in class 2. Thus each sample is a set of zeros and ones. The parametric approach to the problem of two independent samples uses the "two-sample t test" with the test statistic

$$t = \frac{\bar{X} - \bar{Y}}{\sqrt{n_1 S_x^2 + n_2 S_y^2}} \sqrt{\frac{n_1 n_2 (n_1 + n_2 - 2)}{n_1 + n_2}}$$

where $\bar{X}$, $\bar{Y}$, S_x^2, and S_y^2 are the respective sample means and variances. Show that t and T (Equation 1) are related as follows:

$$t^2 = \frac{(n_1 + n_2 - 2)T}{n_1 + n_2 - T}$$

or its equivalent

$$T = \frac{(n_1 + n_2)t^2}{n_1 + n_2 - 2 + t^2}$$

As a result of this relationship, a test that rejects H_0 for large T is equivalent to a test that rejects H_0 for large t^2, if their critical regions coincide.

3. Another way of looking at the test in this section is as follows. O_{11} is a binomial random variable with mean $n_1 p_1$ and variance $n_1 p_1 (1 - p_1)$ and is approximately normal for moderate n_1. Therefore O_{11}/n_1 is approximately normal, with mean p_1 and variance $p_1 (1 - p_1)/n_1$. By the same token, O_{21}/n_2 is approximately normal for moderate n_2, with mean p_2 and variance $p_2 (1 - p_2)/n_2$. Because of independence between O_{11} and O_{21} the random variable

$$X = \frac{O_{11}/n_1 - O_{22}/n_2 - (p_1 - p_2)}{[p_1(1 - p_1)/n_1 + p_2(1 - p_2)/n_2]^{\frac{1}{2}}}$$

is also approximately normal.

(a) Find the mean and variance of X.
(b) Show that if H_0: $p_1 = p_2$ is true and if p_1 and p_2 are estimated using $(O_{11} + O_{21})/(n_1 + n_2)$, then $X = T_1$ in this section.
(c) Justify the preceding statement which says "X is also approximately normal."

4.2. THE $r \times c$ CONTINGENCY TABLE

As an immediate generalization of the 2×2 contingency table of the previous section, we have the contingency table with r rows and c columns, called the $r \times c$ contingency table. This contingency table may be used, as in the previous section, to present a tabulation of the data contained in several samples, where the data represent at least a nominal scale of measurement, and to test the hypothesis that the probabilities do not differ from sample to sample. Another use for the $r \times c$ contingency table is with the single sample, where each element in the sample may be classified into one of r different categories according to one criterion and at the same time, into one of c different categories according to a second criterion. Both of these applications are treated the same in the statistical analysis, but basic differences between the two applications justify separate discussions of the two situations. A third application, similar to the other two, will also be discussed.

First we will consider the extension of the application presented in the previous section. Now, instead of only two samples, we have r samples, where each sample is tabulated in one of the r rows. Instead of each sample furnishing two categories (formerly called class 1 and class 2), we now consider c categories, corresponding to the c columns. Thus the entry in the (i, j) cell (ith row and jth column) is the number of observations from the ith sample that belong to the jth category.

The Chi-Square Test for Differences in Probabilities, $r \times c$

DATA. There are r populations in all, and one random sample is drawn from each population. Let n_i represent the number of observations in the ith sample (from the ith population) for $1 \le i \le r$. Each observation in each sample may be classified into one of c different categories. Let O_{ij} be the number of observations from the ith sample that fall into category j, so

(1) $$n_i = O_{i1} + O_{i2} + \cdots + O_{ic} \qquad \text{for all } i$$

The data are arranged in the following $r \times c$ contingency table.

	Class 1	Class 2	Class c $\cdots$		Totals
Population 1	O_{11}	O_{12}	$\cdots$	O_{1c}	n_1
Population 2	O_{21}	O_{22}	$\cdots$	O_{2c}	n_2
$\cdots$	$\cdots$	$\cdots$	$\cdots$	$\cdots$	$\cdots$
Population r	O_{r1}	O_{r2}	$\cdots$	O_{rc}	n_r
Totals	C_1	C_2	$\cdots$	C_c	N

The total number of observations from all samples is denoted by N.

(2) $$N = n_1 + n_2 + \cdots + n_r$$

The number of observations in the jth column is denoted by C_j. That is, C_j is the total number of observations in the jth category, or class, from all samples combined.

(3) $$C_j = O_{1j} + O_{2j} + \cdots + O_{rj}, \qquad \text{for } j = 1, 2, \ldots, c$$

ASSUMPTIONS

1. Each sample is a random sample.
2. The outcomes of the various samples are all mutually independent (particularly among samples, because independence within samples is part of the first assumption).
3. Each observation may be categorized into exactly one of the c categories or classes.

HYPOTHESES. Let the probability of a randomly selected value from the ith population being classified in the jth class be denoted by p_{ij}, for $i = 1, 2, \ldots, r$, and $j = 1, 2, \ldots, c$.

H_0: All of the probabilities in the same column are equal to each other (i.e., $p_{1j} = p_{2j} = \cdots = p_{rj}$, for all j)

H_1: At least two of the probabilities in the same column are not equal to each other (i.e., $p_{ij} \neq p_{kj}$ for some j, and for some pair i and k)

Note that it is not necessary to stipulate the various probabilities. The null hypothesis merely states the the probability of being in class j is the same for all populations, no matter what that probability might be (and no matter which category we are considering).

TEST STATISTIC. The test statistic T is given by

(4)
$$T = \sum_{i=1}^{r} \sum_{j=1}^{c} \frac{(O_{ij} - E_{ij})^2}{E_{ij}}, \qquad \text{where} \qquad E_{ij} = \frac{n_i C_j}{N}$$

While the term O_{ij} represents the observed number in cell (i, j), the term E_{ij} represents the *expected* number of observations in cell (i, j), if H_0 is really true. That is, if H_0 is true the number of observations in cell (i, j) should be close to the ith sample size n_i multiplied by the proportion C_j/N of all observations in category j.

An equivalent expression for T, more suited for machine computations, is given by

(5)
$$T = \sum_{i=1}^{r} \sum_{j=1}^{c} \frac{O_{ij}^2}{E_{ij}} - N$$

If there are only two rows ($r = 2$) then the test statistic may be computed using one of the following simpler forms

(6)
$$T = \frac{1}{\frac{n_1}{N}\left(1 - \frac{n_1}{N}\right)} \sum_{j=1}^{c} \frac{\left(O_{1j} - \frac{n_1 C_j}{N}\right)^2}{C_j}$$

or, for machine computations,

(7)
$$T = \frac{1}{\frac{n_1}{N}\left(1 - \frac{n_1}{N}\right)} \left(\sum_{j=1}^{c} \frac{O_{1j}^2}{C_j} - \frac{n_1^2}{N} \right)$$

Similarly, if $c = 2$, the corresponding equations are

(8)
$$T = \frac{1}{\frac{C_1}{N}\left(1 - \frac{C_1}{N}\right)} \sum_{i=1}^{r} \frac{\left(O_{i1} - \frac{C_1 n_i}{N}\right)^2}{n_i}$$

and

(9)
$$T = \frac{1}{\frac{C_1}{N}\left(1 - \frac{C_1}{N}\right)} \left(\sum_{i=1}^{r} \frac{O_{i1}^2}{n_i} - \frac{C_1^2}{N} \right)$$

If both r and c equal 2, T reduces to

(10)
$$T = \frac{N(O_{11}O_{22} - O_{12}O_{21})^2}{n_1 n_2 C_1 C_2}$$

which is the same T as presented in the previous section.

DECISION RULE. Because of the difficulties involved in tabulating the exact distribution of T, the approximation based on the large sample distribution (where the E_{ij}s are large) is used to find the critical region. The critical region of approximate size α corresponds to values of T larger than $x_{1-\alpha}$, the $1-\alpha$ quantile of a chi-square random variable with $(r-1)(c-1)$ degrees of freedom obtained from Table A2. Reject H_0 if T exceeds $x_{1-\alpha}$. Otherwise accept H_0.

Because the asymptotic distribution is used, the approximate value for α, as found here, is a good approximation to the true value of α if the E_{ij}s are fairly large. However, if some of the E_{ij}s are small, the approximation may be very poor. Cochran (1952) states that if any E_{ij} is less than 1 or if more than 20% of the E_{ij} are less than 5, the approximation may be poor. This seems to be overly conservative according to unpublished studies by various researchers, including students of Oscar Kempthorne and students of B. L. van der Waerden, and an article by Roscoe and Byars (1971). If r and c are not too small, I feel that the E_{ij}s may be as small as 1.0 without endangering the validity of the test. If some of the E_{ij}s are too small, several categories may be combined to eliminate the E_{ij}s that are too small. Just which categories should be combined is a matter of judgment. Generally, categories are combined only if they are similar in some respects, so that the hypotheses retain their meaning.

Example 1. A sample of students randomly selected from private high schools and a sample of students randomly selected from public high schools were given standardized achievement tests with the following results.

	Test Scores				
	0–275	276–350	351–425	426–500	Totals
Private School	6	14	17	9	46
Public School	30	32	17	3	82
Totals	36	46	34	12	128

To test the null hypothesis that the distribution of test scores is the same for private and public high school students, the test for differences in probabilities is used. A critical region of approximate size $\alpha = .05$ corresponds to values of T greater than 7.815, obtained from the chi-square distribution in Table A2 with $(r-1)(c-1) = (2-1)(4-1) = 3$ degrees of freedom.

The values of E_{ij} are computed using Equation 4 and are given as follows.

Column	1	2	3	4
E_{ij}: Row 1	12.9	16.5	12.2	4.3
Row 2	23.1	29.5	21.8	7.7

Note that the E_{ij} satisfy Cochran's criteria. Also note that the row and column sums for the E_{ij} are always the same as those for the O_{ij}. This may be used as a check on the calculations.

For the cell in row 1, column 1, we have

$$\frac{(O_{ij} - E_{ij})^2}{E_{ij}} = \frac{(O_{11} - E_{11})^2}{E_{11}} = \frac{(6 - 12.9)^2}{12.9} = \frac{47.61}{12.9} = 3.69$$

A similar calculation is made for each cell and the result, using Equation 4 for purposes of illustration, is

$$T = 3.69 + .38 + 1.89 + 5.14$$
$$+ 2.06 + .21 + 1.06 + 2.87$$
$$= 17.3$$

Since 17.3 is greater than 7.815, the null hypothesis is rejected. In fact, the null hypothesis could have been rejected using a level of significance as small as .001, so

$$\hat{\alpha} \cong .001$$

The conclusion is that test scores are distributed differently among public and private high school students

In Example 1 the data (the test scores before grouping) possessed at least an ordinal scale of measurement, a stronger scale than the nominal scale of measurement considered to be more appropriate for the test used. Actually a more powerful nonparametric test based on ranks—the Mann–Whitney test presented in the next chapter—could have been used on the test scores before they were grouped. However, the data were sufficient to result in a clear cut decision using this test, and so the more powerful (and more tedious) test was not needed in this case.

Not all of the calculations in Example 1 were required to determine that H_0 could be rejected. By inspection a shrewd observer might have determined that cells $(1, 1)$ and $(1, 4)$ might yield the largest contributions to the test statistic, and computations for those two cells alone indicate that the test statistic is at least $3.69 + 5.14 = 8.83$, already in the region of rejection.

☐ *Theory.* The exact distribution of T in the $r \times c$ case may be found in exactly the same way as it was found in the previous section for the 2×2 case. That is, the row totals (sample sizes) are held constant, and then all possible contingency tables having those same row totals are listed, and their probabilities are calculated. The column totals may vary freely from one table to the next, but the row totals may not change. This is the essential difference between this application of the contingency table and the next to be described. In the next application the row totals are not fixed and, therefore, a greater number of different contingency tables are possible. The only requirement is that the total number of observations N

remains the same for all tables. Also, a third variation will be presented, sometimes known as Fisher's exact test. In that application the row totals and the column totals are all fixed and do not vary from table to table. The number of possible tables is greatly reduced, and the exact distribution is then much easier to find.

In all three applications of contingency tables in this section, the asymptotic distribution of T is the same, namely chi-square with $(r-1)$ $c-1$ degrees of freedom. Therefore this distribution is used to provide an approximate value for α, so that exact tables are not needed. The asymptotic distribution is derived in Cramér (1946).

The second application of the $r \times c$ contingency table involves a single random sample of size N, where each observation may be classified according to two criteria. There are r categories (rows) resulting from the first criterion, and c categories (columns) resulting from the second criterion. Each observation is classified according to both criteria and thus ends up being assigned to a particular cell in the $r \times c$ contingency table. The cell entries represent the number of observations belonging to that cell. A nominal scale of measurement is all that is required, although higher scales may be used. The hypothesis tested is one of independence; loosely stated, the null hypothesis says that the rows and columns represent two independent classification schemes. A more precise description is now given.

The Chi-Square Test for Independence

DATA. A random sample of size N is obtained. The observations in the random sample may be classified according to two criteria. Using the first criterion each observation is associated with one of the r rows, and using the second criterion each observation is associated with one of the c columns. Let O_{ij} be the number of observations associated with row i and column j simultaneously. The cell counts O_{ij} may be arranged in an $r \times c$ contingency table.

Column	1	2	3	$\cdots$	c	Totals
Row 1	O_{11}	O_{12}	O_{13}	$\cdots$	O_{1c}	R_1
2	O_{21}	O_{22}	O_{23}	$\cdots$	O_{2c}	R_2
$\cdots$	$\cdots$	$\cdots$	$\cdots$	$\cdots$	$\cdots$	$\cdots$
r	O_{r1}	O_{r2}	O_{r3}	$\cdots$	O_{rc}	R_r
Totals	C_1	C_2	C_3	$\cdots$	C_c	N

The total number of observations in row i is designated by R_i, (instead of n_i as the previous test, to emphasize that the row totals are now random rather than fixed), and in column j by C_j. The sum of the numbers in all of the cells is N.

ASSUMPTIONS

1. The sample of N observations is a random sample. (Each observation has the same probability as every other observation of being classified in row i and column j, independently of the other observations.)
2. Each observation may be classified into exactly one of r different categories according to one criterion and into exactly one of c different categories according to a second criterion.

HYPOTHESES

H_0: The event "an observation is in row i" is independent of the event "that same observation is in column j," for all i and j

By the definition of independence of events, H_0 may be stated as follows.

$$H_0: P(\text{row } i, \text{ column } j) = P(\text{row } i) \cdot P(\text{column } j), \qquad \text{for all } i, j$$

The negation of H_0 is conveniently stated as

$$H_1: P(\text{row } i, \text{ column } j) \neq P(\text{row } i) \cdot P(\text{column } j) \qquad \text{for some } i, j$$

TEST STATISTIC. Let E_{ij} equal $R_i C_j / N$. Then the test statistic is given by

$$(11) \qquad T = \sum_{i=1}^{r} \sum_{j=1}^{c} \frac{(O_{ij} - E_{ij})^2}{E_{ij}}$$

or, for machine calculation,

$$(12) \qquad T = \sum_{i=1}^{r} \sum_{j=1}^{c} \frac{O_{ij}^2}{E_{ij}} - N$$

where the summation is taken over all cells in the contingency table. More convenient equations for the cases where $r = 2$ and/or $c = 2$ may be obtained by using Equations 6 to 10 and replacing n_i by R_i, the new notation for row total.

DECISION RULE. Reject H_0 if T exceeds the $1 - \alpha$ quantile of a chi-square random variable with $(r-1)(c-1)$ degrees of freedom, obtained from Table A2. The approximate level of significance is then α. For more discussion of the decision rule, see the previous test of this section. The same discussion of the decision rule applies to this test also.

Example 2. A random sample of students at a certain university were classified according to the college in which they were enrolled and also according to whether they graduated from a high school in the state or out of the state. The results were put into a 2×4 contingency table.

	Engineering	Arts and Sciences	Home Economics	Other	Totals
In State	16	14	13	13	56
Out of State	14	6	10	8	38
Totals	30	20	23	21	94

In order to test the null hypothesis that the college in which each student is enrolled is independent of whether high school training was in state or out of state, the chi-square test for independence is selected. The rejection region corresponds to values of T greater than 7.815, the .95 quantile of a chi-square random variable with $(r-1)(c-1)=3$ degrees of freedom, obtained from Table A2. Therefore α is approximately .05.

Using either Equation 11 or the more convenient Equation 6, T is computed for these data, resulting in

$$T = 1.55$$

Therefore H_0 is accepted. From Table A2 we may say that $\hat{\alpha}$ exceeds .25.

☐ *Theory.* The exact distribution of T may be found in the manner described earlier in this section and is illustrated here for the relatively simple case where $N=4$. Let p_{ij} be the probability of an observation being classified in row i and column j (cell i, j). (Note that this p_{ij} is not the same as the p_{ij} of the previous test. Here the sum of the p_{ij} in all cells is one. In the previous test the p_{ij} in each row added to unity.) Then the probability of the particular outcome

	Column	
	1	2
Row 1	a	b
Row 2	c	d

N

is found using the multinomial distribution, to be

$$(13) \qquad \frac{N!}{a!\,b!\,c!\,d!} (p_{11})^a (p_{12})^b (p_{21})^c (p_{22})^d$$

because the number of ways N objects can result in the preceding cell counts is given by the multinomial coefficient $N!/a!\,b!\,c!\,d!$, and each result has probability

$$(14) \qquad (p_{11})^a (p_{12})^b (p_{21})^c (p_{22})^d$$

The maximum size of the upper tail of T, when H_0 is true, is found by setting all of the p_{ij}s equal to each other, 1/4 in this case (we will not prove this). Therefore α is found by computing

$$(15) \qquad P\left(\begin{array}{|c|c|}\hline a & b \\\hline c & d \\\hline\end{array}\right) = \frac{N!}{a!\,b!\,c!\,d!}\left(\frac{1}{4}\right)^N$$

for each possible arrangement. For $N=4$ there are 35 different contingency tables. These are listed in Figure 1 along with their probabilities

T = 0

Outcome	Probability
4 0	$(1/4)^4$
0 0	
0 4	$(1/4)^4$
0 0	
0 0	$(1/4)^4$
0 4	
0 0	$(1/4)^4$
4 0	
3 1	$4(1/4)^4$
0 0	
0 3	$4(1/4)^4$
0 1	
0 0	$4(1/4)^4$
1 3	
1 0	
3 0	
1 3	$4(1/4)^4$
0 0	
0 1	$4(1/4)^4$
0 3	
0 0	$4(1/4)^4$
3 1	
3 0	$4(1/4)^4$
1 0	
2 2	$6(1/4)^4$
0 0	
0 2	$6(1/4)^4$
0 2	
0 0	$6(1/4)^4$
2 2	
2 0	$6(1/4)^4$
2 0	
1 1	$24(1/4)^4$
1 1	

Total = 84/256

T = 4/9

Outcome	Probability
2 1	$12(1/4)^4$
1 0	
1 2	$12(1/4)^4$
0 1	
0 1	$12(1/4)^4$
1 2	
1 0	$12(1/4)^4$
2 1	

Total = 48/256

T = 4/3

Outcome	Probability
2 1	$12(1/4)^4$
0 1	
0 2	$12(1/4)^4$
1 1	
1 0	$12(1/4)^4$
1 2	
1 1	$12(1/4)^4$
2 0	
2 0	$12(1/4)^4$
1 1	
1 2	$12(1/4)^4$
1 0	
1 1	$12(1/4)^4$
0 2	
0 1	$12(1/4)^4$
2 1	

Total = 96/256

T = 4

Outcome	Probability
3 0	$4(1/4)^4$
0 1	
0 3	$4(1/4)^4$
1 0	
1 0	$4(1/4)^4$
0 3	
0 1	$4(1/4)^4$
3 0	
2 0	$6(1/4)^4$
0 2	
0 2	$6(1/4)^4$
2 0	

Total = 28/256

Figure 1

and the corresponding values of T. As before, we define zero divided by zero to be zero.

Figure 1 shows the exact distribution of T, when all p_{ij}s equal 1/4, to be

$$P(T = 0) = 84/256 = .33$$
$$P(T = 4/9) = 48/256 = .19$$
$$P(T = 4/3) = 96/256 = .37$$
$$P(T = 4) = 28/256 = .11$$

If the critical region corresponds to the largest value of T, $T = 4$, then $\alpha = .11$. This compares with $\alpha = .125$ for a similar situation discussed in the previous section. A comparison of the preceding distribution, where only N is fixed, with the distribution derived in the previous section where the row totals are also fixed, shows that in this case the distribution of T is more complicated to obtain because of the many more possible tables. Also, additional values of T are now possible, and the probability distribution is altered somewhat.

☐ Even though the exact distributions of T under the two applications differ somewhat, the asymptotic distributions are both chi-square with $(r-1)(c-1)$ degrees of freedom.

In the third application of the contingency table, not only are the row totals fixed, as in the first application, but the column totals are also fixed. Thus the exact distribution of T is easier to find than in both applications previously introduced. However, easier or not, the exact distribution is still too complicated for practical purposes, unless extensive tables or a computer are available. The chi-square approximation is recommended for finding the critical region and α.

The Chi-Square Test with Fixed Marginal Totals

DATA. The data are summarized in an $r \times c$ contingency table, as in the two previous applications, except that the row and column totals are determined beforehand, and are therefore fixed, not random.

Column	1	2	$\cdots$	Totals	
Row 1	O_{11}	O_{12}	$\cdots$	O_{1c}	$n_1.$
2	O_{21}	O_{22}	$\cdots$	O_{2c}	$n_2.$
$\cdots$	$\cdots$	$\cdots$	$\cdots$	$\cdots$	$\cdots$
r	O_{r1}	O_{r2}	$\cdots$	O_{rc}	$n_r.$
Totals	$n_{.1}$	$n_{.2}$	$\cdots$	$n_{.c}$	N

The row and column totals are denoted by $n_{i.}$ and $n_{.j}$ respectively, to emphasize the fact that they are given and not random. The total number of observations is N.

ASSUMPTIONS
1. Each observation is classified into exactly one cell.
2. The observations are observations on a random sample. Each observation has the same probability of being classified into cell (i, j) as any other observation.
3. The row and column totals are given, not random.

HYPOTHESES. The hypotheses may be either of the two sets of hypotheses introduced in the two previous applications in this section, under the condition that the row and column totals are fixed. Or the hypotheses may be tailored to fit the particular experimental situation. Usually the hypotheses are variations of the independence hypotheses of the previous test. See Examples 3 and 4 for particular modifications that are dictated by the experiment.

TEST STATISTIC. Let $E_{ij} = n_{i.} n_{.j}/N$ be the expected number of observations in cell (i, j). Then the test statistic, as before, is given by

$$(16) \qquad T = \sum_{i=1}^{r} \sum_{j=1}^{c} (O_{ij} - E_{ij})^2/E_{ij}$$

where the summation is over all rc cells. If $r = 2$ or $c = 2$ the special equations 6 to 10 may be used, with the appropriate change in notation.

DECISION RULE. Reject H_0 if T exceeds the $1 - \alpha$ quantile of a chi-square random variable with $(r-1)(c-1)$ degrees of freedom, obtained from Table A2. The approximate level of significance is then α. For more details, see the decision rule given in the first test of this section. All of the comments made there are equally valid here also.

Example 3. The chi-square test with fixed marginal totals may be used to test the hypothesis that two random variables X and Y are independent. Starting with a scatter diagram of 24 points, which represent independent observations on the bivariate random variable (X, Y), a contingency table may be constructed. The x-coordinate of each point is the observed value of X and the y-coordinate is the observed value of Y in each observation on (X, Y). Assume the observed pairs (X, Y) are mutually independent. We wish to test

H_0: X and Y are independent of each other

against the alternative hypothesis of dependence.

To form the contingency table so that all E_{ij}s are equal, we note that 3 and 4 both are factors of the sample size 24. Therefore we divide the points into 3 rows of 8 points each, and 4 columns of 6 points each, using dotted lines as in Figure 2. (It is recommended that if the E_{ij}s are small, they should be very nearly equal to each other. One way of accomplishing this is by having equal row totals and equal column totals.) The resulting contingency table of counts is given as follows.

Column	1	2	3	4	Totals
Row 1	0	4	4	0	8
2	2	1	2	3	8
3	4	1	0	3	8
Totals	6	6	6	6	24

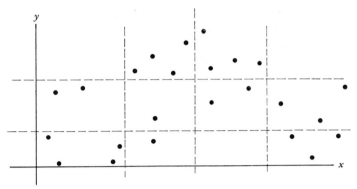

Figure 2

The critical region of approximate size .05 corresponds to values of T greater than 12.59, the .95 quantile of a chi-square random variable with $(r-1)(c-1) = (2)(3) = 6$ degrees of freedom, obtained from Table A2.

The test statistic is evaluated using Equation 16, and $E_{ij} = (6)(8)/24 = 2$.

$$T = \sum_{i=1}^{r} \sum_{j=1}^{c} \frac{(O_{ij} - E_{ij})^2}{E_{ij}} = \sum_{i=1}^{3} \sum_{j=1}^{4} \frac{(O_{ij} - 2)^2}{2}$$

(17) $= 14$

Because T exceeds 12.59, H_0 is rejected, and we conclude that X and Y are not independent. H_0 could have been rejected at a level of significance as small as .03, so $\hat{\alpha} = .03$.

Example 4. A psychologist asks a subject to learn 25 words. The subject is given 25 blue cards, each with one word on it. Five of the words are nouns, 5 are adjectives, 5 are adverbs, 5 are verbs, and 5 are prepositions. She must pair these blue cards with 25 white cards, each with one word on it and also containing the different parts of speech, 5 words each. The subject is allowed 5 minutes to pair the cards (1 white card with each blue card) and 5 minutes to study the pairs thus formed. Then she is asked to close her eyes, and the words on the white cards are read to her one by one. When each word is read to her, she tries to furnish the word on the blue card associated with the word read.

The psychologist is not interested in the number of correct words but, instead, in examining the pairing structure to see if it represents an ordering of some sort. The hypotheses are as follows.

H_0: There is no organization of pairs according to parts of speech

against the alternative

H_1: The subject tends to pair particular parts of speech on the blue cards with particular parts of speech (not necessarily the same) on the white cards

The pairings are summarized in a 5×5 contingency table.

	Noun	Adjective	Adverb	Verb	Preposition	Totals
Noun		3			2	5
Adjective	4	1				5
Adverb				5		5
Verb			5			5
Preposition	1	1			3	5
Totals	5	5	5	5	5	25

The chi-square test with fixed marginal totals is selected because the experimenter feels that large values of T indicate H_1 is true. The marginal totals represent the number of words in each category, which was fixed in advance of the actual experiment. The critical region of approximate size .05 corresponds to values of T greater than 26.30, the .95 quantile of a chi-square random variable with $(r-1)(c-1) = (4)(4) = 16$ degrees of freedom, obtained from Table A2. The observed value of T is obtained using Equation 16.

$$E_{ij} = \frac{(5)(5)}{25} = 1 \quad \text{for all } i \text{ and } j$$

$$T = \sum_{i=1}^{5} \sum_{j=1}^{5} \frac{(O_{ij} - 1)^2}{1}$$

(18)
$$= 66$$

Because $T = 66$, H_0 is soundly rejected in favor of H_1. The critical level $\hat{\alpha}$ is less than .001.

☐ *Theory.* The exact distribution of T is found in a manner similar to the method described in the two previous applications of this section, and once in the previous section. We will describe the method in the 2×2 case.

Let A represent the event the following contingency table is obtained

	Class 1	Class 2	Total
Row 1	x_1	$n_1 - x_1$	n_1
Row 2	x_2	$n_2 - x_2$	n_2

where the row totals are considered to be fixed. Then the probability of A, when H_0 is true, was computed in the previous section and was found to be given by Equation 4.1.4

(19)
$$P(A) = \binom{n_1}{x_1}\binom{n_2}{x_2} p^{x_1 + x_2}(1-p)^{N - x_1 - x_2}$$

Now let B be the event the column totals are $(x_1 + x_2)$ and $(N - x_1 - x_2)$ for columns 1 and 2, respectively Since the probability of being in class 1 is denoted by p, we have a situation where the binomial distribution applies, so

$$(20) \qquad P(B) = \frac{N!}{(x_1 + x_2)! \, (N - x_1 - x_2)!} \, p^{x_1 + x_2} (1 - p)^{N - x_1 - x_2}$$

We are interested in the probability of the event A given that the column totals are fixed, that is, given B is true. Because of the definition of conditional probability,

$$(21) \qquad P(A \mid B) = \frac{P(AB)}{P(B)}$$

we can find $P(A \mid B)$ by finding $P(AB)$, because $P(B)$ is given by Equation 20.

Consider the fact that if A occurs (we obtain the 2×2 table specified by the event A), B automatically occurs (the column totals are $x_1 + x_2$ and $N - x_1 - x_2$). The event AB occurs if and only if the event A occurs. So we have

$$(22) \qquad P(AB) = P(A)$$

where $P(A)$ is given by Equation 19. Thus, by dividing Equation 19 by Equation 20, we get

$$(23) \qquad P(A \mid B) = \frac{\binom{n_1}{x_1}\binom{n_2}{x_2}}{\binom{N}{x_1 + x_2}}$$

Therefore the probability for each 2×2 table, given that the row totals and the column totals are fixed, is given by Equation 23, which may be recognized as the probability of the hypergeometric distribution. Note that we would have obtained the same result if we had fixed first the column totals and then the row totals.

If the row totals and column totals all equal 2, there are three possible contingency tables.

Table	Probability	T
$\begin{array}{|c|c|}\hline 2 & 0 \\\hline 0 & 2 \\\hline\end{array}$	$\dfrac{\binom{2}{2}\binom{2}{0}}{\binom{4}{2}} = 1/6$	4
$\begin{array}{|c|c|}\hline 1 & 1 \\\hline 1 & 1 \\\hline\end{array}$	$\dfrac{\binom{2}{1}\binom{2}{1}}{\binom{4}{2}} = 2/3$	0

0	2
2	0

$$\frac{\binom{2}{0}\binom{2}{2}}{\binom{4}{2}} = 1/6 \qquad\qquad 4$$

Because the probability distribution of T is unique, H_0 is simple in this application.

Fixed row totals and fixed column totals greatly reduce the number of contingency tables possible, and so the exact distribution of T is more feasible in this case than in the previous two applications. When $r = 2$ and $c = 2$, the test is known as "Fisher's exact test," and extensive exact tables of probabilities are available (Finney, 1948). Programming Fisher's exact test is discussed by Robertson (1960).

For r and c in general, the exact probability of the table

Column	1	2	$\cdots$	c	Totals
Row 1	O_{11}	O_{12}	$\cdots$	O_{1c}	$n_{1.}$
2	O_{21}	O_{22}	$\cdots$	O_{2c}	$n_{2.}$
$\cdots$	$\cdots$	$\cdots$	$\cdots$	$\cdots$	$\cdots$
r	O_{r1}	O_{r2}	$\cdots$	O_{rc}	$n_{r.}$
Totals	$n_{.1}$	$n_{.2}$	$\cdots$	$n_{.c}$	N

with fixed marginal totals is given by

(24)
$$\text{probability} = \frac{\begin{bmatrix} n_{1.} \\ O_{1i} \end{bmatrix}\begin{bmatrix} n_{2.} \\ O_{2i} \end{bmatrix}\cdots\begin{bmatrix} n_{r.} \\ O_{ri} \end{bmatrix}}{\begin{bmatrix} N \\ n_{.i} \end{bmatrix}}$$

where the multinomial coefficients are as defined by Rule 3 in Section 1.1.

Presumably, because there are fewer values for T in this third case, the chi-square approximation is not as good as in the first and second cases discussed in this section, and the approximation is probably the best in the second case.

☐   The contingency tables of this section and Section 4.1 could be called *two-way contingency tables* because the observations are classified two ways, by rows and by columns. An immediate extension may be made to include the situation where observations are classified according to three or more criteria, and thus the data are presented in the form of a three-(or more) way contingency table. Of course, such a presentation of data becomes awkward on ordinary paper and is therefore not often used. The next few remarks are made just in case such a presentation becomes necessary and the corresponding test is desired.

For convenience in extending the chi-square contingency table test, the two-way test statistic is rewritten as

$$(25) \qquad T = \sum_{i,j} \frac{\left[O_{ij} - N \dfrac{R_i}{N} \dfrac{C_j}{N} \right]^2}{N \dfrac{R_i}{N} \dfrac{C_j}{N}}$$

which has $(r-1)(s-1)$ degrees of freedom. In a three-way contingency table with r rows, s columns, and t blocks, denote the block totals by B_k, $k = 1, 2, \ldots, t$, to correspond to the row totals R_i and the column totals C_j. Let N still denote the total number of observations. Then

$$(26) \qquad R_i = \sum_{j,k} O_{ijk}$$

$$(27) \qquad C_j = \sum_{i,k} O_{ijk}$$

$$(28) \qquad B_k = \sum_{i,j} O_{ijk}$$

where O_{ijk} represents the total number of observations classified in row i, column j, and block k. Then E_{ijk}, the expected number of observations in row i, column j, and block k, assuming the null hypothesis of row-column-block independence is true, may be estimated from

$$(29) \qquad E_{ijk} = N \frac{R_i}{N} \frac{C_j}{N} \frac{B_k}{N}$$

and the test statistic may be computed using

$$(30) \qquad T = \sum_{i,j,k} \frac{(O_{ijk} - E_{ijk})^2}{E_{ijk}}$$

where the summation is over all $r \cdot s \cdot t$ cells. The test statistic is then tested for significance using the chi-square distribution with $rst - r - s - t + 2$ degrees of freedom. The extension of the test to contingency tables of any dimension should be apparent. So-called "loglinear models" have been used successfully to analyze multidimensional contingency tables and are discussed in the final section of this chapter. More detailed discussion of the analysis of multidimensional contingency tables is found in Goodman (1970), Ireland, Ku, and Kullback (1969), Ku, Varner, and Kullback (1971), Kullback (1971), Goodman (1971), Koch, Johnson, and Tolley (1972), Darroch (1974), and Halperin et al. (1977). A short book on analyzing contingency tables has been written by Maxwell (1961). The important subject of estimation in contingency tables is addressed by Feinberg (1970), McNeil and Tukey (1975), and Quade and Salama (1975). If some of the data are only partially classified, see Chen and Fienberg (1974) or Hocking and Oxspring (1974). Contingency tables where one or both categories have a natural ordering are discussed in Section 5.2, as well as by Williams and Grizzle (1972), Simon (1974), and Clayton (1974).

The exact distribution of the 2×3 contingency table test statistic is given for equal column totals and fixed row totals by Bennett and Nakamura (1963, 1964) and is discussed by Healy (1969). Ireland and Kullback (1968) give a different test for contingency tables with given row and column totals. The power of chi-square tests for contingency tables is examined by Chapman and Meng (1966).

An excellent and readable survey article on contingency tables is one by Mosteller (1968). Haynam and Leone (1965) give an approximation to the exact distribution of T. Misclassification of data is the subject of an article by Mote and Anderson (1965). Tables with small or zero cell frequencies are discussed by Ku (1963) and Sugiura and Otake (1968). See Goodman (1964, 1968) and Bhapkar and Koch (1968) for information concerning tests for interaction. A class of bivariate contingency-type distributions is discussed by Plackett (1965), Mardia (1967b), and Steck (1968). Other methods for examining contingency tables are given by Ishii (1960), Gregory (1961), Claringbold (1961), Kullback, Kupperman, and Ku (1962), Diamond (1963), Mielke and Siddiqui (1965), Hoeffding (1965), Gart (1966), and Chacko (1966). The many recent papers on contingency tables illustrate the usefulness and versatility of this type of analysis; see, for example, Elston (1970), Crowley and Breslow (1975), Light and Margolin (1971), Margolin and Light (1974), and Shuster and Downing (1976). Mantel and Haenzel (1959) present a useful procedure for testing homogeneity in a row-column distributions on N independent $2 \times r$ tables. Other applications of contingency tables are given in the following sections of this chapter.

EXERCISES

1. Test whether the following observations indicate a dependence between the two variables observed: (3.6, 13), (4.7, 19), (1.4, 9), (5.5, 15), (4.8, 27), (4.3, 14), (3.0, 6), (4.2, 11), (6.0, 24), (6.8, 26), (4.1, 18), (3.2, 9), (4.0, 8), (1.9, 6), (.4, 7), (4.9, 14), (5.6, 18), and (5.6, 20). Which test of this section is being used?

2. One horse was selected at random from each of 80 races and categorized according to post position (the position assigned to the horse for the start of the race) and the position in which the horse crossed the finish line (first, second, etc.).

		Finish			
		1	2	3	Other
Post Position	1–4	8	6	8	16
	5–9	3	6	5	28

Is the horse's position at the end of the race dependent on post position? Which test of this section is being used?

3. In another study, all of the horses in all of the races for 3 days were classified by post position and by the order in which they finished.

		Finish			
		1	2	3	Other
Post	1–4	15	14	15	52
Position	5–9	9	10	9	72

Is the horse's position at the end of the race dependent on post position? Which test of this section is being used?

4. Four professors are teaching large classes in introductory statistics. At the end of the semester, they compare grades to see if there are significant differences in their grading policies.

Professor	A	B	C	D	Grade F	WP	WF
Smith	12	45	49	6	13	18	2
Jones	10	32	43	18	4	12	6
White	15	19	32	20	6	9	7

Are these differences significant? Which test are you using? Are the grades assigned by Professors Jones and White significantly different? How would the results be interpreted?

PROBLEMS

1. Show that the two forms of T given by Equations 4 and 5 are equivalent.
2. Show that if $r = 2$ and $c = 2$, T may be given by Equation 10.
3. A different method of analyzing contingency table uses the statistic

$$T' = 2 \sum_{i=1}^{r} \sum_{j=1}^{c} O_{ij} \ln (O_{ij}/E_{ij})$$

instead of T, where ln refers to natural logarithm, found on most calculators. Otherwise the two test procedures are exactly the same. Use T' in Exercise 3 to see if the result of the analysis is similar to the result using T. (The two tests are not equivalent in general, even though they may produce similar results in particular cases.) Note that T' may also be written as

$$T' = 2 \sum_{i=1}^{r} \sum_{j=1}^{c} O_{ij} \ln O_{ij} + 2N \ln N - 2 \sum_{i=1}^{r} n_i \ln n_i - 2 \sum_{j=1}^{c} C_j \ln C_j$$

which is easier to use with a calculator.

4.3. THE MEDIAN TEST

The median test is designed to examine whether several samples came from populations having the same median. Actually, the median test is not new to this chapter; it is merely a special application of the chi-square test with fixed marginal totals introduced in the previous section. It is a very useful application, however, and we consider it worth special treatment.

To test whether several c populations have the same median, a random sample is drawn from each population. (The scale of measurement is at least ordinal, or else the term "median" is without meaning.) A $2 \times c$ contingency table is constructed, and the two entries in the ith column are the numbers of observations in the ith sample that are above and below the grand median (the median of all observations combined). The usual chi-square test is then applied to the contingency table.

The Median Test

DATA. From each of c populations a random sample of size n_i is obtained, $i = 1, 2, \ldots, c$. The combined sample median is determined; that is, the number that is exceeded by about half of the observations in the entire array of N ($= n_1 + n_2 + \cdots + n_c$) sample values is determined. This is called the *grand median*. Let O_{1i} be the number of observations in the ith sample that exceed the grand median and let O_{2i} be the number in the ith sample that are less than or equal to the grand median. Arrange the frequency counts into a $2 \times c$ contingency table as follows.

Sample	1	2	...	c	Totals
> Median	O_{11}	O_{12}	...	O_{1c}	a
≤ Median	O_{21}	O_{22}	...	O_{2c}	b
Totals	n_1	n_2	...	n_c	N

Let a equal the total number of observations above the grand median in all samples and let b equal the total number of values less than or equal to the grand median. Then $a + b = N$, the total number of observations.

ASSUMPTIONS

1. Each sample is a random sample.
2. The samples are independent of each other.
3. The measurement scale is at least ordinal.
4. If all populations have the same median, all populations have the same probability p of an observation exceeding the grand median.

HYPOTHESES

H_0: All c populations have the same median
H_1: At least two of the populations have different medians

TEST STATISTIC. The test statistic is obtained by a slight rearrangement of the test statistic given in the previous section, Equation 4.2.6, for the special case of two rows.

$$(1) \qquad T = \frac{N^2}{ab} \sum_{i=1}^{c} \frac{\left(O_{1i} - \frac{n_i a}{N}\right)^2}{n_i}$$

If a calculator is being used, the following form is more convenient.

$$(2) \qquad T = \frac{N^2}{ab} \sum_{i=1}^{c} \frac{O_{1i}^2}{n_i} - \frac{Na}{b}$$

If $a \simeq b$, as it should be unless there are many values equal to the grand median, the following simplification of the test statistic may be used.

$$(3) \qquad T = \sum_{i=1}^{c} \frac{(O_{1i} - O_{2i})^2}{n_i}$$

The simplified form for T given in Equation 3 is exact if $a = b$. It is only approximate otherwise.

DECISION RULE. The exact distribution of T is difficult to tabulate, so the large sample approximation is used to approximate the distribution of T. (See the discussion of the theory, later in this section, for the exact distribution of T.) The critical region of approximate size α corresponds to values of T greater than $x_{1-\alpha}$, the $(1-\alpha)$ quantile of a chi-square random variable with $c-1$ degrees of freedom, obtained from Table A2. If T exceeds $x_{1-\alpha}$, reject H_0. Otherwise accept H_0.

If some of the sample sizes n_i are too small, the preceding approximation may not be very accurate (the true α may vary considerably from the approximate α determined above). The same rule given in the previous section may be used as a rule of thumb. That is, the approximation may not be satisfactory if more than 20% of the n_is are less than 10 or if any of the n_is are less than 2. (This is almost equivalent to the rule stated in terms of the expected cell frequencies E_{ij} in the previous section.) If most of the n_is are about equal to each other, an exception to the rule may be made, and n_is as small as 2 may be allowed; however, the number of samples, c, should not be small in this case.

Example 1. Four different methods of growing corn were randomly assigned to a large number of different plots of land and the yield per acre was computed for each plot.

	Method		
1	*2*	*3*	*4*
83	91	101	78
91	90	100	82
94	81	91	81
89	83	93	77
89	84	96	79
96	83	95	81
91	88	94	80
92	91		81
90	89		
	84		

In order to determine whether there is a difference in yields as a result of the method used, the median test was employed because it was felt that a difference in population medians could be interpreted as a difference in the value of the method used. The hypotheses may be stated as follows.

H_0: All methods have the same median yield per acre

H_1: At least two of the methods differ with respect to the median yield per acre

A quick count reveals there are 34 observations in all, so the average of the seventeenth and eighteenth smallest observations is the grand median. It is not necessary to order all 34 observations to find the grand median. Instead, a count of the numbers in the 70s and 80s reveals that 18 values are less than 90. Thus the average of the two highest yields in the 80s is the grand median and, by inspection, is seen to be 89. Then, for each method (sample), the number of values that exceed 89 and the number that are less than or equal to 89 are recorded in the following form.

Method	1	2	3	4	Totals
>89	6	3	7	0	16
≤89	3	7	0	8	18
Totals	9	10	7	8	34

The sample sizes are fairly small, but they are approximately equal, so we may ignore the fact that three of them are less than 10 and use the chi-square approximation. The critical region corresponds to values of T greater than 7.815, the .95 quantile of a chi-square random variable with $c - 1 = 3$ degrees of freedom, obtained from Table A2. T is computed using Equation 1.

$$T = \frac{(34)^2}{(16)(18)} \left\{ \frac{\left[6 - \frac{(9)(16)}{34} \right]^2}{9} + \cdots + \frac{\left[0 - \frac{(8)(16)}{34} \right]^2}{8} \right\}$$

$$= 4.01(.34 + .29 + 1.97 + 1.78)$$

(4) $$= 17.6$$

Use of the more convenient Equation 3 gives

$$T = \tfrac{9}{9} + \tfrac{16}{10} + \tfrac{49}{7} + \tfrac{64}{8}$$

(5) $= 17.6$

which is identical to the rounded-off value previously obtained.

Because the T of 17.6 exceeds the critical value 7.815, H_0 is rejected. Inspection of Table A2 shows the critical level $\hat{\alpha}$ to be slightly less than .001.

If the median test leads to rejection of the null hypothesis and it is desired to further inspect the samples to determine which population medians are different from each other, any subgroup of two or more populations may be analyzed using the median test, until the differences have been isolated. However, such "sorting out" of the populations by repeatedly using the same test on subgroups of the original data laways distorts the true level of significance of all tests but the first. Such repeated testing procedures are for one's personal satisfaction or for use as an objective "yardstick" for separating the various populations, but they cannot receive the same interpretation as may legitimately be given to the first, overall test. For a further discussion of repeated testing procedures see Gabriel (1966) or Knoke (1976).

In Example 1 the experiment has been arranged in a so-called "completely randomized design," which assumes that the different methods are assigned to the different plots in some random manner (or a manner equivalent to a random manner). The usual parametric method of analyzing the data is called a "one-way analysis of variance."

The median test may be extended to become a "quantile test" for testing the null hypothesis that several populations have the same quantile, for any particular quantile chosen, merely by altering the data section of the test so that the observations are classified as being above or not above the grand quantile for the entire array of values. The remainder of this test remains the same, except that the approximation Equation 3 will seldom be applicable. The exact distribution of T, as given in the theory to follow, remains the same. An extension of the median test to allow for two-stage sampling is given by Wolfe (1977a).

☐ *Theory.* The row totals a and b are fixed, as in the third test of the previous section, because of the objective set of rules that is used to determine which observations are to be counted in the upper or lower cells of the contingency table. For example, if the test is an "upper quartile" test, then a is about $N/4$ and b is about $3N/4$, with allowances for tied, or equal, sample values. Thus the exact distribution of T is a conditional distribution that depends on the row and column totals. The probability of obtaining the table

(6)

O_{11}	O_{12}	$\cdots$	O_{1c}	a
O_{21}	O_{22}	$\cdots$	O_{2c}	b
n_1	n_2	$\cdots$	n_c	

with the column totals fixed, is the product of the binomial probabilities

(7)
$$P\left(\frac{O_{1i}}{O_{2i}}\right) = \binom{n_i}{O_{1i}} p^{O_{1i}} (1-p)^{O_{2i}}; \qquad i = 1, 2, \ldots, c$$

where p is the probability of an observation exceeding the grand median. Now H_0 merely states that all populations have the same median, and this does not necessarily imply that all populations have the same probability p of exceeding the grand (sample) median. On the other hand, H_0 does not preclude this latter situation, and the two statements (H_0, and the previous situation) are very similar in intent. To find the distribution of T when H_0 is true, we need to require that the probabilities of exceeding the grand median be the same for all populations. This is why the fourth assumption was placed in the model.

Because the samples are independent of each other, we can obtain the joint probability by multiplication, using Equation 7;

(8)
$$P\left(\begin{array}{cccc} O_{11} & O_{12} & \cdots & O_{1c} \\ O_{21} & O_{22} & \cdots & O_{2c} \end{array}\right) = \binom{n_1}{O_{11}} \binom{n_2}{O_{12}} \cdots \binom{n_c}{O_{1c}} p^a (1-p)^b$$

where

$$a = O_{11} + O_{12} + \cdots + O_{1c}$$

and

$$b = O_{21} + O_{22} + \cdots + O_{2c}$$

The probability of the event in Equation 8, given the row totals a and b, is found by dividing the probability in Equation 8 by the probability of getting the row totals a and b, as in the latter part of the previous section. The result is

$$P\left(\begin{array}{ccccc} O_{11} & O_{12} & \cdots & O_{1c} & a \\ O_{21} & O_{22} & \cdots & O_{2c} & b \\ n_1 & n_2 & \cdots & n_c & N \end{array}\right) = \frac{\binom{n_1}{O_{11}} \binom{n_2}{O_{12}} \cdots \binom{n_c}{O_{1c}}}{\binom{N}{a}}$$

which can be written in terms of multinomial coefficients as

$$\text{(10)} \qquad\qquad \text{Probability} = \frac{\begin{bmatrix} a \\ O_{1i} \end{bmatrix}\begin{bmatrix} b \\ O_{2i} \end{bmatrix}}{\begin{bmatrix} N \\ n_i \end{bmatrix}}$$

in agreement with Equation 24 of Section 4.2.

Thus the exact distribution of T can be, but almost never is, found using Equation 9 or 10. Instead, the chi-square distribution with $(c-1)$ degrees of freedom is used (because the number of rows is two), as in the previous section.

The preceding median test may be extended so that more complex experiments may be analyzed. Because of the cumbersome notation involved in the extension of the median test, we will introduce the test by presenting an example of its application.

An Extension of the Median Test

Example 2. Four different fertilizers are used on each of six different fields, and the entire experiment is replicated using three different types of seed. The yield per acre is calculated at the conclusion of the experiment under each of the $(4)(6)(3) = 72$ different conditions with the following results.

		Seed 1				Seed 2				Seed 3			
						Fertilizer							
		1	2	3	4	1	2	3	4	1	2	3	4
Field	1	80.5	90.1	87.0	88.0	79.1	87.0	82.6	81.5	85.4	92.3	92.0	89.3
	2	87.0	83.4	89.1	90.3	77.6	82.0	81.4	87.9	89.2	90.1	90.2	93.6
	3	86.1	82.4	91.0	86.1	84.1	80.6	89.0	80.4	90.0	88.1	87.2	90.8
	4	82.1	84.9	84.4	83.1	83.3	79.5	86.3	83.1	83.4	85.3	94.3	87.6
	5	79.3	87.1	92.2	90.8	76.6	86.2	84.0	87.4	87.1	86.3	88.4	93.7
	6	84.2	89.3	85.3	84.7	81.0	84.1	88.1	85.0	82.3	92.9	95.1	82.9

To test the null hypothesis

H_0: There is no difference in median yields due to the different fertilizers

let $x_{i_1 i_2 i_3}$ denote the observed yield using fertilizer i_1 in field i_2 with seed i_3. For example, x_{213} is the yield using fertilizer 2 in field 1 with seed 3, which is 92.3. Then x_{213} is compared with the median of x_{113}, x_{213}, x_{313}, and x_{413}, the four yields obtained under identical circumstances except for fertilizers (which H_0 claims to have no effect). Thus x_{213} is compared with the median of 85.4, 92.3, 92.0, and 89.3, which is

$$(\tfrac{1}{2})(89.3 + 92.0) = 90.65$$

If x_{213} exceeds 90.65, it is replaced in the table by a one; otherwise it is replaced by a zero.

Similarly, each $x_{i_1 i_2 i_3}$ is compared with the median of $x_{1 i_2 i_3}, x_{2 i_2 i_3}, \ldots, x_{c i_2 i_3}$, the observations obtained under similar conditions, except for the c different fertilizers. In our example each yield is compared with the median of the yields in the same row (field) and same block (seed) and replaced by one or zero according to whether it exceeds or does not exceed its respective median. The results are as follows.

		Seed 1				Seed 2				Seed 3			
		1	2	3	4	1	2	3	4	1	2	3	4
Field	1	0	1	0	1	0	1	1	0	0	1	1	0
	2	0	0	1	1	0	1	0	1	0	0	1	1
	3	0	0	1	0	1	0	1	0	1	0	0	1
	4	0	1	1	0	1	0	1	0	0	0	1	1
	5	0	0	1	1	0	1	0	1	0	0	1	1
	6	0	1	1	0	0	0	1	1	0	1	1	0

Let O_j be the number of fields in which fertilizer j was used and where the yield exceeded its respective median. Then O_j is the total number of "ones" under fertilizer j in the preceding tables. The O_j are given in the following table for $j = 1, 2, \ldots, c$.

Fertilizer	1	2	3	4	Total
O_j = Number of "Ones"	3	8	14	10	$a = 35$
Number of "Zeros"	15	10	4	8	$b = 37$
	$n_1 = 18$	$n_2 = 18$	$n_3 = 18$	$n_4 = 18$	$N = 72$

The usual median test is then applied to this table. Using Equation 3, we obtain

$$T = \frac{(144 + 4 + 100 + 4)}{18}$$

(11) $= 14.0$

A comparison of this value of T with the .95 quantile of a chi-square random variable with $c - 1 = 3$ degrees of freedom, obtained from Table A2 as $x_{.95} = 7.815$, results in rejection of H_0. The critical level $\hat{\alpha}$ in this experiment is about .004.

EXERCISES

1. Test the hypothesis that the following samples were obtained from populations having the same medians.

Sample 1: 35, 42, 42, 30, 15, 31, 29, 29, 17, 21
Sample 2: 34, 38, 26, 17, 42, 28, 35, 33, 16, 40
Sample 3: 17, 29, 30, 36, 41, 30, 31, 23, 38, 30
Sample 4: 39, 34, 22, 27, 42, 33, 24, 36, 29, 25

2. A number of oil leases were auctioned to the highest bidder. Each lease received one or more sealed bids. Test the hypothesis that the leases which eventually became producers of oil have the same median number of bids as the leases which never produced oil. A random sample of each type of lease is given below.

Number of Bids on Each Lease

| Producers | 6, | 3, | 1, | 14, | 8, | 9, | 12, | 1, | 3, | 2, | 1, | 7, |
| Nonproducers | 6, | 2, | 1, | 1, | 3, | 1, | 2, | 4, | 8, | 1, | 2 | |

3. Do the experimental results of Example 2 indicate a difference among seeds?

4. Do the experimental results of Example 2 indicate a difference among fields?

PROBLEMS

1. Show that if a equals b, then Equation 1 becomes Equation 3.

2. Show that Equation 1 is the same as Equation 4.2.16 when r equals 2.

3. The usual parametric test for the design in this section (the one-way layout) assumes that each observation has a normal distribution instead of being merely a zero or one, depending on whether it is below or above the median. If the observations in each sample are called 0s when they are below the grand median and 1s if they are equal to or above the grand median, the statistic for the previous parametric test, computed on the 0s and 1s, simplifies to

$$F = \frac{\left(\sum_{j=1}^{c} \frac{O_{ij}^2}{n_j} - \frac{a^2}{N}\right)(N-c)}{\left(a - \sum_{j=1}^{c} \frac{O_{ij}^2}{n_j}\right)(c-1)}$$

Show that F may be written as the following function of T,

$$F = \frac{T(N-c)}{(N-T)(c-1)}$$

and that therefore rejecting H_0 for large T is equivalent to rejecting H_0 for large F.

4.4. MEASURES OF DEPENDENCE

The contingency table is a convenient form for examining data to see if there is some sort of dependence inherent in the data. The particular type of dependence revealed by a contingency table is a row-column dependence. If the different rows represent samples from different populations and the columns represent different categories of classification of the data from the samples, a row-column dependence is synonymous with a functional dependence of the probabilities of being in the various categories on the population from which the sample was obtained. Similarly, if the observations from one random sample are classified into rows and columns according to each of two different criteria, a row-column dependence has an obvious interpretation as a dependence between the two criteria of classification.

Suppose that instead of testing hypotheses, as we have been doing so far in this chapter, we merely wish to express the degree of dependence shown in a particular contingency table. Ideally, we would like to be able to express the degree of dependence in some simple form, and in a form that easily conveys to other people the exact degree of dependence exhibited by the table.

As a first approach, we could use the test statistic of the previous sections

(1)
$$T = \sum_{i=1}^{r} \sum_{j=1}^{c} \frac{(O_{ij} - E_{ij})^2}{E_{ij}}$$

as a measure of dependence, with the philosophy, "If it is good enough to test for dependence, it is good enough to measure dependence." The use of T seems satisfactory as far as convenience and simplicity are concerned. However, in order to convey the degree of dependence to other people, the number of degrees of freedom should also be stated along with the T value because, without knowing the number of degrees of freedom, it is not possible to tell the degree of dependence conveyed by the value of T. Even if the number of degrees of freedom is known, the nonexpert must consult a chi-square table in order to interpret T.

One widely used measure of dependence is the *critical level*, the smallest level of significance that would result in rejection of the null hypothesis in a hypothesis test involving T. Instead of the experimenter reporting the T value and the number of degrees of freedom so that the reader may consult a chi-square table to determine how large T is, the experimenter could do all this for the reader and find the quantile x_p (of the appropriate chi-square distribution) closest to T. The experimenter then simply reports

(2)
$$\hat{\alpha} = 1 - p$$

where T is the pth quantile of the appropriate chi-square distribution. If $\hat{\alpha}$ is small (close to zero) much dependence is shown. If $\hat{\alpha}$ is large (close to 1.0) the interpretation is that an extreme amount of independence is shown. Values of $\hat{\alpha}$ around 0.5 would be commonplace if there were no row-column dependence.

The preceding measure of dependence is also called simply the *chi-square probability*, but this term may cause confusion because it may create the impression that the number being furnished is p instead of $1 - p$.

The Critical Level as a Measure of Dependence

Example 1. In Example 4.2.1 the contingency table was as follows.

	Scores				
	0–276	276–350	351–425	426–500	Totals
Private School	6	14	17	9	46
Public School	30	32	17	3	82
Totals	36	46	34	12	128

For this contingency table T was found to equal 17.3. Now, 17.3 is approximately the .999 quantile of a chi-square random variable with 3 degrees of freedom, so $p = .999$, and we have

$$\hat{\alpha} = 1 - p$$
$$= .001$$

as computed in the example. Such a small value of the critical level indicates a great degree of dependence between the type of school and the test scores.

Another approach to the problem of providing an easily interpreted measure of dependence consists of modifying the value of T in Equation 1 in such a way that the result does not depend as much on the number of degrees of freedom as T does. One such modification considers dividing T by the maximum value T may attain. We know by now that large values of T arise from contingency tables that have a pronounced unevenness among the cell counts. By examining extremely uneven contingency tables we may find, by trial and error, that T is greatest (for a given number of rows r and columns c and total sample size N) when there are zeros in every cell except for one cell in each row and in each column. (If r does not equal c some rows or columns may be all zeros.) That is T is a maximum in a contingency table resembling the following.

Column	1	2	3	4	5	Totals
Row 1	3	0	0	0	0	3
2	0	3	0	0	0	3
3	0	0	3	0	0	3
Totals	3	3	3	0	0	9

For this table, $T = 18$, after 0/0 is defined as 0 or, equivalently, after omitting columns (and rows) with all zero cells.

In general, the maximum value of T is $N(q-1)$, where q is either r or c, whichever is smaller, and N is the total number of observations. Division of T

by its maximum gives

$$(3) \qquad R_1 = \frac{T}{N(q-1)}$$

where q is the smaller of r and c. R_1 is close to 1.0 if the table indicates a strong row-column dependence and close to 0 if the numbers across each row are in the same proportions to each other as the column totals are to each other. This measure was suggested by Cramér (1946, p. 443).

Cramér's Contingency Coefficient

Example 2. In the previous example the 2×4 contingency table furnished a value of $T = 17.3$. Because $N = 128$ and $q = 2$, we have R_1 given by

$$R_1 = \frac{T}{N(q-1)} = \frac{17.3}{128}$$
$$= .135$$

This rather small value of R_1 might be interpreted as being indicative of little or no dependence, but such an interpretation is incorrect, as we have already seen in the previous example. It is true that the measure of dependence is slight when compared with the possibility of total dependence, but it is large when compared with the possibility of no dependence. In general R_1 has the desirable feature of being between 0 and 1.0 at all times, but it has the undesirable feature of depending on r and c for its interpretation. The larger r and c are, the larger T tends to be, and division by $(q-1)$ only partially offsets this tendency.

Two other coefficients are sometimes used. The first is called *Pearson's coefficient of mean square contingency* by Yule and Kendall (1950, p. 53) and is given as

$$(4) \qquad R_2 = \sqrt{\frac{T}{N+T}}$$

We stated that the maximum value of T is $N(q-1)$, and so the maximum value of R_2 is

$$(5) \qquad R_2(\text{max}) = \sqrt{\frac{N(q-1)}{N+N(q-1)}} = \sqrt{\frac{q-1}{q}}$$

which is close to one in many cases. The smallest possible value of T is zero, and so

$$(6) \qquad 0 \le R_2 \le \sqrt{\frac{q-1}{q}} < 1.0$$

R_2 is also called the *contingency coefficient* by McNemar (1962, p. 198) and Siegel (1956, p. 196).

Pearson's Contingency Coefficient

Example 3. In the contingency table of the two previous examples we have $T = 17.3$ and $N = 128$, so

$$R_2 = \sqrt{\frac{T}{N+T}} = \sqrt{\frac{17.3}{128 + 17.3}}$$
$$= .345$$

We present a third measure of dependence, R_3, also atributed to Pearson (by Cramér, 1946, p. 282) and also called the *mean-square contingency* (by Yule and Kendall, 1950, p. 53). R_3 is defined as

(7) $$R_3 = \frac{T}{N}$$

From the preceding discussions we may conclude that

$$0 \le R_3 \le q - 1$$

and that knowledge of r and c is necessary in order to interpret accurately the degree of dependence from the value of R_3.

Pearson's Mean-Square Contingency Coefficient

Example 4. For the same contingency table used in the previous example we have

$$R_3 = \frac{17.3}{128} = .135$$

Finally, we just mention *Tschuprow's coefficient*, given by Yule and Kendall (1950), as

(8) $$R_4 = \sqrt{\frac{T}{N\sqrt{(r-1)(c-1)}}}$$

The choice of a measure of dependence is largely a personal decision, motivated primarily by local traditions instead by statistical considerations. See Stuart (1953) for further discussion.

For the 2×2 contingency table,

	Column 1	2	
Row 1	a	b	r_1
Row 2	c	d	r_2
	c_1	c_2	N

the preceding measures simplify somewhat. We know from previous sections that T reduces to

(9)
$$T = \frac{N(ad - bc)^2}{r_1 r_2 c_1 c_2}$$

Therefore R_3 and R_1 (because $q = 2$) reduce to

(10)
$$R_3 = \frac{T}{N} = \frac{(ad - bc)^2}{r_1 r_2 c_1 c_2}$$

and

(11)
$$R_1 = \frac{T}{N(q-1)} = \frac{T}{N} = \frac{(ad - bc)^2}{r_1 r_2 c_1 c_2} = R_3$$

R_2 may be written as

(12)
$$R_2 = \sqrt{\frac{T}{N+T}} = \sqrt{\frac{(ad - bc)^2}{r_1 r_2 c_1 c_2 + (ad - bc)^2}}$$

In a fourfold contingency table, unlike the general $r \times c$ contingency table, it is sometimes meaningful to distinguish between a positive association and a negative association, such as when the two criteria of classification have corresponding categories.

Example 5. Forty children are classified according to whether their mothers have dark hair or light hair and as to whether their fathers have dark or light hair. The results may show a positive association (positive correlation)

		Father		
		Dark	Light	
Mother	Dark	28	0	28
	Light	5	7	12
		33	7	40

or a negative association (negative correlation)

		Father		
		Dark	Light	
Mother	Dark	21	7	28
	Light	12	0	12
		33	7	40

according to whether $(ad - bc)$ is positive or negative. A lack of association (zero correlation) is indicated by the following.

		Father		
		Dark	Light	
Mother	Dark	23	5	28
	Light	10	2	12
		33	7	40

If the type of association is of interest, care must be taken to set up the table so that a and d represent the numbers of similar classifications (dark-dark, and light-light), while b and c represent the numbers of unlike classifications (dark-light and light-dark). One measure of association that preserves direction is the *phi coefficient*, given by

$$(13) \qquad R_5 = \frac{ad - bc}{\sqrt{r_1 r_2 c_1 c_2}}$$

which may vary from $+1$, when all items are classified in the "alike" cells (both b and c equal zero), to -1, when all items are classified as "unlike" (both a and d equal zero). The phi coefficient is merely the square root of R_3 (see Equation 10), with the sign of $(ad - bc)$ being preserved. One reason for the popularity of the phi coefficient is because it is a special case of the *Pearson product moment correlation coefficient* (presented in the next chapter), computed by representing the classes by numbers.

The Phi Coefficient

Example 6. For the first table in Example 5 we have

$$
\begin{aligned}
a &= 28 & r_1 &= 28 \\
b &= 0 & r_2 &= 12 \\
c &= 5 & c_1 &= 33 \\
d &= 7 & c_2 &= 7
\end{aligned}
$$

so that R_5 is computed as

$$(14) \qquad R_5 = \frac{ad - bc}{\sqrt{r_1 r_2 c_1 c_2}} = \frac{(28)(7) - 0}{\sqrt{(28)(12)(33)(7)}}$$
$$= .703$$

For the second table in Example 5,

$$R_5 = \frac{(21)(0) - (7)(12)}{\sqrt{(28)(12)(33)(7)}}$$

(15) $= -.302$

which reflects the negative association of hairtypes.

Other measures of association for the four-fold contingency table include one proposed by Yule and Kendall (1950, p. 30)

(16) $$R_6 = \frac{ad - bc}{ad + bc}$$

and one proposed by Ives and Gibbons (1967)

(17) $$R_7 = \frac{(a+d) - (b+c)}{a+b+c+d}$$

There is no end to the possible measures that may be defined. One's choice of a coefficient is solely a result of personal preferences.

Sometimes the question arises, "How can I test the null hypothesis of independence, using R_1 (or R_2, R_3, etc) as a test statistic?" The answer is that you can find the exact small sample distribution of any of these measures in the same laborious manner that the exact small sample distribution of T was found in Section 4.2. Therefore, theoretically a test may be devised. But it is much easier, and just as effective, to use the tests presented in Sections 4.1 and 4.2 for the same hypotheses.

In particular, the coefficients

$$R_1 = \frac{T}{N(q-1)}$$

$$R_2 = \sqrt{\frac{T}{N+T}}$$

$$R_3 = \frac{T}{N}$$

and

$$R_4 = \sqrt{\frac{T}{N\sqrt{(r-1)(c-1)}}}$$

will all be "too large" whenever T is "too large," because they increase or decrease when T increases or decreases. The tests of Section 4.2 use T as a test statistic, and when T is significant we may conclude that R is significant.

A one-tailed test, appropriate only for the 2×2 contingency table, may be based on the phi coefficient R_5. Because of the relationship

(18)
$$R_5 = \pm \sqrt{\frac{T}{N}}$$

where the sign is determined by $(ad - bc)$, we may conclude that

(19)
$$\pm \sqrt{T} = \sqrt{N} \cdot R_5$$

and therefore $\sqrt{N} \cdot R_5$ is approximately normally distributed. This is because T is approximately chi-square distributed with 1 degree of freedom (see Theorem 1.5.3). Then the null hypothesis

(20) H_0: There is no positive correlation

may be rejected if $\sqrt{N} \cdot R_5$ is too large (exceeds $x_{1-\alpha}$ from Table A1, for a level of α), and the null hypothesis

(21) H_0: There is no negative correlation

may be rejected if $\sqrt{N} \cdot R_5$ is too small (smaller than x_α from Table A1, for a level of α).

A one-tailed test based on the phi coefficient is illustrated in the following example.

Example 7. In order to see if seat belts help prevent fatalities, records of the last 100 automobile accidents to occur along a particular highway were examined. These 100 accidents involved 242 persons. Each person was classified as using or not using seat belts when the accident occurred and as injured fatally or a survivor.

		Injured Fatally?		
		Yes	No	Totals
Wearing Seat Belts?	Yes	7	89	96
	No	24	122	146
	Totals	31	211	242

The statement we wish to prove, "Seat belts help prevent fatalities," becomes the alternative hypothesis. The null hypothesis may be stated as

       H_0: Seat belts do not help prevent fatalities

or, more correctly, as

H_0: There is no negative correlation between wearing seat belts and being killed in an automobile accident

This example is slightly different than the situation we have been describing in that the two criteria of classification do not have the same corresponding classes. ("Yes" and "no" mean one thing in the rows and another in the columns.) Therefore we need to stop and assess the situation. If H_1 were true we would expect b and c to be larger than a and d. Therefore the inequality

$$ab - bc < 0$$

4.4. Measures of Dependence   **187**

tends to support H_1 and also causes R_5 to be negative. So we reject H_0 if $\sqrt{N}\cdot R_5$ is less than -1.645, the .05 quantile of the standard normal distribution given in Table A1. In this example the test statistic,

$$\sqrt{N}\cdot R_5 = \frac{\sqrt{N}(ad-bc)}{\sqrt{r_1 r_2 c_1 c_2}}$$

$$= \frac{\sqrt{242}[(7)(122)-(89)(24)]}{\sqrt{(96)(146)(31)(211)}}$$

$$= -2.08$$

is less than -1.645, so H_0 is rejected. We may conclude that the use of seat belts is associated with fewer fatalities. (Whether the relationship is a causal relationship, stated by H_1, remains an open question.) The critical level is found from Table A1 to be about .02.

Several other measures of association between variables that are classified according to two criteria are introduced in classical papers by Goodman and Kruskal (1954, 1959, 1963). A partial coefficient is introduced by Davis (1967).

EXERCISES

1. One hundred married couples were interviewed, and the husband and wife were asked separately for their first choice for the next U.S. president, with the following results.

		Wife's Choice		
		A	B	Other
Husband's Choice	A	12	22	6
	B	25	21	4
	Other	3	7	0

Compute the following.

(a) T (b) $\hat{\alpha}$
(c) R_1 (d) R_2
(e) R_3 (f) R_4

2. Fifty factory workers reported to the nurse complaining of soreness due to arthritis. Twenty-five of them were given aspirin, and the rest were given a placebo without their knowledge. One hour later each was asked if the pill they took helped them to feel better. Seventeen in the aspirin group and 12 in the placebo group said it did.
(a) Use R_5 to see if there is a positive correlation between taking aspirin and feeling better.

(b) Compute R_6.

(c) Compute R_7.

3. A traffic study was conducted for a short time on a well-traveled city street. Of the 64 cars observed, 16 were exceeding the limit and 48 were not. Also, 24 of the drivers had passengers and the rest did not. Twelve of the speeders were driving alone. Assume that the observed traffic behaves the same as a random sample of all traffic would.

(a) Use R_5 to see if there is a positive correlation between speeding and driving alone.

(b) Compute R_6.

(c) Compute R_7.

4. A certain type of insect that is found in lakes in the southwestern United States is studied to see if the chromosomal structure is significantly different among states. The number of insects of various chromosomal types are recorded as follows.

Type	Texas	New Mexico	Arizona	California
A	54	72	83	96
B	20	6	18	6
C	17	3	12	0
D	0	12	14	1
E	0	10	0	0

Compute the following.

(a) T	(b) $\hat{\alpha}$
(c) R_1	(d) R_2
(e) R_3	(f) R_4

PROBLEMS

1. Show that T equals $N(q-1)$ for the following contingency table (here $r < c$).

	1	2	$\cdots$	r	$r+1$	$\cdots$	c
1	$\dfrac{N}{r}$	0	$\cdots$	0	0	$\cdots$	0
2	0	$\dfrac{N}{r}$	$\cdots$	0	0	$\cdots$	0
$\cdots$	$\cdots$	$\cdots$	$\cdots$	$\cdots$	$\cdots$	$\cdots$	$\cdots$
r	0	0	$\cdots$	$\dfrac{N}{r}$	0	$\cdots$	0

2. Think of another $r \times c$ contingency table, other than the one in Problem 1, that you would suspect of having a large value of T. Compute T for your contingency table. Is it greater than $N(q-1)$?

3. Prove the identities:

(a) $R_2{}^2 = \dfrac{R_3}{1+R_3}$.

(b) $R_3 = \dfrac{R_2{}^2}{1-R_2{}^2}$.

(c) When $r = 2$ and $c = 2$, $R_3 = R_5{}^2 = R_1$.

4. Show that the phi coefficient is merely Pearson's product moment correlation coefficient

$$ r = \frac{\sum\limits_{i=1}^{N} (X_i - \bar{X})(Y_i - \bar{Y})}{\left[\sum\limits_{i=1}^{N} (X_i - \bar{X})^2 \sum\limits_{j=1}^{N} (Y_j - \bar{Y})^2 \right]^{1/2}} $$

computed on the pairs (X_i, Y_i) where (X_i, Y_i) equals $(0, 0)$, $(0, 1)$, $(1, 0)$, or $(1, 1)$, depending on which cell each observation belongs to. Then show that the same result holds true if the numbers 0 and 1 just used are replaced by any two numbers p and q.

4.5. THE CHI-SQUARE GOODNESS-OF-FIT TEST

Often the hypotheses being tested are statements concerning the unknown probability distribution of the random variable being observed. Examples include "The median is 4.0" and "The probability of being in class 1 is the same for both populations." More comprehensive hypotheses than the ones we have been examining would include "The unknown distribution function is the normal distribution function with mean 3.0 and variance 1.0" or "The distribution function of this random variable is the binomial, with parameters $n = 10$ and $p = .2$" These latter hypotheses are more comprehensive because they include statements concerning all of the quantiles simultaneously, rather than just the median, and all of the probabilities simultaneously instead of an isolated statement about some of the probabilities. Hypotheses such as these may be tested with a "goodness-of-fit" test, that is, with a test designed to compare the sample obtained with the type of sample one would expect from the hypothesized distribution to see if the hypothesized distribution function "fits" the data in the sample.

The oldest and best-known goodness-of-fit test is the chi-square test for goodness of fit, first presented by Pearson (1900).

The Chi-Square Test for Goodness of Fit

DATA. The data consist of N independent observations of a random variable X. These N observations are grouped into c classes, and the numbers of observations in each class are presented in the form of a $1 \times c$ contingency table.

	Class	1	2	$\cdots$	c	Total
Observed Frequencies		O_1	O_2	$\cdots$	O_c	N

Let O_j denote the number of observations in class j, for $j = 1, 2, \ldots, c$.

ASSUMPTIONS

1. The sample is a random sample.
2. The measurement scale is at least nominal.

HYPOTHESES. Let $F(x)$ be the true but unknown distribution function of X, and let $F^*(x)$ be some completely specified distribution function, the hypothesized distribution function. [If $F^*(x)$ is completely specified except for several unknown parameters that must be estimated from the sample, this test requires a slight modification, as described later in the Comment.]

$$H_0: F(x) = F^*(x) \quad \text{for all } x$$
$$H_1: F(x) \neq F^*(x) \quad \text{for at least one } x$$

The hypotheses may be stated in words.

H_0: The distribution function of the observed random variable is $F^*(x)$
H_1: The distribution function of the observed random variable is different than $F^*(x)$

TEST STATISTIC. Let p_j^* be the probability of a random observation on X being in class j, under the assumption that $F^*(x)$ is the distribution function of X. Then define E_j as

$$(1) \qquad E_j = p_j^* N, \qquad j = 1, 2, \ldots, c$$

where E_j represents the expected number of observations in class j when H_0 is true. The test statistic T is given by

$$(2) \qquad T = \sum_{j=1}^{c} \frac{(O_j - E_j)^2}{E_j}$$

An equivalent expression, more convenient for use with a desk calculator, is

$$(3) \qquad T = \sum_{j=1}^{c} \frac{O_j^2}{E_j} - N$$

If some of the E_js are small, the asymptotic chi-square distribution (described in the following material) may not be appropriate, but just how small the E_js may become is not clear. While Cochran (1952) suggests that none of the E_j should be less than 1 and no more than 20% should be smaller than 5, more recent results indicate that this rule can be relaxed. Yarnold (1970) says "if the number of classes s is 3 or more, and if r denotes the number of expectations less than 5, then the minimum expectation may be as small as $5r/s$." Slakter (1973) feels that the number of classes can exceed the number of observations, which means the average expected value can be less than 1. The user may wish to combine some cells with this discussion in mind if many of the E_js are small.

DECISION RULE. The exact distribution of T is difficult to use, so the large sample approximation is used. The approximate distribution of T, valid for large samples, is the chi-square distribution with $(c-1)$ degrees of freedom. (If parameters are being estimated from the sample, fewer degrees of freedom may be used. See the next Comment.) Therefore the critical region of approximate size α corresponds to values of T greater than $x_{1-\alpha}$, the $(1-\alpha)$ quantile of a chi-square random variable with $(c-1)$ degrees of freedom, obtained from Table A2. Reject H_0 if T exceeds $x_{1-\alpha}$. Otherwise accept H_0.

We may always be quite sure that the true distribution function is never exactly the same as the hypothesized distribution function. However, in many cases we are looking for a good approximation to the true distribution function, and this test provides a means of justifying the use of $F^*(x)$ as the good approximation by the acceptance of H_0. We realize that in any goodness of fit test H_0 will be rejected if the sample size is large enough. For this reason T is often used as a *measure* of goodness of fit.

Example 1. The following is a random sample of size 20, after being ordered from smallest to largest.

16.7	18.8	24.0	35.1	39.8
17.4	19.3	24.7	35.8	42.1
18.1	22.4	25.9	36.5	43.2
18.2	22.5	27.0	37.6	46.2

We wish to test the null hypothesis

H_0: This random sample represents observations on a normally distributed random variable, with mean 30 and variance 100

against the alternative hypothesis

H_1: The distribution function is other than as described by H_0

We arbitrarily decide to form four classes with equal expected cell counts. Four classes are formed from the three quartiles $w_{.25}$, $w_{.50}$, and $w_{.75}$ of the standard normal distribution obtained from Table A1, which are converted to the quartiles x_p of the hypothesized normal distribution by applying

Theorem 1.5.1.

$$x_p = \mu + \sigma w_p$$

(4)
$$= 30 + 10 w_p$$

The quartiles x_p are

$$x_{.25} = 30 + 10(-.6745) = 23.255$$

$$x_{.50} = 30$$

$$x_{.75} = 36.745$$

Class 1 contains all observations less than or equal to 23.255. Class 2 contains those observations between 23.255 and 30 including 30. The other two classes are formed in the same way. The data may now be classified.

	Class 1 $(-\infty, 23.26]$	Class 2 $(23.26, 30]$	Class 3 $(30, 36.75]$	Class 4 $(36.75, \infty)$	Total
Observed Frequency	8	4	3	5	20
Expected Frequency	5	5	5	5	

The decision rule is to reject H_0 at $\alpha = .05$ if T exceeds 7.815, the .95 quantile of a chi-square random variable with $c - 1 = 3$ degrees of freedom, obtained from Table A2. The test statistic is computed using Equation 2.

$$T = \frac{(8-5)^2}{5} + \frac{(4-5)^2}{5} + \frac{(3-5)^2}{5} + \frac{(5-5)^2}{5}$$

(5)
$$= 2.8$$

The decision is to accept H_0. The critical level is well above .25.

The following is an example where $F^*(x)$ is a discrete distribution function.

Example 2. A certain computer program is supposed to furnish random digits. If the program is accomplishing its purpose, the computer prints out digits (2, 3, 7, 4, etc.) that seem to be observations on independent and identically distributed random variables, where each digit 0, 1, 2, . . . , 8, 9 is equally likely (probability 0.1) to be obtained. One way of testing

H_0: The numbers appear to be random digits

against the alternative of nonrandomness is by separating a long sequence of the digits into groups of size 10 and then counting the number of even digits in each group. The number of even digits in a group of 10 digits should have a binomial distribution with $n = 10$ and $p = .5$ (because the probability of each digit being even is 5/10) if H_0 is true.

Suppose the following digits are generated and grouped into blocks of 10 digits each.

1578748416	4705188926	6936349612
4653843213	0282868892	3928057043
5101259393	9837006785	3011679938
7122863085	6528271107	2956427027
2671728075	9759178719	9373309535
8363265100	2546793732	2212122529
9453087720	3976759377	9593511031
5605373242	1819898287	3872181027
3494768396	9296177240	8620774591
4659773922	9246724287	8326143939

In the first block there are 5 even digits (8, 4, 8, 4, 6). In the second block there are 6 even digits (4, 0, 8, 8, 2, 6). In each block the number of even digits is counted. The results are summarized in the following table.

The number of blocks containing j even digits, for j equal to

0	1	2	3	4	5	6	7	8	9	10	Total
0	4	1	1	3	14	5	1	0	1	0	30

Each block consists of 10 independent trials (digits) where each trial has probability .5 of resulting in "even" rather than "odd." Therefore, as in Example 1.3.5, the probability of obtaining exactly j even numbers is $\binom{10}{j}(1/2)^{10}$. Thirty repetitions of the experiment, represented by the 30 blocks, result in

$$(6) \qquad E_j = p_j^* N = 30 \binom{10}{j}\left(\frac{1}{2}\right)^{10}$$

from Equation 1. The values of p_j^* are obtained from Table A3, $n = 10$, $p = .5$, by subtracting successive entries to obtain the difference.

The expected number of blocks containing j even digits, for j equal to

	0	1	2	3	4	5	6	7	8	9	10
$F^*(j)$	.0010	.0107	.0547	.1719	.3770	.6230	.8281	.9453	.9893	.9990	1.0000
$p^*(j)$	.0010	.0097	.0440	.1172	.2051	.2460	.2051	.1172	.0440	.0097	.0010
Ej	.030	.291	1.320	3.516	6.153	7.330	6.153	3.516	1.320	.291	.030

Therefore, as a test of randomness, we are actually using a goodness of fit test where $F^*(j)$ is the binomial distribution with $n = 10$ and $p = .5$. If H_0 is true, the observed numbers should agree well with the E_j. If H_0 is not true, some types of nonrandomness will result in poor agreement with $F^*(x)$, while other types of nonrandomness will not be detectable using this test. No test for nonrandomness is consistent against all types of nonrandomness.

Because some of the E_js are small, the categories $j = 0$, 1, 2, and 3 are combined into one class, and so are the categories $j = 7$, 8, 9, and 10. The result is as follows.

Class:	$j \leq 3$	$j = 4$	$j = 5$	$j = 6$	$j \geq 7$	Total
O_j = Observed Number	6	3	14	5	2	30
E_j = Expected Number	5.157	6.153	7.380	6.153	5.157	30

Now all E_j exceed 5.0. The number of classes is now five, so the critical region of approximate size .05 corresponds to values of T greater than 9.488, the .95 quantile of a chi-square random variable with 4 degrees of freedom, obtained from Table 2. The test statistic T is computed using Equation 2.

$$T = \frac{(6-5.157)^2}{5.157} + \frac{(3-6.153)^2}{6.153} + \frac{(14-7.380)^2}{7.380} + \frac{(5-6.153)^2}{6.153}$$
$$+ \frac{(2-5.157)^2}{5.157}$$

(7) $= 9.84$

The observed value of T is greater than the critical value 9.488, so H_0 is rejected. The critical level $\hat{\alpha}$ is about .04. We conclude that the digits produced by the computer program do not seem to be random.

COMMENT. If $F^*(x)$ is completely specified except for a number k of parameters, it is first necessary to estimate the parameters and then to proceed with the test as just outlined. The only change is in the distribution of T, which now may be approximated using a chi-square distribution with $c - 1 - k$ degrees of freedom. That is, 1 degree of freedom is subtracted for each parameter estimated. However, subtraction of degrees of freedom is a privilege accorded only when the parameters are estimated in the proper manner. For example, in a goodness-of-fit test with four classes, H_0 is usually rejected (at $\alpha = .05$) if T exceeds 7.815 (see Table A2). However, if one parameter is estimated from the data before the test is applied, the hypothesized distribution has already been modified so that it will fit the data better. [This is true if the estimate is a "good" estimate. A poor estimate may be used deliberately to result in a poor fit, but then the goodness-of-fit test is no longer valid. Chase (1972) discusses the chi-square test when the parameters are estimated independently of the data.]

The goodness-of-fit test will then be more likely to result in acceptance of H_0; the test becomes conservative and therefore less powerful. We would like to enlarge the critical region so that α again becomes .05 and the test regains some (or all) of the power that was lost. If we subtract 1 degree of freedom, using 2 degrees of freedom instead of 3, the critical region is enlarged and H_0 is rejected if T exceeds 5.991 instead of 7.815 as before. The question is, "Are we justified in subtracting one degree of freedom, as we did?"

Cramér (1946, p. 424; or see Birnbaum, 1962, p. 258) shows that 1 degree of freedom may be subtracted for each parameter estimated by the *minimum chi-square method*. The minimum chi-square method simply involves using the value of the parameter that results in the smallest value of the test statistic, for the given observations. From a practical standpoint this means trying all possible values of the parameter, or all possible combinations if several parameters are unknown, computing a set of E_js and then T for each value of the parameter, and selecting the value of the parameter that results in the smallest T. However, such a procedure is impractical. Therefore Cramér also presents a more usable *modified minimum chi-square method*. Even that procedure is tedious, and so Cramér and Birnbaum, in their examples, actually use a modification of the modified minimum chi-square method, which asymptotically still permits subtracting 1 degree of freedom for each parameter estimated. The method eventually used consists of estimating the k unknown parameters by computing the first k sample moments of the grouped data. (Each observation is assumed to be at the midpoint of its class interval, where all intervals containing observations are of finite length.) Then the first k population moments are set equal to the first k sample moments of the grouped data, and the resulting k equations are solved simultaneously for the k unknown parameters. The following example should help to clarify the above procedure.

Example 3. Fifty two-digit numbers were drawn at random from a telephone book, and the chi-square test for goodness of fit is used to see if they could have been observations on a normally distributed random variable. The numbers, after being arranged in order from the smallest to the largest, are as follows.

23	23	24	27	29	31	32	33	33	35
36	37	40	42	43	43	44	45	48	48
54	54	56	57	57	58	58	58	58	59
61	61	62	63	64	65	66	68	68	70
73	73	74	75	77	81	87	89	93	97

The null hypothesis is

H_0: These numbers are observations on a normally distributed random variable

The normal distribution has two parameters (Definition 1.5.3), both of which are unspecified by H_0, and must be estimated before the goodness of fit test may be applied. For illustration, the procedure is divided into steps.

Step 1. Divide the observations into intervals of finite length. We arbitrarily choose the intervals 20 to 40, 40 to 60, 60 to 80, and 80 to 100, not including the upper limit of each interval.

Interval	20 to 40	40 to 60	60 to 80	80 to 100	Total
Number of Observations	12	18	15	5	50

Step 2. Estimate μ and σ with the sample mean $\bar{X}$ and sample standard deviation S of the grouped data. The 12 observations in the interval 20 to 40 are treated as if they all equal the middle point 30. The 18 observations from 40 to 60 are all considered to be 50, and so on. These are the numbers used for computing $\bar{X}$ and S, using the equations of Definition 2.2.3.

$$\bar{X} = \frac{1}{N} \sum_{i=1}^{N} X_i$$

$$= \frac{1}{50}[12(30) + 18(50) + 15(70) + 5(90)]$$

(8) $$= 55.2$$

$$S = \sqrt{S^2} = \sqrt{\frac{1}{N} \sum_{i=1}^{N} X_i^2 - \bar{X}^2}$$

$$= \{\frac{1}{50}[12(30)^2 + 18(50)^2 + 15(70)^2 + 5(90)^2] - (55.2)^2\}^{\frac{1}{2}}$$

(9) $$= 18.7$$

Therefore our estimates of μ and σ are 55.2 and 18.7, respectively.

Step 3. Using the estimated parameters from Step 2, compute the E_js for the groups in Step 1 and for the "tails."

Class Boundaries b_j	$(b_j - \bar{X})/S = x_p$	$F(x_p)$	Interval	$p_j{}^*$
$b_1 = 20$	-1.88	.03	<20	.03
$b_2 = 40$	$-.813$	.21	20 to 40	.18
$b_3 = 60$	$+.256$	.60	40 to 60	.39
$b_4 = 80$	$+1.33$	.91	60 to 80	.31
$b_5 = 100$	$+2.40$	.99	80 to 100	.08
			≥ 100	.01

To find the hypothesized probabilities of being in the various classes, when the hypothesized distribution is the normal distribution with mean 55.2 and standard deviation 18.7, the class boundaries (column 1 in the table) are considered to be the quantiles of the hypothesized distribution. These quantiles are converted to quantiles of a standard normal random variable (column 2) by Equation 1.5.3 in order to find out *which* quantile the boundries represent (column 3). Subtraction of the items in column 3 then yields the probabilities $p_j{}^*$ of being in the various intervals under the hypothesized distribution. The E_js equal $50p_j{}^*$, from Equation 1, and are given below.

Class	<20	20–40	40–60	60–80	80–100	≥ 100
Expected Number E_j	1.5	9.0	19.5	15.5	4	0.5
Observed Number O_j	0	12	18	15	5	0

Because of the small E_js, the first and last cells are combined with the cells adjacent to them.

Class	<40	40–60	60–80	≥80
Expected Number E_j	10.5	19.5	15.5	4.5
Observed Number O_j	12	18	15	5

Step 4. Compute T. The test statistic is now computed using Equation 2.

$$T = \frac{(12-10.5)^2}{10.5} + \frac{(18-19.5)^2}{19.5} + \frac{(15-15.5)^2}{15.5} + \frac{(5-4.5)^2}{5}$$

(10) $= .395$

The critical region of size .05 corresponds to values of T greater than 3.841, the .95 quantile of a chi-square random variable with $c - 1 - k = 4 - 1 - 2 = 1$ degree of freedom. Therefore H_0 is accepted, with a critical level $\hat{\alpha}$ well above .25.

Usually a modification called Sheppard's correction is used when the variance is being estimated from grouped data and when the interior intervals are of equal width, say h. Sheppard's correction consists of subtracting $h^2/12$ from S^2 in order to obtain a better estimate of variance. In this example $h = 20$ (the width of each interval), so $(20)^2/12 = 33.33$ could have been subtracted in Step 2 before extracting the square root. The result is $S = 17.8$, a smaller estimate for σ. This smaller estimate of σ results in a larger value of T in this example and, since our objective is to obtain the smallest possible value for T, the correction was not used. In most situations we can expect a smaller T when the correction is used.

Another peculiarity of this example is the fact that a smaller value of T (.279) may be obtained by using $\bar{X} = 55.04$ and $S = 19.0$ as estimates of μ and σ. These estimates are the sample moments obtained from the original observations, before grouping. No matter how they are obtained, the estimates to use are the estimates that result in the smallest value of T. The procedure described in this example can be relied on to provide a value of T not far from its minimum value in most cases. Therefore it is the recommended procedure.

In the preceding example the test statistic just happened to be smaller when μ and σ were estimated using the sample mean and standard deviation based on the original, rather than the grouped, observations. This procedure may be used and is even recommended by Yule and Kendall (1950), but it is usually inferior in results to the other method, which uses the grouped data (Chernoff and Lehmann, 1954).

Theory. If the null hypothesis completely specifies the hypothesized distribution function, once the classes are defined there is a known probability p_i

associated with each class if H_0 is true. The probability of any particular arrangement $O_1, O_2, \ldots, O_c$ of the N sample values is then given by the multinomial probability distribution,

$$(11) \qquad P(O_1, O_2, \ldots, O_c \mid N) = \frac{N!}{O_1! O_2! \cdots O_c!} p_1^{O_1} p_2^{O_2} \cdots p_c^{O_c}$$

which is an immediate extension of the binomial distribution to c classes instead of only two classes. With the probability function given by Equation 11 the probability distribution of T may be determined, although the calculation becomes laborious for large N and c. There seems to be no theory developed to find the exact distribution of T when several parameters are first estimated from the sample. Therefore the large sample approximation is both practical and necessary in order to apply this goodness-of-fit test. The theory behind the large sample chi-square approximation may be found in Cramér (1946).

Usage of the chi-square goodness-of-fit test with small expected frequencies is discussed by Slakter (1966, 1968), Dahiya (1971), Dahiya and Gurland (1972, 1973), and Pahl (1969). Exact tables when all $E_i s = 1$ appear in Zahn and Roberts (1971). If the sample is grouped according to time of observation instead of numerical value, the usage of the test may require some modification (Putter, 1964). Chernoff (1967) discusses the adjustment of the degrees of freedom when parameters are estimated. The test is discussed further by Molinari (1977) and Hewett and Tsutakawa (1972) and compared with other goodness-of-fit tests by Holst (1972), Cohen and Sackrowitz (1975), and Horn (1977).

EXERCISES

1. Test the following data to see if they could have come from a population whose values are uniformly distributed between .0000 and .9999.

.4755	.5233	.5440	.5456	.9056
.2186	.7500	.2484	.5101	.8283
.5112	.5484	.5758	.3607	.4352
.3826	.6454	.9145	.3943	.5381
.5758	.8620	.6687	.3979	.5646
.4274	.5482	.3007	.4438	.4102
.4295	.5926	.6521	.6328	.5689
.7297	.3768	.8403	.2925	.2113
.8757	.4403	.4993	.3900	.5166
.8230	.8522	.8312	.7979	.4632
.8432	.4004	.4295	.9763	.5590
.4396	.2595	.3003	.3003	.5836
.5337	.8008	.4887	.2172	.9329
.5498	.3686	.4067	.5274	.4579
.9096	.4995	.2172	.6793	

2. A die was cast 600 times with the following results.

Occurrence	1	2	3	4	5	6
Frequency	87	96	108	89	122	98

Is the die balanced?

3. Test the data in Exercise 1 to see if they might have come from a normal population.

4. Without the aid of books or tables, attempt to write 300 *random* digits. Then apply the test of randomness described in Example 2 to see if you are a good random digit generator.

4.6. COCHRAN'S TEST FOR RELATED OBSERVATIONS

Sometimes the use of a treatment, or condition, results in one of two possible outcomes. For example, the response to a salesperson's technique may be classified as "sale" or "no sale," or a certain treatment may result in "success" or "failure." Of course, if the several treatments, c in number, are each applied in several different independent trials, the results may be given in the form of a $2 \times c$ contingency table, where one row represents the number of successes and the other row represents the number of failures, and the null hypothesis of no treatment differences may be tested using a chi-square contingency table test, as described in Section 4.2. However, it is often possible to detect more subtle differences between treatments, that is, increase the power of the test, by applying all c treatments independently to the same blocks, such as by trying all c sales techniques on each of several persons in an experimental situation and then recording for each person the results of each technique. Thus each block, or person, acts as its own control, and the treatments are more effectively compared with each other. Such an experimental technique is called "blocking," and the experimental design is called a "randomized complete block design." If the treatment result may be classified into one of two categories, the following test, proposed by Cochran (1950), may be an appropriate method of analysis.

The Cochran Test

DATA. Each of c treatments is applied independently to each of r blocks, or subjects, and the result of each treatment application is recorded as either 1 or 0, to represent "success" or "failure," or any other dichotomization of the possible treatment results. The results are then given in the form of a table with r rows representing the blocks and c columns representing the c treatments, with entries that are either zeros or ones. Let R_i represent the row totals, $i = 1, 2, \ldots, r$, and let C_j represent the column totals, $j = 1, 2, \ldots, c$. Then the data appear as follows, where the X_{ij} are either 0 or 1, and N represents the total number of ones in the table.

Treatment Blocks	1	2		c	*Row Totals*
1	X_{11}	X_{12}	$\cdots$	X_{1c}	R_1
2	X_{21}	X_{22}	$\cdots$	X_{2c}	R_2
$\cdots$	$\cdots$	$\cdots$	$\cdots$	$\cdots$	$\cdots$
r	X_{r1}	X_{r2}	$\cdots$	X_{rc}	R_r
Column Totals	C_1	C_2		C_c	$N=$ Grand Total

ASSUMPTIONS

1. The blocks were randomly selected from the populations of all possible blocks.
2. The outcomes of the treatments may be dichotomized in a manner common to all treatments within each block, so the outcomes are listed as either "0" or "1."

HYPOTHESES

H_0: The treatments are equally effective
H_1: There is a difference in effectiveness among treatments

In more mathematical terms, let

(1) $$p_{ij} = P(X_{ij} = 1); \qquad i = 1, \ldots, r; \qquad j = 1, \ldots, c$$

Then equal effectiveness among treatments implies

(2) $$p_{i1} = p_{i2} = \cdots = p_{ic}, \qquad \text{for each } i \text{ from 1 to } r$$

That is, for each block the probability of a treatment being a success does not depend on which treatment is being used. Then the hypotheses may be restated as follows.

H_0: $p_{i1} = p_{i2} = \cdots = p_{ic}$, for each i from 1 to r
H_1: $p_{ij} \neq p_{ik}$ for some j and k, and for some i

TEST STATISTIC. The test statistic T may be written as

(3) $$T = \sum_{j=1}^{c} \frac{c(c-1)\left(C_j - \dfrac{N}{c}\right)^2}{\sum\limits_{i=1}^{r} R_i(c - R_i)}$$

or, equivalently, as

(4) $$T = c(c-1) \frac{\sum\limits_{j=1}^{c} \left(C_j - \dfrac{N}{c}\right)^2}{\sum\limits_{i=1}^{r} R_i(c - R_i)}$$

The following form is more suitable for machine computation.

(5)
$$T = \frac{c(c-1) \sum\limits_{j=1}^{c} C_j^2 - (c-1)N^2}{cN - \sum\limits_{i=1}^{r} R_i^2}$$

DECISION RULE. The exact distribution of T is difficult to tabulate, so the large sample approximation is used instead. The number of blocks r is assumed to be large. Then the critical region of approximate size α corresponds to all values of T greater than $x_{1-\alpha}$, the $(1-\alpha)$ quantile of a chi-square random variable with $(c-1)$ degrees of freedom, obtained from Table A2. If T exceeds $x_{1-\alpha}$, reject H_0. Otherwise accept the null hypothesis of no differences in the effectiveness of the various treatments.

Example 1. Each of three basketball enthusiasts had devised his own system for predicting the outcomes of collegiate basketball games. Twelve games were selected at random, and each sportsman presented a prediction of the outcome of each game. After the games were played, the results were tabulated, using 1 for successful prediction and 0 for unsuccessful prediction.

	Sportsman			
Game	1	2	3	Totals
1	1	1	1	3
2	1	1	1	3
3	0	1	0	1
4	1	1	0	2
5	0	0	0	0
6	1	1	1	3
7	1	1	1	3
8	1	1	0	2
9	0	0	1	1
10	0	1	0	1
11	1	1	1	3
12	1	1	1	3
Totals	8	10	7	25

The assumptions of the Cochran test were met, because the games (blocks) were selected at random from among all college basketball games being played. Therefore the Cochran test was used to test the null hypothesis

H_0: Each sportsman is equally effective in his ability to predict the outcomes of the basketball games

The test statistic is computed using Equation 4.

$$T = c(c-1) \frac{\sum\limits_{j=1}^{c} \left(C_j - \frac{N}{c}\right)^2}{\sum\limits_{i=1}^{r} R_i(c - R_i)}$$

$$= \frac{(3)(2)[(-\frac{1}{3})^2 + (\frac{5}{3})^2 + (-\frac{4}{3})^2]}{2+2+2+2+2}$$

(6) $\qquad = 2.8$

The critical region of approximate size .05 corresponds to values of T greater than 5.99, the .95 quantile of a chi-square random variable with 2 degrees of freedom obtained from Table A2. Therefore, in this example H_0 is accepted and we conclude that no significant differences among prediction methods were detected. H_0 could have been rejected using an α of about .25, so $\hat{\alpha} = .25$.

☐ *Theory.* Each X_{ij}, as defined, is a random variable that follows the point binomial distribution (the binomial distribution with $n = 1$) with parameter p_{ij}. The column total, C_j, defined by

(7) $\qquad\qquad C_j = \sum\limits_{i=1}^{r} X_{ij}$

is therefore a random variable also. If the p_{ij} in each column were the same, C_j would follow a binomial distribution, but even if H_0 is assumed to be true the p_{ij} down the column may differ from each other. The hypothesis merely states that the p_{ij}s across each row are equal to each other but may vary from one block (row) to another. However, because the random variable C_j is the sum of r independent random variables, the central limit theorem still applies, and for large r the distribution of C_j is approximately normal. This implies that the distribution function of

$$\frac{C_j - E(C_j)}{\sqrt{\text{Var}(C_j)}}$$

may be approximated by the standard normal distribution function, and, according to Theorem 1.5.3, the sum

(8) $\qquad \sum\limits_{j=1}^{c} \left[\frac{C_j - E(C_j)}{\sqrt{\text{Var}(C_j)}}\right]^2 = \sum\limits_{j=1}^{c} \frac{[C_j - E(C_j)]^2}{\text{Var}(C_j)}$

may be approximated by the chi-square distribution with c degrees of freedom. However, the parameters $E(C_j)$ and $\text{Var}(C_j)$ are unknown, and the following method of estimating those parameters is shown by Blomqvist (1951) to result in the loss of 1 degree of freedom.

The mean of C_j may be estimated by the sample mean

(9)
$$\frac{1}{c} \sum_{j=1}^{c} C_j = \frac{N}{c} = \text{estimate of } E(C_j)$$

The same estimate is used for the mean of every C_j. The variance of C_j equals the sum of the variances of the X_{ij} in the jth column,

(10)
$$\text{Var}(C_j) = \sum_{i=1}^{r} \text{Var}(X_{ij})$$

because of the block to block independence of the X_{ij} (see also Theorem 1.4.3). The variance of X_{ij} is given by Equation 1.4.8 as

(11)
$$\text{Var}(X_{ij}) = p_{ij}(1 - p_{ij})$$

Under H_0 the probability of a "success" p_{ij} is the same for all columns within a block, and therefore it is natural to estimate p_{ij} by the average number of successes in row i, R_i/c. That is,

(12)
$$\text{estimate of } p_{ij} = R_i/c$$

and, from Equation 11,

(13)
$$\text{estimate of Var}(X_{ij}) = \frac{R_i}{c}\left(1 - \frac{R_i}{c}\right)$$

However, such an estimate tends to be too small and is improved by multiplication by $c/(c-1)$. Then $\text{Var}(X_{ij})$ is estimated by

(14)
$$\text{estimate of Var}(X_{ij}) = \frac{R_i(c - R_i)}{c(c-1)}$$

and $\text{Var}(C_j)$ is estimated, from Equation 10, as

(15)
$$\text{estimate of Var}(C_j) = \frac{1}{c(c-1)} \sum_{i=1}^{r} R_i(c - R_i)$$

which does not depend on j and so is used for all C_js. Substitution of the estimates for $E(C_j)$, Equation 9, and $\text{Var}(C_j)$, Equation 15, into Equation 8 gives

$$T = \sum_{j=1}^{c} \frac{\left(C_j - \dfrac{N}{c}\right)^2}{\displaystyle\sum_{i=1}^{r} \dfrac{R_i(c - R_i)}{c(c-1)}}$$

(16)
$$= c(c-1) \frac{\displaystyle\sum_{j=1}^{c}\left(C_j - \dfrac{N}{c}\right)^2}{\displaystyle\sum_{i=1}^{r} R_i(c - R_i)}$$

which provides some insight into the use of the chi-square distribution with $(c-1)$ degrees of freedom for the test statistic T.

Cochran's test is considered by Berger and Gold (1973) and Bhapkar and Somes (1977). Patil (1975) discusses the exact distribution of the test statistic. For another model in which Cochran's test is valid, see Fleiss (1965). The asymptotic approximation is considered by Tate and Brown (1970).

COMMENT. If only two treatments are being considered, such as "before" and "after" observations on the same block, with r blocks, the experimental situation is the same as that analyzed by the McNemar test for significance of changes. That is, in each situation the null hypothesis is that the proportion of the population in class one is the same using treatment 1 (before) as it is using treatment 2 (after). Thus it appears that if $c = 2$, the experimenter has a choice of using the Cochran test or the McNemar test. In fact, there is no choice because if $c = 2$, the Cochran test is identical with the McNemar test (Section 3.5), as shown in the following.

For $c = 2$ the Cochran test statistic reduces to

$$T = 2\,\frac{\left(C_1 - \dfrac{C_1 + C_2}{2}\right)^2 + \left(C_2 - \dfrac{C_1 + C_2}{2}\right)^2}{\displaystyle\sum_{i=1}^{r} R_i(2 - R_i)}$$

$$= 2\,\frac{\left(\dfrac{C_1}{2} - \dfrac{C_2}{2}\right)^2 + \left(\dfrac{C_2}{2} - \dfrac{C_1}{2}\right)^2}{\displaystyle\sum_{i=1}^{r} R_i(2 - R_i)}$$

(17)
$$= \frac{(C_1 - C_2)^2}{\displaystyle\sum_{i=1}^{r} R_i(2 - R_i)}$$

If a block has ones in both columns, then $R_i = 2$ and $R_i(2 - R_i) = 0$. Similarly, if both columns have zeros, then $R_i = 0$ and $R_i(2 - R_i) = 0$. If there is a change from zero to one or one to zero in a given row, then $R_i = 1$ and $R_i(2 - R_i) = 1$. Thus the denominator of Equation 17 is merely the total number of rows that go from 0 to 1 or 1 to 0, which is $b + c$ in the notation of the McNemar test. Also, C_1 is the total number of ones in column one, or "before," which is $c + d$ in the notation of the McNemar test. Similarly, $C_2 = b + d$. Therefore we have

$$C_1 - C_2 = c + d - b - d$$
$$= c - b$$

and Equation 17 becomes

$$T = \frac{(c - b)^2}{b + c} = \frac{(b - c)^2}{b + c}$$

which is identical with the form of McNemar's test statistic given in Equation 3.5.1. Both the McNemar test statistic and the Cochran test statistic with $c = 2$ are approximated by a chi-square random variable with 1 degree of freedom.

EXERCISES

1. The relative effectiveness of two different sales techniques was tested on 12 volunteer housewives. Each housewife was exposed to each sales technique and asked to buy a certain product, the same product in all cases. At the end of each exposure, each housewife rated the technique with a 1 if she felt she would have agreed to buy the product and a 0 if she probably would not have bought the product.

								Housewife					
		1	2	3	4	5	6	7	8	9	10	11	12
Technique 1		1	1	1	1	1	0	0	0	1	1	0	1
Technique 2		0	1	1	0	0	0	0	0	1	0	0	1

(a) Use Cochran's test.
(b) Rearrange the data and use McNemar's test in the large sample form suggested by Equation 3.5.1.
(c) Ignore the blocking effect in this experiment and treat the data as if 24 different housewives were used. Analyze the data using the test for differences in probabilities given in Section 4.1.

2. On a ship, 12 groups with three sailors in each group were chosen in a random manner, where the sailors in each group did similar work and were in the same division aboard ship. In a random manner the sailors in each group were given treatment 1, 2, or 3, no two sailors from the same group receiving the same treatment. Treatment 1 was a "flu shot," treatment 2 was a "flu pill," and treatment 3 was a promise of 2 weeks extra leave if they did not catch the flu. As each sailor reported to sick bay with the flu, a report to the experimenter was made. At the end of the winter, these were the results.

Group	Sailors with the Flu (by Treatment Number)
1	2
2	1, 2
3	1, 2, 3
4	2, 3
5	2
6	None
7	1, 2
8	1, 2
9	1
10	2
11	1, 2, 3
12	2

Do these results indicate a significant difference between the various treatments?

3. In an attempt to compare the relative power of three statistical tests, 100 sets of artificial data were generated using a computer. On each set of data the three statistical tests were used, with $\alpha = .05$, and the decision to accept or reject H_0 was recorded. The results were as follows.

Test 1	Test 2	Test 3	Number of Sets of Data
Accept	Accept	Accept	26
Accept	Accept	Reject	6
Accept	Reject	Accept	12
Reject	Accept	Accept	4
Reject	Reject	Accept	18
Reject	Accept	Reject	5
Accept	Reject	Reject	7
Reject	Reject	Reject	22

Is there a difference in the power of the three tests when applied to populations from which the simulated data were obtained?

PROBLEMS

1. Suppose that instead of just one observation in every treatment-block combination, we now have m independent observations in each cell. Let C_j, R_i, and N represent the treatment sum, row sum, and overall sum as before. Then justify comparing the statistic

$$T' = mc(c-1) \frac{\sum\limits_{j=1}^{c} \left(C_j - \frac{N}{c}\right)^2}{\sum\limits_{i=1}^{r} R_i(mc - R_i)}$$

with the chi-square distribution, $c-1$ degrees of freedom, in the same way the distribution of T is justified in this section.

2. The usual parametric test for the design in this section assumes that the observations are made on normal random variables instead of point binomial random variables and uses the "F statistic." If the F statistic is computed on the 0s and 1s, it simplifies to

$$F = (r-1) \frac{c \sum\limits_{j=1}^{c} C_j^2 - N^2}{rcN - r \sum\limits_{i=1}^{r} R_i^2 - c \sum\limits_{j=1}^{c} C_j^2 + N^2}$$

Show that this F is the following function of T

$$F = \frac{(r-1)T}{r(c-1) - T}$$

and that rejecting H_0 for large T is equivalent to rejecting H_0 for large F.

4.7. SOME COMMENTS ON LOGLINEAR MODELS

The methods described in this chapter are not the only methods available for analyzing contingency tables. They may be summarized by saying that they use the test statistic

$$(1) \qquad T_1 = \sum_{\substack{\text{all} \\ \text{cells}}} \frac{(O_i - E_i)^2}{E_i}$$

where O_i is the observed count in cell i and E_i is the expected count in cell i. A different method of analysis, called the log likelihood ratio test, employs the statistic

$$(2) \qquad T_2 = 2 \sum_{\substack{\text{all} \\ \text{cells}}} O_i \ln \left(\frac{O_i}{E_i} \right)$$

instead of T_1, where ln refers to natural logarithms, easily obtained on most calculators. The statistic T_2 is compared with the chi-square distribution, just as for T_1 with the same number of degrees of freedom as used for T_1. Although the two statistics T_1 and T_2 have the same asymptotic distribution, their values for a particular contingency table will differ, possibly by a large amount. The choice of whether to use T_1 or T_2 depends largely on the user's preference.

Another popular method of analysis is the "loglinear models." This method works well in analyzing contingency tables with three or more dimensions (three-way or more) if the proper computer programs are available. Analysis by hand computation is not recommended. The same statistics T_1 and T_2 just given are used with loglinear models; the difference is in the method used for obtaining the E_is. Usually iterative methods are used, which require a computer.

The name loglinear models arises for the following reason. In a two-way contingency table the null hypothesis of independence may be expressed as

$$H_0: p_{ij} = p_{i+} \cdot p_{+j}, \qquad \text{all } i \text{ and } j$$

where p_{ij} is the probability of an observation being classified in cell (i, j) and where p_{i+} and p_{+j} are the row and column marginal probabilities. Taking the logarithm of p_{ij} gives

$$H_0: \log p_{ij} = \log p_{i+} + \log p_{+j}$$

which is a linear equation. The test then amounts to a test of whether or not the model for the logarithms of the cell probabilities is a linear function of the logarithms of the marginal probabilities. A complete presentation of loglinear models and their analyses is found in Bishop, Fienberg, and Holland (1975). An elementary treatment of the subject by Ku and Kullback (1974) or Lee (1978) is recommended for beginners. See also a recent book by Fienberg (1977).

The interested reader may pursue the topic of loglinear models by reading articles by Bishop (1969, 1971), Fienberg (1970, 1972), Fienberg and Larntz (1976), Chen and Fienberg (1976), Grizzle, Starmer, and Koch (1969), Koch and Reinfurt (1971), Grizzle and Williams (1972), Koch, Imrey, and Reinfurt (1972), Wagner (1970), Odoroff (1970), Goodman (1971), Gart (1972), Haberman (1973), and Read (1977).

4.8. REVIEW PROBLEMS FOR CHAPTERS 3 AND 4

1. In a Danish research project to see if alcoholism is hereditary, five psychiatrists studied men who had been separated from their biological parents since early infancy. Fifty-five of the men had one parent who had been diagnosed as an alcoholic, and 10 of these 55 were found to be alcoholic. These were compared with 78 men whose biological parents had no history of alcoholism, and 4 of these 78 were alcoholic. The study found that "significantly more" of the first group had a history of drinking problems. What would your statistical analysis look like? (Associated Press, February 21, 1973.)

2. Prior to an election, a random sample of 200 voters were asked which candidate they preferred. Candidate A was preferred by 85 people, candidate B by 111 voters, and 4 voters were undecided. How would you predict the election results? (Discuss the items that need discussion.)

3. Several nonfreshman students at a community college were asked some questions, including how they felt about legalizing marijuana, with the following (fictitious) results.

Student Number	Sex	Commuting Distance from Home to College	Political Party Preference	Marijuana Question[a]	Freshman G.P.A.
1	M	32	N	1	2.66
2	M	10	D	1	3.18
3	M	28.5	R	1	2.15
4	M	3.5	R	2	1.61
5	M	4	D	4	1.54
6	F	7	N	5	2.12
7	M	3.5	R	3	1.35
8	M	10	N	4	2.26
9	F	6	R	4	2.70
10	M	32	D	4	2.84
11	M	22.5	D	4	2.60
12	M	7	D	1	1.13
13	M	6.5	N	4	0.81
14	M	5	D	1	3.11
15	M	35	R	1	2.47
16	M	5.5	D	5	3.15

[a] 1 = strongly disagree, . . . , 5 = strongly agree.

Student Number	Sex	Commuting Distance from Home to College	Political Party Preference	Marijuana Question[a]	Freshman G.P.A.
17	M	26.5	D	4	2.33
18	F	24	N	5	2.46
19	M	32	D	5	3.59
20	F	5	R	1	2.00
21	M	5.5	R	1	2.90
22	M	11.5	N	5	3.26
23	M	9.5	R	4	2.71
24	M	25.5	O	3	2.22
25	M	15	R	1	3.00
26	F	9	N	4	2.06
27	M	15	D	1	1.75
28	M	24	R	3	2.42

[a] $1 =$ strongly disagree, ..., $5 =$ strongly agree.

(a) Test the hypothesis that political party preference is independent of the attitude toward legalizing marijuana.
(b) Does the G.P.A. appear to be related to the commuting distance?
(c) Estimate the median commuting distance.
(d) Estimate the percent of female students.
(e) Does political party preference appear to be independent of whether students are male or female?
(f) Test the hypothesis that male and female students have the same freshman G.P.A.

4. Ten students were given a special course, and for each student a precourse test score X and a postcourse test score Y were recorded as follows:

					Student Number					
	1	2	3	4	5	6	7	8	9	10
X	92	80	74	85	71	68	81	82	80	91
Y	94	86	72	91	70	77	89	91	86	94

(a) Do the postcourse scores seem to be significantly better than the precourse scores?
(b) Let $p =$ probability of a student's postcourse score being better than the precourse score. Find a confidence interval for p.
(c) Find a confidence interval for the median precourse score.
(d) Is the upper quartile of Y significantly greater than 75?

5. Consider the following experiment in pathology to determine the effectiveness of trying to detect whether ethionine was in an animal's diet by measuring, in an autopsy, the amount of iron absorbed by the liver. Thirty-four animals are randomly placed into one of two groups; 17 received ethionine in their diets and 17 did not. The animals are paired (1 from each group) and fed the same amounts within each pair. After a period of time, each liver was extracted and treated with radioactive iron in a solution that is either warm (37°C) or cool (25°C). The data consist of the amount of iron absorbed by the various livers.

	Warm			Cool	
Pair	Ethionine	None	Pair	Ethionine	None
1	2.59	1.40	9	6.77	4.71
2	1.54	1.51	10	4.97	1.60
3	3.68	2.49	11	1.46	0.67
4	1.96	1.74	12	0.96	0.71
5	2.94	1.59	13	5.59	5.21
6	1.61	1.36	14	9.56	5.12
7	1.23	3.00	15	1.08	0.95
8	6.96	4.81	16	1.58	1.56
			17	8.09	1.68

(a) Do livers from ethionine-fed animals seem to absorb more iron in warm solutions than livers from the other group? How about in cool solutions?

(b) Does the cool solution (rather than the warm solution) significantly enhance the absorption of iron in the livers of the ethionine-fed animals? How about the animals in the other group?

(c) Find a two-sided tolerance limit for the amount of iron absorbed by the liver of an ethionine-fed animal when treated with a cool solution of radioactive iron.

(d) Does there seeem to be a correlation between the amounts of iron absorbed in the livers of the 2 animals in the same pair?

6. An experiment consists of sampling the air in 20 Eskimo homes in Bethel, Alaska, a native village on the Kuskokwima River in southwestern Alaska. Ten of the homes were new homes built as part of a housing development project; the other 10 were standard houses in Bethel. The objective is to compare the old houses with the new houses to see if there is a difference in the number of bacterial colonies per cubic foot of air. The measurements in the houses, as they were estimated from streptocel plates, were as follows.

Old House Number	Bacterial Colonies per Cubic foot	New House Number	Bacterial Colonies per Cubic foot
1	37.0	1N	1.0
2	2.6	2N	5.3
3	48.6	3N	3.4
4	47.8	4N	2.3
5	99.3	5N	5.1
6	1.4	6N	38.7
7	2.3	7N	5.0
8	3.1	8N	50.6
9	3.0	9N	1.6
10	0.3	10N	22.7

(a) Analyze the data.

(b) Find a confidence interval for the median bacterial count in the old houses.

7. Dr. G. Noether has suggested the following test for trend. Group the sequence of observations into nonoverlapping groups of three adjacent observations. Let T

equal the number of monotonic groups (either increasing or decreasing). For example:

$$\underbrace{42, 44, 63,}_{\text{Increasing}} \quad \underbrace{61, 44, 52,}_{} \quad \underbrace{73, 72, 46,}_{\text{Decreasing}} \quad \underbrace{48, 42, 53}_{} \quad .$$

For these numbers $T = 2$.

(a) What is the distribution of T under the null hypothesis of independent and identically distributed random variables?

(b) How does one find the critical region?

8. A random sample consisting of 12 typewriters owned by the College of Business Administration showed that 8 were IBM typewriters. A random sample consisting of 36 typewriters owned by the School of Medicine showed that 30 were IBM typewriters.

(a) Is this difference significant?

(b) Find a confidence interval for the overall proportion of typewriters owned by the School of Medicine that are IBM typewriters.

9. The intramural department would like to add javelin throwing to its curriculum and must decide how many yard markers are needed in the practice field to mark the length of throw. They decide to select randomly several students to throw the javelin and mark the field only between the shortest throw and the longest from that group of students. How many students should they select so that they can be 90% sure that at least 95% of the students will be throwing their javelins within the marked boundaries?

10. A certain broker noted the number of bonds he sold each month for a 24-month period.

	January	February	March	April	May	June
1972	12	16	14	18	18	14
1973	19	22	20	17	18	20

July	August	September	October	November	December
10	21	12	18	17	17
20	16	16	21	24	25

Does this record indicate an increasing trend?

11. To see if a chimpanzee can learn to recognize letters, five different letters are placed randomly on five buttons. When the light goes on and the chimpanzee presses the letter E, he gets a banana, which he likes. Each day five trials are run, and the letters are changed randomly after each trial. The experiment continues for 6 days, with the following results.

Number of presses until E					
Trial Number	1	2	3	4	5
Monday	6	4	4	2	3
Tuesday	7	8	6	1	3
Wednesday	4	2	1	2	3
Thursday	1	4	3	2	2
Friday	5	2	3	1	2
Saturday	4	2	1	2	1

(a) Does the chimpanzee seem to improve during the course of the five trials within each day?
(b) Does the chimpanzee seem to be improving through the week?
(c) If the chimpanzee is pressing the buttons randomly, the number of presses should follow the geometric distribution given by $P(X = k) = (.2)(.8)^{k-1}$, for $k = 1, 2, 3. \ldots$ Test the hypothesis that the chimpanzee is pressing the buttons randomly.

CHAPTER 5

Some Methods Based on Ranks

PRELIMINARY REMARKS

Most of the statistical procedures introduced in the previous chapters can be used on data that have a nominal scale of measurement. In Chapter 3 several statistical methods were presented for analyzing data that were naturally dichotomous, that is, the zero-one or success-failure type of data. In Chapter 4 the discussion centered around the analysis of data that may be classified according to two or more different criteria and into two or more separate classes by each criterion. All of those procedures may also be used where more than nominal information concerning the data is available but, for various reasons such as speed and ease of calculation, abundance of data, or the particular interpretation desired of the data, some of the information contained in the data is disregarded and the data are reduced to nominal-type data for analysis. Such a loss of information usually results in a corresponding loss of power. In this chapter several statistical methods are presented that utilize more of the information contained in the data, if the data have at least an ordinal scale of measurement.

Data may be nonnumeric ("good, better, best") or numeric (7.36, 4.91, etc.). If data are nonnumeric but are ranked as in ordinal-type data, the methods of this chapter are often the most powerful ones available. If data are numeric and, furthermore, are observations on random variables that have the normal distribution so that all of the assumptions of the usual parametric tests are met, the loss of efficiency caused by using the methods of this chapter is surprisingly small. In those situations the relative efficiency of tests using only the ranks of the observations is frequently about .95, depending on the situation.

The rank tests of this chapter are valid for all types of populations, whether continuous, discrete, or mixtures of the two. Earlier results in nonparametric statistics required the assumption of continuous random variables in order for the tests based on ranks to be valid. Recent results by Conover (1973a) and others have shown that the continuity assumption is not necessary. It can be replaced by the trivial assumption that $P(X = x) < 1$ for each x. Since it is unlikely that any experimenter will be sampling from a population consisting entirely of a single number, we will not list this assumption for the tests in this chapter.

Thus data with many ties (two observations are said to be tied if they equal each other) may be analyzed using rank tests if the data are ordinal. A word of caution should be offered here; the so-called "exact tables" for small samples are exact only when there are no ties in the data. Otherwise they are approximate. Exact tables for a given set of ties can be obtained in the same way as in the case of no ties, but it is not practical to have a set of tables for each possible configuration of ties. If there are extensive ties in the data, the large sample approximation and not the small sample tables in this book should be used.

A convenient method for arranging observations in increasing order is the stem-and-leaf method presented by Tukey (1977). Perhaps the simplest way of explaining the stem-and-leaf method is by way of an example. Suppose a class of 28 students obtained the following scores on an exam.

74	63	88	69	81	91	75
82	91	87	77	86	86	87
96	84	93	73	74	93	78
70	84	90	97	79	89	93

The tens digit in each score is the stem in this case. There are four different stems: 6, 7, 8, and 9. The units digit is considered the leaf. First the stems are listed.

$$9$$
$$8$$
$$7$$
$$6$$

Then each leaf is written to the right of the appropriate stem. That is, the first score, 74, is written as a 4 next to the 7 stem. Each score is written in this manner, with the following result.

9	6 1 3 0 7 1 3 3
8	2 4 4 8 7 1 6 6 9 7
7	4 0 7 3 4 9 5 8
6	3 9

A picture of the distribution of exam scores immediately emerges. But, more important for our purposes, the scores may be arranged from smallest to largest quite easily now. In this way the ranks may be assigned to the observations. This simple stem-and-leaf method can make the methods in this chapter easier to use.

5.1. TWO INDEPENDENT SAMPLES

The test presented in this section is known as the Mann–Whitney test and also as the Wilcoxon test. Equivalent forms of the same test appeared in the literature under various names, probably partly because of the intuitive appeal of the test procedure. Although primarily a two-sample test, the Mann–Whitney test may be applied in many different situations other than the usual two-sample situation.

The usual two-sample situation is one in which the experimenter has obtained two samples from possibly different populations and wishes to use a statistical test to see if the null hypothesis that the two populations are identical can be rejected. That is, the experimenter wishes to detect differences between the two populations on the basis of random samples from those populations. If the samples consist of ordinal-type data, the most interesting difference is a difference in the locations of the two populations. Does one population tend to yield larger values than the other population? Are the two medians equal? Are the two means equal?

An intuitive approach to the two-sample problem is to combine both samples into a single ordered sample and then assign ranks to the sample values from the smallest value to the largest, without regard to which population each value came from. Then the test statistic might be the *sum* of the ranks assigned to those values from one of the populations. If the sum is too small (or too large), there is some indication that the values from that population tend to be smaller (or larger, as the case may be) than the values from the other population. Hence the null hypothesis of no differences between populations may be rejected if the ranks associated with one sample tend to be larger than those of the other sample.

Ranks may be considered preferable to the actual data for several reasons. First, if the numbers assigned to the observations have no meaning by themselves but attain meaning only in an ordinal comparison with the other observations, the numbers contain no more information than the ranks contain. Such is the nature of ordinal data. Second, even if the numbers have meaning but the distribution function is not a normal distribution function, the probability theory is usually beyond our reach when the test statistic is based on the actual data. The probability theory of statistics based on ranks is relatively simple and does not depend on the distribution in many cases. A third reason for preferring ranks is that the A.R.E. of the Mann–Whitney test is never too bad when compared with the two-sample t test, the usual parametric counterpart. And yet the contrary is not true; the A.R.E. of the t test compared to the

Mann–Whitney test may be as small as zero, or "infinitely bad." So the Mann–Whitney test is a safer test to use.

The Mann–Whitney Test

DATA. The data consist of two random samples. Let $X_1, X_2, \ldots, X_n$ denote the random sample of size n from population 1 and let $Y_1, Y_2, \ldots, Y_m$ denote the random sample of size m from population 2. Assign the ranks 1 to $n+m$. Let $R(X_i)$ and $R(Y_j)$ denote the rank assigned to X_i and Y_j for all i and j. For convenience, let $N = n + m$.

If several sample values are exactly equal to each other (tied), assign to each the average of the ranks that would have been assigned to them had there been no ties (see Example 1).

ASSUMPTIONS

1. Both samples are random samples from their respective populations.
2. In addition to independence within each sample, there is mutual independence between the two samples.
3. The measurement scale is at least ordinal.

HYPOTHESES.

A. (Two-Tailed Test) Let $F(x)$ and $G(x)$ be the distribution functions corresponding to populations 1 and 2, respectively, and of X and Y, respectively. Then the hypotheses may be stated as follows.

$$H_0: F(x) = G(x) \quad \text{for all } x$$
$$H_1: F(x) \neq G(x) \quad \text{for some } x$$

In many real situations any difference between distributions implies that $P(X < Y)$ is no longer equal to 1/2. Therefore the following set of hypotheses is often used instead of the above.

$$H_0: P(X < Y) = \tfrac{1}{2}$$
$$H_1: P(X < Y) \neq \tfrac{1}{2}$$

B. (One-Tailed Test)

$$H_0: P(X < Y) \leq \tfrac{1}{2}$$
$$H_1: P(X < Y) > \tfrac{1}{2}$$

C. (One-Tailed Test)

$$H_0: P(X < Y) \geq \tfrac{1}{2}$$
$$H_1: P(X < Y) < \tfrac{1}{2}$$

The Mann–Whitney test is unbiased and consistent when testing the preceding hypotheses involving $P(X<Y)$. However, the same is not always true for the following hypotheses, which are sometimes stated instead of these forms. To insure that the test remains consistent and unbiased for the following hypotheses, it is sufficient to add another assumption to the previous model.

Assumption 4. If there is a difference between population distribution functions, that difference is a difference in the location of the distribution. That is, if $F(x)$ is not identical with $G(x)$, then $F(x)$ is identical with $G(x+c)$, where c is some constant.

Then the hypotheses are stated in terms of the means of X and Y, if they exist.

A. (Two-Tailed Test)

$$H_0: E(X)=E(Y)$$
$$H_1: E(X)\neq E(Y)$$

B. (One-Tailed Test)

$$H_0: E(X)\geq E(Y)$$
$$H_1: E(X)<E(Y)$$

C. (One-Tailed Test)

$$H_0: E(X)\leq E(Y)$$
$$H_1: E(X)>E(Y)$$

TEST STATISTIC. If there are no ties, or just a few ties, the sum of the ranks assigned to population 1 can be used as a test statistic.

(1)
$$T=\sum_{i=1}^{n} R(X_i)$$

If there are many ties, subtract the mean from T and divide by the standard deviation to get

(2)
$$T_1=\frac{T-n\frac{N+1}{2}}{\sqrt{\frac{nm}{N(N-1)}\sum_{i=1}^{N} R_i^2-\frac{nm(N+1)^2}{4(N-1)}}}$$

where $\sum R_i^2$ refers to the sum of the squares of all N of the ranks or average ranks actually used in both samples.

DECISION RULE. Use decision rule A, B, or C, depending on whether the hypothesis of interest is classified as A, B, or C. The quantiles w_p of T are given in Table A7 for p equal to .001, .005, .01, .025, .05, and .10. The upper quantiles are not given but may be computed by subtraction from $n(N+1)$. That is,

$$(3) \qquad\qquad w_{1-p} = n(N+1) - w_p$$

As an alternative to using upper quantiles, the statistic T', defined as

$$(4) \qquad\qquad T' = n(N+1) - T$$

may be used with the lower quantiles whenever an upper quantile test is desired.

The approximate quantiles of T_1 are given in Table A1, the normal distribution. In the following set of rules, substitute T_1 for T if T_1 is actually being used.

A. (Two-Tailed Test) Reject H_0 at the level of significance α if T is less than the $\alpha/2$ quantile $w_{\alpha/2}$ or if T is greater than the $1-\alpha/2$ quantile $w_{1-\alpha/2}$. Accept H_0 if T is between or equal to the two quantiles.

B. (One-Tailed Test) Small values of T indicate that H_1 is true. Therefore reject H_0 at a level of significance α if T is less than the αth quantile w_α. Accept H_0 if T is greater than or equal to w_α.

C. (One-Tailed Test) Large values of T, or small values of T', indicate that H_1 is true. Therefore reject H_0 at the level of significance α if T is greater than $w_{1-\alpha}$ or, (equivalently), if T' is less than w_α. Accept H_0 if T is less than or equal to $w_{1-\alpha}$.

Example 1. The senior class in a particular high school had 48 boys. Twelve boys lived on farms and the other 36 lived in town. A test was devised to see if farm boys in general were more physically fit than town boys. Each boy in the class was given a physical fitness test in which a low score indicates poor physical condition. The scores of the farm boys (X_i) and the town boys (Y_i) are as follows.

X_i: Farm Boys				Y_i: Town Boys			
14.8	10.6	12.7	16.9	7.6	2.4	6.2	9.9
7.3	12.5	14.2	7.9	11.3	6.4	6.1	10.6
5.6	12.9	12.6	16.0	8.3	9.1	15.3	14.8
6.3	16.1	2.1	10.6	6.7	6.7	10.6	5.0
9.0	11.4	17.7	5.6	3.6	18.6	1.8	2.6
4.2	2.7	11.8	5.6	1.0	3.2	5.9	4.0

Neither group of boys is a random sample from any population. However, it is reasonable to assume that these scores resemble hypothetical random samples from the populations of farm and town boys in that age group, at least for similar localities. The other assumptions of the model seem to be reasonable, such as independence between groups. Therefore the Mann–Whitney test is selected to test

H_0: Farm boys do not tend to be more fit, physically, than town boys

H_1: Farm boys tend to be more fit than town boys

The null hypothesis could also be stated as $H_0\colon P(X<Y)\geq 1/2$, or $H_0\colon E(X)\leq E(Y)$, according to set C of hypotheses.

The scores are ranked as follows.

X	Y	Rank	X	Y	Rank	X	Y	Rank
	1.0	1		6.2	17		11.3	33
	1.8	2	6.3		18	11.4		34
	2.1	3		6.4	19		11.8	35
	2.4	4		6.7	20.5⎤	12.5		36
	2.6	5		6.7	20.5⎦		12.6	37
2.7		6	7.3		22		12.7	38
	3.2	7		7.6	23	12.9		39
	3.6	8		7.9	24		14.2	40
	4.0	9		8.3	25		14.8	41.5⎤
4.2		10	9.0		26	14.8		41.5⎦
	5.0	11		9.1	27		15.3	43
	5.6	13⎤		9.9	28		16.0	44
	5.6	13⎟		10.6	30.5⎤	16.1		45
5.6		13⎦		10.6	30.5⎟		16.9	46
	5.9	15	10.6		30.5⎟		17.7	47
	6.1	16		10.6	30.5⎦		18.6	48

There are four groups of tied scores, as indicated by the square brackets. Within each group the ranks that should have been assigned are averaged, and the average rank is assigned instead, as illustrated.

The test is one tailed. The critical region corresponds to large values of T_1. Note that this is not a large number of ties, so it is probably acceptable to use T instead of T_1. Both methods will be compared later in the example. From Table A1 we see that a critical region of size $\alpha = .05$ corresponds to values of T_1 greater than 1.6449.

Here we have $n = 12$, $m = 36$, so $N = 12+36 = 48$. The sum of the ranks assigned to the Xs is

$$T = \sum_{i=1}^{n} R(X_i)$$

$$= 6+10+13+18+22+26+30.5$$
$$+34+36+39+41.5+45 = 321$$

The sum of the squares of all 48 ranks is

$$\sum_{i=1}^{N} R_i^2 = 38{,}016$$

which is slightly less than the sum 38,024 of the squares of all the ranks from

1 to 48 if there had been no ties (using Lemma 1.4.2). Now we can compute T_1.

$$T_1 = \frac{T - n\dfrac{N+1}{2}}{\sqrt{\dfrac{nm}{N(N-1)} \displaystyle\sum_{i=1}^{N} R_i^2 - \dfrac{nm(N+1)^2}{4(N-1)}}}$$

$$= \frac{321 - 13\dfrac{49}{2}}{\sqrt{\dfrac{(12)(36)}{(48)(47)}(38,016) - \dfrac{(12)(36)(49)^2}{4(47)}}}$$

$$= .6431$$

which is not in the critical region, so H_0 is accepted and we conclude that these data do not show that farm boys are more physically fit than town boys. A comparison of $T_1 = .6431$ with Table A1 shows that .6431 is close to the .74 quantile, so that the null hypothesis could have been rejected at an α level of about $1 - .74 = .26$, and therefore $\hat{\alpha} = .26$.

If we had ignored the few ties and used the large sample approximation at the end of Table A7 we would have obtained an approximate .95 quantile for T as

$$w_{.95} = n\frac{N+1}{2} + (1.6449)\sqrt{nm(N+1)/12}$$

$$= 294 + (1.6449)(42)$$

$$= 363.1$$

and H_0 would have been accepted as before. Working backwards from the equation

$$T = n\frac{N+1}{2} + x_{1-\alpha}\sqrt{nm(N+1)/12}$$

to find $x_{1-\alpha}$, and hence $\hat{\alpha}$, gives

$$x_{1-\alpha} = \frac{T - n\dfrac{N+1}{2}}{\sqrt{nm(N+1)/12}} = .6429$$

in essential agreement with T_1, so $\hat{\alpha}$ is again .26.

The next example illustrates a situation in which no random variables are defined explicitly. The pieces of flint are ranked according to hardness by direct comparison with each other. A random variable that assigns a measure of hardness to each piece of flint is conceivable but unnecessary in this case.

Example 2. A simple experiment was designed to see if flint in area A tended to have the same degree of hardness as flint in area B. Four sample pieces of flint were collected in area A and five sample pieces of flint were collected in area B. To determine which of two pieces of flint was harder, the two pieces were rubbed against each other. The piece sustaining less damage was judged the harder of the two. In this manner all nine pieces of flint were ordered according to hardness. The rank 1 was assigned to the softest piece, rank 2 to the next softest, and so on.

Origin of Piece	Rank
A	1
A	2
A	3
B	4
A	5
B	6
B	7
B	8
B	9

The hypothesis to be tested is

H_0: The flints from areas A and B are of equal hardness

against the alternative

H_1: The flints are not of equal hardness

The Mann–Whitney two-tailed test is used where

$$T = \text{sum of the ranks of pieces from area A}$$
$$= 1 + 2 + 3 + 5$$
$$= 11$$

The two-tailed critical region of approximate size .05 corresponds to values of T less than 12 and values of T greater than $(4)(10) - 12 = 28$. Because T in this example is less than 12, the null hypothesis is rejected, and it is concluded that flints from the two areas differ in degree of hardness. Because of the direction of the difference, the further conclusion that the flint in area B is harder than the flint in area A may also be drawn.

The critical level $\hat{\alpha}$ may be considered to be .05 because, of the values for $\hat{\alpha}$ given in Table A7, .05 is the smallest that results in rejection of H_0.

☐ *Theory.* The null distribution of T is found by assuming that X_i and Y_j are identically distributed. This is strictly true only when H_0 is true in the two-tailed test. However, in the one-tailed tests, α is found by maximizing the probability of T falling into the appropriate rejection region, and that probability is a maximum, under H_0, when the two populations have identical distributions. Therefore the distribution of T is found by assuming that the X_i and Y_j are identically distributed, no matter which of the three forms of H_0 is being tested.

If the X_i and the Y_j are independent and identically distributed, every arrangement of the Xs and Ys in the ordered combined sample is equally likely. This is the basic principle behind many rank tests. A formal proof of this statement requires calculus and is therefore beyond the scope of this book. However, the truth of the statement may seem to be intuitively obvious after one attempts to furnish a reason for some arrangements being more probable than others. There is no valid reason for this and, therefore, we can accept the fact that all ordered arrangements are equally likely as an intuitively obvious but unproved (here) statement.

If the X_i and Y_j are independent and identically distributed, the ranks assigned to the X_i in the combined sample should resemble a random selection of n of the integers from 1 to $n+m$. That is, there is no reason why any particular rank should have a better chance than any other rank of being assigned to a value of X_i. Because each number from 1 to $n+m$ is equally likely to be assigned to X_i as its rank and because n different numbers are selected as ranks for the Xs, the probability distribution of T, the sum of the ranks, may be obtained by considering the probability distribution of the sum of n integers selected at random, without replacement, from among the integers from 1 to $n+m$.

The number of ways of selecting n integers from a total number of $n+m$ integers is $\binom{n+m}{n}$, and each way has equal probability according to the basic premise just stated. Hence the probability that $T=k$ may be found by counting the number of different sets of n integers from 1 to $n+m$ that add up to the value k and then dividing that number by $\binom{n+m}{n}$.

For example, if the sample sizes are $n=3$ and $m=4$, the number of ways of selecting three out of seven ranks is

$$\binom{n+m}{n} = \frac{7!}{3!\,4!} = 35$$

The smallest value that T may assume is 6, which occurs if the three ranks 1, 2, 3 are selected. The next value that T may assume is 7, which occurs only one way: 1, 2, 4. The value $T=8$ may be assumed two ways, with the ranks 1, 2, 5 or with 1, 3, 4. Therefore,

$$P(T=6) = \frac{1}{35} \qquad P(T \le 6) = .029$$

$$P(T=7) = \frac{1}{35} \qquad P(T \le 7) = .057$$

$$P(T=8) = \frac{2}{35} \qquad P(T \le 8) = .114$$

$$\text{etc.} \qquad\qquad\qquad \text{etc.}$$

Because T is the sum of the ranks of the nXs, for large n and m the central limit theorem may be applied to obtain an approximate distribution for T. This was done in Example 1.5.7. [The requirement, in Example 1.5.7, that n be less than $(n+m)/2$, is not needed if both n and m are large.] The results of Example 1.5.7 state that T is approximately normal, with mean and variance given by Theorem 1.4.5 as

(5)
$$E(T) = \frac{n(n+m+1)}{2}$$

and

(6)
$$\text{Var}\,(T) = \frac{n(n+m+1)m}{12}$$

Therefore the quantiles of T may be approximated with the aid of Theorem 1.5.1:

(7)
$$w_p = E(T) + x_p\sqrt{\text{Var}\,(Y)}$$

where x_p is the pth quantile of the standard normal distribution. The justification for using the normal approximation for T_1 is similar to the preceding justification for using the normal approximation on T, except that the term $\text{Var}\,(T)$ must be based on the actual ranks and average ranks used in the two samples. The details are deferred until Section 5.3.

The Mann–Whitney test may be used for testing

(8)
$$H_0: E(X) = E(Y) + d, \quad \text{or} \quad E(X) - E(Y) = d$$

where d is some specified number. We simply add the number d to each Y_i and then use the Mann–Whitney test on the original Xs and the newly adjusted Ys.

By collecting all the values of d that would result in acceptance of the preceding H_0, we have a confidence interval for $E(X) - E(Y)$, the difference between the two means. This confidence interval is sometimes more meaningful to an experimenter than merely testing whether the two means are equal. We will now describe a method for obtaining the confidence interval without having to use the Mann–Whitney test over and over again.

Confidence Interval for the Difference Between Two Means

DATA. The data consist of two random samples $X_1, \ldots, X_n$ and $Y_1, \ldots, Y_m$ of sizes n and m, respectively. Let X and Y denote random variables with the same distribution as the X_i and the Y_j, respectively.

ASSUMPTIONS

1. Both samples are random samples from their respective populations.
2. In addition to independence within each sample, there is mutual independence between the two samples.

3. The two population distribution functions are identical except for a possible difference in location parameters. That is, there is a constant d (say) such that $X + d$ has the same distribution function as Y.

Note that no assumption of continuity need be made here. Noether (1967b) shows that if the confidence coefficient of a confidence interval is $1 - \alpha$ when sampling from a continuous population then, for general populations, the true confidence coefficient of the same interval including its end points is at least $1 - \alpha$ and without its end points is at most $1 - \alpha$. We will include the end points.

METHOD. Determine the $\alpha/2$ quantile $w_{\alpha/2}$ for n and m from Table A7 where $(1 - \alpha)$ is the desired confidence coefficient. Note that Table A7 is used even if there are many ties. Then calculate k, given by

$$(9) \qquad k = w_{\alpha/2} - n(n+1)/2$$

From all of the possible pairs (X_i, Y_j), find the k largest differences $X_i - Y_j$ and find the k smallest differences. To find the largest and smallest differences, it is convenient to order each sample first, from smallest to largest. The kth largest difference is the upper limit U and the kth smallest difference is the lower limit L. That is, counting toward the middle of the ordered array of all mn possible differences, the kth differences from each end of the array are the points L and U. Then the confidence interval is given by

$$(10) \qquad P[L \leq E(X) - E(Y) \leq U] \geq 1 - \alpha$$

Example 3. A certain type of batter is to be mixed until it reaches a specified level of consistency. Five batches of the batter are mixed using mixer A, and another five batches are mixed using mixer B. The required times for mixing are given as follows (in minutes).

Mixer A	Mixer B
7.3	7.4
6.9	6.8
7.2	6.9
7.8	6.7
7.2	7.1

A 95% confidence interval is sought for the mean difference in mixing times, more specifically for $E(X) - E(Y)$, where X refers to mixer A and Y refers to mixer B.

For $n = 5$, $m = 5$, $\alpha = .05$, Table A7 yields $w_{.025} = 18$, so $k = 18 - (5)(6)/2 = 3$. The two samples are ordered from smallest to largest.

X_i	Y_j
6.9	6.7
7.2	6.8
7.2	6.9
7.3	7.1
7.8	7.4

Then the largest and smallest differences are found.

Smallest Differences	Largest Differences
$6.9 - 7.4 = -.5$	$7.8 - 6.7 = 1.1$
$6.9 - 7.1 = -.2$	$7.8 - 6.8 = 1.0$
$7.2 - 7.4 = -.2 = L$	$7.8 - 6.9 = .9 = U$

The resulting 95% confidence interval (L, U) for $E(X) - E(Y)$ is $(-.2, .9)$.

Theory. Note that there are mn pairs (X_i, Y_j). Let k denote the number of pairs where $X_i > Y_j$, that is, where $X_i - Y_j > 0$. Then T in Equation 1, the sum of the ranks of Xs, is $k + n(n+1)/2$ (see Problem 1). That is, if no Ys are smaller than any of the Xs, $T = 1 + 2 + \cdots + n = n(n+1)/2$ (from Lemma 1.4.1). The effect of having k pairs (X_i, Y_j) where Y is less than S is to increase T by k units.

The "borderline" value of T, where H_0 is barely accepted, is given in Table A7 as $w_{\alpha/2}$. By subtracting $n(n+1)/2$ from $w_{\alpha/2}$, we find the borderline value of k. Now we want to find the value of d that we can add to the Ys to achieve barely this borderline value of k, that is, so that exactly k of the pairs $(X_i, Y_j + d)$ satisfy $X_i > Y_j + d$, or $X_i - Y_j > d$.

If we add the maximum of all of the differences $X_i - Y_j$ to each of the Ys, obviously none of the Xs will be greater than the adjusted Ys because the Ys are too large. By adding the kth largest difference $X_i - Y_j$ to each of the Ys, we achieve the borderline case: fewer than k pairs satisfy $X_i > Y_j + d$, and at least k pairs satisfy $X_i \geq Y_j + d$. In this way we obtain the largest value d that results in acceptance of $H_0 : E(X) = E(Y) + d$. By reversing the procedure and working from the lower end, we obtain the smallest value of d that results in acceptance of the same hypothesis. This collection of values of d gives us the confidence interval we desire.

COMPARISON WITH OTHER PROCEDURES. The natural procedure to compare with the Mann–Whitney test is the two-sample t test, as mentioned earlier. This version of the t test involves the sample means $\bar{X}$ and $\bar{Y}$ of the two samples in the following formula.

(11)
$$t = \frac{(\bar{X} - \bar{Y})\sqrt{mn(N-2)/N}}{\sqrt{\sum_{i=1}^{n} (X_i - \bar{X})^2 + \sum_{j=1}^{m} (Y_j - \bar{Y})^2}}$$

The value of t is compared with quantiles obtained from Table A25, with $N-2$ degrees of freedom. In order for these quantiles to be accurate the additional assumption must be made that both populations have a normal distribution. With this assumption the t test is the most powerful test. The assumption of normality is difficult to verify, however, and certain types of nonnormal distributions result in very little power when using the t test as compared with the Mann–Whitney test. This is especially true when one or both samples contain unusually large or small observations, called "outliers."

If a computer program for the t statistic is available, it can be used to facilitate calculations in the Mann–Whitney test, especially if there are many ties. Merely compute the t statistic on the ranks $R(X_i)$ and $R(Y_j)$, instead of the data X_i and Y_j, and compare the result with quantiles from Table A25, $N-2$ degrees of freedom. Although this approximation is not quite the same as the usual normal approximation, it is slightly more accurate in most cases. For an even better approximation, find the average of T_1 given by Equation 2 and the t statistic computed on the ranks, and compare this with the average of the two quantiles obtained from Tables A1 and A25. For more details on this method see Iman (1976).

The A.R.E. of the Mann–Whitney test as compared with the t test is computed under the assumption that the distributions of X and Y are identical except for their means. If the populations are normal the A.R.E. is .955, if the populations are uniform the A.R.E. is 1.0, and if the populations have a symmetric distribution (known as the double exponential distribution) the A.R.E. is 1.5. If the two populations differ only in their location parameters the A.R.E. is never lower than .864 but may be as high as infinity (Hodges and Lehmann, 1956).

The median test also may be used for data of this type. The A.R.E. of the Mann–Whitney test relative to the median test is 1.5 for normal populations, 3.0 for uniform distributions, but only 0.75 in the double exponential case. Remember that this is *asymptotic* relative efficiency. For small samples the Mann–Whitney test may have much more power than the median test in the case of double exponential distributions (see Conover, Wehmanen, and Ramsey, 1978). On the other hand, the median test does not require that the populations be identical when H_0 is true. It only requires that they have the same median. Hence the median test may be applied in situations where the Mann–Whitney test is not valid.

The Mann–Whitney test was first introduced for the case $n = m$ by Wilcoxon (1945). Wilcoxon's test was extended to the case of unequal sample sizes by White (1952) and van der Reyden (1952). A test equivalent to Wilcoxon's was also developed independently and introduced by Festinger (1946). Mann and Whitney (1947) seem to be the first to consider unequal sample sizes and to furnish tables suitable for use with small samples. It is largely the work of Mann and Whitney that led to widespread use of the test. Because the test is attributed to various authors, it is the user's perogative as to which name to call it by.

The modification of the Mann–Whitney test to examine differences in dispersion or variance or scale, introduced by Siegel and Tukey in 1960, is similar in principle to an earlier test devised by Freund and Ansari (1957). The relationship between the two tests is described on page 126 of Hájek and Sidák (1967).

More extensive tables for the Mann–Whitney test are given by Verdooren (1963) for n and $m \leq 25$ and by Milton (1964) for $n \leq 20$ and $m \leq 40$. Other tables and a bibliography are found in Jacobson (1963). The distribution of the Mann–Whitney test statistic is discussed by Klotz (1966) and Buckle, Kraft, and van Eeden (1969). Other articles are by Zaremba (1965) and Serfling (1967).

The efficiency of the Mann–Whitney test and other closely related tests is the subject of papers by Chanda (1963), Noether (1963), Haynam and Govindarajulu (1966), McNeil (1967), Shorack (1967), Stone (1967), and Conover and Kemp (1976). Justification for the treatment of ties is given by Conover (1973a). Modifications for sequential testing are given by Alling (1963), Woinsky and Kurz (1969), Bradley, Martin, and Wilcoxon (1965), Bradley, Merchant, and Wilcoxon (1966), Sen and Ghosh (1974), and Spurrier and Hewitt (1976). The problem of testing circular distributions, as discussed by Batschelet (1965), is approached by Beran (1969) and Schach (1969b).

If the two samples are censored (i.e., if some of the largest and/or smallest sample values are not observable) the data may be analyzed with modifications of the Mann–Whitney test, such as those discussed by Gastwirth (1965a), Gehan (1965a, 1965b), Gehan and Thomas (1969), Saw (1966), Basu (1968), Hettmansperger (1968), and Shorack (1968). A rank test for the bivariate two-sample problem is given by Mardia (1967a, 1968). Other two-sample nonparametric tests are presented and discussed by Hudimoto (1959), Haga (1960), Tamura (1963), Potthoff (1963), Wheeler and Watson (1964), Gastwirth (1965b), Bhattacharyya and Johnson (1968), Mielke (1972), and Pettitt (1976). The efficiency of some of these tests is examined by Mikulski (1963), Basu (1967a), Hollander (1967a), and Gibbons and Gastwirth (1970). Other related papers include Hollander, Pledger, and Lin (1974), Bickel and Lehmann (1975), Hettmansperger and Malin (1975), Doksum and Sievers (1976), and Fligner, Hogg, and Killeen (1976).

The method for finding confidence intervals is discussed by Noether (1967a), and a related graphical procedure is described by Moses in Walker and Lev (1953). An algorithm that may be useful when sample sizes are large is given by McKean and Ryan (1977). Other estimates of location differences are discussed by Hodges and Lehmann (1963), Høyland (1965), Rao, Schuster, and Littell (1975), and Switzer (1976). Related papers are by Moses (1965), Govindarajulu (1968), Bauer (1972), Ury (1972), and Kraft and van Eeden (1972).

EXERCISES

1. Test the following data to see if the mean high temperature in Des Moines is higher than the mean high temperature in Spokane.

Des Moines	Spokane
83	78
91	82
94	81
89	77
89	79
96	81
91	80
92	81
90	

2. In a controlled environment laboratory, 10 men and 10 women were tested to determine the room temperature they found to be the most comfortable. The results were as follows.

Men	Women
74	75
72	77
77	78
76	79
76	77
73	73
75	78
73	79
74	78
75	80

Assuming that these temperatures resemble a random sample from their respective populations, is the average comfortable temperature the same for men and women?

3. Seven students were taught algebra using the present method, and six students learned algebra according to a new method. Find a 90% confidence interval for the difference in achievement scores expected from the two methods.

Method	Students' Achievement Scores
Present	68 72 79 69 84 80 78
New	64 60 68 73 72 70

4. Diet A was given to four overweight girls and diet B was given to five other overweight girls, with the following observed weight losses. Find a 90% confidence interval for mean difference in effectiveness of the two diets.

Diet	Weight Losses (pounds)
A	7, 2, −1, 4
B	6, 5, 2, 8, 3

PROBLEMS

1. Let S equal the number of pairs (X_i, Y_j) in which X_i exceeds Y_j (counting ties as one-half). Note that there are mn pairs in all. Show that S and T satisfy the relationship

$$S = T - \frac{n(n+1)}{2}$$

What statistic seems reasonable to use as an estimate of $P(X > Y)$?

2. In the case where $n = 3$, $m = 2$, and H_0 is true, find the exact distribution of T and compare it with Table A7.

3. Compute the two-sample t statistic on the data in Exercise 2 and compare the results with those obtained using the Mann–Whitney test.

5.2. SEVERAL INDEPENDENT SAMPLES

The Mann–Whitney test for two independent samples, presented in Section 5.1, was extended to the problem of analyzing k independent samples, for $k \geq 2$, by Kruskal and Wallis (1952). The experimental situation is one where k random samples have been obtained, one from each of k possibly different populations, and we want to test the null hypothesis that all of the populations are identical against the alternative that some of the populations tend to furnish greater observed values than other populations. The term "greater" applies to observations on random variables, but actually any observations that may be arranged in increasing order according to some property such a quality, value, and the like may be analyzed using the Kruskal–Wallis test in a manner analogous to the analysis of nonnumeric data using the Mann–Whitney test, as in Example 5.1.2.

The experimental design described here is called the *completely randomized* design and was already introduced in Section 4.3, following Example 4.3.1, where the median test was introduced as a possible method of analysis. The Kruskal–Wallis test uses more information contained in the observations than the median test does. That is, the Kruskal–Wallis test statistic is a function of the ranks of the observations in the combined sample, as was true with the Mann–Whitney test, while the median test statistic was dependent only on the knowledge of whether the observations were below or above the grand median. For this reason the Kruskal–Wallis test is usually more powerful than the median test. However, computation of the test statistic involves ranking all of the observations and, therefore, involves more effort than the median test.

The Kruskal–Wallis Test

DATA. The data consist of k random samples of possibly different sizes. Denote the ith random sample of size n_i by $X_{i1}, X_{i2}, \ldots, X_{in_i}$. Then the data may be arranged into columns

Sample 1	Sample 2	$\cdots$	Sample k
$X_{1,1}$	$X_{2,1}$		$X_{k,1}$
$X_{1,2}$	$X_{2,2}$		$X_{k,2}$
$\cdots$	$\cdots$		$\cdots$
X_{1,n_1}	X_{2,n_2}		X_{k,n_k}

Let N denote the total number of observations

$$(1) \qquad N = \sum_{i=1}^{k} n_i$$

Assign rank 1 to the smallest of the totality of N observations, rank 2 to the second smallest, and so on to the largest of all N observations, which receives rank N. Let $R(X_{ij})$ represent the rank assigned to X_{ij}. Let R_i be the sum of the ranks assigned to the ith sample.

$$(2) \qquad R_i = \sum_{j=1}^{n_i} R(X_{ij}) \qquad i = 1, 2, \ldots, k$$

Compute R_i for each sample.

If the ranks may be assigned in several different ways because several observations are equal to each other, assign the average rank to each of the tied observations, as in the previous test of this chapter.

ASSUMPTIONS

1. All samples are random samples from their respective populations.
2. In addition to independence within each sample, there is mutual independence among the various samples.
3. The measurement scale is at least ordinal.
4. Either the k population distribution functions are identical, or else some of the populations tend to yield larger values than other populations do.

HYPOTHESES

H_0: All of the k population distribution functions are identical

H_1: At least one of the populations tends to yield larger observations than at least one of the other populations

Because the Kruskal–Wallis test is designed to be sensitive against differences among means in the k populations, the alternative hypothesis is sometimes stated as follows.

H_1: The k populations do not all have identical means

TEST STATISTIC. The test statistic T is defined as

$$(3) \qquad T = \frac{1}{S^2} \left(\sum_{i=1}^{k} \frac{R_i^2}{n_i} - \frac{N(N+1)^2}{4} \right)$$

where N and R_i are defined by Equations 1 and 2, respectively, and where

$$(4) \qquad S^2 = \frac{1}{N-1} \left(\sum_{\substack{\text{all} \\ \text{ranks}}} R(X_{ij})^2 - N \frac{(N+1)^2}{4} \right)$$

If there are no ties S^2 simplifies to $N(N+1)/12$, and the test statistic reduces to

$$(5) \qquad T = \frac{12}{N(N+1)} \sum_{i=1}^{k} \frac{R_i^2}{n_i} - 3(N+1)$$

If the number of ties is moderate there will be very little difference between Equations 3 and 5, so the simpler Equation 5 may be preferred.

DECISION RULE. If $k = 3$, all of the sample sizes are 5 or less, and there are no ties, the exact quantile may be obtained from Table A8. More extensive exact tables are given in Iman, Quade, and Alexander (1975). When there are ties, or when exact tables are not available, the approximate quantiles may be obtained from Table A2, the chi-square distribution with $k-1$ degrees of freedom. Reject H_0 at the level α if T exceeds the $1-\alpha$ quantile thus obtained.

MULTIPLE COMPARISONS. If, and only if, the null hypothesis is rejected, we may use the following procedure to determine which *pairs* of populations tend to differ. We can say that populations i and j seem to be different if the following inequality is satisfied:

$$(6) \qquad \left|\frac{R_i}{n_i} - \frac{R_j}{n_j}\right| > t_{1-(\alpha/2)}\left(S^2 \frac{N-1-T}{N-k}\right)^{\frac{1}{2}}\left(\frac{1}{n_i} + \frac{1}{n_j}\right)^{\frac{1}{2}}$$

where R_i and R_j are the rank sums of the two samples, $t_{1-\alpha/2}$ is the $(1-\alpha/2)$ quantile of the t distribution obtained from Table A25 with $N-k$ degrees of freedom, S^2 comes from Equation 4, and T comes from Equation 3 or 5. This procedure is repeated for all pairs of populations. The same α level is used here as in the Kruskal–Wallis test.

Example 1. Data from a completely randomized design were given in Example 4.3.1, where four different methods of growing corn resulted in various yields per acre on various plots of ground where the four methods were tried. Ordinarily, only one statistical analysis is used, but here we will use the Kruskal–Wallis test so that a rough comparison may be made with the median test, which previously furnished a critical level $\hat{\alpha}$ of slightly less than .001.

The hypotheses may be stated as follows.

H_0: The four methods are equivalent
H_1: Some methods of growing corn tend to furnish higher yields than others

The observations are ranked from the smallest, 77, of rank 1 to the largest, 101, of rank $N=34$. Tied values receive the average ranks. The ranks of the observations, with the sums R_i, are given next.

			Method				
	1		2		3		4
Observation	Rank	Observation	Rank	Observation	Rank	Observation	Rank
83	11	91	23	101	34	78	2
91	23	90	19.5	100	33	82	9
94	28.5	81	6.5	91	23	81	6.5
89	17	83	11	93	27	77	1
89	17	84	13.5	96	31.5	79	3
96	31.5	83	11	95	30	81	6.5
91	23	88	15	94	28.5	80	4
92	26	91	23			81	6.5
90	19.5	89	17				
		84	13.5				
R_i:	196.5		153.0		207.0		38.5
n_i:	9		10		7		8
$N=34$							

The critical region of approximate size $\alpha = .05$ corresponds to values of T greater than the .95 quantile of chi-square random variable with $k - 1 = 3$ degrees of freedom, which is given in Table A2 as 7.815. (Note that the median test also used the chi-square distribution with $k - 1$ degrees of freedom, so the critical regions of the two tests will seem to be the same although the test statistics are different.)

The value of T obtained using Equation 5 is

$$T = 25.46$$

which clearly leads to rejection of H_0. A rough idea of the power of the Kruskal–Wallis test as compared with the median test may be obtained by comparing the value of the test statistics in both tests. Both test statistics have identical asymptotic distributions, the chi-square distribution with 3 degrees of freedom. However, the value 25.46 attained in the Kruskal–Wallis test is somewhat larger than the value 17.6 computed in the median test.

Because H_0 is rejected, the multiple comparison procedure may be used. We can ignore the few ties and use the simpler form

(7) $$S^2 = N(N+1)/12 = 99.167$$

so that

(8) $$\frac{S^2(N-1-T)}{N-k} = \frac{(99.167)(33-25.464)}{34-4} = 24.911$$

and the remaining calculations are as follows:

Populations	$\left\lvert \dfrac{R_i}{n_i} - \dfrac{R_j}{n_j} \right\rvert$	$2.041(24.911)^{\frac{1}{2}}\left(\dfrac{1}{n_i} + \dfrac{1}{n_j}\right)^{\frac{1}{2}}$
1 and 2	6.533	4.681
1 and 3	7.738	5.134
1 and 4	17.021	4.950
2 and 3	14.271	5.020
2 and 4	10.488	4.832
3 and 4	24.759	5.272

In every case the second column exceeds the third column, so we may state that the multiple comparisons procedure shows every pair of populations to be different.

As a result of recent advances in the theory of rank tests, there should no longer be any hesitation to apply the rank tests of this chapter to situations that have many ties. In fact, the Kruskal–Wallis test is an excellent test to use in a contingency table, where the rows represent ordered categories and the columns represent the different populations, as in the following.

Population	1	2	3	...	k	Row Totals	$\bar{R}_i$ = Average Rank
Category 1	O_{11}	O_{12}	O_{13}	...	O_{1k}	t_1	$(t_1 + 1)/2$
2	O_{21}	O_{22}	O_{23}	...	O_{2k}	t_2	$t_1 + (t_2 + 1)/2$
3	O_{31}	O_{32}	O_{33}	...	O_{3k}	t_3	$t_1 + t_2 + (t_3 + 1)/2$
...	...	...	...	...	...		
c	O_{c1}	O_{c2}	O_{c3}	...	O_{ck}	t_c	$\sum_{i=1}^{c-1} t_i + (t_c + 1)/2$
Column Totals	n_1	n_2	n_3	...	n_k	N = Grand Total	

O_{ij} is the number of observations in population j that fall into the ith category. The average rank for row i is $\bar{R}_i$, which is computed from the row totals, as shown. The difference between this structure and ordinary contingency tables is that the categories (rows) are ordered. That is, all of the observations in row 1 are considered equal to each other but less than the observations in row 2, and so on. To compute the test statistic, the following form is recommended. Let the sum of the ranks in population (column) j be denoted by R_j,

(9)
$$R_j = \sum_{i=1}^{c} O_{ij}\bar{R}_i$$

and compute S^2 from the following equation.

(10)
$$S^2 = \frac{1}{N-1}\left[\sum_{i=1}^{c} t_i\bar{R}_i^2 - N(N+1)^2/4\right]$$

Then the test statistic T is computed by substituting Equations 9 and 10 into Equation 3, as before. Note that Equations 10 and 4 yield the same value of S^2, but Equation 10 is easier to use in this situation. If the null hypothesis is rejected the multiple comparisons procedure may be used, as described, to pinpoint differences where they exist.

Example 2. Three instructors compared the grades they assigned over the past semester to see if some of them tended to give lower grades than others. The null hypothesis is:

H_0: The three instructors grade evenly with each other

and the alternative of interest is

H_1: Some instructors tend to grade lower than others

The grades being examined are as follows.

	Instructor			Row	Average
Grades	1	2	3	Totals	Ranks
A	4	10	6	20	10.5
B	14	6	7	27	34
C	17	9	8	34	64.5
D	6	7	6	19	91
F	2	6	1	9	105
Total number of students	43	38	28	109	

The column rank sums are found from Equation 9.

$$R_1 = 2370.5 \qquad R_2 = 2156.5 \qquad R_3 = 1468$$

As a check on our calculations so far, the sum of the R_j should equal $N(N+1)/2 = 5995$ for $N = 109$, and it does. From Equation 10 we compute $S^2 = 941.71$ and, finally, Equation 3 yields $T = .3209$.

The critical region of size .05, from Table A2 for 2 degrees of freedom, corresponds to all values of T greater than 5.991. The null hypothesis is clearly accepted. None of the instructors can be said to grade higher or lower than the others on the basis of the evidence presented.

☐ *Theory.* The exact distribution of T is found under the assumption that all observations were obtained from the same or identical populations. The method is that of randomization, which was also used in finding the distribution of the Mann–Whitney test statistic. That is, under the preceding assumption, each arrangement of the ranks 1 to N into groups of sizes $n_1, n_2, \ldots, n_k$, respectively, is equally likely, and occurs with probability $n_1! \, n_2! \cdots n_k!/N!$, which is the reciprocal of the number of ways the N ranks may be divided into groups of sizes $n_1, n_2, \ldots, n_k$. The value of T is computed for each arrangement. The probabilities associated with equal values of T are then added to give the probability distribution of T.

For example, if $n_1 = 2$, $n_2 = 1$, and $n_3 = 1$ in the three-sample case, there are 12 equally likely arrangements of the four ranks; thus each arrangement has probability 1/12. The 12 arrangements, with the associated values of T, are given as follows.

	Sample			
Arrangement	1	2	3	T
1	1, 2	3	4	2.7
2	1, 2	4	3	2.7
3	1, 3	2	4	1.8
4	1, 3	4	2	1.8
5	1, 4	2	3	0.3
6	1, 4	3	2	0.3
7	2, 3	1	4	2.7
8	2, 3	4	1	2.7
9	2, 4	1	3	1.8
10	2, 4	3	2	1.8
11	3, 4	1	2	2.7
12	3, 4	2	1	2.7

Therefore the probability function $f(x)$ and the distribution function $F(x)$ are given as follows for $n_1 = 2$, $n_2 = 1$, and $n_3 = 1$.

x	$f(x) = P(T = x)$	$F(x) = P(T \leq x)$
0.3	$2/12 = 1/6$	$1/6$
1.8	$4/12 = 1/3$	$1/2$
2.7	$6/12 = 1/2$	1.0

The large sample approximation for the distribution of T is based on the fact that R_i in Equation 2 is the sum of n_i random variables, and for large n_i the central limit theorem may be used. Thus

$$\frac{R_i - E(R_i)}{\sqrt{\text{Var}(R_i)}}$$

is approximately distributed as a standardized normal random variable when H_0 is true. From Theorem 1.4.5 the mean and variance of R_i may be expressed by

(11)
$$E(R_i) = \frac{n_i(N+1)}{2}$$

and

(12)
$$\text{Var}(R_i) = \frac{n_i(N+1)(N-n_i)}{12}$$

Therefore,

(13)
$$\left[\frac{R_i - E(R_i)}{\sqrt{\text{Var}(R_i)}} \right]^2 = \frac{\{R_i - [n_i(N+1)/2]\}^2}{n_i(N+1)(N-n_i)/12}$$

is approximately distributed as a chi-square random variable with 1 degree of freedom. If the R_i were independent of each other the distribution of the sum

(14)
$$T' = \sum_{i=1}^{k} \frac{\{R_i - [n_i(N+1)/2]\}^2}{n_i(N+1)(N-n_i)/12}$$

could be approximated using the chi-square distribution with k degrees of freedom. However, the sum of the R_is is $N(N+1)/2$, so there is a dependence among the R_is. Kruskal (1952) showed that if the ith term in T' is multipled by $(N-n_i)/N$ for $i = 1, 2, \ldots, k$, then the result

(15)
$$T = \sum_{i=1}^{k} \frac{\{R_i - [n_i(N+1/2)]\}^2}{n_i(N+1)N/12}$$

is asymptotically distributed as a chi-square random variable with $k - 1$ degrees of freedom. Equation 15 is merely a rearrangement of the terms in Equation 5 which originally defined the test statistic T. Therefore we have rationalized the use of the chi-square approximation for the distribution of the Kruskal–Wallis test statistic. □

Kruskal and Wallis (1952) found that for small α (less than about .10) and for selected small values of n_1, n_2, and n_3, the true level of significance is smaller than the stated level of significance associated with the chi-square distribution, which indicates that the chi-square approximation furnishes a conservative test in many if not most situations. Gabriel and Lachenbruch (1969) show that the chi-square approximation is good even though the sample sizes may be small. The chi-square approximation is compared with other approximations by Iman and Davenport (1976).

For two samples the Kruskal–Wallis test is equivalent to the Mann–Whitney test. Recall that in the Mann–Whitney test (Section 5.1) one sample was called $X_1, \ldots, X_n$, while the other was $Y_1, \ldots, Y_m$. The statistic T was defined by Equation 5.1.1 as

(16)
$$T = \sum_{i=1}^{n} R(X_i)$$

the sum of the ranks of the Xs in the combined sample corresponding to R_1 in the Kruskal–Wallis test. The Mann–Whitney two-tailed test consisted of rejecting H_0 if the statistic T was too large or too small. Because T is approximately normal for large sample sizes, one could reject H_0 if the quantity

(17)
$$\frac{T - E(T)}{\sqrt{\mathrm{Var}\,(T)}}$$

is above or below the appropriate standardized normal quantiles or if its square,

(18)
$$\frac{[T - E(T)]^2}{\mathrm{Var}\,(T)}$$

is above the $1 - \alpha$ quantile in a chi-square distribution with 1 degree of freedom, according to Theorem 1.5.3. So the chi-square distribution with 1 degree of freedom could have been used in the Mann–Whitney two-tailed test, with the quantity in Equation 18 as a test statistic. The Kruskal–Wallis test, with two samples, also uses the chi-square distribution with 1 degree of freedom to test the same hypothesis as in the Mann–Whitney two-tailed test and, in fact, the Kruskal–Wallis test statistic is identical to the form of the Mann–Whitney test statistic given in Equation 18. Showing this is left as an exercise for the reader.

Justification of the usage of the rank tests presented thus far in the case of noncontinuous distributions is given by Conover (1973a). The exact distribution of the test statistic when ties are present is discussed by Klotz and Teng (1977). The multiple comparisons procedure is simply the usual parametric procedure, called Fisher's least significant difference, computed on the ranks rather than the data, as described by Conover and Iman (1979).

The usual parametric procedure is called the "one-way analysis of variance," or sometimes simply the one-way F test. The statistic used is given by

(19)
$$F = \frac{\left(\sum_{i=1}^{k} T_i^2 / n_i - C\right) / (k-1)}{\left(\sum_{i=1}^{k} \sum_{j=1}^{n_i} X_{ij}^2 - \sum_{i=1}^{k} T_i^2 / n_i\right) / (N-k)}$$

where T_i is the sum of the observations in the ith sample, and $C = T^2/N$ where T is the total of all of the observations. If the assumptions of the Kruskal–Wallis test are valid and, in addition, if the populations have in common a normal distribution, then the quantiles of the F statistic are given in Table A26. Look in the column for $k_1 = k - 1$ and the row marked $k_2 = N - k$ for N and k given by the experiment. Violation of the normality assumption usually has little effect on the distribution of the F statistic when the null hypothesis is true. However, the power of the F test may be considerably less than the Kruskal–Wallis test for certain types of nonnormality when H_0 is false. Data containing outliers are better suited for the Kruskal–Wallis test, for example.

The A.R.E. of the Kruskal–Wallis test relative to the F test is never less than 0.864 but may be as high as infinity if the distribution functions have identical shapes but differ only in their means. If the populations are normal the A.R.E. is $3/\pi = 0.955$. For uniform distributions the A.R.E. relative to the F test is 1.0; for double exponential distributions it is 1.5. Compared with the median test the A.R.E. of the Kruskal–Wallis test is 1.5, 3.0, and 0.75, respectively, for the three distributions just mentioned.

Rank sum tests similar to the Kruskal–Wallis test have been adapted by Steel (1960), Sherman (1965), and McDonald and Thompson (1967) for making multiple comparisons. Some tables for making multiple comparisons are provided by Tobach, Smith, Rose, and Richter (1967). Procedures for selecting the best populations are described by Rizvi and Sobel (1967), Sobel (1967), Rizvi, Sobel, and Woodworth (1968), and Puri and Puri (1969). Rank tests are presented for censored data by Basu (1967b) and Breslow (1970); for testing against ordered alternatives by Shorack (1967), Odeh (1971, 1972), and Tryon and Hettmansperger (1973); and for analysis of covariance by Puri and Sen (1969a). For other work concerned with rank tests and several independent samples see Sen (1962, 1966), Matthes and Truax (1965), Quade (1966), Crouse (1966), Sen and Govindarajulu (1966), Odeh (1967), Deshpande (1970), and Bhapkar and Deshpande (1968). Analysis of covariance is discussed by Quade (1967). Brunden (1972) considers using ranks to analyze 2×3 contingency tables.

EXERCISES

1. Random samples from each of three different types of light bulbs were tested to see how long the light bulbs lasted, with the following results.

Brand

A	B	C
73	84	82
64	80	79
67	81	71
62	77	75
70		

Do these results indicate a significant difference between brands? If so, which brands appear to differ?

2. Four job training programs were tried on 20 new employees, where 5 employees were randomly assigned to each training program. The 20 employees were then placed under the same supervisor and, at the end of a certain specified period, the supervisor ranked the employees according to job ability, with the lowest ranks being assigned to those employees with the lowest job ability.

Program	Ranks
1	4, 6, 7, 2, 10
2	1, 8, 12, 3, 11
3	20, 19, 16, 14, 5
4	18, 15, 17, 13, 9

Do these data indicate a difference in the effectiveness of the various training programs? If so, which ones seem to be different?

3. The amount of damage to the soil on a farm caused by water and wind is examined for many different farms. At the same time the type of farming practiced on each farm is noted, with the following results.

Number of Farms with:	Type of Farming			
	Minimum Tillage	Contour	Terrace	Other
No damage	17	19	4	21
Slight damage	3	10	4	42
Moderate damage	0	2	2	34
Severe damage	0	0	2	6

Does the type of farming affect the degree of damage? If so, which types of farming are significantly different?

4. Three different types of radios, manufactured by the same company, all carry 1-year guarantees. A record is kept of how many radios needed to be replaced, were repairable, or were not returned under warranty.

	Type		
	A	B	C
Replaced	12	3	6
Repaired	10	8	7
Not Returned	82	96	58

Does there seem to be a significant difference among the reliabilities of the different radio types? If so, which ones seem to be different?

PROBLEMS

1. Show that Equations 3 and 5 are equivalent when there are no ties.
2. Find the exact distribution of the Kruskal–Wallis test statistic when H_0 is true, $n_1 = 3$, $n_2 = 2$, $n_3 = 1$, and there are no ties. Compare your results with the quantiles given in Table A8.
3. In the two-sample case, what are some of the reasons why we might prefer to use the Mann–Whitney test instead of the Kruskal–Wallis test?
4. Show that Equations 10 and 4 are equivalent.
5. Suppose the F statistic in Equation 19 is computed on the ranks $R(X_{ij})$ instead of the observations X_{ij}. Then show that the relationship

$$F = \frac{T/(k-1)}{(N-1-T)/(N-k)}$$

holds between F and T, given by Equation 3. Therefore the test that rejects H_0 for large T is equivalent to the test that rejects H_0 for large F, if F is computed on the ranks.

5.3. A TEST FOR EQUAL VARIANCES

The usual standard of comparison for several populations is based on the means or other measures of location of the populations. However, in some situations the variances of the populations may be the quantity of interest. For example, it has been claimed that the effect of seeding clouds with silver iodide is to increase the variance of the resulting rainfall. Such a claim may be tested by the method presented in this section.

This test for variances is analogous to the tests just presented for means. For example, to test H_0: $E(X) = E(Y)$, the two independent samples were combined, ranked, and the sum of the ranks of the Xs was used as a test statistic. Recall that the variance is defined as the expected value of $(X - \mu)^2$ where μ is the mean of X. Thus to test H_0: $E[(X - \mu_x)^2] = E[(Y - \mu_y)^2]$ it seems reasonable to record the values of $(X_i - \mu_x)^2$ and $(Y_i - \mu_y)^2$ from two independent samples, assign ranks to them, and use the sum of the ranks of the $(X - \mu_x)^2$s as the test statistic. Although this technique could be used, more power is obtained when the ranks are squared first and then summed. This section contains a more accurate description of such a test.

The Squared Ranks Test for Variances

DATA. The data consist of two random samples. Let $X_1, X_2, \ldots, X_n$ denote the random sample of size n from population 1 and let $Y_1, Y_2, \ldots, Y_m$ denote the random sample of size m from population 2. Convert each X_i and Y_j to its absolute deviation from the mean using

(1) $$U_i = |X_i - \mu_1|, \qquad i = 1, \ldots, n$$

and

(2) $$V_j = |Y_j - \mu_2|, \qquad j = 1, \ldots, m$$

where μ_1 and μ_2 are the means for populations 1 and 2. If μ_1 and μ_2 are unknown, use $\bar{X}$ for μ_1 and $\bar{Y}$ for μ_2, and the following test is still approximately valid.

Assign the ranks 1 to $n + m$ to the combined sample of Us and Vs in the usual way. If several values of U and/or V are exactly equal to each other (tied), assign to each the average of the ranks that would have been assigned to them had there been no ties. Let $R(U_i)$ and $R(V_j)$ denote the ranks and average ranks thus assigned. Note that ranking the U_is and V_js achieves the same results and is easier than ranking the values of $(X_i - \mu_1)^2$ and $(Y_j - \mu_2)^2$.

ASSUMPTIONS

1. Both samples are random samples from their respective populations.
2. In addition to independence within each sample there is mutual independence between the two samples.
3. The measurement scale is at least interval.

HYPOTHESES

A. (Two-Tailed Test)

H_0: X and Y are identically distributed, except for possibly different means

H_1: $\text{Var}(X) \neq \text{Var}(Y)$

B. (One-Tailed Test)

H_0: X and Y are identically distributed, except for possibly different means

H_1: $\text{Var}(X) < \text{Var}(Y)$

C. (One-Tailed Test)

H_0: X and Y are identically distributed, except for possibly different means

H_1: $\text{Var}(X) > \text{Var}(Y)$

TEST STATISTIC. If there are no ties the sum of the squares of the ranks assigned to population 1 can be used as the test statistic.

(3) $$T = \sum_{i=1}^{n} [R(U_i)]^2$$

If there are ties subtract the mean from T and divide by the standard deviation to get

(4) $$T_1 = \frac{T - n\overline{R^2}}{\left[\dfrac{nm}{N(N-1)} \sum_{i=1}^{N} R_i^4 - \dfrac{nm}{N-1} (\overline{R^2})^2 \right]^{\frac{1}{2}}}$$

where $N = n + m$, $\overline{R^2}$ represents the average of the squared ranks of both samples combined:

(5)
$$\overline{R^2} = \frac{1}{N}\left\{\sum_{i=1}^{n}[R(U_i)]^2 + \sum_{j=1}^{m}[R(V_j)]^2\right\}$$

and $\sum R_i^4$ represents the sum of the ranks raised to the fourth power:

(6)
$$\sum_{i=1}^{N} R_i^4 = \sum_{i=1}^{n}[R(U_i)]^4 + \sum_{j=1}^{m}[R(V_j)]^4$$

DECISION RULE. Use decision rules A, B, or C, depending on whether the hypothesis of interest is classified as A, B, or C of the preceding. Some exact quantiles for T are given in Table A9 for the case with no ties, and the large sample approximation for T is given at the end of that table. If there are ties approximate quantiles for T_1 may be obtained from Table A1. In case of ties substitute T_1 for T in the following instructions.

A. (Two-Tailed Test) Reject H_0 at the level of significance α if T is less than its $\alpha/2$ quantile or greater than its $1 - \alpha/2$ quantile. Accept H_0 if T is between those two quantiles or equal to one of them.

B. (One-Tailed Test) Reject H_0 at the level of significance α if T is less than its αth quantile, because small values of T indicate that H_1 is true. Accept H_0 if T is greater than or equal to its αth quantile.

C. (One-Tailed Test) Reject H_0 at the level of significance α if T is greater than its $1 - \alpha$th quantile. Accept H_0 if T is less than or equal to its $1 - \alpha$th quantile.

A TEST FOR MORE THAN TWO SAMPLES. If there are three or more samples, this test is modified easily to test the equality of several variances. From each observation subtract its population mean (or its sample mean when μ_i is unknown) and convert the sign of the resulting difference to $+$, as just described for two samples. Rank the combined absolute differences from smallest to largest, assigning average ranks in case of ties, again as described. Compute the sum of the squares of the ranks for each sample, letting $S_1, S_2, \ldots, S_k$ denote the sums for each of the k samples. Thus S_1 corresponds to T in the preceding two-sample case.

H_0: All k populations are identical, except for possibly different means

H_1: Some of the population variances are not equal to each other

The test statistic is

(7)
$$T_2 = \frac{1}{D^2}\left[\sum_{j=1}^{k}\frac{S_j^2}{n_j} - N(\bar{S})^2\right]$$

where n_j = number of observations in sample j

$$N = n_1 + n_2 + \cdots + n_k$$

S_j = the sum of the squared ranks in sample j

$$\bar{S} = \frac{1}{N} \sum_{j=1}^{k} S_j = \text{the average of all the squared ranks}$$

$$D^2 = \frac{1}{N-1} \left[\sum_{i=1}^{N} R_i^4 - N(\bar{S})^2 \right]$$

and $\sum R_i^4$ represents the sum resulting after raising each rank to the fourth power. If there are no ties D^2 and $\bar{S}$ simplify to

(8) $$D^2 = N(N+1)(2N+1)(8N+11)/180$$

and

(9) $$\bar{S} = (N+1)(2N+1)/6$$

The null hypothesis is rejected if T_2 exceeds the $1 - \alpha$ quantile of the chi-square distribution with $k - 1$ degrees of freedom, obtained from Table A2. If H_0 is rejected, multiple comparisons may be made as described in the previous section. In this case the variances of populations i and j are said to differ if the following inequality is satisfied.

(10) $$\left| \frac{S_i}{n_i} - \frac{S_j}{n_j} \right| > t_{1-\alpha/2} \left(D^2 \frac{N-1-T_2}{N-k} \right)^{\frac{1}{2}} \left(\frac{1}{n_i} + \frac{1}{n_j} \right)^{\frac{1}{2}}$$

where $t_{1-\alpha/2}$ is the $1 - \alpha/2$ quantile of the t distribution obtained from Table A25 with $N - k$ degrees of freedom.

Example 1. A food packaging company would like to be reasonably sure that the boxes of cereal it produces do in fact contain at least the number of ounces of cereal stamped on the outside of the box. In order to do this it must set the average amount per box a little above the advertised amount, because the unavoidable variation caused by the packaging machine will sometimes put a little less or a little more cereal in the box. A machine with smaller variation would save the company money because the average amount per box could be adjusted to be closer to the advertised amount.

A new machine is being tested to see if it is less variable than the present machine, in which case it will be purchased to replace the old machine. Several boxes are filled with cereal using the present machine and the amount in each box is measured. The same is done for the new machine to test:

H_0: Both machines have equal variability

versus

H_1: The new machine has a smaller variance

The measurements and calculations are as follows.

Original Measurements		Absolute Deviation		Rank		Squared Rank	
Present (X)	New (Y)	Present (U)	New (V)	Present	New	Present	New
10.8	10.8	.06	.01	4	2 (tie)	16	4
11.1	10.5	.36	.29	10	8	100	64
10.4	11.0	.34	.21	9	7	81	49
10.1	10.9	.64	.11	12	6	144	36
11.3	10.8	.56	.01	11	2 (tie)	121	4
	10.7		.09		5		25
	10.8		.01		2 (tie)		4
$\bar{X} = 10.74$	$\bar{Y} = 10.79$					$T = 462$	

$$T = \text{sum of squared ranks (present)} = 462$$

$$\overline{R^2} = \frac{1}{12}(16 + 100 + \cdots + 25 + 4) = 54$$

$$\sum_{i=1}^{N} R_i^4 = (16)^2 + (100)^2 + \cdots + (25)^2 + (4)^2 = 60{,}660$$

$$T_1 = \frac{462 - 5(54)}{\left[\frac{(5)(7)}{(12)(11)}(60{,}660) - \frac{(5)(7)}{(11)}(54)^2\right]^{\frac{1}{2}}} = 2.3273$$

The preceding hypotheses match set C, because H_1 specifies the new machine (Y) has a smaller variance. The critical region corresponds to values of T_1 greater than 1.6449, the .95 quantile from Table A1, for an approximate α of .05. In this case T_1 exceeds 1.6449, so H_0 is rejected. A comparison of the observed $T_1 = 2.3273$ with the quantiles from Table A1 reveals a critical level $\hat{\alpha}$ of about .01.

Considerable simplification of the computations results whenever none of the values of U are tied with values of V, as in this example. Then ranks rather than average ranks can be used and the exact tables consulted. That is, in this example the only tie is among three values of V, so instead of using $2^2 = 4$ in the column on the far right, the values $1^2 = 1$, $2^2 = 4$, and $3^2 = 9$ can be used where the three tied values occur and the remainder of the test conducted as if there were no ties. The value of T happens to remain unchanged this time, is greater than the .95 quantile 410 from Table A9 for $n = 5$, $m = 7$, and shows $\hat{\alpha}$ to be about .01 as with the approximate test.

☐ *Theory.* Whenever two random variables X and Y are identically distributed except for having different means μ_1 and μ_2, $X - \mu_1$ and $Y - \mu_2$ not only have zero means, but they are identically distributed also. This means $U = |X - \mu_1|$ has the same distribution as $V = |Y - \mu_2|$, and $U^2 = (X - \mu_1)^2$ has the same distribution as $V^2 = (Y - \mu_2)^2$. So random samples of Xs and Ys furnish Us and Vs that are independent and identically distributed. Thus every assignment of ranks to the Us is equally likely, as in the Mann–Whitney test, and the distribution of any function of the ranks can be found as in Section 5.1.

Note that the ranks of the Us are the same as the ranks of the corresponding values of U^2, since $U_1 < U_2$ if and only if $U_1^2 < U_2^2$. Since we are interested in comparing $E(U^2)$ with $E(V^2)$, we should be looking at the ranks of the values of U^2 and V^2; however, it is *equivalent* and *easier* to consider the ranks of the Us and Vs.

Another important difference that distinguishes this rank test from the previous rank tests is that we are using the squared ranks and not the ranks themselves. We say we are using scores instead of ranks, the scores being denoted by a function of R, $a(R)$, which is used in the test statistic instead of the rank R. Let T equal the sum of the scores associated with one sample, as in this test where the scores are $a(R) = R^2$. The distribution of T is found just as in Section 5.1. If the sample sizes are $n = 3$ and $m = 4$, there are 35 ways of selecting 3 out of the 7 ranks. The three ranks 1, 2, 3 have corresponding scores $a(1)$, $a(2)$, and $a(3)$, which in turn yields some number for the test statistic T (depending on which scores are being used) which has probability 1/35. The 35 ways of selecting three ranks out of seven give 35 values of T, possibly not all different, and the probability function of T is then obtained quite simply as in Section 5.1.

To use the large sample normal approximation for T it is necessary to find the mean and variance of T when H_0 is true. We have

$$(11) \qquad T = \sum_{i=1}^{n} a(R_i)$$

where $R_1, \ldots, R_n$ represent the ranks of $U_1, \ldots, U_n$ in the combined sample of Us and Vs. We will find $E(T)$ and $\mathrm{Var}(T)$ for general scores $a(R)$ and then substitute $a(R) = R^2$ at the end.

The mean of T is written as

$$(12) \qquad E(T) = E\left[\sum_{i=1}^{n} a(R_i)\right] = \sum_{i=1}^{n} E[a(R_i)]$$

with the aid of Theorem 1.4.1. Because $P(R_i = j) = 1/N$ for each $j = 1, 2, \ldots, N$, we have

$$(13) \qquad E[a(R_i)] = \sum_{j=1}^{N} a(j) \cdot \frac{1}{N} = \frac{1}{N} \sum_{j=1}^{N} a(j) = \bar{a}$$

say. This is the same for all $i = 1$ to n, so Equation 12 becomes

(14) $$E(T) = n\bar{a}$$

where $\bar{a}$ is the average of all the scores.

For the variance of T we use Theorem 1.4.3 to get

(15) $$\text{Var}(T) = \sum_{i=1}^{n} \text{Var}[a(R_i)] + \sum_{\substack{i=1 \\ i \neq j}}^{n} \sum_{j=1}^{n} \text{Cov}[a(R_i), a(R_j)]$$

where, by the definition of variance,

(16) $$\text{Var}[a(R_i)] = E\{[a(R_i) - \bar{a}]^2\} = \sum_{k=1}^{N} [a(k) - \bar{a}]^2 \cdot \frac{1}{N} = A$$

and where, by the definition of covariance,

$$\text{Cov}[a(R_i), a(R_j)] = E\{[a(R_i) - \bar{a}][a(R_j) - \bar{a}]\}$$

(17) $$= \sum_{\substack{k=1 \\ k \neq l}}^{N} \sum_{l=1}^{N} [a(k) - \bar{a}][a(l) - \bar{a}] \frac{1}{N(N-1)}$$

because $P(R_i = k, R_j = l) = 1/[N(N-1)]$ for all $k \neq l$. The expression in Equation 17 is simplified by adding and subtracting the term where $k = l$;

(18)
$$\text{Cov}[a(R_i), a(R_j)] = \sum_{k=1}^{N} [a(k) - \bar{a}] \sum_{l=1}^{N} [a(l) - \bar{a}] \frac{1}{N(N-1)}$$
$$- \sum_{k=1}^{N} [a(k) - \bar{a}]^2 \frac{1}{N(N-1)}$$

But the first summation equals zero because of the way $\bar{a}$ was defined in Equation 13, so Equation 18 simplifies to

(19) $$\text{Cov}[a(R_i), A(R_j)] = -\frac{1}{N-1} A$$

where A is defined in Equation 16. Now the variance and covariance terms of Equations 16 and 19 are substituted into Equation 15 to get

$$\text{Var}(T) = \sum_{i=1}^{n} A - \sum_{\substack{i=1 \\ i \neq j}}^{n} \sum_{j=1}^{n} \frac{1}{N-1} A$$

$$= nA - n(n-1)\frac{1}{N-1} A$$

$$= \frac{n(N-n)}{N-1} A$$

(20) $$= \frac{nm}{(N-1)N} \sum_{i=1}^{N} [a(i) - \bar{a}]^2$$

because $N - n = m$. These Equations 14 and 20 were used in Sections 5.1 and 5.2 when ties were present and will be useful later in this chapter. For now we are interested in the case where $a(R) = R^2$ for the squared ranks test. In that case $\bar{a}$ is written as $\overline{R^2}$ in Equation 4, and the denominator of Equation 4 is what the square root of Equation 20 becomes when the identity

$$(21) \qquad \sum_{i=1}^{N} [a(i) - \bar{a}]^2 = \sum_{i=1}^{N} [a(i)]^2 - N(\bar{a})^2$$

is employed for ease in computation.

The extension of the two-sample case to the k-sample case is completely analogous to the extension of the two-sample Mann–Whitney test to the k-sample Kruskal–Wallis test. That is, the sums of scores are found for each of the k samples. Call these $S_1, S_2, \ldots, S_k$. The mean and variance of S_i are found from Equations 14 and 20 to be

$$(22) \qquad E(S_i) = n_i \bar{a}$$

and

$$(23) \qquad \text{Var}(S_i) = \frac{n_i(N - n_i)}{(N-1)N} \sum_{i=1}^{N} [a(i) - \bar{a}]^2$$

The terms $[S_i - E(S_i)]^2 / \text{Var}(S_i)$ are multiplied by $(N - n_i)/N$ for $i = 1$ to k as in the Kruskal–Wallis test and are added together to get

$$(24) \qquad T_2 = \sum_{i=1}^{k} \frac{(S_i - n_i \bar{a})^2}{n_i D^2}$$

where

$$(25) \qquad D^2 = \frac{1}{N-1} \sum_{i=1}^{N} [a(i) - \bar{a}]^2 = \frac{1}{N-1} \left\{ \sum_{i=1}^{N} [a(i)]^2 - N(\bar{a})^2 \right\}$$

Equation 24 simplifies to

$$(26) \qquad T_2 = \frac{1}{D^2} \left[\sum_{j=1}^{k} \frac{S_j^2}{n_j} - N(\bar{a})^2 \right]$$

which matches Equation 7 when the scores are the squared ranks. The suggested multiple comparisons procedure is an approximate procedure that becomes exact as the sample sizes get large.

If the populations of X and Y have normal distributions the appropriate statistic to use is the ratio of the two "sample variances,"

$$(27) \qquad F = \frac{\dfrac{1}{n-1} \sum_{i=1}^{n} (X_i - \bar{X})^2}{\dfrac{1}{m-1} \sum_{j=1}^{m} (Y_j - \bar{Y})^2}$$

which has the F distribution. The upper quantiles of F are given in Table A26 in column $k_1 = n - 1$ and row $k_2 = m - 1$. The lower quantiles are not given but may be found by taking the reciprocal of the upper quantile found in column $k_1 = m - 1$ and row $k_2 = n - 1$. Appropriate one-tailed or two-tailed tests are then obtained.

The F test is very sensitive to the assumption of normality, as pointed out by Siegel and Tukey (1960). The true distribution may be symmetric and resemble somewhat the normal distribution, such as the double exponential distribution, and yet the true level of significance may be two or three times as large as it is supposed to be. For this reason the F test is not a very safe test to use unless one is sure that the populations are normal.

If the squared ranks test is used instead of the F test when the populations are normal the A.R.E. is only $15/(2\pi^2) = 0.76$. However, for the double exponential distribution the A.R.E. is 1.08 and for the uniform distribution the A.R.E. is 1.00. These same efficiencies apply to the k-sample case as well. Thus the sensitivity of the F test to the assumption of normality, coupled with its lack of power in some reasonable nonnormal situations, encourages the consideration of a nonparametric test for variances.

The result of using $\bar{X}$ and $\bar{Y}$ in place of the true means of X and Y in the squared ranks test is to make the test approximate rather than exact and the exact distribution of the test statistic dependent on the true population distribution. The test is asymptotically distribution free, however, which means that the approximation becomes exact as the sample sizes get large.

Another popular nonparametric test for the two-sample scale problem is based on the statistic

$$(28) \qquad T = \sum_{i=1}^{n} \left[R(X_i) - \frac{n + m + 1}{2} \right]^2$$

where $R(X_i)$ is the rank of X_i, as in the Mann–Whitney test. This was proposed by Mood (1954). Exact tables are given by Laubscher, Steffens, and DeLange (1968) under the null hypothesis of identical distribution functions. The null distribution tables of a related statistic

$$(29) \qquad T = \sum_{i=1}^{n} [R(X_i) - \bar{R}_x]^2$$

where

$$\bar{R}_x = \frac{1}{n} \sum_{i=1}^{n} R(X_i)$$

are given by Hollander (1963). Ansari and Bradley (1960) discuss the A.R.E. of the Mood test and others.

Further discussion of the squared ranks test and more extensive tables are given by Conover and Iman (1978). A slight variation of the test is examined by Talwar and Gentle (1977). Other tests for scale are considered by Sen

(1963), Puri (1965), Mielke (1967), Duran and Mielke (1968), Shorack (1969), Hwang and Klotz (1975) and Fligner and Killeen (1976) for two samples, and Tsai, Duran, and Lewis (1975) for several samples. Tests designed to detect both location and scale differences are presented by Lepage (1971, 1973, 1977), Mielke (1972), and Duran, Tsai, and Lewis (1976). The correlation between rank tests for scale and rank tests for location is studied by Gibbons (1967) and Hollander (1968). Estimation of scale parameters is considered further by Moses (1963), van Eeden (1964), Basu and Woodworth (1967), Bauer (1972), Laubscher and Odeh (1976), and Bhattacharyya (1977). If the location parameters are unknown and possibly unequal, see Raghavachari (1965a), Puri (1968), and Nemenyi (1969) for modified tests. Further references may be found in an excellent review article by Duran (1976).

EXERCISES

1. A blood bank kept a record of the rate of heartbeats for several blood donors.

Men	Women
58	66
76	74
82	69
74	76
79	72
65	73
74	75
86	67
	68

Is the variation among the men significantly greater than the variation among women?

2. A particular watershed has been built up extensively in recent years, with housing developments, dams, and so forth. A random sample of stream flow rates (cubic feet per minute) for a stream in that watershed is compared with a sample of rates from earlier times to see if the variability has changed.

Present Rates	Past Rates
32	39
36	21
41	58
27	46
35	30
48	22
31	17
28	19

Is there a significant difference in variances?

3. Three different methods of instruction are compared by assigning fifth-grade students at random to three different classrooms. The grade-level attainment (as measured by a standardized exam) of each student is measured at the beginning of the year and again at the end of the year, and the increase for each student is noted.

Method of Instruction	Increase in Attainment
Structured Classes	.7, 1.0, 2.0, 1.4, .5, .8, 1.0, 1.1, 1.9, 1.2, 1.5
Individual Studies	1.7, 2.1, −.4, 0, 1.0, 1.1, .9, 2.3, 1.3, .4, .5
No Walls Classroom	.9, .9, 1.0, 0, .1, −.6, 2.2, −.3, .6, 2.4, 2.5

Does there seem to be a difference in variance associated with the three methods of instruction? If so, which methods appear to differ in variability?

4. An investment class was divided into three groups of students. One group was instructed to invest in bonds, the second in blue chip stocks, and the third in speculative issues. Each student "invested" (on paper only) $10,000 and evaluated the hypothetical profit or loss at the end of 3 months with the following results.

Bonds	Blue Chip	Speculative
146	176	−540
180	110	1052
192	212	642
185	108	−281
153	196	67

Is the difference in variance significant? If so, which groups are significantly different?

PROBLEMS

1. Find the exact distribution of T as given by Equation 3 for $n = 3$ and $m = 4$ and compare it with the quantiles from Table A9.

2. Show that Equation 24 is equivalent to Equation 26.

3. Find the exact distribution for $n = 3$, $m = 4$, for the Mood statistic given by Equation 28 and the Hollander statistic given by Equation 29.

4. Show that the mean of the Mood statistic, given by Equation 28, is $n(N+1) \times (N-1)/12$.

5. Another test for equal variances was devised by Siegel and Tukey (1960). In the ordered combined sample of Xs and Ys, assign rank 1 to the smallest value, rank 2 to the largest value, rank 3 to the second largest value, rank 4 to the second smallest value, rank 5 to the third smallest value, and so on, alternately assigning ranks to the end values two at a time (after the first) and proceeding toward the middle. The test statistic is the sum of the ranks assigned to the sample of Xs.

(a) Justify the use of the Table A7 for this statistic when both populations are identical.

(b) Which tail (upper or lower) of the critical region should be used for the one-sided alternative Var $(X) >$ Var (Y)?

(c) Use an extreme example to show that this test has little power if the two population means are far apart.

6. Show that if the F statistic, as defined by Equation 5.2.19, is computed on the scores $a(i)$ the result can be simplified to the form

$$F = \frac{T_2/(k-1)}{(N-1-T_2)/(N-k)}$$

where T_2 is given in Equation 24 or 26. Note that this mathematical relationship holds for all types of scores.

5.4. MEASURES OF RANK CORRELATION

A measure of correlation is a random variable that is used in situations where the data consist of pairs of numbers, such as the type of data described in Section 3.4. Suppose a *bivariate* random sample of size n is represented by $(X_1, Y_1), (X_2, Y_2), \ldots, (X_n, Y_n)$. We will use (X, Y) when referring to the (X_i, Y_i) in general. That is, the (X_i, Y_i) for $i = 1, 2, \ldots, n$ have identical bivariate distributions, the same bivariate distribution as (X, Y) has.

Examples of bivariate random variables include one where X_i represents the height of the ith man and Y_i represents his father's height, or where X_i represents a test score of the ith individual and Y_i represents her amount of training. The random variables X and Y may even be independent, as might be the case if X_i represented the scoring average of a basketball player and Y_i his current girlfriend's present grade point average.

By tradition, a measure of correlation between X and Y should satisfy the following requirements in order to be acceptable.

1. The measure of correlation should assume only values between -1 and $+1$.
2. If the larger values of X tend to be paired with the larger values of Y, and hence the smaller values of X and Y tend to be paired together, then the measure of correlation should be positive, and close to $+1.0$ if the tendency is strong. Then we would speak of a positive correlation between X and Y.
3. If the larger values of X tend to be paired with the smaller values of Y, and vice versa, then the measure of correlation should be negative and close to -1.0 if the tendency is strong. Then we say that X and Y are negatively correlated.
4. If the values of X seem to be randomly paired with the values of Y, the measure of correlation should be fairly close to zero. This should be the case when X and Y are independent, and possibly some cases where X and Y are not independent. We then say that X and Y are uncorrelated, or have no correlation, or have correlation zero.

The most commonly used measure of correlation is Pearson's product moment correlation coefficient, denoted by r and defined as

(1)
$$r = \frac{\sum_{i=1}^{n} (X_i - \bar{X})(Y_i - \bar{Y})}{\left[\sum_{i=1}^{n} (X_i - \bar{X})^2 \sum_{i=1}^{n} (Y_i - \bar{Y})^2 \right]^{\frac{1}{2}}}$$

where $\bar{X}$ and $\bar{Y}$ are the sample means as defined in Section 2.2. An easier form to use with a calculator is

(2)
$$r = \frac{\sum_{i=1}^{n} X_i Y_i - n\bar{X}\bar{Y}}{\left(\sum_{i=1}^{n} X_i^2 - n\bar{X}^2 \right)^{\frac{1}{2}} \left(\sum_{i=1}^{n} Y_i^2 - n\bar{Y}^2 \right)^{\frac{1}{2}}}$$

If the numerator and denominator in Equation 1 are divided by n, r becomes

(3)
$$r = \frac{\frac{1}{n} \sum_{i=1}^{n} (X_i - \bar{X})(Y_i - \bar{Y})}{\left[\frac{1}{n} \sum_{i=1}^{n} (X_i - \bar{X})^2 \right]^{\frac{1}{2}} \left[\frac{1}{n} \sum_{i=1}^{n} (Y_i - \bar{Y})^2 \right]^{\frac{1}{2}}}$$

which may be easily remembered as the sample covariance in the numerator, and the product of the two sample standard deviations in the denominator.

This measure of correlation may be used with any data of a numeric nature without any requirements concerning the scale of measurement or the type of underlying distribution, although it is difficult to interpret unless the scale of measurement is at least interval. It meets the necessary requirements of an acceptable measure of correlation. However, r is a random variable and, as such, r has a distribution function. Unfortunately, the distribution function of r depends on the bivariate distribution function of (X, Y). Therefore r has no value as a test statistic in nonparametric tests or for forming confidence intervals unless, of course, the distribution of (X, Y) is known.

In addition to this widely accepted r, many other measures of correlation have been invented that satisfy the preceding requirements for acceptability. An excellent and readable survey article by Kruskal (1958) discusses many of these. Some measures of correlation possess distribution functions that do not depend on the bivariate distribution function of (X, Y) if X and Y are independent and, therefore, they may be used as test statistics in nonparametric tests of independence. The measures of correlation selected for presentation here are functions of only the ranks assigned to the observations. They possess distribution functions that are independent of the bivariate distribution function of (X, Y) if X and Y are independent and continuous. They may even be used as measures of correlation on certain types of nonnumeric data.

Spearman's Rho

DATA. The data may consist of a bivariate random sample of size n, $(X_1, Y_1), (X_2, Y_2), \ldots, (X_n, Y_n)$. Let $R(X_i)$ be the rank of X_i as compared with the other X values, for $i = 1, 2, \ldots, n$. That is, $R(X_i) = 1$ if X_i is the smallest of $X_1, X_2, \ldots, X_n$, $R(X_i) = 2$ if X_i is the second smallest, and so on, with rank n being assigned to the largest of the X_i. Similarly, let $R(Y_i)$ equal $1, 2, \ldots,$ or n, depending on the relative magnitude of Y_i as compared with $Y_1, Y_2, \ldots, Y_n$, for each i.

Or the data may consist of nonnumeric observations occurring in n pairs if the observations are such that they can be ranked in the manner just described. The ranking may be based on the quality of the observations ("worst" observation to "best" observation) or according to the degree of preference attached to the observations, and so on.

In case of ties, assign to each tied value the average of the ranks that would have been assigned if there had been no ties, as was done in the Mann–Whitney and Kruskal–Wallis tests.

MEASURE OF CORRELATION. The measure of correlation as given by Spearman (1904) is usually designated by ρ (rho) and, if there are no ties, is defined as

$$(4) \qquad \rho = \frac{\sum_{i=1}^{n}\left[R(X_i) - \frac{n+1}{2}\right]\left[R(Y_i) - \frac{n+1}{2}\right]}{n(n^2 - 1)/12}$$

An equivalent but computationally easier form is given by

$$(5) \qquad \rho = 1 - \frac{6\sum_{i=1}^{n}[R(X_i) - R(Y_i)]^2}{n(n^2 - 1)} = 1 - \frac{6T}{n(n^2 - 1)}$$

where T represents the entire sum in the numerator. These forms are equivalent only if there are no ties. If there are many ties use

$$(6) \qquad \rho = \frac{\sum_{i=1}^{n} R(X_i)R(Y_i) - n\left(\frac{n+1}{2}\right)^2}{\left(\sum_{i=1}^{n} R(X_i)^2 - n\left(\frac{n+1}{2}\right)^2\right)^{\frac{1}{2}}\left(\sum_{i=1}^{n} R(Y_i)^2 - n\left(\frac{n+1}{2}\right)^2\right)^{\frac{1}{2}}}$$

which is simply Pearson's r computed on the ranks and average ranks. If a moderate number of ties is present in the data, Equation 5 is recommended for computational simplicity, since the difference between Equations 5 and 6 will be slight.

If there are no ties in the data, Spearman's ρ is merely what one obtains by replacing the observations by their ranks and then computing Pearson's r on the ranks. This may be seen as follows. If the data are replaced by their ranks, then $\bar{X}$ and $\bar{Y}$ correspond to

$$\overline{R(X)} = \frac{1}{n}\sum_{i=1}^{n} R(X_i) = \frac{1}{n}\sum_{i=1}^{n} i = \frac{1}{n}\frac{n(n+1)}{2}$$

(7)
$$= \frac{n+1}{2}$$

and

(8)
$$\overline{R(Y)} = \frac{n+1}{2}$$

Also, $\sum_{i=1}^{n}(X_i - \bar{X})^2$ and $\sum_{i=1}^{n}(Y_i - \bar{Y})^2$ correspond to

$$\sum_{i=1}^{n}[R(X_i) - \overline{R(X)}]^2 = \sum_{i=1}^{n}\left(i - \frac{n+1}{2}\right)^2$$

$$= \sum_{i=1}^{n}\left[i^2 - i(n+1) + \left(\frac{n+1}{2}\right)^2\right]$$

$$= \frac{n(n+1)(2n+1)}{6} - \frac{n(n+1)^2}{2} + \frac{n(n+1)^2}{4}$$

(9)
$$= \frac{n(n^2-1)}{12}$$

and

(10)
$$\sum_{i=1}^{n}[R(Y_i) - \overline{R(Y)}]^2 = \frac{n(n^2-1)}{12}$$

so that Equation 1 becomes Equation 4, and Pearson's r reduces to Spearman's ρ if the data are replaced by their ranks.

Example 1. Twelve sets of identical twins were given psychological tests to measure their aggressiveness. The emphasis is on examination of the degree of similarily between twins within the same set.

The data were measures of aggressiveness and are given here.

Twin Set	i	1	2	3	4	5	6	7	8	9	10	11	12
First-Born	X_i	86	71	77	68	91	72	77	91	70	71	88	87
Second-Born	Y_i	88	77	76	64	96	72	65	90	65	80	81	72

The first-born twins were ranked among themselves and the second-born twins were ranked among themselves, with the following results.

Twin Set, i	1	2	3	4	5	6	7	8	9	10	11	12
$R(X_i)$	8	3.5	6.5	1	11.5	5	6.5	11.5	2	3.5	10	9
$R(Y_i)$	10	7	6	1	12	4.5	2.5	11	2.5	8	9	4.5
$[R(X_i) - R(Y_i)]^2$	4	12.25	0.25	0	0.25	0.25	16	0.25	0.25	20.25	1	20.25

The five pairs of ties were given the average ranks for each pair.

First the statistic T in Equation 5 is computed.

$$T = \sum_{i=1}^{12} [R(X_i) - R(Y_i)]^2 = 75$$

Then ρ is obtained from Equation 5.

$$\rho = 1 - \frac{6T}{n(n^2 - 1)} = 1 - \frac{6(75)}{12(143)}$$

$$= .7378$$

As a point of interest, the value of ρ obtained using Equation 4 is .7290, and the ρ obtained using Pearson's r on the ranks is .7354, with the differences being due to the use of average ranks for the ties.

HYPOTHESIS TEST. The Spearman rank correlation coefficient is often used as a test statistic to test for independence between two random variables. Actually, Spearman's ρ is insensitive to some types of dependence, so it is better to be specific as to what types of dependence may be detected. Therefore, the hypotheses take the following form.

A. (Two-Tailed Test)

H_0: The X_i and Y_i are mutually independent
H_1: Either (a) there is a tendency for the larger values of X to be paired with the larger values of Y, or (b) there is a tendency for the smaller values of X to be paired with the larger values of Y

B. (One-Tailed Test for Positive Correlation)

H_0: The X_i and Y_i are mutually independent
H_1: There is a tendency for the larger values of X and Y to be paired together

C. (One-Tailed Test for Negative Correlation)

H_0: The X_i and Y_i are mutually independent
H_1: There is a tendency for the smaller values of X to be paired with the larger values of Y, and vice versa

The alternative hypotheses given here and throughout this section state the existence of correlation between X and Y, so that a null hypothesis of "no correlation between X and Y" would be more accurate than the statement of independence between X and Y as just given. Nevertheless, we will persist in using the null hypothesis of independence because it is in widespread usage and it is easier to interpret.

Spearman's ρ may be used as a test statistic for the preceding hypotheses. Table A10 gives the quantiles of ρ under the assumption of independence, the null hypothesis. Then H_0 in B is rejected if ρ is too large (at a level α if ρ

exceeds the $1-\alpha$ quantile), H_0 in C is rejected if ρ is too small, and the two-tailed test involves rejecting H_0 if ρ exceeds the $1-\alpha/2$ quantile or if ρ is less than the $\alpha/2$ quantile.

Instead of Spearman's ρ, it is usually more convenient to use directly the statistic T in Equation 5, where T is defined explicitly as

$$(11) \qquad T = \sum_{i=1}^{n} [R(X_i) - R(Y_i)]^2$$

Note that in the case of extensive ties, ρ should always be used. However, if the number of ties is moderate, the use of T eliminates some of the arithmetic involved in computing ρ. The test in this form is called the Hotelling-Pabst test, probably because of the paper by Hotelling and Pabst (1936) that emphasized the nonparametric nature of Spearman's ρ. Quantiles of T are given in Table A11. Note, however, that T is large when ρ is small, and vice versa. Therefore the H_0 in B is rejected at the level α if T is less than its α quantile. Also, H_0 in C is rejected if T exceeds its $1-\alpha$ quantile.

Example 2. Let us continue with Example 1. Suppose we want to test

H_0: The measures of aggressiveness of two identical twins are mutually independent

against the two-sided alternative

H_1: There is either a positive correlation or a negative correlation between the two measures of aggressiveness

at the .05 level of significance. The .025 (half of .05) quantile of T is given by Table A11, for $n = 12$, as

$$w_{.025} = 120$$

Because $T = 75$ in Example 1 and because 75 is less than 120, the null hypothesis is rejected. H_0 would have been rejected if T had exceeded the .975 quantile also, given by

$$w_{.975} = \tfrac{1}{3}n(n^2 - 1) - w_{.025}$$
$$= \tfrac{1}{3}(12)(143) - 120$$
$$= 452$$

as explained by the footnote in Table A11.

Since ρ had already been computed, it would have been easier to use ρ as a test statistic. The ρ in Example 1 equals .7378, which exceeds the .975 quantile given by Table A10 as .5804.

The approximate critical level $\hat{\alpha}$, the smallest level at which H_0 could have been rejected, is seen from Tables A10 and A11 to be about .01.

The next measure of correlation we are presenting resembles Spearman's ρ in that it is based on the order (ranks) of the observations rather than the numbers themselves, and the distribution of the measure does not depend on the distribution of X and Y if X and Y are independent and continuous. This measure, called Kendall's tau (τ), is usually considered to be more difficult to compute than Spearman's ρ. The chief advantage of Kendall's τ is that its distribution approaches the normal distribution quite rapidly so that the normal approximation is better for Kendall's τ than it is for Spearman's ρ, when the null hypothesis of independence between X and Y is true. Another advantage of Kendall's τ is its direct and simple interpretation in terms of probabilities of observing concordant and discordant pairs, as defined next.

Kendall's Tau

DATA. The data may consist of a bivariate random sample of size n, (X_i, Y_i) for $i = 1, 2, \ldots, n$. Two observations, for example, $(1.3, 2.2)$ and $(1.6, 2.7)$, are called *concordant* if both members of one observation are larger than their respective members of the other observation. Let N_c denote the number of concordant pairs of observations, out of the $\binom{n}{2}$ total possible pairs. A pair of observations, such as $(1.3, 2.2)$ and $(1.6, 1.1)$, are called *discordant* if the two numbers in one observation differ in opposite directions (one negative and one positive) from the respective members in the other observation. Let N_d be the total number of discordant pairs of observations. Pairs with ties between respective members are neither concordant nor discordant. Because the n observations may be paired $\binom{n}{2} = n(n-1)/2$ different ways, the number of concordant pairs N_c plus the number of discordant pairs N_d plus the number of pairs with ties should add up to $n(n-1)/2$.

The data may also consist of nonnumeric observations occurring in n pairs if the observations are such that N_c and N_d just described may be computed. .

MEASURE OF CORRELATION. The measure of correlation proposed by Kendall (1938) is

(12)
$$\tau = \frac{N_c - N_d}{n(n-1)/2}$$

If all pairs are concordant, Kendall's τ equals 1.0. If all pairs are discordant, the value is -1.0. As a measure of correlation, Kendall's τ satisfies the requirements stated at the beginning of this section.

The computation of τ is simplified if the observations (X_i, Y_i) are arranged in a column according to increasing values of X. Then each Y may be compared only with those below it, and the number of concordant and discordant comparisons is easily determined. Also, each pair of observations is considered only once. The procedure is illustrated in the following example.

Example 3. Again we will use the data in Example 1 for purposes of illustration. Arrangement of the data (X_i, Y_i) according to increasing values of X gives the following.

X_i, Y_i	Concordant Pairs Below (X_i, Y_i)	Discordant Pairs Below (X_i, Y_i)
(68, 64)	11	0
(70, 65)	9	0
tie $\{$ (71, 77)	4	4
(71, 80)	4	4
(72, 72)	5	1
tie $\{$ (77, 65)	5	0
(77, 76)	4	1
(86, 88)	2	2
(87, 72)	3	0
(88, 81)	2	0
tie $\{$ (91, 90)	0	0
(91, 96)	0	0
	$N_c = 49$	$N_d = 12$

Kendall's τ is given by

$$\tau = \frac{N_c - N_d}{n(n-1)/2} = \frac{49 - 12}{(12)(11)/2}$$

$$= .5606$$

There is a positive rank correlation between aggression scores in the twins we observed, as measured by Kendall's τ.

HYPOTHESIS TEST. Kendall's τ may also be used as a test statistic to test the null hypothesis of independence between X and Y, with possible one-tailed or two-tailed alternatives as described with Spearman's ρ. Some arithmetic may be saved, however, by using $N_c - N_d$ as a test statistic, without dividing by $n(n-1)/2$ to obtain τ. Therefore we use T as the Kendall test statistic, where T is defined as

(13) $$T = N_c - N_d$$

Quantiles of T are given in Table A12. If T exceeds the $1 - \alpha$ quantile, reject H_0 in favor of the one-sided alternative of positive correlation, at level α. Values of T less than the α quantile lead to acceptance of the alternative of negative correlation.

Example 4. In Example 3 Kendall's τ was computed by first finding the value of

$$T = N_c - N_d = 49 - 12$$

$$= 37$$

In Table A12 the quantiles for a two-tailed test of size $\alpha = .05$ are found, for $n = 12$, to be

$$w_{.975} = 28$$

and

$$w_{.025} = -28$$

For $T = 37$ the null hypothesis of independence is rejected. The critical level is estimated from Table A12 to be about

$$\hat{\alpha} \cong 2(.005) = .01$$

The same data were used for both Spearman's ρ and Kendall's τ in order to compare the two statistics better. It was seen that Spearman's ρ ($\rho = .7378$) was a larger number than Kendall's τ ($\tau = .5606$). However, the two tests using the two statistics (or their equivalents) produced nearly identical results. Both of the preceding statements hold true in most, but not all, situations. Spearman's ρ tends to be larger than Kendall's τ, in absolute value. However, as a test of significance there is no strong reason to prefer one over the other, because both will produce nearly identical results in most cases.

Daniels (1950) proposed the use of Spearman's ρ to test for trend by pairing measurements, called X_i, with the time (or order) at which the measurements were taken. The assumption is that the X_i are mutually independent, and the null hypothesis is that they are identically distributed. The alternative hypothesis is that the distribution of the X_is is related to time so that as time goes on, the X measurements tend to become larger (or smaller). The idea of trend was discussed more fully in Section 3.5, where the Cox and Stuart test for trend was presented. Tests of trend based on Spearman's ρ or Kendall's τ are generally considered to be more powerful than the Cox and Stuart test. It was mentioned in Section 3.5 that the A.R.E. of the Cox and Stuart test for trend, when applied to random variables known to be normally distributed, is about .78 with respect to the test based on the regression coefficient, while the A.R.E. of these tests using Spearman's ρ or Kendall's τ is about .98 under the same conditions, according to Stuart (1956). However, these tests are not as widely applicable as the Cox and Stuart test. For instance, these tests would be inappropriate in Example 3.5.3. These tests are appropriate in Example 3.5.2, and so we use that example to illustrate the use of Spearman's ρ as a test for trend. The procedure using Kendall's τ is similar.

Example 5. In Example 3.5.2, nineteen years of annual precipitation records are given on the following page. The two-tailed test for trend involves rejection of the null hypothesis of no trend if the total of the last column above is too large or too small. The test statistic is given by

$$T = \sum_{i=1}^{19} [R(X_i) - R(Y_i)]^2$$
$$= 1421.5$$

and the quantiles of T, for $\alpha = .05$, are given in Table A11, for $n = 19$, as

$$w_{.025} = 618$$

Precipitation X_i (inches)	Year Y_i	$R(X_i)$	$R(Y_i)$	$[R(X_i)-R(Y_i)]^2$
45.25	1950	12	1	121
45.83	1951	15	2	169
41.77	1952	11	3	64
36.26	1953	6	4	4
45.27	1954	13	5	64
52.25	1955	17	6	121
35.37	1956	2.5	7	20.25
57.16	1957	18	8	100
35.37	1958	2.5	9	42.25
58.32	1959	19	10	81
41.05	1960	9	11	4
33.72	1961	1	12	121
45.73	1962	14	13	1
37.90	1963	7	14	49
41.72	1964	10	15	25
36.07	1965	4	16	144
49.83	1966	16	17	1
36.24	1967	5	18	169
39.90	1968	8	19	121
				Total 1421.5

and

$$w_{.975} = \tfrac{1}{3}(19)(360) - 618$$
$$= 1662$$

As before, H_0 is readily accepted.

☐ *Theory.* The exact distributions of ρ and τ are quite simple to obtain in principle, although in practice the procedure is most tedious for even moderate-sized n. The exact distributions are found under the assumption that X_i and Y_i are independent and identically distributed. Then each of the $n!$ arrangements of the ranks of the X_is paired with the ranks of the Y_is is equally likely. As in the previous sections of this chapter, the distribution functions are obtained simply by counting the number of arrangements that give a particular value of ρ or τ and dividing that number by $n!$ to get the probability of that value of ρ or τ.

A form of the central limit theorem is applied to obtain the large sample approximate distributions, because both ρ and τ are based on the sum of random variables. Both ρ and τ have probability distributions symmetric about zero, so the means are zero for both. The variances are more difficult to obtain and will not be derived here. Division of ρ and τ by their respective variances thus results in a random variable that is approximately distributed as a standard normal random variable for large n. The approximation is considered quite good when used to find the quantiles of τ for $n \geq 8$, but not nearly as good when used to find the quantiles of ρ.
☐

If (X_i, Y_i), $i = 1, \ldots, n$ are independent and identically distributed bivariate normal random variables, both ρ and τ have an asymptotic relative efficiency of $9/\pi^2 = .912$, relative to the parametric test that uses Pearson's r as a test statistic (Stuart, 1954).

Kendall's Partial Correlation Coefficient

The concept of partial correlation is not an easy one to grasp. However, in order to illustrate the manner in which Kendall's τ may be extended to partial correlation, a brief attempt to describe partial correlation will be made.

In a multivariate random variable $(X_1, X_2, \ldots, X_k)$ there may be correlation between X_1 and X_2, between X_2 and X_5, and so forth, and a measure of this correlation might be any of the measures already described. Those measures estimate the total influence (correlation) of one random variable on the other, including the indirect influence felt because the second random variable is correlated not only with the first random variable, but perhaps with a third random variable that is in turn correlated with the first random variable and hence acts as a carrier of indirect influence between the first and second random variables. Sometimes it is desirable to measure the correlation between two random variables, under the condition that the indirect influence due to the other random variables is somehow eliminated. An estimate of this "partial" correlation between X_1 and X_2, say, while the indirect correlation due to $X_3, X_4, \ldots$, and X_n is eliminated, is denoted by $r_{12.34\ldots n}$ when using the extension of Pearson's r, or by $\tau_{12.34\ldots n}$ when using the extension of Kendall's τ. In the simple case where $n = 3$, the partial correlation may be estimated by Pearson's partial correlation coefficient

$$(14) \qquad r_{12.3} = \frac{r_{12} - r_{13} r_{23}}{\sqrt{(1 - r_{13}{}^2)(1 - r_{23}{}^2)}}$$

where r_{ij} is the ordinary Pearson r computed between X_i and X_j, and by Kendall's partial correlation coefficient

$$(15) \qquad \tau_{12.3} = \frac{\tau_{12} - \tau_{13} \tau_{23}}{\sqrt{(1 - \tau_{13}{}^2)(1 - \tau_{23}{}^2)}}$$

where τ_{ij} is the ordinary Kendall's τ computed between X_i and X_j.

Spearman's ρ has also been extended to measure partial correlation in the same way as described for Kendall's τ. An advantage of using the extension of Spearman's ρ is that existing computer programs for finding Pearson's partial correlation coefficient may be used on the ranks instead of the data, and the rank partial correlation coefficients are obtained easily.

The distribution of $r_{12.3}$ depends on the multivariate distribution function of

(X_1, X_2, X_3) and therefore may not be used as a test statistic in a nonparametric test. The distributions of $\tau_{12.3}$ or $\rho_{12.3}$ also depend on the multivariate distribution and therefore are not distribution free except in the case where all three variables are mutually independent. For more on this subject see recent articles by Simon (1977a, 1977b), Agresti (1977), or Wolfe (1977). Kendall (1942) presents a discussion of partial rank correlation.

Another measure of correlation proposed by Kendall for use in another situation is the *coefficient of concordance*. This may be used to measure total correlation when more then two variates are involved. However, the close relationship between Kendall's coefficient of concordance and a test statistic proposed by Friedman makes it advisable to present both of these statistics at the same time, which will be done in Section 5.8.

A comprehensive study of rank correlation is contained in Kendall's (1955) book on the subject. Knight (1966) gives a computer method for calculating Kendall's τ. Best (1973, 1974) presents extended tables for Kendall's τ and even has tables for different cases when ties are present for $n \leq 25$. Spearman's ρ for contingency tables is explained by Stuart (1963). More extensive tables for Spearman's ρ are given by Zar (1972), using some approximate methods that work quite well. Iman and Conover (1978) compare several approximations. A mechanical interpretation of Spearman's ρ is given by Evans (1973).

Usage of rank correlation methods in regression is discussed by Konijn (1961), Adichie (1967a, 1967b), and Sen (1968a) and is the topic of the next two sections of this chapter. Other papers on rank correlation and the concept of dependence are by Aitkin and Hume (1965), Lehmann (1966), Bell and Doksum (1967), Gokhale (1968), Ruymgaart (1973), Choi (1973) and Shirahata (1975, 1976).

EXERCISES

1. A husband and wife who go bowling together kept their scores for 10 lines to see if there was a correlation between their scores. The scores were:

Line	Husband's Score	Wife's Score	Line	Husband's Score	Wife's Score
1	147	122	6	151	120
2	158	128	7	196	108
3	131	125	8	129	143
4	142	123	9	155	124
5	183	115	10	158	123

(a) Compute ρ.
(b) Compute τ.
(c) Test the hypothesis of independence using a two-tailed test based on ρ.
(d) Do the same as in part c for τ.

2. The following is an example of a situation in which τ and ρ yield widely varying estimates of correlation.

X_i	Y_i	X_i	Y_i	X_i	Y_i
−8.7	−0.6	−1.9	−4.7	2.2	3.8
−8.3	−0.8	−1.6	−5.5	4.0	3.5
−8.2	−1.3	−1.3	−5.6	5.6	3.1
−7.2	−1.9	−0.2	−6.0	5.9	2.6
−6.1	−2.0	0.7	4.6	6.2	2.0
−6.0	−2.1	1.3	4.4	6.6	1.2
−4.1	−4.0	1.6	4.2	6.7	0.6
−2.0	−4.6	2.1	3.9	8.1	0.4

(a) Make a rough scatter diagram.
(b) Compute τ.
(c) Compute ρ.
(d) Does either ρ or τ lead to rejection of the null hypothesis that X and Y are independent?

3. A new worker is assigned to a machine that manufacturers bolts. Each day a sample of bolts is examined and the percent defective is recorded. Do the following data indicate a significant improvement over time for that worker?

Day	Percent	Day	Percent	Day	Percent
1	6.1	6	6.1	10	4.6
2	7.5	7	5.3	11	3.0
3	7.7	8	4.5	12	4.0
4	5.9	9	4.9	13	3.7
5	5.2				

(a) Use Spearman's ρ.
(b) Use Kendall's τ.

4. Is there a significant correlation between the age at which a U.S. president was inaugurated for the first time and the age at which he died?

Name	Inaugurated	Died	Name	Inaugurated	Died
Washington	57	67	Buchanan	65	77
J. Adams	61	90	Lincoln	52	56
Jefferson	57	83	A. Johnson	56	66
Madison	57	85	Grant	46	63
Monroe	58	73	Hayes	54	70
J. Q. Adams	57	80	Garfield	49	49
Jackson	61	78	Arthur	50	56
Van Buren	54	79	Cleveland	47	71
Harrison	68	68	Harrison	55	67
Tyler	51	71	McKinley	54	58
Polk	49	53	T. Roosevelt	42	60
Taylor	64	65	Taft	51	72
Fillmore	50	74	Wilson	56	67
Pierce	48	64	Harding	55	57

Name	Inaugurated	Died	Name	Inaugurated	Died
Coolidge	51	60	Eisenhower	62	78
Hoover	54	90	Kennedy	43	46
F. Roosevelt	51	63	L. Johnson	55	64
Truman	60	88			

(a) Use Spearman's ρ.
(b) Use Kendall's τ.

Note that these data do not represent a random sample, but one might assume that they behave as a random sample of all U.S. presidents, past, present and future.

PROBLEMS

1. Show that Equations 4 and 5 are equivalent expressions for ρ.
2. For $n = 5$ what pairing of ranks results in
   (a) $\rho = 1$?
   (b) $\tau = 1$?
   (c) $\rho = -1$?
   (d) $\tau = -1$?
3. Generalize the result of Problem 2 to any value of n in general and show that ρ and τ do, in fact, assume the values indicated.
4. Suppose someone suggests using

$$R = 1 - \frac{\sum_{i=1}^{n} |R(X_i) - R(Y_i)|}{(1/4)n^2}$$

which is sometimes called "Spearman's footrule."
   (a) Under what conditions will $R = 1$?
   (b) Under what conditions will $R = -1$?
5. Find the exact distribution of ρ, τ, and R from Problem 4 for the case where $n = 3$ under the usual assumption of independence.
6. Compute R defined in Problem 4 for the data in Exercise 2. Does R seem to resemble ρ more than τ in its behavior?

5.5. NONPARAMETRIC LINEAR REGRESSION METHODS

This section is related closely to the previous section on rank correlation in that we are examining a random sample $(X_1, Y_1), \ldots, (X_n, Y_n)$ on the bivariate random variable (X, Y). Correlation methods emphasize estimating the degree of dependence between X and Y. Regression methods are used to inspect the relationship between X and Y more closely. One important objective of regression methods is to predict a value of Y in the pair (X, Y) where only the value for X is known, on the basis of information that we can obtain from previous observations (X_1, Y_1) through (X_n, Y_n). For example, if X represents the score on a college entrance examination and Y represents the grade point average of that student 4 years later, observations on past students may help us to predict how well an incoming student will perform in the 4 years of college. Of course, Y is still a random variable, so we cannot expect to determine Y solely from knowing the associated value of X, but knowing X should help us make a better guess concerning Y.

Formally, the regression of Y on X is merely the mean of Y for a given value of X, say x.

Definition 1. The *regression of Y on X* is $E(Y \mid X = x)$. The *regression equation* is $y = E(Y \mid X = x)$.

If the regression equation is known, we can represent the regression on a graph by plotting y as the ordinate and x as the abscissa. But the regression equation is seldom, if ever, known. It is estimated on the basis of past data. For example, if we would like to predict Y when $X = 6$, we could use $E(Y \mid X = 6)$ if we knew it; otherwise, we could use the sample mean or the sample median of several observed values of Y for which X is equal to 6 or close to 6. In this way point estimates and confidence intervals may be formed for $E(Y \mid X = 6)$ using the methods described in Sections 3.2 and 5.7. In order to have enough observations so that the regression of Y on X can be estimated for each value of X, many observations are needed. It is not unusual to have large data sets that contain hundreds or even thousands of observations, in which case the nonparametric methods just mentioned work very nicely.

A more difficult situation arises when we have only a few observations and wish to estimate the regression of Y on X. This is what we will examine in this section. It is helpful to know something about the relationship between $E(Y \mid X = x)$ and x and to be able to use this information when there are only a few observations. First, we will examine the case where $E(Y \mid X = x)$ is a linear function of x; in the next section we will consider a more general situation where $E(Y \mid X = x)$ is a monotonic (either increasing or decreasing) function of x.

The regression of Y on X is said to be linear if the graph of the regression equation is a straight line.

Definition 2. The regression of Y on X is *linear regression* if the regression equation is of the form

$$(1) \qquad\qquad E(Y \mid X = x) = a + bx$$

for some constants a, called the *y-intercept*, and b, called the *slope*.

Usually the constants a and b are unknown and must be estimated from the data. If all of the observations on X and Y are used in estimating a and b, maximum usage is made of the data and a good estimate of $E(Y \mid X = x)$ for each x can be expected. A commonly accepted method for estimating a and b is called the *least squares method.*

Definition 3. The *least squares method* for choosing estimates A and B of a and b in the regression equation $y = a + bx$ is the method that minimizes the sum of squared deviations

(2) $$SS = \sum_{i=1}^{n} [Y_i - (A + BX_i)]^2$$

for the observations $(X_1, Y_1), \ldots, (X_n, Y_n)$.

The idea behind the least squares method is that an estimate of the regression line should be close to the observed values of X and Y because the true regression line is probably close to the observations. Therefore the estimate is selected so that the vertical distance D_i between Y_i and the estimated regression line, which equals $A + BX_i$ directly above or below Y_i, is small when all of the points are considered at once. We cannot merely make the sum of the Ds small because the sum of the Ds could be zero even though the estimated regression line is not at all close to the observations. That is, the absolute values of the distances D could be large, but the positive Ds could cancel the negative Ds, giving a sum of zero. To avoid this, we choose to minimize the sum of squares of the Ds:

(3) $$SS = \sum_{i=1}^{n} D_i^2$$

where

(4) $$D_i = Y_i - (A + BX_i)$$

This usually produces a straight line that agrees well with the data and, therefore, is a reasonable estimate of the true regression line.

Nonparametric Methods for Linear Regression

DATA. The data consist of a random sample (X_1, Y_1), $(X_2, Y_2), \ldots, (X_n, Y_n)$ from some bivariate distribution.

ASSUMPTIONS

1. The sample is a random sample. The methods of this section are valid if the values of X are nonrandom quantities as long as the Ys are independent with identical conditional distributions.
2. The regression of Y on X is linear.

LEAST SQUARES ESTIMATES. The method of least squares furnishes the estimate

(5) $$y = A + Bx$$

of the true regression line $y = a + bx$, where A and B are computed from

(6)
$$B = \frac{n \sum\limits_{i=1}^{n} X_i Y_i - \left(\sum\limits_{i=1}^{n} X_i \right)\left(\sum\limits_{i=1}^{n} Y_i \right)}{n \sum\limits_{i=1}^{n} X_i^2 - \left(\sum\limits_{i=1}^{n} X_i \right)^2}$$

and

(7)
$$A = \bar{Y} - B\bar{X}$$

where $\bar{X}$ and $\bar{Y}$ are the respective sample means.

TESTING THE SLOPE. To test the hypothesis concerning the slope, add the following assumption to Assumptions 1 and 2.

3. The "residual" $Y - E(Y \mid X)$ is independent of X.

Spearman's ρ may be adapted to test the following hypotheses concerning the slope. Let b_0 represent some specified number.

A. (Two-Tailed Test)

H_0: $b = b_0$
H_1: $b \neq b_0$

B. (One-Tailed Test)

H_0: $b = b_0$
H_1: $b > b_0$

C. (One-Tailed Test)

H_0: $b = b_0$
H_1: $b < b_0$

For each pair (X_i, Y_i) compute $Y_i - b_0 X_i = U_i$ (say). Then find the Spearman rank correlation coefficient ρ on the pairs (X_i, U_i), $i = 1, \ldots, n$, as described in Section 5.4. Table A10 gives the quantiles of ρ when H_0 is true and there are no ties. Reject H_0 in part B if ρ is too large (reject at a level α if ρ exceeds the $1 - \alpha$ quantile), reject H_0 in part C if ρ is too small, and reject H_0 in the two-sided hypothesis if ρ exceeds the $1 - \alpha/2$ quantile or is less than the $\alpha/2$ quantile. In case of ties approximate quantiles are obtained, as described in Section 5.4.

A CONFIDENCE INTERVAL FOR THE SLOPE. Assumptions 1, 2, and 3 are used in this procedure also. For each pair of points (X_i, Y_i) and (X_j, Y_j), such that $i < j$ and $X_i \neq X_j$, compute the "two-point slope,"

(8)
$$S_{ij} = \frac{Y_i - Y_j}{X_i - X_j}$$

Let N be the number of slopes computed. Order the slopes obtained and let

$$S^{(1)} \leq S^{(2)} \leq \cdots \leq S^{(N)}$$

denote the ordered slopes.

For a $1-\alpha$ confidence interval, find $w_{1-\alpha/2}$ from Table A12. Let r and s be given by

(9) $$r = \tfrac{1}{2}(N - w_{1-\alpha/2})$$

(10) $$s = \tfrac{1}{2}(N + w_{1-\alpha/2}) + 1$$

Round r downward and s upward to the next integer if they are not already integers. The $1-\alpha$ confidence interval for b is given by the interval $(S^{(r)}, S^{(s)})$. That is,

(11) $$P(S^{(r)} < b < S^{(s)}) \geq 1 - \alpha$$

COMMENT. The confidence interval for the slope is based on Kendall's τ, a completely different concept than the least squares concept; therefore it is possible, although unlikely, for the least squares estimator B for b to be outside the confidence interval for b. This could happen, for instance, when one value of Y much larger or smaller than we would expect it to be, judging from the other observations. Such an outlying observation can "pull" the least squares line up to fit it more closely at the expense of the other observations. In such a case it makes more sense to choose an estimated regression line that passes through the point (sample median of X, sample median of Y), with slope equal to the median of the slopes S_{ij} defined by Equation 8. That is, we could choose our estimators to be

(12) $$B_1 = \text{sample median of the } S_{ij}\text{s} = S_{(\text{median})}$$

and

(13) $$A_1 = Y_{.50} - B_1 X_{.50}$$

where $X_{.50}$ and $Y_{.50}$ refer to the sample medians.

Example 1. Let us again use the data from the previous section. The measure of aggressiveness for the firstborn in a set of twins is denoted by X_i and for the second twin by Y_i. The twelve observations are (86, 88), (71, 77), (77, 76), (68, 64), (91, 96), (72, 72), (77, 65), (91, 90), (70, 65), (71, 80), (88, 81), and (87, 72). These are plotted in Figure 1, along with the least squares regression line

$$y = 9.38 + .857x$$

which is obtained by substituting

$$\sum_{i=1}^{12} X_i = 949 \qquad \bar{X} = 79.1 \qquad \sum_{i=1}^{12} X_i^2 = 75{,}919$$

$$\sum_{i=1}^{12} Y_i = 926 \qquad \bar{Y} = 77.2 \qquad \sum_{i=1}^{12} X_i Y_i = 73{,}976$$

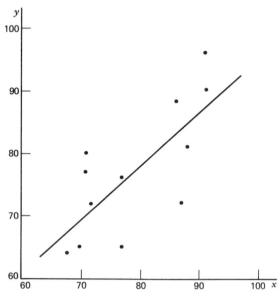

Figure 1. A plot of X_i versus Y_i and the least squares regression line.

into Equations 6 and 7 to obtain $B = .857$ and $A = 9.38$. We may use the regression line as a description of the relationship between Y and X or, more precisely, as an estimate of the conditional mean $E(Y \mid X)$ of Y given X. If the firstborn in an additional set of twins has a score $X = 80$, we can predict the score of the secondborn twin to be about $9.38 + (.857)(80) = 77.9$, or about the same as the firstborn twin.

It seems reasonable to use $b = 1$ in describing the relationship between twins, although the least squares estimate was $B = .857$, somewhat less than 1. To test H_0: $b = 1$ versus H_1: $b \neq 1$, the Spearman rank correlation coefficient is computed between X_i and the residual $U_i = Y_i - (1)X_i$.

	Twin Set i											
	1	*2*	*3*	*4*	*5*	*6*	*7*	*8*	*9*	*10*	*11*	*12*
Firstborn X_i	86	71	77	68	91	72	77	91	70	71	88	87
Residual U_i	2	6	−1	−4	5	0	−12	−1	−5	9	−7	−15
$R(X_i)$	8	3.5	6.5	1	11.5	5	6.5	11.5	2	3.5	10	9
$R(U_i)$	9	11	6.5	5	10	8	2	6.5	4	12	3	1

Equation 5.4.6 is used to compute $\rho = -.1232$, which is not in the two-tailed critical region $|\rho| > .5804$ of size $\alpha = .05$ from Table A10. Therefore the null hypothesis is accepted, and $\hat{\alpha}$ is greater than .20.

For a confidence interval for b, the $N = 63$ pairs (X_i, Y_i) and (X_j, Y_j) with $i < j$ and $X_i \neq X_j$ are used to find the 63 two-point slopes S_{ij} defined by Equation 8. From Table A12 for $n = 12$, $w_{.975}$ is found to be 28, so Equations 9 and 10 yield $r = 17$ and $s = 47$. The seventeenth ordered value of S_{ij} is found to be

$$S^{(17)} = .24$$

while the forty-seventh (or seventeenth from the top) is

$$S^{(47)} = 1.48$$

The 95% confidence interval for b is given by

$$P(.24 < b < 1.48) = .95$$

If for some reason the least squares regression line is considered unsatisfactory, the median value of S_{ij}, which is .89, may be used as the slope in Equation 12, along with $A_1 = 76.5 - (.89)(77) = 7.97$ from Equation 13, for the regression line estimate

$$y = 7.97 + .89x$$

as discussed in the previous comment.

☐ *Theory.* To derive A and B such that SS in Equation 2 is minimized, add and subtract the quantity $(\bar{Y} - B\bar{X})$ inside the brackets to get

(14) $$SS = \sum_{i=1}^{n} [(Y_i - \bar{Y}) - B(X_i - \bar{X}) + (\bar{Y} - B\bar{X} - A)]^2$$

Because of the algebraic identity

(15) $$(c - d + e)^2 = c^2 + d^2 + e^2 - 2cd + 2ce - 2de$$

we can expand Equation 14, using $c = Y_i - \bar{Y}$ and so on, to get

(16) $$SS = \sum_{i=1}^{n} (Y_i - \bar{Y})^2 + B^2 \sum_{i=1}^{n} (X_i - \bar{X})^2 + \sum_{i=1}^{n} (\bar{Y} - B\bar{X} - A)^2$$

$$-2B \sum_{i=1}^{n} (Y_i - \bar{Y})(X_i - \bar{X}) + 2(\bar{Y} - B\bar{X} - A) \sum_{i=1}^{n} (Y_i - \bar{Y})$$

$$-2B(\bar{Y} - B\bar{X} - A) \sum_{i=1}^{n} (X_i - \bar{X})$$

Because $\sum (Y_i - \bar{Y}) = 0$ and $\sum (X_i - \bar{X}) = 0$ by the definition of $\bar{Y}$ and $\bar{X}$, the last two summations equal zero in Equation 16. The third summation is smallest (zero) when

(17) $$A = \bar{Y} - B\bar{X}$$

which gives the least squares solution for A. We are left with the problem of finding the value of B that minimizes the sum of the second and fourth summations, that is, that minimizes

(18) $$B^2 S_x - 2BS_{xy}$$

where

(19)
$$S_x = \sum_{i=1}^{n} (X_i - \bar{X})^2$$

and

(20)
$$S_{xy} = \sum_{i=1}^{n} (X_i - \bar{X})(Y_i - \bar{Y})$$

By adding and subtracting S_{xy}^2/S_x to Equation 18, the sum of the second and fourth summations becomes

$$S_x\left[B^2 - 2B\frac{S_{xy}}{S_x} + \left(\frac{S_{xy}}{S_x}\right)^2 \right] - \frac{S_{xy}^2}{S_x} = S_x\left(B - \frac{S_{xy}}{S_x}\right)^2 - \frac{S_{xy}^2}{S_x}$$

which is obviously a minimum when

(21)
$$B = \frac{S_{xy}}{S_x}$$

in agreement with Equation 6. Note that this reduces the second and fourth summations to $-S_{xy}^2/S_x$, so that the minimum sum of squares is

(22)
$$SS_{min} = \sum_{i=1}^{n} (Y_i - \bar{Y})^2 - \frac{S_{xy}^2}{S_x}$$

$$= (1 - r^2) \sum_{i=1}^{n} (Y_i - \bar{Y})^2$$

where r is the Pearson product moment correlation coefficient given by Equation 5.4.1. Also note that no assumptions regarding the distribution of (X, Y) were made, so the least squares method is distribution free. In fact, the only purpose of assumptions 1 and 2 is to assure us that there is a regression line somewhere to be estimated.

Under assumption 3, the residuals

(23)
$$Y_i - E(Y_i \mid X_i) = Y_i - (a + bX_i)$$

are independent of X_i, so the assumptions of Section 5.4 regarding Spearman's ρ are met. Note that the ranks of $(Y_i - a - bX_i)$, $i = 1$ to n are the same as the ranks of $U_i = (Y_i - bX_i)$, $i = 1$ to n so we can test H_0: $b = b_0$ without knowing a. Just as Spearman's ρ is merely Pearson's r computed on ranks, this test is the rank analogue of computing r on the pairs (X_i, U_i), which is the usual parametric procedure for testing the same null hypothesis, valid with the additional assumption that (X, Y) has the bivariate normal distribution. Under that condition and the condition that the observations on X are equally spaced, the A.R.E. of this procedure is $(3/\pi)^{1/3} = .98$ according to Stuart (1954, 1956); for other distributions the A.R.E. is always greater than or equal to .95 (Lehmann, 1975).

To see the relationship between the slopes S_{ij} and Kendall's τ, note that for any hypothesized slope b_0 we have

$$S_{ij} = \frac{Y_i - Y_j}{X_i - X_j} = \frac{U_i + b_0 X_i - U_j - b_0 X_j}{X_i - X_j}$$

(24)
$$= b_0 + \frac{U_i - U_j}{X_i - X_j}$$

where $U_i = Y_i - b_0 X_i - a$ is the residual of Y_i from the hypothesized regression line $y = a + b_0 x$. The slope S_{ij} is greater than b_0 or less than b_0 according to whether the pair (X_i, U_i) and (X_j, U_j) is concordant or discordant in the sense described in Section 5.4 in the discussion of Kendall's τ. If we use the number of S_{ij}s less than b_0 as our test statistic for determining whether to accept H_0: $b = b_0$, we accept b_0 as long as the number of discordant pairs N_d is not too small or too large. Because N_d is related to the number of concordant pairs N_c by

(25)
$$N_c + N_d = N$$

where N is the total number of pairs, and because the quantiles of $N_c - N_d$ are given by Table 12 if we have the true slope and Assumption 3 of independence, we can say N_d is too small if $N_c - N_d$ is greater than $w_{1-\alpha/2}$ from Table A12. This is equivalent to saying N_d is less than $r = (N - w_{1-\alpha})/2$. In other words, b_0 is acceptable if b_0 is greater than at least r of the S_{ij}s, or $b_0 > S_{(r)}$. The same argument gives an upper bound for b_0, and the confidence interval is obtained. This method, due to Theil (1950), was modified to handle ties by Sen (1968a).

For nonparametric tests applicable to several regression lines see Sen (1972), Adichie (1974, 1975), and Pothoff (1974). Alternative methods of estimating regression coefficients are given by Jureckova (1971, 1977), Huber (1973), and Hettmansperger and McKean (1977). Kalbfleish (1974) discusses ranks in nonlinear models. Further discussions of nonparametric regression appear in Jaeckel (1972), Hollander and Wolfe (1973), Behnen (1976), and Stone (1977).

EXERCISES

1. A driver kept track of the number of miles she traveled and the number of gallons put in the tank each time she bought gasoline.

Miles	Gallons	Miles	Gallons
142	11.1	157	12.5
116	5.7	255	17.9
194	14.2	159	8.8
250	15.8	43	3.4
38	7.5	208	15.2

(a) Draw a diagram showing these points, using gallons as the x-axis.
(b) Estimate a and b using the method of least squares.

(c) Plot the least squares regression line on the diagram of part a.

(d) Suppose the EPA estimated this car's mileage at 18 miles per gallon. Test the null hypothesis that this figure applies to this particular car and driver. (Use the test for slope.)

(e) Find a 95% confidence interval for the mileage of this car and driver.

2. A random sample of American colleges and universities resulted in the following numbers of students and faculty (Spring 1973).

Name	Students	Faculty
American International	2546	129
Bethany Nazarene	1355	75
Carlow	1019	87
David Lipscomb	1858	99
Florida International University	4500	300
Heidelberg	1141	109
Lake Erie	784	77
Mary Hardin Baylor	1063	64
Mt. Angel	267	40
Newberry	753	61
Pacific Lutheran University	3164	190
St. Ambrose	1189	90
Smith	2755	240
Texas Women's University	5602	300
West Liberty State	2697	170
Wofford	988	73

(a) Draw a diagram showing these points using faculty as the x-axis.

(b) Estimate the regression line using the method of least squares.

(c) Plot the least squares regression line on the diagram of part a.

(d) Test the hypothesis that an increase of one faculty member is accompanied by an average increase of 15 students.

(e) Find a confidence interval for the slope.

5.6. METHODS FOR MONOTONE REGRESSION

In Section 5.5 nonparametric methods for linear regression were presented. These may be used in situations such as in Example 5.5.1, where the assumption of linear regression seems reasonable. In other situations it may be unreasonable to assume that the regression function is a straight line, but it may be reasonable to assume that $E(Y|X)$ increases (at least, it does not decrease) as X increases. In such a case we say the regression is *monotonically increasing*. If $E(Y|X)$ becomes smaller as X increases the regression is *monotonically decreasing*. Either case lends itself to the following method.

Nonparametric Methods for Monotonic Regression

DATA. The data consist of a random sample (X_1, Y_1), $(X_2, Y_2), \ldots, (X_n, Y_n)$ from some bivariate distribution.

ASSUMPTIONS

1. The sample is a random sample.
2. The regression of Y on X is monotonic.

AN ESTIMATE OF $E(Y|X)$ AT A POINT. To estimate the regression of Y on X at a particular value of $X = x_0$:

1. Obtain the ranks $R(X_i)$ of the Xs and $R(Y_i)$ of the Ys. Use average ranks in case of ties.
2. Find the least squares regression line on the ranks

(1) $$y = A_2 + B_2 x$$

where

(2) $$B_2 = \frac{\sum_{i=1}^{n} R(X_i)R(Y_i) - n(n+1)^2/4}{\sum_{i=1}^{n} [R(X_i)]^2 - n(n+1)^2/4}$$

and

(3) $$A_2 = (1 - B_2)(n+1)/2$$

3. Obtain a rank $R(x_0)$ for x_0 as follows:
   (a) If x_0 equals one of the observed X_is, let $R(x_0)$ equal the rank of that X_i.
   (b) If x_0 lies between two adjacent values X_i and X_j where $X_i < x_0 < X_j$, interpolate between their respective ranks to get $R(x_0)$.

(4) $$R(x_0) = R(X_i) + \frac{x_0 - X_i}{X_j - X_i}[R(X_j) - R(X_i)]$$

   This "rank" will not necessarily be an integer.
   (c) If x_0 is less than the smallest observed X or greater than the largest observed X, do not attempt to extrapolate. Information on the regression of Y on X is available only within the observed range of X.
4. Substitute $R(x_0)$ for x in Equation 1 to get an estimated rank $R(y_0)$ for the corresponding value of $E(Y|X = x_0)$.

(5) $$R(y_0) = A_2 + B_2 R(x_0)$$

5. Convert $R(y_0)$ into $\hat{E}(Y|X = x_0)$, an estimate of $E(Y|X = x_0)$, by refer-ring to the observed Y_is as follows.
   (a) If $R(y_0)$ equals the rank of one of the observations Y_i, let the estimate $\hat{E}(Y|X = x_0)$ equal that observation Y_i.
   (b) If $R(y_0)$ lies between the ranks of two adjacent values of Y, say Y_i and Y_j where $Y_i < Y_j$, so that $R(Y_i) < R(y_0) < R(Y_j)$, interpolate

between Y_i and Y_j.

(6)
$$\hat{E}(Y \mid X = x_0) = Y_i + \frac{R(y_0) - R(Y_i)}{R(Y_j) - R(Y_i)} (Y_j - Y_i)$$

(c) If $R(y_0)$ is greater than the largest observed rank of Y, let $\hat{E}(Y \mid X = x_0)$ equal the largest observed Y. If $R(y_0)$ is less than the smallest observed rank of Y, let $\hat{E}(Y \mid X = x_0)$ equal the smallest observed Y.

AN ESTIMATE OF THE REGRESSION OF Y ON X. To obtain the entire regression curve consisting of all points that can be obtained in the manner just described, the following procedure may be used.

1. Obtain the end points of the regression curve by using the smallest $X^{(1)}$ and the largest $X^{(n)}$ observations in the preceding procedure to obtain $\hat{E}(Y \mid X = x^{(1)})$ and $\hat{E}(Y \mid X = X^{(n)})$.
2. For each rank of $Y, R(Y_i)$, find the estimated rank of X_i $\hat{R}(X_i)$ from Equation 1.

(7)
$$\hat{R}(X_i) = [R(Y_i) - A_2]/B_2, \qquad i = 1, 2, \dots, n$$

3. Convert each $\hat{R}(X_i)$ to an estimate $\hat{X}_i$ in the manner of the preceding step 5. More specifically:
   (a) If $\hat{R}(X_i)$ equals the rank of some observation X_j, let $\hat{X}_i$ equal that observed value.
   (b) If $\hat{R}(X_i)$ falls between the ranks of two adjacent observations X_j and X_k, where $X_j < X_k$, then use interpolation,

(8)
$$\hat{X}_i = X_j + \frac{\hat{R}(X_i) - R(X_j)}{R(X_k) - R(X_j)} (X_k - X_j)$$

   to get $\hat{X}_i$.
   (c) If $\hat{R}(X_i)$ is less than the smallest observed rank of X or greater than the largest observed rank, no estimate $\hat{X}_i$ is found.
4. Plot each of the points found in step 3 on graph paper, with Y_i as the ordinate and $\hat{X}_i$ as the abscissa. Also plot the two end points found in step 1, with $\hat{E}(Y \mid X)$ as the ordinate and $X^{(1)}$ or $X^{(n)}$ as the abscissa. All of these points should be monotonic, increasing if $B_2 > 0$ and decreasing if $B_2 < 0$.
5. Connect the adjacent points in step 4 with straight lines. This series of connected line segments is the estimate of the regression of Y on X.

Example 1. Seventeen jars of fresh grape juice were obtained to study how long it took for the grape juice to turn into wine as a function of how much sugar was added to the huice. Various amounts of sugar, ranging from none to about 10 pounds, were added to the jars, and each day the jars were checked to see if the transition to wine was complete. At the end of 30 days the experiment was terminated, with three jars still unfermented. An estimate of the regression curve of Y (number of days till fermentation) versus X (pounds of sugar) is desired.

The observations (X_i, Y_i), their ranks $R(X_i)$ and $R(Y_i)$, and the values $\hat{R}(X_i)$ and $\hat{X}_i$ computed from the preceding steps 2 and 3 are given as indicated. Before obtaining $\hat{R}(X_i)$ and $\hat{X}_i$, the least squares coefficients on the ranks are computed from Equations 2 and 3 and substituted into Equation 1 to get the least squares regression line on ranks

$$(9) \qquad\qquad y = 17.4 - (.934)x$$

The end points are obtained by substituting the ranks $R(X^{(1)}) = 1$ and $R(X^{(n)}) = 17$ into Equation 9 to get the estimated ranks 16.47 and 1.52 for $\hat{E}(Y \mid X)$. By interpolating between successive observations on Y, the rank 1.52 converts to $\hat{E}(Y \mid X = 9.8) = 4.01$ from Equation 6. The other value $E(Y \mid X = 0)$ is taken to be ">30" because the estimated rank 16.47 exceeds the largest observed rank of Y. Neither end point is used in the graph because $\hat{R}(X_i) = 1.52$ is greater than the smallest usuable rank obtained from step 2, so that point merely lies on the line segment already formed. The observation ">30" cannot be used as an end point on a line segment.

X_i	Y_i	$R(X_i)$	$R(Y_i)$	$\hat{R}(X_i)$	$\hat{X}_i$
0	>30	1	16	1.50	.25
.5	>30	2	16	1.50	.25
1.0	>30	3	16	1.50	.25
1.8	28	4	14	3.64	1.51
2.2	24	5	13	4.71	2.08
2.7	19	6	12	5.78	2.59
4.0	17	7.5	11	6.85	3.44
4.0	9	7.5	8	10.06	5.62
4.9	12	9	9.5	8.46	4.58
5.6	12	10	9.5	8.46	4.58
6.0	6	11	5	13.28	7.50
6.5	8	12	7	11.13	6.07
7.3	4	13	1.5	17.02	9.80
8.0	5	14	3	15.42	9.01
8.8	6	15	5	13.28	7.50
9.3	4	16	1.5	17.02	9.80
9.8	6	17	5	13.28	7.50

The observations are plotted in Figure 2. The regression curve, consisting of line segments joining successive values of $(\hat{X}_i, Y_i)$, is also plotted in Figure 2. An estimate $\hat{E}(Y \mid X = x_0)$ is obtained easily from Figure 1 by finding the ordinate that corresponds to the abscissa x_0. Note that the "censored" observations ">30" were used in the regression of the ranks, but that portion of the regression curve of the data is not possible to plot.

It is interesting to note how a set of observations, with a regression curve that is obviously nonlinear, is converted to ranks that have a regression curve that seems to be linear. The ranks are plotted in Figure 3 along with Equation 9.

☐ *Theory.* The procedures for monotonic regression are based on the fact that if two variables have a monotonic relationship, their ranks will have a linear relationship. A scattering of the observations around the monotonic regression line should correspond to a scattering of the ranks around their linear regression line. The ranks serve as transformed variables, where the transformation seeks to convert the monotonic regression function to a
☐ linear regression function.

Other methods of handling monotone regression are compared and illustrated by Cryer, Robertson, Wright, and Casady (1972), Casady and Cryer (1976), Hogg (1975), and Iman and Conover (1979). This procedure is explained more fully in Iman and Conover (1979).

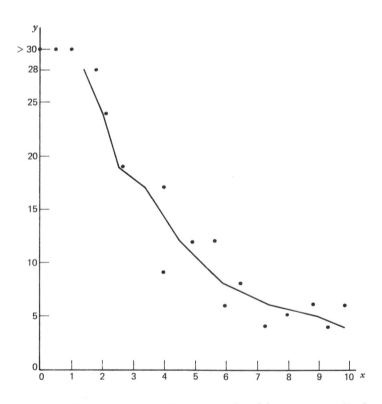

Figure 2. Number of days till fermentation (y) versus pounds of sugar (x), and the estimated monotonic regression curve.

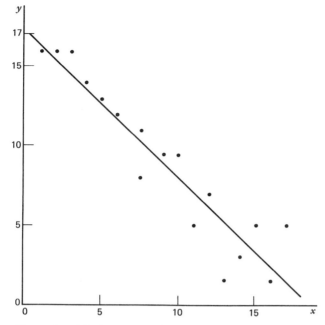

Figure 3. $R(Y_i)$ versus $R(X_i)$ and the least squares regression line.

EXERCISES

1. Dose-response curves such as the following are widely used in biological studies and in the pharmaceutical industry. Suppose that a certain drug (X, measured in milliliters) is administered to guinea pigs to see whether a particular reaction (cancer, diabetes, etc.) occurs. Five guinea pigs are treated at each of several dosage levels of the drug, the percent of the animals showing the reaction is recorded as the Y variable.

X(dosage)	0.5	1.0	1.5	2.0	2.5	3.0	3.5	4.0	4.5	5.0
Y(percent response)	0	0	20	0	40	60	40	80	100	100

(a) Plots the points on a graph. Does the expected value of the response sum to be a linear function of the dosage? A monotonic function?
(b) Estimate $E(Y|X)$ at $X = 3.0$ milliliters.
(c) Estimate $E(Y|X)$ at $X = 3.3$ milliliters.
(d) Estimate the regression of Y on X. Plot the estimated regression curve on the same graph used in part a.

2. Ten companies reported their percent increase in advertising expenses, X, and their percent increase in sales, Y, for last year as compared with the previous year.

					Company					
	1	2	3	4	5	6	7	8	9	10
X (advertising)	4	62	31	−11	47	88	16	−1	74	21
Y (sales)	10	33	39	−14	37	39	18	−8	45	33

(a) Plot the points on a graph. Does expected value of the percent increase in sales seem to be a linear function of the percent increase in advertising? A monotonic function?

(b) Estimate the expected percent increase in sales for a 25% increase in advertising.

(c) Estimate the regression of Y on X. Plot the estimated regression curve on the same graph used in part a.

PROBLEMS

1. Show that the estimates of $E(Y \mid X)$ can never be less than the smallest observed value of Y or greater than the largest observed value of Y. Discuss the advantages or disadvantages of this property relative to the situations described in Exercises 1 and 2.

2. Find the least squares regression line for the data in Exercise 1. Use this regression line to estimate the mean of Y given $X = 0.5$ milliliter. Does this estimate seem reasonable to you?

5.7. THE ONE-SAMPLE OR MATCHED-PAIRS CASE

The rank test of this section deals with the single random sample and the random sample of matched pairs that is reduced to a single sample by considering differences. A matched pair (X_i, Y_i) is actually a single observation on a bivariate random variable. The sign test of Section 3.4 analyzed matched pairs of data by reducing each pair to a plus, a minus, or a tie and applying the binomial test to the resultant single sample. The test of this section also reduces the matched pair (X_i, Y_i) to a single observation by considering the difference

$$(1) \qquad\qquad D_i = Y_i - X_i \qquad \text{for} \qquad i = 1, 2, \ldots, n$$

The analysis is then performed on the D_is as a sample of single observations. Whereas the sign test merely noted whether D_i was positive, negative, or zero, the test of this section notes the sizes of the positive D_is relative to the negative D_is. The model of this section resembles the model used in the sign test. Also, the hypotheses resemble the hypotheses of the sign test. The important difference between the sign test and this test is an additional assumption of *symmetry* of the distribution of differences. Before we introduce the test, we should clarify the meaning of the adjective *symmetric* as it applies to a distribution and discuss the influence of symmetry on the scale of measurement.

Symmetry is easy to define if the distribution is discrete. A discrete distribution is symmetric if the left half of the graph of the probability function is the mirror image of the right half. For example, the binomial distribution is symmetric if $p = 1/2$ (see Figure 4) and the discrete uniform distribution is

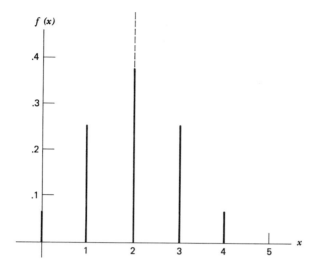

Figure 4. Symmetry in binomial distribution.

always symmetric (see Figure 5). The dotted lines in the figures represent the lines about which the distributions are symmetric.

For other than discrete distributions we are not able to draw a graph of the probability function. Therefore a more abstract definition of symmetry is required, such as the following.

> *Definition 1.* The distribution of a random variable X is *symmetric* about a line $x = c$, for some constant c, if the probability of $X \leq c - x$ equals the probability of $X \geq c + x$ for each possible value of x.

In Figure 4, $c = 2$ and the definition is easily verified for all real numbers x. In Figure 5, $c = 3.5$. Even though we may not know the exact distribution of a random variable, we are often able to say, "It is reasonable to assume that the distribution is symmetric." Such an assumption is not as strong as the assumption of a normal distribution; while all normal distributions are symmetric, not all symmetric distributions are normal.

If a distribution is symmetric, the mean (if it exists) coincides with the median because both are located exactly in the middle of the distribution, at the line of symmetry. One consequence of adding the assumption of symmetry to the model is that any inferences concerning the median are also valid statements for the mean.

A second consequence of adding the assumption of symmetry to the model is that the required scale of measurement is changed from ordinal to interval. With an ordinal scale of measurement, two observations of the random variable need only to be distinguished on the basis of which is larger and which is smaller. It is not necessary to know which one is farthest from the median, such as when the two observations are on opposite sides of the median. If the

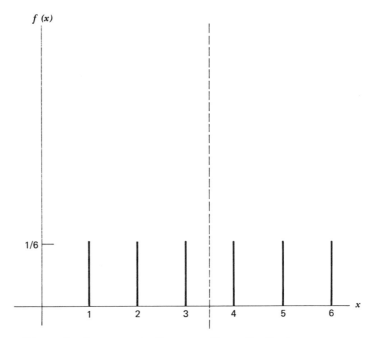

Figure 5.   Symmetry in discrete uniform distribution.

assumption of symmetry is a meaningful one, the distance from the median is a meaningful measurement and, therefore, the distance between two observations is a meaningful measurement. As a result, the scale of measurement is more than just ordinal, it is interval.

A test presented by Wilcoxon (1945) is designed to test whether a particular sample came from a population with a specified median. It may also be used in situations where observations are paired, such as "before" and "after" observations on each of several subjects, to see if the second random variable in the pair has the same median as the first.

The Wilcoxon Signed Ranks Test

DATA. The data consist of n' observations $(x_1, y_1), (x_2, y_2), \ldots, (x_{n'}, y_{n'})$ on the respective bivariate random variables $(X_1, Y_1), (X_2, Y_2), \ldots, (X_{n'}, Y_{n'})$. (For use as a median test with a single sample, see Example 2.) The absolute differences (without regard to sign)

(2) $$|D_i| = |Y_i - X_i| \qquad i = 1, 2, \ldots, n'$$

are then computed for each of the n' pairs (X_i, Y_i).

Omit from further consideration all pairs with a difference of zero (i.e., where $X_i = Y_i$, or $D_i = 0$). Let the number of pairs remaining be denoted by

$n, n \leq n'$. Ranks from 1 to n are assigned to these n pairs according to the relative size of the absolute difference, as follows. The rank 1 is given to the pair (X_i, Y_i) with the smallest absolute difference $|D_i|$; the rank 2 is given to the pair with the second smallest absolute difference; and so on, with the rank n being assigned to the pair with the largest absolute difference.

If several pairs have absolute differences that are equal to each other, assign to each of these several pairs the *average* of the ranks that would have otherwise been assigned [i.e., if the ranks 3, 4, 5, and 6 belong to four pairs, but we do not know which rank to assign to which pair because all four absolute differences are exactly equal to each other, assign the average rank $(1/4)(3+4+5+6) = 4.5$ to each of the four pairs.]

ASSUMPTIONS

1. The distribution of each D_i is symmetric.
2. The D_is are mutually independent.
3. The D_is all have the same median.
4. The measurement scale of the D_is is at least interval.

HYPOTHESES. Let the common median of the D_is be denoted by $d_{.50}$. Then the hypotheses may be stated in several ways, depending on whether the test is one tailed or two tailed.

A. (One-Tailed Test)

$$H_0: d_{.50} \leq 0$$

$$H_1: d_{.50} > 0$$

This alternative hypothesis may be loosely stated as, "The values of the X_is tend to be smaller than the values of the Y_is."

B. (One-Tailed Test)

$$H_0: d_{.50} \geq 0$$

$$H_1: d_{.50} < 0$$

This alternative hypothesis may be loosely stated as, "The values of the X_is tend to be larger than the values of the Y_is."

C. (Two-Tailed Test)

$$H_0: d_{.50} = 0$$

$$H_1: d_{.50} \neq 0$$

If the model is changed slightly, these hypotheses may be broadened considerably. The change consists of adding the assumption, "The (X_i, Y_i), for $i = 1, 2, \ldots, n$, constitute a random (bivariate) sample." This new assumption actually includes Assumptions 2 and 3 of the model. The hypotheses may then be stated as

A.
$$H_0: E(X) \geq E(Y)$$
$$H_1: E(X) < E(Y)$$

B.
$$H_0: E(X) \leq E(Y)$$
$$H_1: E(X) > E(Y)$$

C.
$$H_0: E(X) = E(Y)$$
$$H_1: E(X) \neq E(Y)$$

in place of A, B, and C originally given, if $E(X)$ and $E(Y)$ exist.

TEST STATISTIC. Let R_i, called the signed rank, be defined for each pair (X_i, Y_i) as follows.

(3) R_i = the rank assigned to (X_i, Y_i) if D_i is positive

   R_i = the negative of the rank assigned to (X_i, Y_i) if D_i is negative

The test statistic is defined as follows.

(4)
$$T = \frac{\sum_{i=1}^{n} R_i}{\sqrt{\sum_{i=1}^{n} R_i^2}}$$

In case there are no ties, Equation 4 simplifies to

(5)
$$T = \frac{\sum_{i=1}^{n} R_i}{\sqrt{n(n+1)(2n+1)/6}}$$

with the aid of Lemma 1.4.2. For the case of no ties it is more convenient to use only the positive signed ranks.

(6) $$T^+ = \sum (R_i \text{ where } D_i \text{ is positive})$$

DECISION RULE. The decision rule may be expressed three different ways, corresponding to the three sets of hypotheses A, B, and C. Let w_p be the pth quantile, obtained from Table A13 when using T^+, or approximately given by Table A1 when T is being used (i.e., when there are ties or large samples). If T^+ is being used, substitute T^+ for T in the following.

A. (One-Tailed Test) Large values of T indicate that H_0 is false, so reject H_0 at the level of significance α if T exceeds $w_{1-\alpha}$. Accept H_0 if T is less than or equal to $w_{1-\alpha}$.

B. (One-Tailed Test) Small values of T indicate that H_0 is false, so reject H_0 at the level of significance α if T is less than w_α. Accept H_0 if T is greater than or equal to w_α.

C. (Two-Tailed Test) Reject H_0 at the level of significance α if T exceeds $w_{1-\alpha/2}$ or if T is less than $w_{\alpha/2}$. If T is between $w_{\alpha/2}$ and $w_{1-\alpha/2}$ or equal to either quantile, accept H_0.

Example 1. As in the examples of Section 5.4, 12 sets of identical twins were given psychological tests to measure in some sense the amount of aggressiveness in each person's personality. In the previous sections we used these measurements to compute a measure of agreement between persons in the same twin set. Now we are interested in comparing the twins with each other to see if the firstborn twin tends to be more aggressive than the other. The results are as follows, where the higher score indicates more aggressiveness.

							Twin Set					
	1	2	3	4	5	6	7	8	9	10	11	12
Firstborn X_i	86	71	77	68	91	72	77	91	70	71	88	87
Second twin Y_i	88	77	76	64	96	72	65	90	65	80	81	72
Difference D_i	+2	+6	−1	−4	+5	0	−12	−1	−5	+9	−7	−15
Rank of $|D_i|$	3	7	1.5	4	5.5	—	10	1.5	5.5	9	8	11
R_i	3	7	−1.5	−4	5.5	—	−10	−1.5	−5.5	9	−8	−11

The hypotheses are:

H_0: The firstborn twin does not tend to be more aggressive than the other ($d_{.50} \geq 0$).

H_1: The firstborn twin tends to be more aggressive than the second twin ($d_{.50} < 0$).

These correspond to hypotheses set B. We are assuming that the test scores are accurate measures of the aggressiveness of the individuals. The test statistic is:

$$(7) \qquad T = \frac{\sum R_i}{\sqrt{\sum R_i^2}} = \frac{-17}{\sqrt{505}} = -.7565$$

The critical region (see decision rule B) of size $\alpha = .05$ corresponds to values of T less than -1.6449 (from Table A1). Therefore H_0 is readily accepted. The observed value $-.7565$ corresponds approximately, to the .225 quantile from Table A1, so $\hat{\alpha}$ is about .225.

If we had used T^+ and Table A13 we would have obtained $T^+ = 24.5$ and a critical region corresponding to values of T^+ less than 14. So the same conclusion would have been reached and a similar value for $\hat{\alpha}$ would have been obtained by interpolation between $w_{.20}$ and $w_{.30}$ in Table A13.

The Wilcoxon signed ranks test is equally appropriate as a median test, where the data consist of a single random sample of size n', $X_1, X_2, \ldots, X_{n'}$. Let X be a random variable with the same distribution as the X_i and let m be a specified constant. The hypotheses, corresponding to the preceding hypotheses, sets A, B, and C, are as follows.

A. (One-Tailed Test)

H_0: The median of X is $\geq m$

H_1: The median of X is $< m$

B. (One-Tailed Test)

H_0: The median of X is $\leq m$

H_1: The median of X is $> m$

C. (Two-Tailed Test)

H_0: The median of X equals m

H_1: The median of X is not m

The word "mean" may be substituted for "median" in these hypotheses because of the assumption of symmetry of the distribution of X.

Pairs $(X_1, m), (X_2, m), \ldots, (X_{n'}, m)$ are formed, and the pairs are treated exactly the same as described in the Wilcoxon signed ranks test. The rest of the Wilcoxon test procedure remains unchanged. The following example illustrates the procedure.

Example 2. Thirty observations on a random variable X are obtained in order to test the hypothesis that $E(X)$, the mean of X, is no larger than 30 (hypotheses set B).

$$H_0: \ E(X) \leq 30$$

$$H_1: \ E(X) > 30$$

The observations, the differences $m - X_i$, and the ranks of the pairs are as follows. (The random sample was ordered first, for convenience.)

X_i	$D_i = 30 - X_i$	Rank of $|D_i|$	X_i	$D_i = 30 - X_i$	Rank of $|D_i|$
23.8	+6.2	17	35.9	−5.9	15
26.0	+4.0	11	36.1	−6.1	16
26.9	+3.1	8	36.4	−6.4	18
27.4	+2.6	6	36.6	It is not necessary to compute	
28.0	+2.0	5	37.2	the remaining differences and	
30.3	−0.3	1	37.3	ranks.	
30.7	−0.7	2	37.9		
31.2	−1.2	3	38.2		
31.3	−1.3	4	39.6		
32.8	−2.8	7	40.6		
33.2	−3.2	9	41.1		
33.9	−3.9	10	42.3		
34.3	−4.3	12	42.8		
34.9	−4.9	13	44.0		
35.0	−5.0	14	45.8		

The approximate .05 quantile is obtained from Table A13.

$$w_{.05} \cong \frac{n(n+1)}{4} + x_{.05}\sqrt{n(n+1)(2n+1)/24}$$

$$= \frac{(30)(31)}{4} + (-1.645)\sqrt{(30)(31)(61)/(24)}$$

$$= 232.5 - (1.645)(48.6)$$

$$= 152.6$$

Therefore the critical region of approximate size .05 corresponds to values of the test statistic less than 152.6.

The test statistic is defined by Equation 6. In this case T^+ equals the sum of the ranks associated with the positive D_i.

(8) $$T^+ = 47$$

The small value of T^+ results in rejection of H_0. We conclude that the mean of X is greater than 30.

The approximate critical level is found by solving the equation

(9) $$47 = 232.5 + x_\alpha(48.6)$$

to get

(10) $$x_\alpha = -3.82$$

Table A1 shows that $\hat{\alpha}$ is smaller than .0001.

☐ *Theory.* The model states that all of the differences D_i share a common median, say $d_{.50}$. By the definition of median, whenever the D_is are assumed to have continuous distributions, each D_i has probability .5 of exceeding $d_{.50}$, the median, and probability .5 of being less than $d_{.50}$. If $d_{.50} = 0$ (H_0 in set C), each D_i has probability .5 of being positive and the same of being negative. Because of the symmetric distribution of each D_i, the size of each difference and the resulting rank are independent of whether the difference is above or below the median. (Without symmetry it would be possible for the positive differences to tend to be much larger than the negative differences, or vice versa.)

The purpose of these considerations is to find the distribution of the test statistic T when H_0 is true. First, we will consider the null hypothesis of the two-tailed test. The resulting distribution applies equally well in the one-tailed tests.

Consider n chips numbered from 1 to n, corresponding to the n ranks if there are no ties. Suppose each chip has its number written on one side and the negative of its number on the other side (like 6 and −6). Each chip is tossed into the air so that it is equally likely to land with either side showing, corresponding to the ranks of (X_i, Y_i), which are equally likely to correspond to a positive D_i, in which case R_i of Equation 3 equals the rank, or a negative D_i, in which case R_i is a negative rank. Let T^+ be the sum of the positive numbers showing after all n chips are tossed, corresponding to the definition of T^+ in Equation 6. The probability distribution of T^+ is the same in the game with the chips as it is when H_0 is true, but the game with the chips is easier to imagine.

The sample space in the game with the chips consists of points such· as $(1, 2, 3, -4, -5, 6, 7, \ldots, n)$, simply a reordering of the R_i associated with a set of data like in Example 1. The tosses are independent of each other, so each of the 2^n points has probability $(1/2)^n$. The test statistic T^+ equals the sum of the positive numbers in the sample point. Therefore the probability that T^+ equals any number x is found by counting the points whose positive numbers add to x, then multiplying that count by the probability $(1/2)^n$.

For example, if $n = 8$, $T^+ = 0$ one way (all the positive numbers landed face down), and so $P(T = 0) = (1/2)^8$. $T^+ = 1$ only one way, $T^+ = 2$ only one way, but $T^+ = 3$ two ways, points $(-1, -2, 3, -4, -5, -6, -7, -8)$ and $(1, 2, -3, -4, -5, -6, -7, -8)$. Also, $T^+ = 4$ two ways. That is,

$$P(T^+ = 0) = (1/2)^8 = \frac{1}{256} \qquad P(T^+ \le 0) = .0039$$

$$P(T^+ = 1) = \frac{1}{256} \qquad P(T^+ \le 1) = .0078$$

$$P(T^+ = 2) = \frac{1}{256} \qquad P(T^+ \le 2) = .0117$$

$$P(T^+ = 3) = \frac{2}{256} \qquad P(T^+ \le 3) = .0195$$

$$P(T^+ = 4) = \frac{2}{256} \qquad P(T^+ \le 4) = .0273$$

$$\text{etc.} \qquad\qquad\qquad \text{etc.}$$

The distribution function of T^+ is tabulated in Owen (1962) for $n \le 20$ and in Harter and Owen (1970) for $n \le 50$. A table of selected quantiles for $n \le 100$ is given by McCornack (1965). That table is more extensive than we need here, so the more useful quantiles were selected and are given in Table A13. The use of Table A13 will generally result in a slightly conservative test, because the probability of being less than the pth quantile may be less than p. For example, if $n = 8$, as in the preceding paragraph, the .025 quantile of T^+ is given in Table A13 as 4, while the actual size of the critical region corresponding to values of T^+ less than 4 is .0195. Further results on the exact distribution of T^+ are given by Claypool (1970) and Chow and Hodges (1975).

For thc one-tailed tests, the probability of getting a point in the critical region is a maximum when $d_{.50} = 0$, so this is the situation to be considered. Thus the preceding distribution of T^+ is equally valid when H_0 is true in the one-tailed tests.

To find the conditional distribution of T^+ when there are ties, only the initial step in the discussion is changed. That is, the numbers on the chips must agree with the ranks and midranks assigned to the pairs (X_i, Y_i) in the particular set of data under consideration. Call these ranks and midranks $a_1, a_2, \ldots, a_n$. In Example 1 we have $a_1 = 1.5$, $a_2 = 1.5$, $a_3 = 3$, and so on. For this set of numbers we can find the distribution of T^+. Because there are 11 numbers in Example 1, there are $2^{11} = 2048$ points in the sample space. The smallest 5% of these, about 102 points, constitute the critical region. This is a large number of points to tabulate by hand, so the normal approximation is used.

To use the normal approximation, let S equal the sum of all the R_i. Then, to apply the central limit theorem (set B) from Section 1.5, we need the mean and variance of S when H_0 is true. Note that under H_0,

$$P(R_i = a_i) = \tfrac{1}{2} \qquad \text{and} \qquad P(R_i = -a_i) = \tfrac{1}{2}$$

so that

(11) $$E(R_i) = a_i(\tfrac{1}{2}) + (-a_i)(\tfrac{1}{2}) = 0$$

and

(12) $$\text{Var}(R_i) = a_i^2(\tfrac{1}{2}) + (-a_i)^2(\tfrac{1}{2}) = a_i^2$$

Since the R_is are independent of each other (the tosses of the chips are independent), we can apply Theorems 1.4.1 and 1.4.3 to get

(13) $$E(S) = \sum_{i=1}^{n} E(R_i) = 0$$

and

(14) $$\text{Var}(S) = \sum_{i=1}^{n} \text{Var}(R_i) = \sum_{i=1}^{n} a_i^2$$

But since a_i^2 always equals R_i^2 (the sign always becomes $+$), we can say

(15) $$\text{Var}(S) = \sum_{i=1}^{n} R_i^2$$

and apply the central limit theorem to

(16) $$T = \frac{\sum_{i=1}^{n} R_i}{\sqrt{\sum_{i=1}^{n} R_i^2}}$$

and use the normal distribution as an approximation whenever exact tables are not available. Justification for the treatment of ties is given by Vorlickova (1970) and Conover (1973a).

☐

The method presented here for handling zero differences was suggested by Wilcoxon (1949). Another method of handling zero differences is thoroughly discussed by Pratt (1959). It involves leaving the zero differences in, ranking the $|D_i|$ as described, but treating all of the $D_i = 0$ values as a tie and assigning the average rank in the usual manner. Then R_i is defined as in Equation 3, except that $R_i = 0$ if $D_i = 0$. Then T is computed from Equation 4 and compared with Table A1. Table A13 is not used with Pratt's method when testing hypotheses, but some exact tables are given by Rahe (1974). A comparison by Conover (1973b) shows that each method of handling ties at zero is more powerful than the other in some situations, so there is little reason to prefer one over the other. Pratt's suggestion of retaining the zero differences is incorporated into the following method for finding a confidence interval for $d_{.50}$, the common median of the D_is, which appears in Tukey (1949) and Walker and Lev (1953).

Confidence Interval for the Median Difference

DATA. The data consist of n observations $(x_1, y_2), \ldots, (x_n, y_n)$ on the bivariate random variables $(X_1, Y_1), (X_2, Y_2), \ldots, (X_n, Y_n)$, respectively. Compute the differences

$$D_i = Y_i - X_i$$

For each pair and arrange them in order from the smallest (the most negative) to the largest (the most positive), denoted as follows.

$$D^{(1)} \leq D^{(2)} \leq \cdots \leq D^{(n-1)} \leq D^{(n)}$$

Or, in the usual situation, the data consist of a single sample $D_1, D_2, \ldots, D_n$, arranged in order as shown. We wish to find a confidence interval for the common median of the D_is.

ASSUMPTIONS

1. The distribution of each D_i is symmetric.
2. The D_is are mutually independent.
3. The D_is all have the same median.
4. The measurement scale of the D_is is at least interval.

METHOD. To obtain a $1 - \alpha$ confidence interval, obtain the $\alpha/2$ quantile $w_{\alpha/2}$ from Table A13. (If $w_{\alpha/2} = $ zero no confidence interval may be obtained for that value of α.) Then consider the $n(n + 1)/2$ possible averages $(D_i + D_j)/2$ for all i and j, including $i = j$. The $w_{\alpha/2}$th largest of these averages and the

$w_{\alpha/2}$th smallest of these averages constitute the upper and lower bounds for the $1 - \alpha$ confidence interval. It is not necessary to compute all $n(n + 1)/2$ averages; only the averages near the largest and the smallest need to be computed to obtain a confidence interval.

Example 1 (continued). The 12 values of D_i, arranged in order, are

$$-15, -12, -7, -5, -4, -1, -1, 0, 2, 5, 6, 9$$

To find a 95% confidence interval for the median difference, Table A13 is entered with $n = 12$ to obtain $w_{.025} = 14$. The 14 smallest averages, starting with $(-15 - 15)/2$, are

$$-15, -13.5, -12, -11, -10, -9.5, -9.5, -8.5, -8, -8, -8, -7.5, -6.5, -6.5$$

so the lower bound for the confidence interval is -6.5. The 14 largest averages are

$$9, 7.5, 7, 6, 5.5, 5.5, 5, 4.5, 4, 4, 4, 3.5, 3, 2.5$$

so the upper bound for the confidence interval is 2.5. The 95% confidence interval for the median difference of aggressiveness scores (firstborn twin minus second twin) is

(17) $$P(-6.5 \le d_{.50} \le 2.5) \ge .95$$

Theory. To see the relationship between average differences $(D_i + D_j)/2$ and the rank of the difference, consider the following. The rank of any D_i, say $D_i = 6$ in the previous example, is equal to the number of D_js as close or closer to 0 than $D_i = 6$ is, assuming no ties. By counting the averages between 0 and 6, which involve D_i, we obtain the rank of D_i. (We must be careful to include the average of D_i with itself in the count.) By repeating this for all positive D_i we obtain, as a total count, the test statistic T^+.

A confidence interval for the median $d_{.50}$ of D_i is found by using the Wilcoxon test to test

$$H_0: d_{.50} = m$$

for various values of m. This procedure is equivalent to subtracting the value m from each D_i and testing to see if the median of the new D_is equals zero. But instead of subtracting m from each D_i and then reranking and recomputing T^+, it is easier to look at the averages of the original D_is, counting how many averages are above m (instead of zero, as we did before) and that equals T^+. By working backward, starting with the critical value for T^+ and finding those largest averages, the stopping point is the value of m that would have barely resulted in acceptance of H_0. Thus the bounds for the confidence interval are found.

Noether (1967b) shows that if the continuity assumption is not true, the confidence interval with its end points (U and L) has a confidence coefficient of at least $1-\alpha$, while the interval without its end points has a confidence coefficient of at least $1-\alpha$, Therefore we recommend inclusion of the end points and a statement of the form

$$P(L \leq d_{.50} \leq U) \geq 1-\alpha$$

A discussion of this method of finding confidence intervals is given by Moses (1965). If the sampling is stratified rather than random, see the article by McCarthy (1965). For another type of dependency in the sample, see Høyland (1968). Confidence regions for the case of multivariate random variables are given by Puri and Sen (1968). Sequential sampling methods offer some advantages according to Geertsema (1970) and Srivastava and Sen (1973). Other methods of estimating the center of a distribution are discussed by Schuster and Navarta (1973), Noether (1973), Johns (1974), and Maritz, Wu, and Staudte (1977). See Beran (1977) for a theoretical discussion of robust location estimates.

COMPARISON WITH OTHER PROCEDURES

When one encounters paired observations and wishes to test whether the mean difference is zero, and the scale of measurement is interval as in this section, the first test that usually comes to mind is the "paired t test," also called the "one-sample t test." This test uses the test statistic

(18)
$$t = \frac{\bar{D}}{\sqrt{\dfrac{1}{n(n-1)} \sum_{i=1}^{n} (D_i - \bar{D})^2}}$$

where $\bar{D}$ is the sample mean of the D_is, and compares it with quantiles of the t distribution from Table A25, in the row $k = n-1$. In order for the quantiles from Table A25 to be accurate, an additional assumption of normality must be made. That is, add to the assumptions for the Wilcoxon test the assumption that the D_is are identically distributed normal random variables.

The assumptions of the Wilcoxon test are easier to justify than the assumption of normality. If the data are discrete, we know right away that the distribution is nonnormal because the normal distribution is continuous. If the data have an occasional very large or very small observation, called "outliers," the power of the t test drops considerably and should not be used. Unfortunately, this type of nonnormality is difficult to detect.

If a computer program for the t test is available, it can be used for the Wilcoxon test also. Merely use the R_is instead of the D_is in computing t and compare the result with Table A25, as just described. This approximation is slightly more accurate than the normal approximation described earlier and works well with ties. For an approximation that is better than either of these, use the average of T and t and compare it with the average of the two critical values obtained from Tables A1 and A25. See Iman (1974a) for more details of this method.

The A.R.E. of the Wilcoxon signed ranks test relative to the paired t test is computed under the following restrictions.

1. The bivariate random variables $(X_1, Y_1), \ldots, (X_n, Y_n)$ constitute a random sample.
2. The distribution of X_i is identical with the distribution of Y_i, except for a possible difference in means.

Under these conditions the A.R.E. may range from $108/125 = .864$ up to infinity, but with the surprising insurance feature of never being less than .864. Therefore the Wilcoxon test never can be too bad, but it can be infinitely good as compared with the usual parametric test under these circumstances.

Under the further restriction that the differences D_i have the normal distribution, the A.R.E. is $3/\pi = .955$. If we instead assume that the differences D_i have the uniform distribution, the A.R.E. is 1.0. For a distribution known as the double exponential distribution, the A.R.E. is 1.5.

Under the preceding restrictions the sign test (Section 3.4) may be used to test the same hypotheses as the Wilcoxon test. Then the A.R.E. of the sign test, relative to the Wilcoxon test, is as follows.

Assumed Distribution	*A.R.E.*
Normal	$\frac{2}{3}$
Uniform	$\frac{1}{3}$
Double exponential	$\frac{4}{3}$

It may be surprising to some that the sign test may be more powerful than the Wilcoxon test under some circumstances. The Wilcoxon A.R.E., compared to the paired t test, is $3/2$ for the double exponential distribution and, therefore, multiplication gives the A.R.E. of the sign test, relative to the paired t test, as

$$(\tfrac{4}{3})(\tfrac{3}{2}) = 2$$

For the double exponential distribution the sign test has twice the asymptotic efficiency as the paired t test. For other investigations of power and efficiency, see Klotz (1963, 1965), Arnold (1965), Noether (1967a), and Kraft and van Eeden (1972).

The Wilcoxon signed ranks test is sometimes called a test of symmetry. Schuster (1975) and Rao, Schuster, and Littell (1975) discuss estimation with symmetric distributions, while Rothman and Woodroofe (1972) present an alternative test for symmetry. Symmetry tests for bivariate random variables are discussed by Bell and Haller (1969), Hollander (1971), and Bhattacharyya, Johnson, and Neave (1971). Extensions to multivariate random variables are examined by Bennett (1965) and Sen and Puri (1967). Hollander (1970) adapts the Wilcoxon test to test for parallelism of two regression lines. Adaptations to sequential sampling are presented by Miller (1970), Weed, Bradley, and Govindarajulu (1974), Sen and Ghosh (1974), Reynolds (1975), and Spurrier and Hewett (1976). For other papers related to this section see Groeneveld

(1972) and Bickel and Lehmann (1975). A test proposed by Walsh (1949) is identical to the Wilcoxon test for n less than 7, but not for n of 7 or more. An application of the Wilcoxon test to problems involving circular distributions, as presented by Batschelet (1965), is given by Schach (1969a).

EXERCISES

1. A random sample consisting of 20 people who drove automobiles was selected to see if alcohol affected reaction time. Each driver's reaction time was measured in a laboratory before and after drinking a specified amount of a beverage containing alcohol. The reaction times in seconds were as follows.

Subject	Before	After	Subject	Before	After
1	.68	.73	11	.65	.72
2	.64	.62	12	.59	.60
3	.68	.66	13	.78	.78
4	.82	.92	14	.67	.66
5	.58	.68	15	.65	.68
6	.80	.87	16	.76	.77
7	.72	.77	17	.61	.72
8	.65	.70	18	.86	.86
9	.84	.88	19	.74	.72
10	.73	.79	20	.88	.97

   Does alcohol affect reaction time?

2. A grocer wishes to see whether the median number of items bought on each sale could be considered to be 10, so he observes 12 customers at the checkout counter.

Customer	Number of Items	Customer	Number of Items
1	22	7	15
2	9	8	26
3	4	9	47
4	5	10	8
5	1	11	31
6	16	12	7

   Test $H_0: d_{.50} = 10$ using the Wilcoxon test. Which assumptions of the model are violated in this problem?

3. Test the data of Example 3.5.3 to see if there is a tendency for the observations in the second year to be less than the observations in the first year.

4. Each member of a girls' basketball team was given a brief warmup period and then told to shoot 25 free throws. The number X of goals was recorded. Then the team was given an extensive workout and, after a brief rest period, was told to shoot another 25 free throws each. The number Y of successful attempts was again recorded. Do the data indicate that the percentages tend to drop when the players are tired?

Player	1	2	3	4	5	6	7	8	9	10	11	12
X_i (before)	18	12	7	21	19	14	8	11	19	16	8	11
Y_i (after)	16	10	8	23	13	10	8	13	9	8	8	5

5. A candidate for political office realizes that she will maximize her vote-getting ability if she adopts the median position of her constituents. Therefore she devises a questionnaire and distributes it to 15 voters (who hopefully resemble a random sample). The results of the questionnaire are scored from one extreme (0) to the other (10).

Voter	Score	Voter	Score	Voter	Score
1	6.7	6	9.3	11	8.8
2	4.2	7	8.9	12	5.4
3	4.1	8	7.4	13	6.1
4	2.3	9	7.4	14	6.0
5	6.1	10	9.3	15	4.9

Find a 90% confidence interval for the median score. Based on the procedure of this section, what is your point estimate of the median score?

6. An emergency rescue squad is responsible for accidents occurring in a long narrow lake. They wish to build a permanent station in the spot that will minimize the total distance they will have to travel going to accidents in the future. That spot should be at the median (by distance from some point of reference) spot at which accidents will occur. Assuming that the accidents occurred thus far resemble a random sample of all accidents yet to occur, the distances (from the dam) are measured.

Accident	Distance (miles)	Accident	Distance (miles)
1	7.1	8	6.1
2	4.4	9	2.2
3	3.9	10	6.7
4	2.2	11	4.9
5	4.2	12	7.3
6	3.4	13	0.3
7	1.1	14	7.6

What is a 95% confidence interval for the optimal distance from the dam for the station?

PROBLEMS

1. Find the probability distribution of the test statistic T^+ of this section for $N=5$. (Assume H_0 is true in the two-tailed test.)

2. Compare the two forms of the t statistic, one given by Equation 18 of this section and the other given by Equation 5.12.1. Show that these two forms are algebraically equivalent.

3. Suppose that the t statistic is computed on the signed ranks R_i instead of the differences D_i. Show that this statistic is the following function of T,

$$t_R = \frac{T}{\left(\dfrac{n}{n-1} - \dfrac{1}{n-1}\, T^2\right)^{\frac{1}{2}}}$$

as stated in Equation 5.12.4. Also show that as T increases, t_R increases and, therefore, the test that rejects H_0 for large T is equivalent to the test that rejects H_0 for large t_R.

4. Compute the paired t test statistic on the data of Exercise 4 and compare the results with those of the Wilcoxon test. (Use Table A25 with row $k = 11$; zeros are not discarded in the paired t test.)

5.8. SEVERAL RELATED SAMPLES

In Section 5.2 we presented the Kruskal–Wallis rank test for several independent samples, which is an extension of the Mann–Whitney test for two independent samples introduced in Section 5.1. In this section we consider the problem of analyzing several *related* samples, which is an extension of the problem of matched pairs, or two related samples, examined in the previous section. First, we will present a test that is an extension of the Wilcoxon signed-ranks test to the case of several related samples. Then we will present the Friedman test for the same situation. The Friedman test is an extension of the sign test of Sections 3.4 and 3.5 and requires fewer assumptions than the first test. The Friedman test is the better-known test for this experimental situation, but it has less power in some situations.

The problem of several related samples arises in an experiment that is designed to detect differences in k possibly different treatments, $k \geq 2$. The observations are arranged in *blocks*, which are groups of k experimental units similar to each other in some important respects, such as k puppies that are littermates and therefore may tend to respond to a particular stimulus more similarly than would randomly selected puppies from various litters. The k experimental units within a block are matched randomly with the k treatments being scrutinized, so that each treatment is administered once and only once within each block. In this way the treatments may be compared with each other without an excess of unwanted effects confusing the results of the experiment. The total number of blocks used is denoted by b, $b > 1$.

The experimental arrangement described here is usually called a *randomized compete block design*. This design may be compared with the *incomplete* block design described in the next section, in which the blocks do not contain enough experimental units to enable all the treatments to be applied in all the blocks, and so each treatment appears in some blocks but not in others. Examples of randomized complete block designs are as follows.

1. *Psychology.* Five litters of mice, with four mice per litter, are used to examine the relationship between environment and aggression. Each litter is considered to be a block. Four different environments are designed. One mouse from each litter is placed in each environment, so that the four mice from each litter are in four different environments. After a suitable length of time, the mice are regrouped with their littermates and are ranked according to degree of aggressiveness.
2. *Home economics.* Six different types of bread dough are compared to see which bakes the fastest by forming three loaves with each type of dough. Three different ovens are used, and each oven bakes the six different types of bread at the same time. The ovens are the blocks and the doughs are the treatments.
3. *Environmental engineering.* One experimental unit may form a block if the different treatments may be applied to the same unit without leaving residual effects. Seven different men are used in a study of the effect of color schemes on work efficiency. Each man is considered to be a block and spends some time in each of three rooms, each with its own type of color scheme. While in the room, each man performs a work task and is measured for work efficiency. The three rooms are the treatments.

By now the reader should have some idea of the nature of a randomized complete block design. The usual parametric method of testing the null hypothesis of no treatment differences is called the two-way analysis of variance. The following nonparametric method depends only on the ranks of the observations within each block and the ranks of the block to block sample ranges. Therefore it may be considered a two-way analysis of variance on ranks. We prefer to follow the convention of naming the test after its inventor, Dana Quade (1972, 1979).

The Quade Test

DATA. The data consist of b mutually independent k-variate random variables $(X_{i1}, X_{i2}, \ldots, X_{ik})$, called b blocks, $i = 1, 2, \ldots, b$. The random variable X_{ij} is in block i and is associated with treatment j. The b blocks are arranged as follows.

Block	Treatment 1	2	$\cdots$	k
1	X_{11}	X_{12}	$\cdots$	X_{1k}
2	X_{21}	X_{22}	$\cdots$	X_{2k}
3	X_{31}	X_{32}	$\cdots$	X_{3k}
$\cdots$	$\cdots$	$\cdots$	$\cdots$	$\cdots$
b	X_{b1}	X_{b2}	$\cdots$	X_{bk}

Let $R(X_{ij})$ be the rank, from 1 to k, assigned to X_{ij} within block (row) i. That is, for block i the random variables $X_{i1}, X_{i2}, \ldots, X_{ik}$ are compared with each other and the rank 1 is assigned to the smallest observed value, the rank 2 to the second smallest, and so on to the rank k, which is assigned to the largest observation in block i. Ranks are assigned in all of the b blocks. Use average ranks in case of ties.

The next step again uses the original observations X_{ij}. Ranks are assigned to the blocks themselves according to the size of the sample range in each block. The sample range within block i is the difference between the largest and the smallest observations within that block.

$$(1) \qquad \text{Range in block } i = \underset{j}{\text{maximum}}\, \{X_{ij}\} - \underset{j}{\text{minimum}}\, \{X_{ij}\}$$

There are b sample ranges, one for each block. Assign rank 1 to the block with the smallest range, rank 2 to the second smallest, and so on to the block with the largest range, which gets rank b. Use average ranks in case of ties. Let $Q_1, Q_2, \ldots, Q_b$ be the ranks assigned to blocks $1, 2, \ldots, b$, respectively.

Finally, the block rank Q_i is multiplied by the difference between the rank within block i, $R(X_{ij})$ and the average rank within blocks, $(k+1)/2$, to get the product S_{ij}, where

$$(2) \qquad S_{ij} = Q_i\left[R(X_{ij}) - \frac{k+1}{2}\right]$$

is a statistic that represents the relative size of each observation within the block, adjusted to reflect the relative significance of the block in which it appears.

Let S_j denote the sum for each treatment:

$$(3) \qquad S_j = \sum_{i=1}^{b} S_{ij}$$

for $j = 1, 2, \ldots, k$.

ASSUMPTIONS

1. The b k-variate random variables are mutually independent. (The results within one block do not influence the results within the other blocks.)
2. Within each block the observations may be ranked according to some criterion of interest.
3. The sample range may be determined within each block so that the blocks may be ranked.

(*Note.* This test is valid even if there are many ties in the rankings.)

HYPOTHESES

H_0: Each ranking of the random variables within a block is equally likely (i.e., the treatments have identical effects).

H_1: At least one of the treatments tends to yield larger observed values than at least one other treatment.

TEST STATISTIC. For convenience, first calculate the term

(4)
$$A_1 = \sum_{i=1}^{b} \sum_{j=1}^{k} S_{ij}^2$$

where S_{ij} is given by Equation 2. This is called the "total sum of squares." If there are no ties, A_1 simplifies to

(5)
$$A_1 = b(b+1)(2b+1)k(k+1)(k-1)/72$$

Next calculate the term

(6)
$$B_1 = \frac{1}{b} \sum_{j=1}^{k} S_j^2$$

where S_j is given by Equation 3. This is called the "treatment sum of squares." The test statistic is

(7)
$$T_1 = \frac{(b-1)B_1}{A_1 - B_1}$$

If $A_1 = B_1$, consider that point to be in the critical region and calculate the critical level as $\hat{\alpha} = (1/k!)^{b-1}$.

DECISION RULE. Reject the null hypothesis at the level α if T_1 exceeds the $1-\alpha$ quantile of the F distribution as given in Table A26 with $k_1 = k-1$ and $k_2 = (b-1)(k-1)$. Actually, the F distribution only approximates the exact distribution of T_1, but exact tables are not available at this time. As b becomes large, the F approximation comes closer to being exact.

MULTIPLE COMPARISONS. *Only if* the preceding procedure results in rejection of the null hypothesis are multiple comparisons made. Treatments i and j are considered different if the inequality

(8)
$$|S_i - S_j| > t_{1-\alpha/2} \left[\frac{2b(A_1 - B_1)}{(b-1)(k-1)} \right]^{\frac{1}{2}}$$

is satisfied, where S_i, S_j, A_1, and B_1 are given previously, and where $t_{1-\alpha/2}$ is obtained from Table A25 with $(b-1)(k-1)$ degrees of freedom. This comparison is made for all pairs of treatments, using the same α used in the Quade test.

Example 1. Seven stores are selected for a marketing survey. In each store five different brands of a new type of hand lotion are placed side by side. At the end of a week, the number of bottles of lotion sold for each brand is tabulated, with the following results.

Numbers of customers (rank within stores)
Brand

Store	A	B	C	D	E
1	5 (2)	4 (1)	7 (3)	10 (4)	12 (5)
2	1 (2.5)	3 (5)	1 (2.5)	0 (1)	2 (4)
3	16 (2)	12 (1)	22 (3.5)	22 (3.5)	35 (5)
4	5 (4.5)	4 (2.5)	3 (1)	5 (4.5)	4 (2.5)
5	10 (3.5)	9 (2)	7 (1)	13 (5)	10 (3.5)
6	19 (2)	18 (1)	28 (3)	37 (4)	58 (5)
7	10 (5)	7 (2.5)	6 (1)	8 (4)	7 (2.5)

The observations are ranked from 1 to 5 within each store, with average ranks assigned when there are ties. These ranks $R(X_{ij})$ appear in parentheses.

Next, the sample range within each store is computed by subtracting the smallest observation from the largest. In store 1 the sample range is $12-4=8$. These sample ranges are listed next, along with the ranks Q_i of the sample ranges, and the products

$$S_{ij} = Q_i[R(X_{ij}) - (k+1)/2]$$

Store Number	Sample Range	Rank Q_i	Brand A	$S_{ij} = Q_i[R(X_{ij}) - 3]$ B	C	D	E
1	8	5	−5	−10	0	+5	+10
2	3	2	−1	+4	−1	−4	+2
3	23	6	−6	−12	+3	+3	+12
4	2	1	+1.5	−0.5	−2	+1.5	−0.5
5	6	4	+2	−4	−8	+8	+2
6	40	7	−7	−14	0	+7	+14
7	4	3	+6	−1.5	−6	+3	−1.5
			$S_j =$ −9.5	−38	−14	+23.5	+38

From Equation 4,

$$A_1 = \sum_{i=1}^{7}\sum_{j=1}^{5} S_{ij}^2 = (-5)^2 + (-10)^2 + \cdots = 1366.5$$

which is slightly less than the more easily obtained value 1400, from Equation 5, applicable if there had been no ties. Equation 6 yields

$$B_1 = \frac{1}{7}\sum_{j=1}^{5} S_j^2 = \frac{1}{7}[(-9.5)^2 + (-38)^2 + \cdots] = 532.4$$

which gives, when substituted into Equation 7, the test statistic

$$T_1 = \frac{6(532.4)}{1366.5 - 532.4} = 3.83$$

This value of T_1 is greater than 2.78, the .95 quantile of the F distribution with $k_1 = 4$ and $k_2 = 24$, obtained from Table A26; therefore the null hypothesis is rejected at $\alpha = .05$. In fact, perusal of Table A26 shows $\hat{\alpha}$ to be slightly less than .025. Some brands seem to be preferred over others by the store customers.

Because the null hypothesis is rejected multiple comparisons are made. From Equation 8, two treatments are considered different if the difference between their sums $|S_i - S_j|$ exceeds

$$t_{1-\alpha/2}\left[\frac{2b(A_1 - B_1)}{(b-1)(k-1)}\right]^{\frac{1}{2}} = 2.064\left[\frac{14(834.1)}{24}\right]^{\frac{1}{2}} = 45.53$$

where $t_{1-\alpha/2} = t_{.975}$ is obtained from Table A25 and $(b-1)(k-1) = 24$ degrees of freedom. Thus the brands that may be considered different from each other are brand A and E, brands B and D, brands B and E, and brands C and E.

Another test for the randomized complete blocks design is the following test, named after its inventor, noted economist Milton Friedman. Where the previous test is a multisample extension of the Wilcoxon signed ranks test, the Friedman test is an extension of the sign test. That is, for $k = 2$ treatments, the Friedman test is equivalent to the sign test. Therefore the power of the Friedman test for a small number of treatments may be less than the power of the previous test in many situations of interest. If the number of treatments is five or more, this test appears to be more powerful than the Quade test, according to an unpublished study by R. L. Iman and the author.

The Friedman test may be preferred when comparisons among the different blocks are not possible. The test is easier to execute than the previous test and is therefore simpler to understand.

The Friedman Test

DATA. Find the ranks within blocks $R(X_{ij})$ as described in the previous test and then sum the ranks for each treatment to obtain R_j where:

$$(9) \qquad R_j = \sum_{i=1}^{b} R(X_{ij})$$

for $j = 1, 2, \ldots, k$. Note that no comparisons between different blocks need to be made in this test.

ASSUMPTIONS. The assumptions are the same as the first two assumptions of the previous test. The third assumption is not needed because no comparisons are made between blocks.

HYPOTHESES. The hypotheses are the same as in the previous test.

TEST STATISTICS. First calculate the term A_2, given by

(10)
$$A_2 = \sum_{i=1}^{b} \sum_{j=1}^{k} [R(X_{ij})]^2$$

If there are no ties A_2 simplifies to

(11)
$$A_2 = \frac{bk(k+1)(2k+1)}{6}$$

Then calculate the term

(12)
$$B_2 = \frac{1}{b} \sum_{j=1}^{k} R_j^2$$

where R_j is given by Equation 9. The test statistic is

(13)
$$T_2 = \frac{(b-1)[B_2 - bk(k+1)^2/4]}{A_2 - B_2}$$

If $A_2 = B_2$, consider that point to be in the critical region and compute the critical level as $\hat{\alpha} = (1/k!)^{b-1}$.

DECISION RULE. Reject the null hypothesis at the level α if T_2 exceeds the $1-\alpha$ quantile of the F distribution as given in Table A26 with $k_1 = k-1$ and $k_2 = (b-1)(k-1)$. Actually, the F distribution only approximates the exact distribution of T_2, but the approximation is fairly good and improves as b gets large.

COMMENT. The usual form of the Friedman test compares the statistic

(14)
$$T_3 = \frac{(k-1)[bB_2 - b^2k(k+1)^2/4]}{A_2 - bk(k+1)^2/4}$$

with quantiles of the chi-square distribution, $k-1$ degrees of freedom. Recent studies by Iman and Davenport (1979) show the F approximation to T_2 as clearly superior to the previous chi-square approximation, so that is what we recommend here.

MULTIPLE COMPARISONS. The following method for comparing individual treatments may be used only if the Friedman test results in rejection of the null hypothesis. Treatments i and j are considered different if the following inequality is satisfied.

(15)
$$|R_j - R_i| > t_{1-\alpha/2} \left[\frac{2b(A_2 - B_2)}{(b-1)(k-1)} \right]^{\frac{1}{2}}$$

where R_i, R_j, A_2, and B_2 are given previously and where $t_{1-\alpha/2}$ is the $1-\alpha/2$ quantile of the t distribution given by Table A25 with $(b-1)(k-1)$ degrees of freedom. The value for α is the same one used in the Friedman test.

Example 2. Twelve homeowners are randomly selected to participate in an experiment with a plant nursery. Each homeowner was asked to select four fairly identical areas in his yard and to plant four different types of grasses, one in each area. At the end of a specified length of time each homeowner was asked to rank the grass types in order of preference, weighing important criteria such as expense, maintenance and upkeep required, beauty, hardiness, wife's preference, and so on. The rank 1 was assigned to the least preferred grass and the rank 4 to the favorite. The null hypothesis was that there is no difference in preferences of the grass types, and the alternative was that some grass types tend to be preferred over others. Each of the 12 blocks consists of four fairly identical plots of land, each receiving care of approximately the same degree of skill because the four plots are presumably cared for by the same homeowner. The results of the experiment are as follows.

		Grass		
Homeowner	1	2	3	4
1	4	3	2	1
2	4	2	3	1
3	3	1.5	1.5	4
4	3	1	2	4
5	4	2	1	3
6	2	2	2	4
7	1	3	2	4
8	2	4	1	3
9	3.5	1	2	3.5
10	4	1	3	2
11	4	2	3	1
12	3.5	1	2	3.5
R_j (totals)	38	23.5	24.5	34

First A_2 is computed by squaring each $R(X_{ij})$ and summing to get $A_2 = 356.5$ as the total sum of squares. Note that if there had been no ties Equation 11 could have been used to get $A_2 = 360$. This is always too large when ties are present. Equation 12 gives

$$B_2 = \frac{1}{12}[(38)^2 + (23.5)^2 + (24.5)^2 + (34)^2] = 312.71$$

for the treatment sum of squares. Substitution of A_2 and B_2 into Equation 13 gives

$$T_2 = \frac{11(312.71 - 300)}{356.5 - 312.71} = \frac{139.81}{43.79} = 3.19$$

The critical region of approximate size $\alpha = .05$ corresponds to all values of T_2 greater than 2.90, the .95 quantile of the F distribution with $k_1 = 3$,

$k_2 = 33$, obtained from Table A26. Therefore the null hypothesis is rejected. We may conclude that there is a tendency for some types of grass to be preferred over others. The critical level is about .04, which is obtained by interpolation in Table A26. This means that the null hypothesis could have been rejected at a significance level as small as $\alpha = .04$. For multiple comparisons $t_{.975}$ with $(11)(3) = 33$ degrees of freedom is found to be 2.036, by interpolation in Table A25. From Equation 15 we have

$$t_{.975} \left[\frac{2b(A_2 - B_2)}{(b-1)(k-1)} \right]^{\frac{1}{2}} = 11.49$$

Any two grasses whose rank sums are more than 11.49 units apart may be regarded as unequal. Therefore grass 1 may be considered better than grasses 2 and 3. No other differences are significant.

☐ *Theory.* The exact distribution of T_1, T_2, and T_3 is found under the assumption that each ranking within a block is equally likely, which is the null hypothesis. There are $k!$ possible arrangements of ranks $R(X_{ij})$ within a block and, therefore, $(k!)^b$ possible arrangements of ranks in the entire array of b blocks. The preceding statements imply that each of these $(k!)^b$ arrangements is equally likely under the null hypothesis. Therefore the probability distributions of T_1, T_2, and T_3 may be found for a given number of samples k and blocks b, merely by listing all possible arrangements of ranks and by computing T_1, T_2, or T_3 for each arrangement.

For example, if $k = 2$ and $b = 3$, there are $(2!)^3 = 8$ equally likely arrangements of the ranks, which are listed next along with their associated values of T_2 and T_3. We will consider T_1 later.

		Arrangements						
Blocks	1	2	3	4	5	6	7	8
1	2 1	2 1	2 1	1 2	1 2	1 2	2 1	1 2
2	2 1	2 1	1 2	2 1	1 2	2 1	1 2	1 2
3	2 1	1 2	2 1	2 1	2 1	1 2	1 2	1 2
Probability	$\frac{1}{8}$	$\frac{1}{8}$	$\frac{1}{8}$	$\frac{1}{8}$	$\frac{1}{8}$	$\frac{1}{8}$	$\frac{1}{8}$	$\frac{1}{8}$
Value of T_2	∞	$\frac{1}{4}$	$\frac{1}{4}$	$\frac{1}{4}$	$\frac{1}{4}$	$\frac{1}{4}$	$\frac{1}{4}$	∞
Value of T_3	3	$\frac{1}{3}$	$\frac{1}{3}$	$\frac{1}{3}$	$\frac{1}{3}$	$\frac{1}{3}$	$\frac{1}{3}$	3

Therefore the probability distribution of T_3 is given by $P(T_3 = \frac{1}{3}) = \frac{3}{4}$ and $P(T_3 = 3) = \frac{1}{4}$ under H_0. The probability distribution of T_2 is given by $P(T_2 = \frac{1}{4}) = \frac{3}{4}$ and $P(T_2 = \infty) = \frac{1}{4}$.

To examine the behavior of T_1 under the null hypothesis we again start out with the eight equally likely arrangements of ranks $R(X_{ij})$, as just given. The average rank 1.5 is subtracted from each rank and, for the moment, we consider the case where the block ranks are given by $Q_1 = 1$, $Q_2 = 2$, $Q_3 = 3$. The resulting arrays of S_{ij} are given here.

		Arrangements		
Blocks	1	2	3	4
1	$-.5+.5$	$-.5+.5$	$-.5+.5$	$+.5-.5$
2	$-1+1$	$-1+1$	$+1-1$	$-1+1$
3	$-1.5+1.5$	$+1.5-1.5$	$-1.5+2.5$	$-1.5+1.5$
Conditional Probability	$\frac{1}{8}$	$\frac{1}{8}$	$\frac{1}{8}$	$\frac{1}{8}$
Value of T_1	12	0	$\frac{4}{19}$	$1\frac{3}{13}$

		Arrangements		
Blocks	5	6	7	8
1	$+.5-.5$	$+.5-.5$	$-.5+.5$	$+.5-.5$
2	$+1-1$	$-1+1$	$+1-1$	$+1-1$
3	$-1.5+1.5$	$+1.5-1.5$	$+1.5-1.5$	$+1.5-1.5$
Conditional Probability	$\frac{1}{8}$	$\frac{1}{8}$	$\frac{1}{8}$	$\frac{1}{8}$
Value of T_1	0	$\frac{4}{19}$	$1\frac{3}{13}$	12

The probability for each value of T_1 is:

$$\tfrac{1}{8} \cdot P(Q_1 = 1, Q_2 = 2, Q_3 = 3)$$

because $\frac{1}{8}$ represents the conditional probability for that value of T_1, given the assignment of ranks Q_1, Q_2, and Q_3. Suppose a different assignment of ranks Q_1, Q_2, Q_3 is considered, say $Q_1 = 2$, $Q_2 = 1$, $Q_3 = 3$. Then the reader may easily verify, by listing the eight arrangements of values of S_{ij} as we just did, that again we observe the same eight values of T_1, and each of these eight values has probability

$$\tfrac{1}{8} \cdot P(Q_1 = 2, Q_2 = 1, Q_3 = 3)$$

By considering all six (3!) permutations of ranks for Q_1, Q_2, and Q_3, we arrive at the total probability for each value of T_1: $\frac{1}{8}$. Thus, for purposes of calculating the null distribution of T_1, only the one case given here, $Q_i = i$ for $i = 1, 2, 3$, must be considered. The probability distribution of T_1 is obtained by collecting identical values of T_1 to get

$$P(T_1 = 0) = \tfrac{1}{4}, \qquad P(T_1 = \tfrac{4}{19}) = \tfrac{1}{4}, \qquad P(T_1 = 1\tfrac{3}{13}) = \tfrac{1}{4}, \qquad P(T_1 = 12) = \tfrac{1}{4}$$

The approximations to the distributions of T_1, T_2, and T_3 that use the F or chi-square distributions are justified using the central limit theorem. Some of the details are beyond the scope of this book, so the entire development of the asymptotic distributions is omitted. The reader is referred to Quade (1972, 1979) or Lawler (1978) for T_1, Iman and Davenport (1979) for T_2, and Friedman (1937) for T_3.

The preceding distributions of T_1, T_2, and T_3 were obtained, as always, under the assumption that H_0 is true. If H_0 is false, the treatment sums S_j and R_j may be expected to vary widely from their mean values 0 and $b(k+1)/2$, respectively. This causes all three statistics to tend to increase in size. Therefore the decision rule is to reject H_0 when T_1, T_2, or T_3 is large.

The parametric procedure for analyzing data from the randomized complete block design is valid only if the data are normally distributed with the same variance. The null hypothesis is that the random variables within the same block have the same mean. The test statistic is

(16)
$$F = \frac{(b-1)SSB}{SST - SSB - SSR}$$

where

(17)
$$SSB = \frac{1}{b} \sum_{j=1}^{k} T_j^2 - \frac{T^2}{bk}$$

(18)
$$SSR = \frac{1}{k} \sum_{i=1}^{b} \left(\sum_{j=1}^{k} X_{ij} \right)^2 - \frac{T^2}{bk}$$

(19)
$$SST = \sum_{i=1}^{b} \sum_{j=1}^{k} X_{ij}^2 - \frac{T^2}{bk}$$

(20)
$$T_j = \sum_{i=1}^{b} X_{ij}$$

and

(21)
$$T = \sum_{i=1}^{b} \sum_{j=1}^{k} X_{ij}$$

The statistic in Equation 16 is compared with quantiles from the F distribution in Table A26, $k_1 = k - 1$, $k_2 = (b-1)(k-1)$.

If this same F statistic is computed on the ranks $R(X_{ij})$ instead of the data, the statistic is the same as T_2. If the F statistic is computed on the weighted ranks S_{ij}, the result is the statistic T_1. The adjustment for ties is automatically incorporated in the F statistic. There is a direct relationship between T_2 and T_3:

(22)
$$T_2 = \frac{(b-1)T_3}{b(k-1) - T_3}$$

The method of making multiple comparisons described here is merely the parametric procedure, known as Fisher's least significant difference (LSD) method, but it is computed on the number S_{ij} or on the ranks $R(X_{ij})$ instead of on the data.

For two samples ($k = 2$) the A.R.E. of the Friedman test relative to the usual parametric t test is the same as that of the sign test, $2/\pi = 0.637$, in situations where the t test is the most powerful test. For k samples the A.R.E. of the Friedman test relative to the usual parametric F test is dependent on the number of samples k and equals $(0.955)k/(k+1)$ if the populations are normal, $k/(k+1)$ if the populations are uniform, and $3k/2(k+1)$ if the populations are double exponential. In fact, the A.R.E. of the Friedman test relative to the popular F test never falls below $(0.864)k/(k+1)$ under purely translation-type alternative hypotheses. The A.R.E. of the Friedman test is discussed more fully in Noether (1967a).

For $k=2$ the A.R.E. of the Quade test relative to the usual parametric t test is the same as the Wilcoxon signed ranks test, $3/\pi = 0.955$, when the distributions are normal. As with the Wilcoxon test, the A.R.E. relative to the t test never falls below 0.864 but may be as high as infinity. The A.R.E. of this test for more than two samples has not yet been found.

The following discussion shows that the Friedman test statistic is closely related to some other popular nonparametric statistics. This discussion is limited to the case when there are no ties merely for simplicity. A similar comparison can be made for the case when ties are present.

THE RELATIONSHIP WITH KENDALL'S COEFFICIENT OF CONCORDANCE. A statistic W called Kendall's coefficient of concordance was introduced independently by Kendall and Babington-Smith (1939) and Wallis (1939). It may be used in the same situation where Friedman's test statistic is applicable, although it was probably intended primarily as a measure of "agreement in rankings" in the b blocks rather than as a test statistic. Using the same notation as before, Kendall's W is defined as

(23)
$$W = \frac{12}{b^2 k(k+1)(k-1)} \sum_{j=1}^{k} \left[R_j - \frac{b(k+1)}{2} \right]^2$$

If there is perfect agreement in the rankings in all b blocks, treatment 1 receives the same rank in all b blocks, treatment 2 receives the same rank in all b blocks, and so on, and the resulting value of W is 1.0. If there is "perfect disagreement" among rankings, the values of R_j will either be equal or very nearly equal to each other and their mean, and W will be 0 or very close to 0.

A comparison of Kendall's W with Friedman's T_3 of Equation 14 reveals the relationship

(24)
$$W = \frac{T_3}{b(k-1)}$$

Thus W is a simple modification of the Friedman test statistic, and any hypothesis test that uses W as a test statistic may be conducted by computing T_3 instead of W. If T_3 exceeds its $1-\alpha$ quantile, W exceeds its own $1-\alpha$ quantile.

THE RELATIONSHIP WITH SPEARMAN'S ρ. Spearman's ρ, defined by Equation 5.4.4, may be computed between any two blocks, say block i and block m, by considering the two blocks as two samples and the two ranks under each treatment as being a pair of related ranks. If Spearman's ρ computed between blocks i and m is denoted by ρ_{im}, conversion of the equation for

Spearman's ρ, Equation 5.4.4, into the preceding notation gives

(25)
$$\rho_{im} = \frac{\sum_{j=1}^{k} \{R(X_{ij}) - [(k+1)/2]\}\{R(X_{mj}) - [(k+1)/2]\}}{k(k+1)(k-1)/12}$$

The average value of Spearman's ρ, averaged over all pairs of blocks, bears a direct relationship with Friedman's test statistic, as we will now demonstrate.

Let ρ_a denote the average value of Spearman's ρ. There are $b(b-1)$ values of ρ_{im} to be averaged, counting both ρ_{im} and ρ_{mi}, even though $\rho_{im} = \rho_{mi}$ by symmetry. To compute the average ρ, we will sum over all i and m and then subtract those ρ_{im} where i equals m; that is, we will subtract the values of ρ where a block is paired with itself. In those b cases ρ_{im} equals 1. Thus the average ρ may be expressed as

(26)
$$\rho_a = \frac{1}{b(b-1)} \left(\sum_{i=1}^{b} \sum_{m=1}^{b} \rho_{im} - b \right)$$

The random variable ρ_a equals 1 if there is "perfect agreement" among rankings, in the sense previously described, because then each ρ_{im} equals 1. If there is disagreement among rankings ρ_a will be smaller than 1 and may even assume negative values. However, it is not possible for ρ_a to be as small as -1 except in the special case where there are only two blocks, $b = 2$.

The two preceding equations may be combined and simplified as follows. Substitution of Equation 25 into Equation 26 gives

$$\rho_a = \frac{1}{b(b-1)} \left(\sum_{i=1}^{b} \sum_{m=1}^{b} \frac{\sum_{j=1}^{k} \{R(X_{ij}) - [(k+1)/2]\}\{R(X_{mj}) - [(k+1)/2]\}}{k(k+1)(k-1)/12} - b \right)$$

$$= \frac{12}{b(b-1)k(k+1)(k-1)} \sum_{i=1}^{b} \sum_{m=1}^{b} \sum_{j=1}^{k} \left[R(X_{ij}) - \frac{k+1}{2} \right]$$

(27)
$$\times \left[R(X_{mj}) - \frac{k+1}{2} \right] - \frac{1}{b-1}$$

This expression is summed first over i and then over m. Also, the fact that

$$R_j = \sum_{i=1}^{b} R(X_{ij})$$

is used to simplify Equation 27 as follows. Summation over i leaves

$$\rho_a = \frac{12}{b(b-1)k(k+1)(k-1)} \sum_{m=1}^{b} \sum_{j=1}^{k} \left[R_j - \frac{b(k+1)}{2} \right]$$

(28)
$$\times \left[R(X_{mj}) - \frac{k+1}{2} \right] - \frac{1}{b-1}$$

and the summation over m leaves

(29) $$\rho_a = \frac{12}{b(b-1)k(k+1)(k-1)} \sum_{j=1}^{k} \left[R_j - \frac{b(k+1)}{2} \right]^2 - \frac{1}{b-1}$$

A comparison of the previous equation for ρ_a with the definition of Friedman's test statistic T_3, Equation 14, reveals the relationship

(30) $$\rho_a = \frac{T}{(b-1)(k-1)} - \frac{1}{b-1}$$

Thus the quantiles for the average Spearman's ρ may be easily obtained from the quantiles of the Friedman test statistic.

The preceding relationship between the Friedman test statistic and Spearman's ρ illustrates that the Friedman test may be used as a test for linear dependence in the two-sample case where Spearman's ρ was applicable. Spearman's ρ has the advantage of being tabulated for small samples, although the exact distribution of Friedman's test statistic could be easily obtained from the distribution of Spearman's ρ. Both tests would be equivalent, and therefore both would have an A.R.E. of .912 when compared with the usual parametric test using Pearson's r as a test statistic in the situation where Pearson's r is appropriate.

AN EXTENSION TO THE CASE OF SEVERAL OBSERVATIONS PER EXPERIMENTAL UNIT. If there are several (m) observations for each treatment in each block instead of only one observation per experimental unit as before, the null hypothesis of no differences among treatments may be tested by slightly modifying Friedman's procedure. The observations within each block are ranked as before, with the exception that the ranks range from 1 to mk. The sum of ranks R_j is defined, as before, as the sum of ranks assigned to all observations involving treatment j. Let the observations in block i using treatment j be denoted by $X_{ij1}, X_{ij2}, \ldots, X_{ijm}$. The mean of R_j becomes

$$E(R_j) = \sum_{i=1}^{b} \sum_{n=1}^{m} E[R(X_{ijn})] = \sum_{i=1}^{b} \sum_{n=1}^{m} \frac{mk+1}{2}$$

(31) $$= \sum_{i=1}^{b} \frac{m(mk+1)}{2} = \frac{bm(mk+1)}{2}$$

The variance of R_j is found with the aid of Theorem 1.4.5, in which n is replaced by m and N is replaced by mk.

$$\text{Var}(R_j) = \sum_{i=1}^{b} \text{Var}\left[\sum_{n=1}^{m} R(X_{ijn}) \right] = \sum_{i=1}^{b} \frac{m(mk+1)(mk-m)}{12}$$

(32) $$= \frac{bm^2(mk+1)(k-1)}{12}$$

If there are ties in the data, Var (R_j) is given by

(33)
$$\text{Var}\,(R_j) = \frac{m(k-1)}{k(mk-1)} \left[\sum_{\substack{\text{all} \\ \text{ranks}}} R(X_{ijn})^2 - mkb(mk+1)^2/4 \right]$$

which is the same for all values of j.

The test statistic

(34)
$$T_4 = \sum_{j=1}^{k} \frac{(k-1)}{k} \frac{[R_j - E\,(R_j)]^2}{\text{Var}\,(R_j)}$$

is used here. The mean and variance of R_j are given above in Equations 31 and 32 or in Equation 33. The chi-square tables with $k-1$ degrees of freedom are used as before. Multiple comparisons use the inequality

(35)
$$|R_j - R_i| > t_{1-\alpha/2} \left\{ \frac{2kb(mk-1)\,\text{Var}\,(R_j)}{(k-1)(mbk-k-b+1)} \left[1 - \frac{T_4}{b(mk-1)} \right] \right\}^{\frac{1}{2}}$$

where Var (R_j) is given by Equation 32 or 33, T_4 is given by Equation 34, and $t_{1-\alpha/2}$ is obtained from Table A25 with $mbk - k - b + 1$ degrees of freedom.

Some references on rank sum tests for two-way layouts include Page (1963), Hollander (1967b), and Pirie (1974) if the alternative hypothesis specifies an ordering of treatment effects, and Dunn (1964) and McDonald and Thompson (1967) for multiple comparisons. Other methods of analysis are suggested by Doksum (1967), Puri and Sen (1967), Sen (1968b), and Lemmer, Stoker, and Reinach (1968). Asymptotic efficiency is studied by Mehra and Sarangi (1967) and Sen (1967a). Small-sample efficiency is studied by Gilbert (1972). Patel and Hoel (1973) present a nonparametric test for interaction. Extensions to the multivariate case are considered by Gerig (1969, 1975). Koch (1970) discusses a split-plot variation with several observations per cell. Li and Schucany (1975) and Schucany and Beckett (1976) consider measuring the concordance between two sets of blocked rankings. For a complete presentation of an "aligned rank" procedure, which applies to the two-way layout and has a higher A.R.E. than the Friedman test, consult Lehmann (1975). The interested reader should also see Section 5.12.

EXERCISES

1. A survey was taken of all seven hospitals in a particular city to obtain the number of babies born over a 12 month period. This time period was divided into the four seasons to test the hypothesis that the birth rate is constant over all four seasons.

The results of the survey are as follows:

Number of Births

Hospital	Winter	Spring	Summer	Fall
A	92	112	94	77
B	9	11	10	12
C	98	109	92	81
D	19	26	19	18
E	21	22	23	24
F	58	71	51	62
G	42	49	44	41

(a) Analyze these data using the Quade test.
(b) Analyze these data using the Friedman test.
(c) Can you account for the wide discrepancy in the results of the two tests?

2. Twelve randomly selected students are involved in a learning experiment. Four lists of words are made up by the experimenter. Each list contains 20 pairs of words, but different methods of pairing are used on the four lists. Each student is handed a list, given 5 minutes to study it, and then examined on his or her ability to remember the words. This procedure is repeated for all four lists for each student, the order of the lists being rotated from one student to the next. The examination scores are as follows (20 is perfect).

Student

List	1	2	3	4	5	6	7	8	9	10	11	12
1	18	7	13	15	12	11	15	10	14	9	8	10
2	14	6	14	10	11	9	16	8	12	9	6	11
3	16	5	16	12	12	9	10	11	13	9	9	13
4	20	10	17	14	18	16	14	16	15	10	14	16

Are some lists easier to learn than others?
(a) Use the Quade test.
(b) Use the Friedman test.

3. Rework Example 2 using T_3 and compare critical levels.

PROBLEMS

1. Show that for $k = 2$ the statistic T_1 is a function of the Wilcoxon signed ranks statistic given by Equation 5.7.4 and, therefore, the two tests are equivalent. (*Hint*. First show that Q_i is equal to the absolute value of R_i given by Equation 5.7.3.)

2. Show that for $k = 2$ the Friedman test is equivalent to the two-tailed sign test (large sample approximation).

5.9. THE BALANCED INCOMPLETE BLOCK DESIGN

In the randomized complete block design described at the beginning of Section 5.8 every treatment is applied in every block. However, it is sometimes impractical or impossible for all of the treatments to be applied to each block, especially when the number of treatments is large and the block size is limited. For example, if 20 different foods are to be tasted, each judge (block) may find it quite difficult to rank accurately all 20 foods in order of preference. But if each judge tastes only 5 foods and then four times as many judges are used (or each judge is used four times), the judging may be easier and more accurate. Those experimental designs in which not all treatments are applied to each block are called incomplete block designs. Furthermore, if the design is balanced so that (1) every block contains k experimental units, (2) every treatment appears in r blocks, and (3) every treatment appears with every other treatment an equal number of times, the design is called a *balanced incomplete block design*.

Durbin (1951) presented a rank test that may be used to test the null hypothesis of no differences among treatments in a balanced incomplete block design. Parametric methods of analyzing data obtained using a balanced incomplete block design exist and are based on certain normality assumptions that will not be explained here. The Durbin test may be preferred over the parametric test if the normality assumptions are not met, if an easier method of analysis is desired, or if the observations consist merely of ranks. The Durbin test reduces to the Friedman test if the number of treatments equals the number of experimental units per block. If the third condition just given is not completely satisfied, the Durbin test is still valid in most situations.

The Durbin Test

DATA. We will use the following notation.

t = the number of treatments to be examined.
k = the number of experimental units per block ($k < t$).
b = the total number of blocks.
r = the number of times each treatment appears ($r < b$).
λ = the number of blocks in which the ith treatment and the jth treatment appear together. (λ is the same for all pairs of treatments.)

The data are arrayed in a balanced incomplete block design, just defined. Let X_{ij} represent the result of treatment j in the ith block, if treatment j appears in the ith block.

Rank the X_{ij} within each block by assigning the rank 1 to the smallest observation in block i, the rank 2 to the second smallest, and so on to the rank k, which is assigned to the largest observation in block i, there being only k observations within each block. Let $R(X_{ij})$ denote the rank of X_{ij} where X_{ij} exists.

Compute the sum of the ranks assigned to the r observed values under the

jth treatment and denote this sum by R_j. Then R_j may be written as

$$(1) \qquad R_j = \sum_{i=1}^{b} R(X_{ij})$$

where only r values of $R(X_{ij})$ exist under treatment j; therefore only r ranks are added to obtain R_j.

If the observations are nonnumeric but such that they are amenable to ordering and ranking within blocks according to some criterion of interest, the ranking of each observation is noted and the values R_j for $j = 1, 2, \ldots, t$ are computed as described.

If the ranks may be assigned in several different ways because of several observations being equal to each other, we recommend assigning the average of the disputed ranks to each of the tied observations. This procedure changes the null distribution of the test statistic, but the effect is negligible if the number of ties is not excessive

ASSUMPTIONS

1. The blocks are mutually independent of each other.
2. Within each block the observations may be arranged in increasing order according to some criterion of interest. (A moderate number of ties may be tolerated.)

HYPOTHESES

H_0: Each ranking of the random variables within each block is equally likely (i.e., the treatments have identical effects)
H_1: At least one treatment tends to yield larger observed values than at least one other treatment

TEST STATISTIC. The Durbin test statistic is defined as

$$(2) \qquad T = \frac{12(t-1)}{rt(k-1)(k+1)} \sum_{j=1}^{t} \left[R_j - \frac{r(k+1)}{2} \right]^2$$

and may be written in the equivalent machine form

$$(3) \qquad T = \frac{12(t-1)}{rt(k-1)(k+1)} \sum_{j=1}^{t} R_j^2 - 3\frac{r(t-1)(k+1)}{k-1}$$

DECISION RULE. Reject the null hypothesis at the approximate level of significance α if the Durbin test statistic T exceeds the $(1-\alpha)$ quantile of a chi-square random variable with $t-1$ degrees of freedom, obtained from Table A2.

Example 1. Suppose an ice cream manufacturer wants to test the taste preferences of several people for her seven varieties of ice cream. She asks each person to taste three varieties and to rank them 1, 2, and 3, with the rank 1 being assigned to the favorite variety. In order to design the experiment so that each variety is compared with every other variety an equal number of times, a Youden square layout given by Federer (1963) is used. Seven people are each given three varieties to taste, and the resulting ranks are as follows.

				Variety			
Person	1	2	3	4	5	6	7
1	2	3		1			
2		3	1		2		
3			2	1		3	
4				1	2		3
5	3				1	2	
6		3				1	2
7	3		1				2
R_j =	8	9	4	3	5	6	7

In this experiment,

$t = 7$ = total number of varieties
$k = 3$ = number of varieties compared at one time
$b = 7$ = number of people (blocks)
$r = 3$ = number of times each variety is tasted
$\lambda = 1$ = number of times each variety is compared with each other variety

Therefore the design is a balanced incomplete block design, and the Durbin test may be used to test the null hypothesis that no variety of ice cream tends to be preferred over any other variety of ice cream.

The critical region of approximate size $\alpha = .05$ corresponds to all values of T greater than 12.59, which is the .95 quantile of a chi-square random variable with $t - 1 = 6$ degrees of freedom, obtained from Table A2. The value of the Durbin test statistic is found as follows.

$$T = \frac{12(t-1)}{rt(k-1)(k+1)} \sum_{j=1}^{t} \left[R_j - \frac{r(k+1)}{2} \right]^2$$

$$= \frac{(12)(6)}{(3)(7)(2)(4)} [(8-6)^2 + (9-6)^2 + \cdots + (7-6)^2]$$

$$= 12$$

which is not in the critical region, so H_0 is accepted. However, the critical level is quite small and is estimated by interpolation to be about .065.

☐ *Theory.* The theoretical development of the Durbin test is very similar to that of the Friedman test. That is, the exact distribution of the Durbin test statistic is found under the assumption that each arrangement of the k ranks within a block is equally likely because of no differences between treatments. There are $k!$ equally likely ways of arranging the ranks within each block, and there are b blocks. Therefore each arrangement of ranks over the entire array of b blocks is equally likely and has probability

$1/(k!)^b$ associated with it, because there are $(k!)^b$ different arrays possible. The Durbin test statistic is calculated for each array and then the distribution function is determined, just as it was for the Friedman test statistic in the previous section.

The exact distribution is not practical to find in most cases, so the distribution of the Durbin test statistic is approximated by the chi-square distribution with $t-1$ degrees of freedom, if the number of repetitions r of each treatment is large. The justification for this approximation is as follows.

If the number r of repetitions of each treatment is large, the sum of the ranks, R_j, under the jth treatment is approximately normal, according to the central limit theorem. Therefore the random variable

$$\frac{R_j - E(R_j)}{\sqrt{\mathrm{Var}\,(R_j)}}$$

has approximately a standard normal distribution. As in the previous section, if the R_j were independent, the statistic

(4)
$$T' = \sum_{j=1}^{t} \frac{[R_j - E(R_j)]^2}{\mathrm{Var}\,(R_j)}$$

could be considered as the sum of t independent, approximately chi-square, random variables and the distribution of T' then could be approximated with a chi-square distribution with t degrees of freedom. But the R_j are not independent. Their sum is fixed as

(5)
$$\sum_{j=1}^{t} R_j = \frac{bk(k+1)}{2}$$

so that the knowledge of $t-1$ of the R_j enables us to state the value of the remaining R_j. Durbin (1951) shows that multiplication of T' by $(t-1)/t$ results in a statistic that is approximately chi-square with $t-1$ degrees of freedom, with the form

(6)
$$T = \frac{t-1}{t} T' = \frac{t-1}{t} \sum_{j=1}^{t} \frac{[R_j - E(R_j)]^2}{\mathrm{Var}\,(R_j)}$$

It only remains to find the mean and variance of R_j in order to transform Equation 6 into the usual form given by Equation 2.

The sum of ranks R_j is the sum of independent random variables $R(X_{ij})$.

(7)
$$R_j = \sum_{i=1}^{b} R(X_{ij})$$

Each $R(X_{ij})$, where it exists, is a randomly selected integer from 1 to k. Therefore the mean and variance of $R(X_{ij})$ are given by Theorem 1.4.5 as

(8)
$$E[R(X_{ij})] = \frac{k+1}{2}$$

and

(9) $$\mathrm{Var}\,[R(X_{ij})] = \frac{(k+1)(k-1)}{12}$$

Then the mean and variance of R_j are easily found to be

(10) $$E(R_j) = \sum_{i=1}^{b} E[R(X_{ij})] = \frac{r(k+1)}{2}$$

and

(11) $$\mathrm{Var}\,(R_j) = \sum_{i=1}^{b} \mathrm{Var}\,[R(X_{ij})] = \frac{r(k+1)(k-1)}{12}$$

The mean and variance of R_j given here are substituted into the Durbin test statistic given by Equation 6 to obtain

$$
\begin{aligned}
T &= \frac{t-1}{t} \sum_{j=1}^{t} \frac{[R_j - r(k+1)/2]^2}{r(k+1)(k-1)/12} \\
(12)\qquad &= \frac{12(t-1)}{rt(k+1)(k-1)} \sum_{j=1}^{t} \left[R_j - \frac{r(k+1)}{2} \right]^2
\end{aligned}
$$

which is in the same form as given in the explanation of the Durbin test.

The chi-square approximation is based on the assumption that the number r of repetitions of each treatment is reasonably large. In practice the approximation is used even if r is as small as 3 or 2, out of sheer necessity. The stated α level is probably not very accurate in those circumstances.

The Durbin test has been generalized to the case where some experimental units may contain several observations by Benard and van Elteren (1953). Noether (1967a) also discusses the Durbin test and its generalizations and shows that the A.R.E. of the Durbin test relative to its parametric counterpart is the same as that of the Friedman test relative to its parametric counterpart. See the preceding section for details. The case of paired comparisons ($k = 2$) is discussed by Puri and Sen (1969b).

MULTIPLE COMPARISONS. The following method may be used for comparing pairs of treatments *if the null hypothesis is rejected.* Consider two treatments i and j different if their rank sums satisfy the inequality

(13) $$|R_j - R_i| > t_{1-\alpha/2} \left\{ \frac{r(k+1)(k-1)[bk(t-1) - tT]}{6(t-1)(bk - t - b + 1)} \right\}^{\frac{1}{2}}$$

where $t_{1-\alpha/2}$ is obtained from Table A25, $bk - t - b + 1$ degrees of freedom.

AN ADJUSTMENT FOR TIES. If there are a large number of ties the simplest method of adjusting for ties is to use a regular computer program for the balanced incomplete block design if one is available. The data consist of the ranks and average ranks obtained as described in this section.

If such a program is not available, find the sample variance

(14)
$$S_i^2 = \frac{1}{k} \sum_{j=1}^{k} \left[R(X_{ij}) - \frac{k+1}{2} \right]^2, \qquad i = 1, 2, \ldots, b$$

for the actual ranks and average ranks in each block. In blocks having no ties, S_i^2 will equal $(k-1)(k+1)/12$, as given in Equation 9. Then compute

(15)
$$\text{Var}(R_j) = \sum_{i=1}^{b} \text{Var}(R_{ij}) = \sum_{r \text{ blocks}} S_i^2$$

where the sum is taken only in those blocks that make a contribution to the rank sum R_j. The variance of R_j may be different from column to column. These values of $\text{Var}(R_j)$ are used in Equation 6 along with the usual $E(R_j)$ given by Equation 10 to compute T. The rest of the procedure remains unchanged. The same multiple comparisons procedure as given here may be used as an approximation.

EXERCISES

1. Seven types of automobile tires are being tested for durability. It is felt that the best test is to see how the tires perform under actual driving conditions. However, only four tires may be compared at a time because only four-wheeled vehicles are available for testing. Therefore the experiment is designed using a balanced incomplete block design. Each of seven drivers is given four tires that are placed on each car in a random order and rotated regularly during the experiment. The tires are replaced when necessary, and ranks are assigned to the original tires according to the order of replacement.

Driver	Tire Type						
	1	2	3	4	5	6	7
1			3		1	4	2
2	1			3		4	2
3	2	1			3		4
4	1	2	4			3	
5		1	4	3			2
6	2		4	1	3		
7		1		2	3	4	

Do the results indicate a significant difference in durability? (First examine the experiment to be sure it follows a balanced incomplete block design.) If there is a significant difference in durability, use the multiple comparisons procedure to determine which tire types are better than others.

2. A experiment is designed to determine which of five scents tends to be the most attractive to coyotes, for purposes of predator control. The experimenter has observed that the presence of more than three scents at a time tends to confuse the coyotes and produce inconsistent results. Therefore three scents at a time are placed in separate areas of a large pen. One coyote at a time is released into the pen, and the amount of time (seconds) the coyote spends at each scent is recorded.

The scents are rotated according to a balanced incomplete block design, with the following results.

Coyote	Scent 1	2	3	4	5
1	12	23		14	
2		17	2		2
3	16		1	6	
4		42		10	0
5	8		6		1
6	22	31			0
7	28	16	4		
8	15			7	4
9		67	5	18	
10			6	16	1

Is this a balanced incomplete block design? Are there significant differences between scents? If so, which scents are better than others?

PROBLEMS

1. Show that $kb = rt$. (*Hint.* Count the number of observations in two different ways.)
2. Show that $\lambda = r(k-1)/(t-1)$. (*Hint.* First note that any particular treatment occurs in r blocks. Then count the number of units in which the treatment does not appear in those r blocks and count them in two different ways.)

5.10. TESTS WITH A.R.E. OF 1 OR MORE

The tests described in this section all share one property in common. The A.R.E. of each of these tests is 1 when compared with the usual parametric test in situations where the parametric test is appropriate. If the normality assumptions underlying the parametric test are not satisfied, under certain conditions that are easily met, the A.R.E. is always greater than 1 and may be as high as infinity. It seems to be a rather strong statement to say that the tests of this section are always at least as good as the usual parametric tests, such as the t test and the F test, as measured by their asymptotic relative efficiencies. However it is true. Remember that A.R.E. is only one way to compare tests, although admittedly it is probably the most universally accepted method of comparison. Relative efficiency, without the word "asymptotic," is also a method of comparison. It compares the sample sizes required for two tests to have the same power under identical conditions where the sample sizes are finite. On the basis of relative efficiencies, the tests in this section may be better or worse than their usual parametric counterparts, depending on the circumstances. It is not possible to examine all of the possible circumstances, and that is why the A.R.E. is usually used to compare tests.

Unlike the previous sections of this chapter, no new experimental situations are introduced in this section. We have already introduced nonparametric methods, based on ranks, of handling the one-way layout in Section 5.2, correlation in Section 5.4, and the randomized complete block design in Section 5.8. Those methods are widely accepted, reasonably powerful, and not too difficult to administer. By comparison, the methods of this section are usually slightly more powerful, but they are also slightly more difficult to administer. The assumptions behind these tests are practically identical to the assumptions underlying the earlier tests. Indeed, these tests are basically rank tests with a little dressing to improve the A.R.E. The user may decide whether to use these tests or the previously introduced tests; there is no solid statistical basis for preferring some to others.

The first tests we will describe are based on a simple idea suggested by van der Waerden (1952/1953). Instead of using the ranks of the observations as the basis for all of our computations, suppose we use other numbers that more nearly resemble observations from a normal distribution; in particular, instead of the rank k, suppose we use the $k/(N+1)$ quantile of a standard normal distribution, for $k = 1, 2, \ldots, N$ where N is the sample size. These quantiles, sometimes called *normal scores*, are readily available from Table A1. For example, suppose a random sample of five observations, arranged from smallest to largest, is given by 7.3, 7.7, 9.2, 12.0, and 26.4. Note that the three smallest observations are close together, the fourth observation is somewhat larger, and the largest observation is more than twice as large as any of the others. Replacement of these observations by their ranks 1, 2, 3, 4, and 5 amounts to a transformation of the nonsymmetrical original numbers into very symmetric, evenly spaced, somewhat "uniformly distributed" numbers. In the previous sections of this chapter we explained how the same kind of analysis customarily performed on the original observations can also be performed on the ranks. Now, following van der Waerden's suggestion, we transform the ranks into normal scores by replacing the rank k with the $k/(N+1)$ quantile from the normal distribution given in Table A1. Thus rank 1 is replaced by $w_{1/6} = w_{.167} = -0.9661$, rank 2 is replaced by $w_{2/6} = w_{.333} = -0.4316$, and so on. Then, instead of performing the analysis on the ranks, we analyze the normal scores: -0.9661, -0.4316, 0.0000, 0.4316, and 0.9661. In general, these normal scores will be symmetrically distributed around zero and will be spread out much the same as the "perfect normal sample" would be although, of course, there is no such thing as a "perfect normal sample." The result of using normal scores is a nonparametric test that has the same asymptotic efficiency as the parametric test when the population is really normal and a larger asymptotic efficiency when the population is nonnormal.

We start out by showing how the normal scores may be used as a modification of the Kruskal–Wallis test of Section 5.2 for testing equality among k populations. The two-sample problem is special case that relates to the Mann–Whitney test of Section 5.1.

The van der Waerden (Normal Scores) Test for Several Independent Samples

---

DATA. The data consist of k random samples of possibly unequal sample sizes. Denote the ith sample, of size n_i, by $X_{i1}, X_{i2}, \ldots, X_{in_i}$. Let N denote the total number of observations. Rank all N values from rank 1 to rank N, as explained in the Kruskal–Wallis test, using average ranks in case of ties as usual. Let $R(X_{ij})$ denote the rank of X_{ij}.

Convert each rank R into the $R/(N+1)$ quantile of a standard normal random variable obtained from Table A1. For brevity called these quantiles "normal scores" and denote them by A_{ij}.

(1) $$A_{ij} = w_{R(X_{ij})/(N+1)} = \text{the } \frac{R(X_{ij})}{N+1} \text{ th quantile from Table A1}$$

For convenience in obtaining the normal scores, round each value of $R(X_{ij})/(N+1)$ off to three decimal places before consulting Table A1. Find the average score

(2) $$\bar{A}_i = \frac{1}{n_i} \sum_{j=1}^{n_i} A_{ij} . \qquad i = 1, 2, \ldots, k$$

for each of the k samples and the variance

(3) $$S^2 = \frac{1}{N-1} \sum_{\substack{all \\ scores}}^{n_i} A_{ij}^{2}$$

Note that the overall mean equals zero if there are no ties and is essentially zero even if there are many ties, so the overall mean may be ignored when computing the variance.

ASSUMPTIONS. The assumptions here are the same as in the Kruskal–Wallis test.

HYPOTHESES. As in the Kruskal–Wallis test, we have:

H_0: All of the k population distribution functions are identical
H_1: At least one of the populations tends to yield larger observations than at least one of the other populations

TEST STATISTIC. The test statistic T_1 is defined as

(4) $$T_1 = \frac{1}{S^2} \sum_{i=1}^{k} n_i (\bar{A}_i)^2$$

where $\bar{A}_i$ and S^2 are given by Equations 2 and 3, respectively.

DECISION RULE. Reject H_0 at the level α if T_1 exceeds the $1 - \alpha$ quantile of a chi-square random variable with $k - 1$ degrees of freedom, given in Table

A2. Note that this is only an approximation, but it is good enough for most practical applications.

MULTIPLE COMPARISONS. If the null hypothesis is rejected we can say populations i and j seem to be different if the inequality

(5)
$$|\bar{A}_i - \bar{A}_j| > t_{1-\alpha/2}\left(S^2\frac{N-1-T_1}{N-k}\right)^{\frac{1}{2}}\left(\frac{1}{n_i}+\frac{1}{n_j}\right)^{\frac{1}{2}}$$

is satisfied, where $t_{1-\alpha/2}$ is the $1-\alpha/2$ quantile of the t distribution with $N-k$, degrees of freedom, obtained from Table A25, and the other terms are defined previously. This procedure is usually repeated for all pairs i and j. The same value for α as used before is also used for multiple comparisons.

Example 1. The same example that was used to illustrate the Kruskal–Wallis test in Section 5.2 and the median test in Section 4.3 will also be used here for ease in comparing these methods.

Four methods of growing corn resulted in the following observations and their ranks.

Method	1			2			3			4	
Obser-		Normal	Obser-		Normal	Obser-		Normal	Obser-		Normal
vation	Rank	Score	vation	Rank	Score	vation	Rank	Score	vation	Rank	Score
83	11	−0.4845	91	23	0.4043	101	34	1.8957	78	2	−1.5805
91	23	0.4043	90	19.5	0.1434	100	33	1.5805	82	9	−0.6526
94	28.5	0.8927	81	6.5	−0.8927	91	23	0.4043	81	6.5	−0.8927
89	17	−0.0351	83	11	−0.4845	93	27	0.7421	77	1	−1.8957
89	17	−0.0351	84	13.5	−0.2898	96	31.5	1.2816	79	3	−1.3658
96	31.5	1.2816	83	11	−0.4845	95	30	1.0669	81	6.5	−0.8927
91	23	0.4043	88	15	−0.1789	94	28.5	0.8927	80	4	−1.2055
92	26	0.6526	91	23	0.4043				81	6.5	−0.8927
90	19.5	0.1434	89	17	−0.0351						
			84	13.5	−0.2898						
	Average										
	score $\bar{A}_i$: 0.3582			−0.1703			1.1234			−1.1723	

The ranks are converted to normal scores in the following manner. The total sample size is $N=34$, so each rank is divided by $N+1=35$ and rounded off to three decimal places. The first observation has rank 11, and 11/35 equals .314. The .314 quantile from Table A1 is −0.4845, as noted.

The average score for each method of growing corn is also given before. The variance is computed using Equation 3 by squaring each of the 34 normal scores, summing them, and dividing by $N-1=33$. The result is $S^2=.8447$. In general, S^2 will always be slightly less than 1.0. The observed value of $T_1=25.1840$ is much greater than the .95 quantile of a chi-square random variable with $k-1=3$ degrees of freedom, which is 7.815 from Table A2. Therefore the null hypothesis is clearly rejected. The critical level is less than .001, as it was with the median test (Example 4.3.1, where $T=17.6$) and the Kruskal–Wallis test (Example 5.2.1, where $T=25.46$).

The multiple comparisons procedure uses the .975 quantile of the t distribution with 30 $(= 34 - 4)$ degrees of freedom, which is given in Table A25 as 2.042. The results of the computations are as follows.

$|\bar{A}_i - \bar{A}_j|$		$t_{.975}\left(S^2 \dfrac{N-1-T_1}{N-k}\right)^{\frac{1}{2}}\left(\dfrac{1}{n_i} + \dfrac{1}{n_j}\right)^{\frac{1}{2}}$
$i = 1, j = 2$	.5286	.4401
$i = 1, j = 3$	.7652	.4828
$i = 1, j = 4$	1.5305	.4655
$i = 2, j = 3$	1.2937	.4721
$i = 2, j = 4$	1.0020	.4544
$i = 3, j = 4$	2.2957	.4958

In each case the average scores are far enough apart to result in the conclusion that the two populations seem to be different. Note that these are the same conclusions that were reached using the Kruskal–Wallis test in Section 5.2. These two tests will agree in their conclusions quite often, but not always. To avoid ambiguous situations, either one test or the other, but not both, should normally be used.

It should be clear by now that the normal scores are used in the same way the ranks were: as numbers used to replace the original observations. The analysis on the normal scores is analogous to the analysis on the ranks. Exact tables could be given, but we do not present them here. Instead we rely on the large sample approximation for all sample sizes, large or small.

The test of the van der Waerden type for the one-sample problem, as an analogue to the Wilcoxon signed ranks test, is mentioned by van Eeden (1963). Let R_i represent the signed rank of Section 5.7, defined by Equation 5.7.4. Instead of using the signed ranks R_i, use the $\frac{1}{2}[1 + R_i/n + 1)]$th quantile from the normal distribution given in Table A1. Note that n is the number of nonzero differences obtained from the data. Call these signed normal scores and denote them by A_i. Note that A_i will have the same sign as R_i. Then the test statistic

(6)
$$T_2 = \frac{\sum\limits_{i=1}^{n} A_i}{\sqrt{\sum\limits_{i=1}^{n} A_i^2}}$$

is compared with quantiles from the standard normal distribution as an approximate test of the same hypothesis tested by the Wilcoxon signed ranks test and under the same assumptions.

Example 2. For comparison purposes we will use the same data given in Example 5.7.1 to test

H_0: The firstborn twin does not tend to be more aggressive than the other

versus

H_1: The firstborn twin tends to be more aggressive than the second twin

The data are as follows.

Twin Set	Firstborn X_i	Secondborn Y_i	Difference D_i	Rank of $|D_i|$	Signed Rank R_i	Signed Normal Score A_i
1	86	88	+2	3	3	0.3186
2	71	77	+6	7	7	0.8134
3	77	76	−1	1.5	−1.5	−0.1560
4	68	64	−4	4	−4	−0.4316
5	91	96	+5	5.5	5.5	0.6098
6	72	72	0	—	—	—
7	77	65	−12	10	−10	−1.3852
8	91	90	−1	1.5	−1.5	−0.1560
9	70	65	−5	5.5	−5.5	−0.6098
10	71	80	+9	9	9	1.1503
11	88	81	−7	8	−8	−0.9661
12	87	72	−15	11	−11	−1.7279

The test statistic T_2, defined by Equation 6, equals

(7)
$$T_2 = \frac{\sum\limits_{i=1}^{n} A_i}{\sqrt{\sum\limits_{i=1}^{n} A_i^2}} = \frac{-2.5405}{8.9027} = -0.8514$$

which corresponds to a one-tailed critical level of .197, from Table A1, in reasonable agreement with the results of the Wilcoxon signed ranks test which had $T = -.7565$ and $\hat{\alpha} = .225$.

The two-sample test for equal variances using normal scores was introduced by Klotz (1962). The test begins like the van der Waerden test for two samples but, in the end, the square of the normal scores is used rather than the normal scores themselves in the statistic

(8)
$$T_3 = \frac{\sum\limits_{i=1}^{n} A_i^2 - \frac{n}{N} \sum\limits_{i=1}^{N} A_i^2}{\left\{ \frac{nm}{N(N-1)} \left[\sum\limits_{i=1}^{N} A_i^4 - \frac{1}{N} \left(\sum\limits_{i=1}^{N} A_i^2 \right)^2 \right] \right\}^{\frac{1}{2}}}$$

where the normal scores are A_i, the sample sizes are m and n, and $N = n + m$ denotes the combined sample size. Then T_3 is compared with normal quantiles from Table A1. If the two samples come from populations with different means, the means (if they are known) are subtracted from the respective observations before the initial ranks are assigned.

Example 3. Refer to Example 5.3.1 for details of this example and a comparison with the squared ranks test. A new machine is being tested to see if it is less variable than the present machine. So the null hypothesis

H_0: Both machines have the same variability

is being tested against the one-sided alternative

H_1: The new machine has a smaller variance

We adjust the data by subtracting the sample means, because the population means are unknown. The result is an approximate test, just as in the squared ranks test.

X_i	$X_i - \bar{X}$	Ranks	$\dfrac{R_i}{N+1}$	Normal Score A_i	A_i^2
10.8	.06	8	.615	0.2924	.0855
11.1	.36	11	.846	1.0914	1.0392
10.4	−.34	2	.154	−1.0194	1.0392
10.1	−.64	1	.077	−1.4255	2.0321
11.3	.56	12	.923	1.4255	2.0321
Y_i	$Y_i - \bar{Y}$				
10.8	.01	6	.462	−0.0954	.0091
10.5	−.29	3	.231	−0.7356	.5411
11.0	.21	10	.769	0.7356	.5411
10.9	.11	9	.692	0.5015	.2515
10.8	.01	6	.462	−0.0954	.0091
10.7	−.09	4	.308	−0.5015	.2515
10.8	.01	6	.462	−0.0954	.0091

The sum of the A_i^2 from the first sample (from the present machine) is the basic measure of variability used in the Klotz test. To check its level of significance, we subtract its mean and divide by its standard deviation under the null hypothesis to get

(9)
$$T_3 = \frac{6.2280 - 3.2669}{1.2629} = 2.3447$$

(see Equation 8), which is compared with Table A1 to get a one-sided critical level of about $\hat{\alpha} = .01$, similar to the results using the squared ranks test.

To use the normal scores in regression and correlation, the normal scores replace the ranks of the X_is; then, in a separate step, the ranks of the Y_is are replaced by normal scores. If there are no ties the same set of normal scores is used for the X variable as is used for the Y variable, just as the same set of ranks 1 to n is used with each variable. The Pearson product moment correlation coefficient is computed on the normal scores (see Equation 5.4.2 for the coefficient). In this case it simplifies to

(10)
$$\rho = \frac{\sum\limits_{i=1}^{n} A_i B_i}{\sum\limits_{i=1}^{n} A_i^2}$$

(A_i and B_i represent the normal scores assigned to X_i and Y_i, respectively) because the mean scores are zero. Equation 10 may be used with ties unless the ties are quite extensive, in which case the safest procedure is to revert to computing Equation 5.4.2 on the actual normal scores used. Methods described in Sections 5.4 to 5.6 may be used on these scores, but we do not present the details here.

For the two-way layout, recall that the Friedman test of Section 5.8 used a ranking of the observations within each block. Normal scores may be substituted for these ranks in the usual manner. Let A_{ij} represent the normal score assigned to the variable X_{ij} in block i, treatment j, and let A_j be the sum of the scores in treatment j, analogous to $R(X_{ij})$ and R_j in the Friedman test. Then the test statistic

$$(11) \qquad T_4 = \frac{k-1}{S^2} \left(\sum_{j=1}^{k} A_i{}^2 \right)$$

where

$$(12) \qquad S^2 = \sum_{\substack{\text{all} \\ \text{scores}}} A_{ij}{}^2$$

is compared with the chi-square distribution, $k-1$ degrees of freedom, given by Table A2, as with the Friedman test. The rest of the details are the same as in the discussion of the Friedman test. Multiple comparisons are made as described in Section 5.8, except that the preceding values of T_4 and S^2 are used whenever appropriate.

By now the pattern of using normal scores instead of ranks should be clear. The result is a slightly higher A.R.E. relative to the best parametric test. The A.R.E. relative to the rank tests presented in previous sections may be greater than 1 or less than 1, depending on the particular situation. Other scores may be used instead of normal scores to achieve identical A.R.E.s with the normal scores tests just described. Two of these types of scores are called "random normal deviates" and "expected normal scores." We will now describe them briefly.

RANDOM NORMAL DEVIATES. One way to replace a random sample $X_1, \ldots, X_n$ from an arbitrary distribution, with numbers that seem to have come from a normal distribution, is to obtain somehow a group of n numbers that seem to have come from a normal population and to replace the smallest observation from the original sample with the smallest number from this pseudonormal sample, the second smallest with the second smallest, and so on. That is, the original observation of rank k is replaced by the number that has rank k in the pseudo normal sample. Note that only the ranks of the original observations need to be known in order to accomplish this replacement, so the resulting statistical procedures are rank tests. The pseudonormal sample may be obtained from tables of such numbers, such as in the book *A Million Random Digits with 100,000 Normal Deviates* by the Rand Corporation

(1955), or by using computer programs specifically designed to produce such numbers. These numbers are called "random normal deviates," although in actuality they are not random in the true sense of the word. They are deliberately generated numbers that seem to resemble a random sample from a standard normal distribution.

For example, the random sample we used earlier to illustrate normal scores was 7.3, 7.7, 9.2, 12.0, 26.4. A group of five numbers from a table of normal deviates is .026, -1.388, 2.338, 1.066, $-.173$. The smallest of these, -1.388, replaces the 7.3. The next smallest, $-.173$, replaces 7.7, and so on. From this point these new numbers are used in much the way as the normal scores, or the ranks, were. Of course, someone else who analyzes the same data will most likely select a different five numbers to work with, and the resulting analysis will be slightly (or not so slightly, sometimes) different. This results in the unpleasant situation of two statisticians using the same test to analyze the same data but coming up with conflicting conclusions. For this reason these procedures are seldom, if ever, used in practice. However, they are very interesting to study from a theory point of view because their A.R.E. is the same as that of the normal scores tests.

The principle of using random normal deviates is explained more fully by Bell and Doksum (1965). Earlier mention of the method appears in an article by Durbin (1961), the last problem in Fraser (1957), and in an article by Ehrenberg (1951).

EXPECTED NORMAL SCORES. One way of looking at the normal deviates procedures is to think of the actual order statistics $X^{(i)}$ as being replaced by order statistics $Z^{(i)}$ from a normal distribution. The next type of scores we consider uses the mean of the $Z^{(i)}$s, $E(Z^{(i)})$, instead of the order statistics themselves. These "expected normal scores" are well-defined numbers that are commonly available in tables such as Fisher and Yates (1957), Pearson and Hartley (1962), and Owen (1962). Therefore the unpleasant variability connected with using the $Z^{(i)}$s themselves, as random normal deviates, is eliminated. This type of procedure is still based only on the ranks of the observations and is therefore a rank test. Fisher and Yates (1975) suggest using these exact scores instead of the original data and then applying the usual parametric procedures to these expected normal scores as a nonparametric procedure. The A.R.E. of these methods is the same as that of the normal scores procedures and the random normal deviates procedures. A more complete presentation of this variation is given by Bradley (1968).

☐ *Theory.* The same reasoning we used to find the exact distribution of the test statistics in previous sections is used here, with minor modifications. One obvious modification is that instead of working with ranks 1, 2, 3, ..., we are working now with other numbers which we will call $a(1), a(2), a(3), \ldots$. These numbers $a(i)$ represent the normal scores or expected normal scores or any other set of numbers chosen independently of the data. Another modification to our previous tests is that these new numbers have different means and variances than the ranks have, and these means and variances need to be determined.

As one example of the method used to find the exact distribution, consider two independent samples $X_1, X_2, \ldots, X_n$ of size n and $Y_1, Y_2, \ldots, Y_m$ of size m, as in the Mann-Whitney test. Under the null hypothesis of identical distributions, the rank of X_1 is equally likely to be any of the ranks from 1 to $n+m$. Therefore the score assigned to X_1 is equally likely to be any of the scores from $a(1)$ to $a(n+m)$. The same holds true for $X_2, X_3, \ldots, Y_1, Y_2, Y_3$, and so on. Thus there are $\binom{n+m}{n}$ ways of selecting n ranks as belonging to the Xs and each of the $\binom{n+m}{n}$ ways is equally likely and has probability $1/\binom{n+m}{n}$. This also means that there are $\binom{n+m}{n}$ ways of selecting n of the scores $a(1)$ to $a(n+m)$, and each of these ways has probability $1/\binom{n+m}{n}$. This enables the null distribution of any statistic based on the scores (or ranks) assigned to the Xs (or Ys) to be found by the same counting methods used earlier in this chapter.

To be more specific, let $n=2$ and $m=3$, and suppose the scores we are using are the normal scores.

$$a(1) = -0.9661$$
$$a(2) = -0.4316$$
$$a(3) = 0.0000$$
$$a(4) = 0.4316$$
$$a(5) = 0.9661$$

The possible ranks for X_1 and X_2, their associated scores, and the sum of scores are given as follows

$R(X_1), R(X_2)$	Scores (A_1, A_2)	Sum	Probability
(1, 2)	(−0.9661, −0.4361)	−1.3977	0.1
(1, 3)	(−0.9661, 0.0000)	−0.9661	0.1
(1, 4)	(−0.9661, 0.4316)	−0.5345	0.1
(1, 5)	(−0.9661, 0.9661)	0.0000	0.1
(2, 3)	(−0.4316, 0.0000)	−0.4316	0.1
(2, 4)	(−0.4316, 0.4316)	0.0000	0.1
(2, 5)	(−0.4316, 0.9661)	0.5345	0.1
(3, 4)	(0.0000, 0.4316)	0.4316	0.1
(3, 5)	(0.0000, 0.9661)	0.9661	0.1
(4, 5)	(0.4316, 0.9661)	1.3977	0.1

Thus the distribution function of the sum of scores can be found. In a similar manner, the distribution function of any of the statistics presented in this section can be found. However, we do not present tables of the exact distributions, but suggest using the approximate distributions instead for these tests.

In order to find the mean and variance of the sum of ranks in the two sample situation, the same procedure shown in Section 5.3 is used. The result there, which is equally valid in this situation, is summarized in Equation 5.3.14 for the mean and Equation 5.3.20 for the variance. A more complete discussion of the use of scores instead of ranks is given by Hajek and Sidak (1967). There the method of choosing the best scores for each particular situation is described, and the complete theory is presented. The theory is well beyond the scope of this text, but Hajek and Sidak is strongly recommended reading for any statistician.

A discussion of random normal numbers may be found in Marsaglia (1968) or Lewis (1975), which are only two of the many references available on the subject. Jogdeo (1966) shows that the relative efficiency of the random normal deviates procedures is less than one for some fixed alternatives. Ramsey (1971) examines the small sample power of some of these two-sample tests, while large sample efficiencies are considered by Raghavachari (1965b), Thompson, Govindarajulu, and Doksum (1967), Bhattacharyya (1967), Stone (1968), and Gokhale (1968). Some variations of these tests are presented by Bradley, Patel, and Wackerly (1971) for the multivariate case, Johnson and Mehrotra (1972) for censored data, and Pirie and Hollander (1972) for ordered alternatives in the randomized block design. A broader prespective for these methods may be obtained by consulting Lehmann (1975) or Hogg (1976).

EXERCISES

1. Work Exercise 5.2.1 using normal scores instead of ranks and compare the results of the two methods.

2. Work Exercise 5.2.3 using normal scores instead of ranks and compare the results of the two methods.

3. Work Exercise 5.7.1 using normal scores instead of ranks and compare the results of the two methods.

4. Work Exercise 5.7.3 using normal scores instead of ranks and compare the results of the two methods.

5. Use the Klotz test on the data in Exercise 5.3.1 and compare the results of the two methods.

6. Use the Klotz test in Exercise 5.3.2.

7. Use normal scores in Exercise 5.4.1 to compute the correlation coefficient as given by Equation 10. How does this coefficient compare in size with Spearman's and Kendall's coefficients? Compare $\rho\sqrt{n-1}$ with the quantiles in Table A1 for significance, and test the hypothesis of independence. Compare with the results of Exercise 5.4.1.

8. Use normal scores in Exercise 5.4.3 to test for trend. Compare $\rho\sqrt{n-1}$ with the quantiles of Table A1 for significance. How do these results compare with the results of Exercise 5.4.3?

PROBLEMS

1. Find the exact distribution of the statistic given by Equation 6 for $n = 5$.

2. Find the exact distribution of the Klotz statistic given by Equation 8 for $n = 2$, $m = 3$.

3. Obtain 34 random normal deviates from a table of random numbers or from a computer program that generates random normal numbers. Rank them from smallest to largest and then use them in Example 1 instead of normal scores. How do the results of this Bell–Doksum procedure compare with the results of the van der Waerden and Kruskal–Wallis tests?

5.11. FISHER'S METHOD OF RANDOMIZATION

In the previous section we introduced a variety of ways of obtaining nonparametric tests. Each method consisted of using a set of scores $a(1)$ to $a(N)$ in place of the ranks 1 through N. Some suggested scores included quantiles from a standard normal distribution, numbers that seem to be a random sample from a standard normal distribution, and the expected values of order statistics from a standard normal distribution. We mentioned that any numbers whatsoever may be used as scores, but that some types of numbers resulted in more power against particular alternative hypotheses.

Suppose that in our search to find a "good" set of scores to use in place of the ranks from 1 to N we decide to use the numbers that actually occurred in the sample. These are convenient numbers to use, since there are exactly N of them and they are readily available. We use these numbers as scores just as we used the normal scores, for example, to obtain nonparametric tests in the manner described in the previous section. But is this choice of scores likely to have as much power as the normal scores? Apparently so, according to studies by Lehmann and Stein (1949), Hoeffding (1952), and many others, who find that these procedures have an A.R.E. of 1.0 when compared to the most powerful parametric tests in some situations. So these scores compare favorably not only with normal scores, but with any other type of test for those situations. Why, then, doesn't everybody use the data as scores in hypothesis tests?

The major disadvantage of this suggested procedure is that the test becomes very tedious to perform. This is because it is not possible to make tables of the critical regions or to present quantiles of the test statistics, since the scores are different in each test. Therefore each time a test of this type is performed, the critical region must be determined specifically for the set of data observed. Each different sample means a different set of scores and a different critical region. And even though the asymptotic distribution of the test statistic is, under easily met conditions, one of the standard distributions such as normal or chi-square, the use of the asymptotic distribution as an approximation may not be accurate for some types of scores. At least when ranks, normal scores, or expected normal scores are used, we know what the scores are and the accuracy of each approximation may be determined. In those cases the asymptotic distribution works well as an approximation. But when the set of scores changes from one sample to the next, it is impossible to measure the accuracy of any asymptotic distribution. So, in short, exact critical values may be obtained for each case after some effort (considerable effort when sample sizes are not small). Methods for finding approximate critical values are available but may not be accurate.

The idea of using the data themselves as scores is credited to Fisher (1935), and the resulting tests are traditionally known as randomization tests. Although our presentation may lead one to think that randomization tests are a third generation of nonparametric tests, after rank tests and tests using other scores, randomization tests actually preceded these other tests in time. Randomization tests may be used in any of the situations for which we have described rank tests. We will describe in more detail the randomization tests for two independent samples and for matched pairs, with examples of each, to clarify how these tests may be used. The first test is analogous to the Mann–Whitney test of Section 5.1.

The Randomization Test for Two Independent Samples

DATA. The data of two random samples $X_1, X_2, \ldots, X_n$ and $Y_1, Y_2, \ldots, Y_m$ of sizes n and m, respectively.

ASSUMPTIONS

1. Both samples are random samples from their respective populations.
2. In addition to independence within each sample there is mutual independence between the two samples.
3. The measurement scale is at least interval.
4. Either the two population distribution functions are identical, or one population has a larger mean that the other. (Without this assumption the test is still valid but might lack consistency.)

HYPOTHESES. Only the two-tailed test is presented; the one-tailed tests may be surmised by direct analogy with the Mann–Whitney test in Section 5.1.

$$H_0: E(X) = E(Y)$$
$$H_1: E(X) \neq E(Y)$$

TEST STATISTIC. The test statistic T_1 is the sum of the X observations:

(1)
$$T_1 = \sum_{i=1}^{n} X_i$$

DECISION RULE. Reject H_0 at the level α if either $T_1 > w_{1-\alpha/2}$ or $T_1 < w_{\alpha/2}$, where the quantiles w_p are found as follows.

Consider the observed values of X_i and Y_j as merely a group of $n + m$ numbers, and consider the ways in which n of those numbers may be selected. There are $\binom{n+m}{n}$ such ways. To find the p quantile w_p, consider the $\binom{n+m}{n}(p)$

selections that yield the smallest sums, which sum we will call T_1. The largest T_1 thus obtained is w_p.

As before, if $\binom{n+m}{n}(p)$ is not an integer, round upward to the next higher integer. If $\binom{n+m}{n}(p)$ is integer valued, w_p is the average of the largest T_1 thus obtained and the T_1 that would result from considering $\binom{n+m}{n}(p)+1$ selections.

The critical value $\hat{\alpha}$ is obtained by counting the number of ways n of the $n+m$ observations may be selected so that their sum is smaller (or larger if the observed T_1 is in the upper tail) than, or equal to, the observed T_1 from the data. This number is doubled, because the test is two tailed and divided by $\binom{n+m}{n}$ to get $\hat{\alpha}$.

Example 1. Suppose that a random sample yielded X_is of 0, 1, 1, 0, and -2 and an independent random sample of Y_js gave 6, 7, 7, 4, -3, 9, and 14. The null hypothesis

$$H_0: E(X) = E(Y)$$

is tested against

$$H_1: E(X) \neq E(Y)$$

at $\alpha = .05$, with the randomization test for two independent samples.

One sample is of size $n = 5$ and the other of size $m = 7$, so there are $\binom{12}{5} = 792$ ways of forming a subset containing 5 of the 12 numbers. Not all of the subsets are distinguishable because there are two 0s, two 1s, and two 7s, so we will distinguish between identical numbers with the aid of subscripts such as $0_1, 0_2, 1_1$, and so on. Because $(792)(.025) = 19.8$, we need to find the 20 groups that yield the lowest values of T_1 to obtain $w_{.025}$. These groups of numbers and the corresponding values of T_1 are as follows.

Observations	T_1	Observations	T_1
$-3, -2, 0_1, 0_2, 1_1$	-4	$-3, -2, 0_2, 1_2, 4$	0
$-3, -2, 0_1, 0_2, 1_2$	-4	$-3, -2, 1_1, 1_2, 4$	1
$-3, -2, 0_1, 1_1, 1_2$	-3	$-3, -2, 0_1, 0_2, 6$	1
$-3, -2, 0_2, 1_1, 1_2$	-3	$-3, -2, 0_1, 1_1, 6$	2
$-3, -2, 0_1, 0_2, 4$	-1	$-3, -2, 0_1, 1_2, 6$	2
$-3, 0_1, 0_2, 1_1, 1_2$	-1	$-3, -2, 0_2, 1_1, 6$	2
$-2, 0_1, 0_2, 1_1, 1_2$	0	$-3, -2, 0_2, 1_2, 6$	2
$-3, -2, 0_1, 1_1, 4$	0	$-3, 0_1, 0_2, 1_1, 4$	2
$-3, -2, 0_1, 1_2, 4$	0	$-3, 0_1, 0_2, 1_2, 4$	2
$-3, -2, 0_2, 1_1, 4$	0	$-3, -2, 0_1, 0_2, 7_1$	2

The largest T_1 thus obtained is

$$w_{.025} = 2$$

It is not necessary to find $w_{.975}$ even though this is a two-tailed test, because the observed T_1 is in the lower tail. The observed value of T_1 from the data is

$$T_1 = \sum_{i=1}^{5} X_i = 0+1+1+0-2 = 0$$

which is less than $w_{.025} = 2$, so H_0 is rejected. In fact, H_0 could have been rejected at the level

$$\hat{\alpha} = \frac{2(11)}{792} = .028$$

because there are 11 possible arrangements of the numbers that yield values less than or equal to 0.

The previous randomization test is typical of randomization tests in general. Instead of the statistic T_1, the usual two-sample t statistic could have been used and computed for each of the 20 groups in each tail of the distribution, as was done in the example. This additional labor is unnecessary however, because the t statistic is a monotonic function of T_1, as mentioned in Section 5.1, so the 20 groups that provide the largest values of T_1 will be the same groups that provide the largest values of t. We mention this so that the extension of randomization tests to other cases, such as the one-way layout, two-way layout, or tests for correlation, becomes more obvious. The usual statistic for the situation, or any monotonic function of that statistic that is easier to compute, is used to determine the "most extreme" arrangements of the data and thus the critical region for the statistic being used. Because of the difficult calculations involved, the actual critical region is sometimes not found, but only the critical level $\hat{\alpha}$ is found, especially if $\hat{\alpha}$ is close to zero and only a few arrangements of the data need to be considered.

The randomization test for matched pairs follows a slightly different pattern than the other randomization tests, and we will now present it in detail. This test is analogous to the Wilcoxon signed ranks test of Section 5.7. According to Kempthorne and Doerfler (1969), the randomization test for matched pairs is always to be preferred over the Wilcoxon test or the sign test.

The Randomization Test for Matched Pairs

DATA. The data consist of observations on n' bivariate random variables $(X_1, Y_1), (X_2, Y_2), \ldots, (X_{n'}, Y_{n'})$. Omit from further consideration all pairs

(X_i, Y_i) whose difference $Y_i - X_i$ is zero and let the remaining number of pairs be denoted by n. Denote the nonzero differences $Y_i - X_i$ by $D_1, D_2, \ldots, D_n$.

ASSUMPTIONS

1. The distribution of each D_i is symmetric.
2. The D_is are mutually independent.
3. The D_is all have the same median.
4. The measurement scale of the D_is is at least interval.

HYPOTHESES. Only the two-tailed version of this test is presented, although the one-tailed versions may be obtained by comparison with the Wilcoxon signed ranks test of Section 5.7. Let the common median of the D_is be denoted by $d_{.50}$.

$$H_0: d_{.50} = 0$$
$$H_1: d_{.50} \neq 0$$

TEST STATISTIC. The test statistic T_2 equals the sum of the positive differences.

(2) $$T_2 = \sum D_i \qquad \text{only for those } D_i > 0$$

DECISION RULE. Reject H_0 at the level α if $T_2 > w_{1-\alpha/2}$ or if $T_2 < w_{\alpha/2}$, where the quantiles $w_{1-\alpha/2}$ and $w_{\alpha/2}$ are found as follows.

Consider only the absolute values of the D_is, $|D_i|$, without regard for whether they were originally positive or negative. There are 2^n ways of assigning $+$ or $-$ signs to the set of absolute differences obtained, that is, we might assign $+$ signs to all n of the $|D_i|$, or we might assign a $+$ to $|D_1|$ but $-$ signs to $|D_2|$ to $|D_n|$, and so on. To find the p quantile w_p, $0 \leq p \leq 1$, first find the $(2^n)(p)$ assignments of signs that give the smallest values for T_2 the sum of the "positive" absolute differences. [If $(2^n)(p)$ is not an integer, use the next larger integer.] The largest value of T_2 thus obtained is the p quantile w_p of T_2 under the null hypothesis. [If $(2^n)(p)$ is an integer, use the average of the largest value of T_2 thus obtained, and the largest value of T_2 possible if $(2^n)(p) + 1$ arrangements had been considered instead, according to our usual convention.]

The preceding method of finding w_p works for all values of p from 0 to 1, but in practice it should be used only for small values of p, such as $p = \alpha/2$. For large values of p such as $p = 1 - \alpha/2$ the relationship

(3) $$w_{1-\alpha/2} = \sum_{i=1}^{n} |D_i| - w_{\alpha/2}$$

should be used. The relationship in Equation 3 is apparent if one considers that for every assignment of signs that results in a small value of T_2 a complete reversal of signs (pluses replaced by minuses, and vice versa) results in a large value of T_2. The latter value of T_2, the sum of the "positive" $|D_i|$s, plus the former value of T_2, the sum of the now "negative" $|D_i|$s, add up to the sum of all of the $|D_i|$s, as indicated by Equation 3.

The critical value $\hat{\alpha}$ is obtained by counting the number of assignments of signs that result in a smaller (or larger, if the observed $T_2 > 1/2 \sum_{i=1}^{n} |D_i|$) value of T_2, or the same value for T_2, as the one obtained from the data. This number is doubled and divided by 2^n to get $\hat{\alpha}$.

Example 2. Suppose that eight matched pairs resulted in the following differences: $-16, -4, -7, -3, 0, +5, +1, -10$. The zero is discarded, and we have

$$D_1 = -16, D_2 = -4, D_3 = -7, D_4 = -3, D_5 = +5, D_6 = +1, D_7 = -10$$

and $n = 7$. The null hypothesis

$$H_0: d_{.50} = 0$$

is tested against the alternative

$$H_1: d_{.50} \neq 0$$

using the randomization test at the level $\alpha = .05$.

The quantile $w_{.025}$ is found by considering the 4 [because $(2^7)(.025) = 3.2$] ways of assigning signs that result in the lowest sum of the "positive" absolute differences. These are given as follows.

Assignment of Signs	$\sum$ "positive" $|D_i|$
$-16, -4, -7, -3, -5, -1, -10$	$T_2 = 0$
$-16, -4, -7, -3, -5, +1, -10$	$T_2 = 1$
$-16, -4, -7, +3, -5, -1, -10$	$T_2 = 3$
$-16, -4, -7, +3, -5, +1, -10$	$T_2 = 4$
(also $-16, +4, -7, -3, -5, -1, -10$ gives	$T_2 = 4$)

The largest of these T_2 values is 4, so

$$w_{.025} = 4$$

From Equation 3 we have

$$w_{.975} = \sum_{i=1}^{7} |D_i| - w_{.025}$$

$$= 46 - 4 = 42$$

The value of the test statistic obtained from the data is

$$T_2 = \sum \text{positive } D_i$$

$$= 5 + 1 = 6$$

which is neither less than 4 nor greater than 42, so H_0 is accepted.

The critical level $\hat{\alpha}$ is found by listing the assignments of signs that result in $T_2 \leq 6$, in addition to the five just listed.

Assignment of Signs	$\sum$ *"Positive"* $\|D_i\|$
$-16, +4, -7, -3, -5, +1, -10$	$T_2 = 5$
$-16, -4, -7, -3, +5, -1, -10$	$T_2 = 5$
$-16, -4, -7, -3, +5, +1, -10$	$T_2 = 6$

Thus there are eight arrangements that give values of T_2 less than or equal to the observed value of 6. Because this is a two-tailed test, this number is doubled, and $\hat{\alpha}$ is given by

$$\hat{\alpha} = \frac{2(8)}{2^7} = \frac{16}{128} = .125$$

☐   *Theory.* The theory behind the randomization tests is partially explained by the method of finding the critical region. In the test for two independent samples, for instance, it is obvious that we are considering each selection of n X observations to be equally likely, from the $n + m$ observations available. It just remains to explain why we may consider the selections to be equally likely and why we are working with the observations themselves as our "sample space," so to speak.

The selections may be considered to be equally likely because of the null hypothesis, which states (along with the assumptions) that the Xs and the Ys are all independent and identically distributed. Therefore the Xs should have no more of a tendency to be low than the Ys have, or to be high, or to be in the middle. Given any group of $m + n$ numbers, whether they be observations or not, each subgroup of n of those numbers is just as likely to be the n values of X as any other subgroup of n of those numbers, because the numbers that are not Xs have to be Ys and the overall probability attached to that group of numbers does not depend on which numbers are called Xs and which numbers are called Ys. Now, if the Xs are distributed differently than the Ys, it will matter which numbers are called Xs and which are called Ys but, for purposes of finding a critical region of size α, we restrict our consideration to identically distributed random variables. So that is the intuitive argument for considering each selection of n observations as Xs to be equally likely.

That also leads to the second question, "Why are we working with the observations themselves as our sample space?" We explained before that any set of $m + n$ numbers satisfies the "equally likely" criterion. But in a testing situation we need to identify the $m + n$ numbers used with the $m + n$ observations obtained. In a rank test the $m + n$ numbers used are the integers from 1 to $m + n$, and they are matched one for one with the observations by assigning ranks to the observations. In this case we are using the observations themselves as the numbers. This eliminates the problem of which numbers to assign to which observations, which occurs in the rank tests when ties confuse the ranking procedure. By using the observations themselves as the numbers, it is easy to identify one of the selections of n numbers as the one actually obtained in the data. Then, with the aid of the test statistic, all selections more extreme than the one obtained may also be identified, counted, and used to compute the critical level $\hat{\alpha}$.

The critical region is thus determined for individual subsets of the sample space, such as for the subset of all outcomes that have the same numerical values as the observed values in the data. These subsets are mutually exclusive, cover the entire sample space (given *any* set of observations we can find the critical region for that subset of the sample space), and each subset has a critical region of size α relative to the size of the entire subset. So the overall size of all the critical regions combined is also α, which shows that the test is indeed a valid one.

The principal difference between the test for two independent samples and the test for matched pairs is that in the test for matched pairs the assumption of symmetry is used to justify the change of algebraic signs without changing the probability. If a difference D_i can be a $+6$, say, then it can be a -6 with the same probability when its distribution is symmetric about zero. Again, it does not matter which numbers are used. The Wilcoxon test used ranks. We use the observations themselves as numbers so that we may easily identify one of the assignments of signs as corresponding to the one actually obtained.

☐

The randomization test for matched pairs is discussed by Fisher (1935). The randomization test for two independent samples is presented by Pitman (1937/1938) along with a randomization test for correlation and an analysis of variance test.

A randomization test for multivariate data is presented by Chung and Frazer (1958). Further discussions of randomization tests may be found in articles by Welch (1937), Scheffé (1943), Moses (1952), Smith (1953), and Kempthorne (1955). A paper on multisample permutation tests is by Sen (1967b). Useful approximations to the distributions of the test statistics are discussed by Cleroux (1969). Other recent papers on Fisher's randomization tests are by Tsutakawa and Yang (1974), Oden and Wedel (1975), Boyett and Shuster (1977) and Soms (1977).

EXERCISES

1. One random sample furnished the values 1, 4, 3, -2, and 0, and the other gave 4, 6, 4. Can we conclude that the two population means are unequal?

2. A random sample of eight adults were asked how old they were when they went on their first date. The three men responded with ages 15, 17, 16, while the five women answered 12, 14, 15, 10, and 12. Test the hypothesis that the average is the same for both sexes against the alternative that girls tend to be younger on the occasion of their first date.

3. The number of customers served by each of two salespersons is observed for each hour. The differences $Y_i - X_i$, are noted, where Y_i and X_i represent the number of

customers served by each salesperson. Test whether the median difference $Y_i - X_i$ may be considered to be zero, where the observed differences are $+7$, $+3$, $+2$, $+8$, and -1.

4. Two highway patrolmen kept track of the numbers of traffic tickets they wrote, Y_i and X_i, for 7 days. The paired observations on (X_i, Y_i) are $(17, 14)$, $(15, 14)$, $(12, 15)$, $(9, 7)$, $(17, 16)$, $(18, 18)$, and $(14, 10)$. Is the median of $Y_i - X_i$ zero?

PROBLEMS

1. Suppose someone suggests subtracting a constant from all of the observations in the randomization test for two independent samples to make the calculations easier, such as subtracting 10 from each observation in Exercise 2 before analyzing the data. Does this affect the results of the test? Explain. Would division of the observations by a constant affect the results?

2. Would the results of the randomization test for matched pairs be affected by subtracting a constant from all of the observations or by division of the data by a constant? Explain.

3. In the randomization test for correlation the critical region is determined, as in the rank correlation tests, by assuming that each pairing of Xs with Ys is equally likely, where the data consist of a bivariate sample (X_i, Y_i), $i = 1, 2, \ldots, n$. Explain how to find the p quantile w_p of the test statistic $T_3 = \sum_{i=1}^{n} X_i Y_i$ under the null hypothesis of independence between the Xs and the Ys.

5.12. SOME COMMENTS ON THE RANK TRANSFORMATION

Most of the nonparametric procedures presented in this chapter are examples of procedures that arise by applying the rank transformation to the data (i.e., replacing the data by their ranks) and then using the usual parametric procedure, but on the ranks instead of on the data. The most obvious use of the rank transformation is in Section 5.4, where the usual product moment correlation coefficient known as Pearson's r is applied to ranks and called Spearman's ρ. In Section 5.6 the usual least squares regression line is found for the ranks rather than for the original data.

Other nonparametric procedures, such as the Wilcoxon signed ranks test, the Mann–Whitney test, and the Kruskal–Wallis test, are less obvious applications of the rank transformation. Let us examine the Wilcoxon signed ranks test. This test applies to a random sample of differences $D_1, D_2, \ldots, D_n$ to test the null hypothesis that the mean $E(D_i) = 0$. The usual parametric procedure assumes that the D_is are a random sample from a normal distribution and rejects the null hypothesis when the statistic

$$(1) \qquad t = \frac{\sum\limits_{i=1}^{n} D_i}{\left[\dfrac{n}{n-1} \sum\limits_{i=1}^{n} D_i^2 - \dfrac{1}{n-1}\left(\sum\limits_{i=1}^{n} D_i\right)^2\right]^{\frac{1}{2}}}$$

is too large or too small. This is called the one-sample t test and was mentioned earlier in Section 5.7. For the Wilcoxon signed ranks test, the D_is are replaced by signed ranks (see Section 5.7 for further details) R_1 to R_n, and the null hypothesis is rejected when

$$(2) \qquad T = \frac{\sum\limits_{i=1}^{n} R_i}{\sqrt{\sum\limits_{i=1}^{n} R_i^2}}$$

is too large or too small. The rank transformation procedure would suggest computing the statistic t on the signed ranks to obtain

$$(3) \qquad t_R = \frac{\sum\limits_{i=1}^{n} R_i}{\left[\dfrac{n}{n-1} \sum\limits_{i=1}^{n} R_i^2 - \dfrac{1}{n-1} \left(\sum\limits_{i=1}^{n} R_i \right)^2 \right]^{\frac{1}{2}}}$$

A closer comparison of Equations 2 and 3 reveals that Equation 3 is actually a function of Equation 2. That is, by dividing the numerator and denominator of Equation 3 by $\sqrt{\sum_{i=1}^{n} R_i^2}$, Equation 3 becomes

$$(4) \qquad t_R = \frac{T}{\left(\dfrac{n}{n-1} - \dfrac{1}{n-1} T^2 \right)^{\frac{1}{2}}}$$

where T is given by Equation 2. Note that as T gets larger, t_R gets larger, and as T gets smaller, t_R gets smaller. This means that a test that rejects the null hypothesis when T is too large or too small achieves identical results as the test that rejects the null hypothesis when t_R is too large or too small. So the two tests, the Wilcoxon signed ranks test and the rank transformation procedure, are exactly equivalent. To find the exact .95 quantile of t_R merely substitute the exact .95 quantile of T into Equation 4. The result will not be exactly the same as the .95 quantile from the t distribution (Table A25) with $n-1$ degrees of freedom, but the latter value serves as a good approximation to the exact value when the exact value is not known.

Just as the Wilcoxon signed ranks test is equivalent to the one-sample t test on the rank transformed data, so is the Mann–Whitney test equivalent to the two-sample t test after the rank transformation. The two-sample t statistic, given by Equation 5.1.11, when computed on the ranks of the Xs and Ys, becomes

$$(5) \qquad t_R = \frac{T_1}{\sqrt{\dfrac{N-1}{N-2} - \dfrac{1}{N-2} T_1^2}}$$

where T_1 is the Mann–Whitney test statistic given by Equation 5.1.2. Again, it can be seen from Equation 5 that if T_1 gets larger, t_R gets larger and as T_1 gets smaller, so does t_R. So tests based on T_1 and t_R are equivalent tests.

In Section 5.2 the Kruskal–Wallis test statistic T is given by Equation 5.2.3, and the F statistic used in the one-way analysis of variance is given by Equation 5.2.19. The F statistic computed on the ranks of the observations is a function of T:

$$(6) \qquad F_R = \frac{T/(k-1)}{(N-1-T)/(N-k)}$$

and because F_R increases or decreases as T increases or decreases, the rank transformation procedure is equivalent to the Kruskal–Wallis test.

These examples and others in this chapter, such as the test for slope in Section 5.5, the Friedman test, and the Durbin test, illustrate the idea of applying the usual parametric test statistic or its equivalent to the ranks of the observations to get a nonparametric procedure with high efficiency in most situations. Of course, the trick in each case is to rank the observations in such a way that all possible rankings are equally likely under the null hypothesis. This technique has been used successfully by Worsley (1977) in cluster analysis and by Shirley (1977) for contrasting increasing dose levels of a treatment.

In situations where such a method of ranking is not possible, or quite difficult, the principle of the rank transformation may still be useful. Two areas of statistics in which the rank transformation is useful even though it does not result in nonparametric procedures are the areas of experimental design and multiple regression.

To analyze an experimental design using the rank transformation, first rank all of the observations together from smallest to largest and then apply the usual analysis of variance to the ranks. The result is a procedure that is only conditionally distribution free. However, it is "robust," which means that the true level of significance is usually fairly close to the approximate level of significance used in the test, no matter what the underlying population distribution might be. The resulting procedure usually has good efficiency (Iman, 1974b and Conover and Iman, 1976). The recommended procedure in experimental designs for which no nonparametric test exists is to use the usual analysis of variance on the data and then to use the same procedure on the rank transformed data. If the two procedures give nearly identical results the assumptions underlying the usual analysis of variance are likely to be reasonable and the regular parametric analysis valid. When the two procedures give substantially different results, the analysis on ranks is probably more accurate than the analysis on the data and should be preferred. In such cases the experimenter may want to take a closer look at the data and to look especially for outliers (observations that are unusually large compared with the bulk of the data) or very nonsymmetric distributions. These aberrations in the data affect the analysis of the data to a great extent by changing the level of significance and decreasing the power, but the analysis on the ranks is not affected nearly as much. The rank transformation is used in experimental designs by Crouse (1967), Lemmer and Stoker (1967), Crouse (1968), Macdonald (1971), Scheirer, Ray, and Hare (1976), and Hamilton (1976).

In multiple regression each variable is ranked separately, as in the bivariate regression procedure of Section 5.5. Then the usual regression methods are applied to the ranks. The result is a robust regression method that is not sensitive to outliers or nonnormal distributions to the extent that the regular regression methods on the data are affected. As before, the recommended procedure is to analyze the data, analyze the ranks, and interpret the results in the light of both analyses. Prediction of the dependent variable may be accomplished as in Section 5.5 by predicting the rank from a regression equation and interpolating among the known values of the dependent variable.

Application of the rank transformation in discriminant analysis results in methods that are both simple to use and powerful in classifying observations. Briefly, each variable is ranked separately, and the popular linear discriminant function or quadratic discriminant function is computed on the ranks. A more complete discussion of this method and extensive Monte Carlo power comparisons are given by Conover and Iman (1978 and 1980).

Other areas of statistics are just as fertile for application of the rank transformation. These methods are usually not distribution free. However, they seem to be more robust and often more powerful than the standard procedures are when the assumptions behind the standard procedures are not reasonable. See Hettmansperger and McKean (1978) for a general discussion of the use of ranks. Other robust procedures, not necessarily based on ranks, are receiving a lot of attention currently. Some important references for these robust procedures include Huber (1972) and Hogg (1977). These methods are discussed by Labovitz (1970) and Allan (1976). The paper by Kim (1975) contains many references on the subject.

5.13. REVIEW PROBLEMS FOR CHAPTER 5

1. The state highway commission wants to buy a good grade of paint for painting lines on the highways. The choice has narrowed to two brands of paint. Twenty stripes are painted across a short stretch of highway. Ten of these stripes are painted with Brand A and the other 10 stripes with Brand B, in a random order. The stripes are inspected after 6 months and ranked according to wear. The results are as follows.

Ranks

| Brand A | 2, | 3, | 4, | 6, | 8, | 9, | 10, | 12, | 13, | 14 |
| Brand B | 1, | 5, | 7, | 11, | 15, | 16, | 17, | 18, | 19, | 20 |

Is there a significant difference between Brands A and B?

2. A local university conducted a study to see if height was related to monthly earnings. Graduates of the university were paired in such a way that the two

members in each pair had similar I.Q. scores, grade point averages, degrees, major and minor areas of study, and marital status, except one person was short and the other was tall. Ten pairs were studied, and their monthly earnings were as follows.

					Pairs					
	1	2	3	4	5	6	7	8	9	10
Short person	1080	1160	1200	1020	1110	1120	1080	1180	1090	1100
Tall Person	1130	1205	1130	1000	1110	1090	1130	1240	1160	1090

Does height seem to be related to earnings?

3. Ten golfers agreed to test a new type of golf ball in a tournament. Five golfers were selected at random from the 10 to try the new type of ball, and the other 5 were supplied with the old type of ball. The results after four rounds were as follows.

Scores with New Ball	295	301	288	290	289
Scores with Old Ball	302	306	292	306	314

(a) Do these results provide convincing evidence that the new type of ball tends to produce lower scores?

(b) What other statistical methods could you have used in part a? What are the main advantages and disadvantages of each, including the method you used?

4. While waiting for a customer, a caddy saw eight golfers finish their round of golf, pay their caddies, and leave. He estimated the age of each golfer and noted how much they paid their caddies.

				Golfer				
	1	2	3	4	5	6	7	8
Age (Estimated)	32	30	33	41	43	47	28	30
Amount Paid	2.50	2.80	2.25	2.90	3.20	3.50	2.20	2.35

(a) Does there seem to be a tendency for older golfers to pay their caddies more?

(b) What other statistical methods could you have used in part a? What are the main advantages and disadvantages of each, including the method you used?

5. Two horse trainers are comparing the results of their last five horses to see which trainer is best at teaching a horse to run faster. The first trainer gave the following times for running a quarter mile.

			Horse		
	1	2	3	4	5
Before Training	26.3	24.1	27.6	25.3	26.8
After Training	23.3	22.0	24.1	22.8	23.0

The second trainer gave the results for his last five horses.

			Horse		
	1	2	3	4	5
Before Training	25.4	26.2	24.0	26.0	27.7
After Training	23.6	23.9	21.8	23.6	25.7

Test the hypothesis that the two trainers are equally adept at training a horse to run faster.

6. To see if any further training was necessary, one particular horse was clocked on a quarter mile distance each morning for 10 consecutive days with the following results.

Day	1	2	3	4	5	6	7	8	9	10
Speed (Seconds)	22.2	22.8	21.0	21.4	22.4	21.9	22.0	22.6	21.8	21.1

Do these results indicate that the horse is still improving?

7. Eighteen high school students, selected at random, were given a conduct rating X, where $X = 10$ represents a perfect score, and an achievement rating Y, where $Y = 20$ represents satisfactory achievement in each of 20 areas.

Student

	1	2	3	4	5	6	7	8	9
X	1.8	8.9	8.3	4.0	8.8	9.2	9.5	8.1	5.3
Y	11	17	16	10	16	17	20	16	11

Student

	10	11	12	13	14	15	16	17	18
X	7.3	7.7	6.8	7.9	8.8	9.9	9.0	9.3	9.2
Y	14	15	12	14	17	20	18	19	18

(a) Is there a significant positive correlation between X and Y?
(b) Find the least squares regression line.
(c) Find a 95% confidence interval for the slope of the least squares regression line.
(d) Estimate the regression curve $E(Y|X)$ using rank regression.
(e) Draw a graph, showing the data points, the least squares regression line, and the estimate of the monotonic regression curve using rank regression. Which estimate of the regression seems to agree better with the data?

8. A random sample of men and women resulted in the following measurements of height (inches):

Men	Women
$X_1 = 70.1$	$Y_1 = 62.2$
$X_2 = 67.8$	$Y_2 = 64.7$
$X_3 = 71.6$	$Y_3 = 65.3$

Test the null hypothesis that the heights of men and women are identically distributed against the alternative hypothesis that men tend to be taller than women. Let the test statistic T equal the *largest rank* assigned to the Y values, where the ranks 1 to 6 are assigned to the combined sample of both men and women in order of increasing height.
(a) Find the probability distribution of T under the null hypothesis.
(b) Find, and sketch a graph of, the distribution function of T under the null hypothesis.
(c) Obtain a reasonable critical region and find the level of significance.
(d) Test the null hypothesis using the preceding test.
(e) Test the null hypothesis using any other nonparametric test you have learned or can invent.

9. At the beginning of the year a first-grade class was randomly divided into two groups. One group was taught to read using a uniform method, where all students progressed from one stage to the next at the same time, following the teacher's direction. The second group was taught to read using an individual method, where each student progressed at his or her own rate according to a programmed workbook, under supervision of the teacher. At the end of the year each student was given a reading ability test, with the following results.

First Group				Second Group			
227	55	184	174	209	271	63	19
176	234	147	194	14	151	184	127
252	194	88	248	165	235	53	151
149	247	161	206	171	147	228	101
16	99	171	89	292	99	271	179

(a) Test the null hypothesis that there is no difference in the two teaching methods against the alternative that the two population means are different.

(b) Test the null hypothesis of equal variances against the alternative that the variance of the second population is greater than the variance of the population that used the uniform method of learning to read.

10. A certain grade school has 121 students. The first semester the number of students with their number of absences was summarized as follows.

										More Than
Number of Absences	0	1	2	3	4	5	6	7	8	8
Number of Students	54	32	10	4	5	5	3	0	1	7
(Boys)	28	15	4	2	2	2	1	0	0	3
(Girls)	26	17	6	2	3	3	2	0	1	4

(a) Discuss the concepts of "target population" and "sampled population" as they apply to this problem.

(b) Are girls less likely than boys to have perfect attendance records?

(c) Do girls in general tend to have more absences than boys?

11. Seven judges were asked to rank the five finalists in a local beauty pageant. The ranks went from 1 for the best to 5 for the worst. The results were as follows.

			Girl		
Judge	A	B	C	D	E
1	5	2	1	3	4
2	5	1	2	3	4
3	3	1	2	4	5
4	2	3	4	1	5
5	3	1	2	5	4
6	4	1	2	3	5
7	4	2	3	1	5

May the null hypothesis of random assignment of ranks be rejected?

12. A hairbrush marketing company wishes to choose one of five different styles of hairbrushes to market. As part of their analysis, they test the consumer preference on these brushes by selecting 10 girls from senior high school. Each of these 10 girls is given two styles of hairbrush to use for 1 month and is then asked to report their preferences. The results are as follows. For simplicity the different styles of hairbrushes are called A, B, C, D, and E.

Alice prefers B over A
Betty prefers A over C
Charlene prefers D over A
Donna prefers E over A
Ellen prefers B over C

Fawn prefers B over D
Greta prefers E over B
Heather prefers D over C
Inga prefers E over C
Jean prefers E over D

Is there a significant difference in preferences?
If so, which brushes are significantly different?

13. The rate of return on investment in several common stocks over a certain period of time is figured by taking the market price of each stock at the end of the time period plus any dividends that were paid during the time period and dividing the result by the price of the stock at the beginning of the time period. The rate of return is recorded here for several stocks over nine 3-month periods. Does there seem to be a significant difference in the rate of return for the different stocks?

			Stock		
	A	B	C	D	E
Period 1	1.022	1.018	1.031	1.009	1.018
2	.996	.998	1.021	.981	.992
3	1.001	.993	.998	1.010	1.008
4	1.064	1.073	1.020	1.051	1.061
5	1.013	1.009	1.026	1.042	1.000
6	1.113	1.126	1.088	1.141	1.103
7	.998	.992	1.012	1.002	.977
8	.993	1.004	1.010	.998	.987
9	1.061	1.020	.999	1.031	1.040

14. In another part of the same study as in Problem 13, the total rate of return over the nine-quarter period was calculated for 40 stocks. These 40 stocks were selected to represent four different types of industry, 10 stocks in each type of industry.

		Twenty-Seven-Month Rate of Return			
Industry Type		A	B	C	D
	1	1.062	1.060	1.101	1.003
Stocks in	2	1.021	1.001	.981	1.067
Alphabetical	3	1.000	1.124	1.173	1.084
Order	4	1.316	.961	1.126	1.049
	5	1.177	1.054	1.002	1.056
	6	1.289	1.048	.964	1.012
	7	1.405	1.113	1.142	1.008
	8	1.566	1.147	1.226	1.051
	9	1.304	1.067	1.184	1.058
	10	1.111	1.073	1.098	1.042

Does there seem to be a significant difference in the rate of return for stocks in the four types of industry?

15. A rural appraiser has kept a record of prices paid for all plots of land, 20 acres or more, sold in the vicinity of a certain town for the last year. She has reduced the data from each sale to two variables, X = the distance from the city limits and Y = the price per acre.

	1	2	3	4	5	6	7	Parcel 8	9	10	11	12	13	14	15
X(Miles)	12.1	4.8	13.9	1.6	17.4	7.5	19.9	21.8	2.4	5.8	2.3	12.8	25.6	8.8	7.3
Y(Dollars per Acre)	280	590	163	530	157	394	177	110	620	492	761	210	115	245	334

She is asked to suggest a fair market price for a parcel of land located 4.4 miles from the city limits. Taking into consideration only the preceding information, what should be the price per acre?

CHAPTER 6

Statistics of the Kolmogorov–Smirnov Type

PRELIMINARY REMARKS

In Chapter 2 the empirical distribution function was introduced as a function based on a random sample that may be used to estimate the true distribution function of the population. If we want to see if two or more samples are governed by the same unknown distribution, it seems natural to compare the empirical distribution functions of those samples to see if they look somewhat similar. To be precise, however, some measure of disparity between or among these functions is needed. Kolmogorov and Smirnov developed statistical procedures that use the maximum vertical distance between these functions as a measure of how well the functions resemble each other. Their methods and other methods that use the same idea are presented in this chapter.

6.1. THE KOLMOGOROV GOODNESS-OF-FIT TEST

We will begin this chapter with a test for goodness of fit that was introduced by Kolmogorov (1933). This test is perhaps the most useful of the tests in this chapter, partly because it furnishes us with an alternative, designed for ordinal data, to the chi-square test for goodness of fit introduced in Section 4.5, which was designed for nominal type data, and partly because the Kolmogorov test statistic enables us to form a "confidence band" for the unknown distribution function, as we will explain in this section.

A test for goodness of fit usually involves examining a random sample from some unknown distribution in order to test the null hypothesis that the unknown distribution function is in fact a known, specified function. That is,

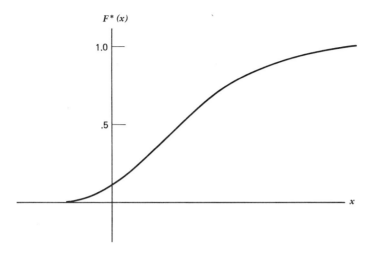

Figure 1.   A hypothesized distribution function.

the null hypothesis specifies some distribution function $F^*(x)$, perhaps graphically as in Figure 1, or perhaps as a mathematical function that may be graphed. A random sample $X_1, X_2, \ldots, X_n$ is then drawn from some population and is compared with $F^*(x)$ in some way to see if it is reasonable to say that $F^*(x)$ is the true distribution function of the random sample.

One logical way of comparing the random sample with $F^*(x)$ is by means of the empirical distribution function $S(x)$, which was defined by Definition 2.2.1 as the fraction of X_is that are less than or equal to x for each x, $-\infty < x < +\infty$. We learned in Section 2.2 that the empirical distribution function $S(x)$ is useful as an estimator of $F(x)$, the unknown distribution function of the X_is. So we can compare the empirical distribution function $S(x)$ with the hypothesized distribution function $F^*(x)$ to see if there is good agreement. If there is not good agreement, then we may reject the null hypothesis and conclude that the true but unknown distribution function, $F(x)$, is in fact not given by the function $F^*(x)$ in the null hypothesis.

But what sort of test statistic can we use as a measure of the discrepancy between $S(x)$ *and* $F^*(x)$? One of the simplest measures imaginable is the largest distance between the two graphs $S(x)$ and $F(x)$, measured in a vertical direction. This is the statistic suggested by Kolmogorov (1933). That is, if $F^*(x)$ is given by Figure 1 and a random sample of size 5 is drawn from the population, the empirical distribution function $S(x)$ may be drawn on the same graph along with $F^*(x)$, as shown in Figure 2. If $F^*(x)$ and $S(x)$ are as given the maximum vertical distance between the two graphs occurs just before the third step of $S(x)$. This distance is about 0.5 in Figure 2; therefore the Kolmogorov statistic T equals 0.5 in this case. Large values of T as determined by Table A14 lead to rejection of $F^*(x)$ as a reasonable approximation to the unknown true distribution function $F(x)$.

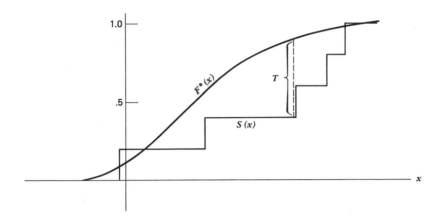

Figure 2.   The hypothesized distribution function $F^*(x)$, the empirical distribution function $S(x)$, and Kolmogorov's statistic T.

The Kolmogorov test may be preferred over the chi-square test for goodness of fit if the sample size is small; the Kolmogorov test is exact even for small samples, while the chi-square test assumes that the number of observations is large enough so that the chi-square distribution provides a good approximation às the distribution of the test statistic. There is controversy over which test is the more powerful, but the general feeling seems to be that the Kolmogorov test is probably more powerful than the chi-square test in most situations. For further comparisons see a paper by Slakter (1965).

The title of this chapter is "Statistics of the Kolmogorov–Smirnov Type." Statistics that are functions of the maximum vertical distance between $S(x)$ and $F^*(x)$ are considered to be Kolmogorov-type statistics. Statistics that are functions of the maximum vertical distance between two empirical distribution functions are of the Smirnov type. This entire chapter is concerned with statistics that are determined only by the vertical distances between distribution functions, either hypothesized or empirical distribution functions.

The Kolmogorov Goodness-of-Fit Test

DATA. The data consist of a random sample $X_1, X_2, \ldots, X_n$ of size n associated with some unknown distribution function, denoted by $F(x)$.

ASSUMPTIONS

1. The sample is a random sample.

HYPOTHESES. Let $F^*(x)$ be a completely specified hypothesized distribution function.

A. (Two-Sided Test)

$$H_0: F(x) = F^*(x) \quad \text{for all } x \text{ from } -\infty \text{ to } +\infty$$
$$H_1: F(x) \neq F^*(x) \quad \text{for at least one value of } x$$

B. (One-Sided Test)

$$H_0: F(x) \geq F^*(x) \quad \text{for all } x \text{ from } -\infty \text{ to } +\infty$$
$$H_1: F(x) < F^*(x) \quad \text{for all least one value of } x$$

C. (One-Sided Test)

$$H_0: F(x) \leq F^*(x) \quad \text{for all } x \text{ from } -\infty \text{ to } +\infty$$
$$H_1: F(x) > F^*(x) \quad \text{for at least one value of } x$$

TEST STATISTIC. Let $S(x)$ be the empirical distribution function based on the random sample $X_1, X_2, \ldots, X_n$. The test statistic is defined differently for the three different sets of hypotheses, A, B, and C.

A. (Two-Sided Test) Let the test statistic T be the greatest (denoted by "sup" for supremum) vertical distance between $S(x)$ and $F^*(x)$. In symbols we say

$$(1) \qquad T = \sup_x |F^*(x) - S(x)|$$

which is read "T equals the supremum, over all x, of the absolute value of the difference $F^*(x) - S(x)$."

B. (One-Sided Test) Denote this test statistic by T^+ and let it equal the greatest vertical distance attained by $F^*(x)$ above $S(x)$. That is,

$$(2) \qquad T^+ = \sup_x [F^*(x) - S(x)]$$

which is similar to T except that we consider only the greatest difference where the function $F^*(x)$ is above the function $S(x)$.

C. (One-Sided Test) For this test use the test statistic T^-, defined as the greatest vertical distance attained by $S(x)$ above $F^*(x)$. Formally this becomes

$$(3) \qquad T^- = \sup_x [S(x) - F^*(x)]$$

DECISION RULE. Reject H_0 at the level of significance α if the appropriate test statistic, T, T^+, or T^- exceeds the $1-\alpha$ quantile $w_{1-\alpha}$ as given by Table A14. This table is exact only if $F^*(x)$ is continuous; otherwise these quantiles lead to a conservative test (Noether, 1967a). For a method of finding the exact critical level when $F^*(x)$ is discrete, see the instructions following Example 1.

Quantiles are provided for use in two-sided tests at $\alpha = .20, .10, .05, .02,$ and $.01$ and for one-sided tests at α values of $.10, .05, .025, .01,$ and $.005$. The tables are exact for $n \leq 20$ in the two-sided test. For the one-sided test and for

$n > 20$ in the two-sided test, the tables provide good approximations that are exact in most cases. The approximation for $n > 40$ is based on the asymptotic distribution of the test statistics and is not very accurate until n becomes large.

Example 1. A random sample of size 10 is obtained: $X_1 = 0.621$, $X_2 = 0.503$, $X_3 = 0.203$, $X_4 = 0.477$, $X_5 = 0.710$, $X_6 = 0.581$, $X_7 = 0.329$, $X_8 = 0.480$, $X_9 = 0.554$, $X_{10} = 0.382$. The null hypothesis is that the distribution function is the uniform distribution function whose graph is given in Figure 3. The mathematical expression for the hypothesized distribution function is

$$
\begin{aligned}
F^*(x) &= 0 & \text{if} \quad & x < 0 \\
&= x & \text{if} \quad & 0 \le x < 1 \\
&= 1 & \text{if} \quad & 1 \le x
\end{aligned}
$$

(4)

Formally, the hypotheses are given by

$$H_0: F(x) = F^*(x) \qquad \text{for all } x$$
$$H_1: F(x) \ne F^*(x) \qquad \text{for at least one } x$$

where $F(x)$ is the unknown distribution function common to the X_is and $F^*(x)$ is given by Equation 4.

The two-sided Kolmogorov test for goodness of fit is used. The critical region of size $\alpha = 0.05$ corresponds to values of T greater than the .95 quantile 0.409, obtained from Table A14 for $n = 10$. The value of T is obtained by graphing the empirical distribution function $S(x)$ on top of the hypothesized distribution function $F^*(x)$, as shown in Figure 4. The largest vertical distance separating the two graphs in Figure 4 is 0.290, which occurs at $x = 0.710$ because $S(0.710) = 1.000$ and $F^*(0.710) = 0.710$. In other words,

$$
\begin{aligned}
T &= \sup_x |F^*(x) - S(x)| \\
&= |F^*(0.710) - S(0.710)| \\
&= 0.290
\end{aligned}
$$

Since $T = 0.290$ is less than 0.409, the null hypothesis is accepted. The critical level $\hat{\alpha}$ is seen, from Table A14, to be somewhat larger than .20.

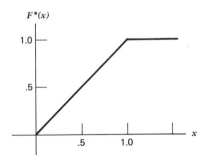

Figure 3. The hypothesized distribution function.

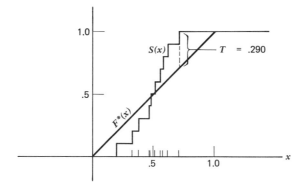

Figure 4.   Graphs of $F^*(x)$ and $S(x)$, with T.

If we had wished to test the null hypothesis

$$H_0 : F(x) \geq F^*(x) \qquad \text{for all } x$$

against the one-sided alternative

$$H_1 : F(x) < F^*(x) \qquad \text{for some } x$$

the test statistic T^+ would have been used. The decision rule is to reject H_0 at $\alpha = 0.05$ if T^+ exceeds the .95 quantile for a one-sided test, 0.369, as given by Table A14 for $n = 10$. The value for T^+ in this case is computed just to the left of the second jump of $S(x)$.

$$T^+ = \sup_x [F^*(x) - S(x)]$$

$$= F^*(0.3289) - S(0.3289)$$

$$= 0.3289 - 0.100$$

$$= 0.2289$$

To be more precise, we should say that $T^+ = 0.228999\ldots$, which is rounded off to 0.229. The end result is the same.

A one-sided test in the other direction would have resulted in

$$T^- = \sup_x [S(x) - F^*(x)]$$

$$= S(0.710) - F^*(0.710)$$

$$= 1.000 - 0.710$$

$$= 0.290$$

The two-sided test is the appropriate test for this situation. The one-sided tests were presented merely to show how their test statistics are evaluated. In general, of course, the two-sided test statistic T always equals the larger of the two one-sided test statistics T^+ and T^-.

A METHOD OF OBTAINING THE EXACT CRITICAL LEVEL WHEN $F^*(x)$ IS DISCRETE. If the hypothesized distribution function $F^*(x)$ is discrete and the conservative approximation for the critical level obtained from Table A14 is not satisfactory, the exact critical level may be obtained for a particular observed value of the test statistic. This computational procedure may be accomplished by hand for sample sizes of about 5 or less. A computer is recommended for larger sample sizes. For sample sizes larger than 30 or 40 the calculations become tricky, even on a computer. The labor may prove worthwhile, however, because the exact critical values for discrete distributions are often only about one-third as large as their approximations from Table A14.

A. (Two-Sided Test) Let t be the observed value of the test statistic T. Compute $P(T^+ \geq t)$ and $P(T^- \geq t)$ as described in parts B and C that follow, using t instead of t^+ and t^-. Then

$$(5) \qquad\qquad P(T \geq t) \doteq P(T^+ \geq t) + P(T^- \geq t)$$

is an approximation that is very close to the true critical level in most cases, unless t is small. The error is on the conservative side.

B. (One-Sided Test) Let t^+ denote the observed value of T^+.

Step 1. Compute the probabilities f_j for $0 \leq j < n(1-t^+)$ by drawing a horizontal line with ordinate $1 - t^+ - j/n$ directly on a graph of $F^*(x)$. Then $f_j = 1 - t^+ - j/n$ unless the horizontal line intersects $F^*(x)$ at a jump, in which case f_j equals the height of $F^*(x)$ at the bottom of the jump. One of the horizontal lines may intersect $F^*(x)$ directly at the top of a jump; in this event f_j equals the ordinate of the horizontal line.

Step 2. Compute the constants $e_0, e_1, \ldots,$ from the recursive relationship $e_0 = 1$ and

$$(6) \qquad\qquad e_k = 1 - \sum_{j=0}^{k-1} \binom{k}{j} f_j^{k-i} e_j \qquad k \geq 1$$

for all k such that $f_k > 0$ in Step 1. Note that these constants are of the form

$e_0 = 1$

$e_1 = 1 - f_0$

$e_2 = 1 - f_0^2 - 2f_1 e_1$

$e_3 = 1 - f_0^3 - 3f_1^2 e_1 - 3f_2 e_2$

$e_4 = 1 - f_0^4 - 4f_1^3 e_1 - 6f_2^2 e_2 - 4f_3 e_3$

$e_5 = 1 - f_0^5 - 5f_1^4 e_1 - 10f_2^3 e_2 - 10f_3^2 e_3 - 5f_4 e_4$

etc.

Step 3. Compute the critical level

(7)
$$P(T^+ \geq t^+) = \sum_{j=0}^{[n(1-t^+)]} \binom{n}{j} f_j^{n-i} e_j$$

from the f_j and e_j of Steps 1 and 2.

C. (One-Sided Test) Let t^- denote the observed value of T^-.

Step 1. Compute the probabilities c_j for $0 \leq j < n(1-t^-)$ as follows. Draw a horizontal line with the ordinate $t^- + j/n$ directly on a graph of $F^*(x)$. Then $c_j = 1 - t^- - j/n$ unless the horizontal line intersects $F^*(x)$ at a jump of $F^*(x)$. In that case $c_j = 1.0$ minus the height of $F^*(x)$ at the top of the jump. One of the horizontal lines may intersect $F^*(x)$ exactly at the bottom of a jump, in which event $c_j = 1.0$ minus the ordinate of that line.

Step 2. Compute the constants $b_0, b_1, \ldots$, from the recursive relationship $b_0 = 1$ and

(8)
$$b_k = 1 - \sum_{j=0}^{k-1} \binom{k}{j} c_j^{k-1} b_j \qquad k \geq 1$$

for all k such that $c_k > 0$ in Step 1. These constants follow the same pattern as the e_ks in part B, with the f_is replaced by c_is.

Step 3. Compute the critical level

(9)
$$P(T^- \geq t^-) = \sum_{j=0}^{[n(1-t^-)]} \binom{n}{j} c_j^{n-i} b_j$$

from the c_j and b_j of Steps 1 and 2.

The following example illustrates the method of computing the exact critical level when $F^*(x)$ is discrete.

Example 2. Let $F^*(x)$ be the discrete uniform distribution with equal probabilities $1/5$ at the five points $x = 1, 2, 3, 4, 5$. Suppose a random sample of size 10 with the (ordered) values 1, 1, 1, 2, 2, 2, 3, 3, 3, 3, is drawn from some population and the null hypothesis is that $F^*(x)$ is the population distribution function. The greatest distance between $F^*(x)$ and $S(x)$ occurs at $x = 3$ (see Figure 5), so the test statistic for the two-sided Kolmogorov test becomes

(10)
$$T = \sup_x |F^*(x) - S(x)| = 0.4 = t$$

To find the critical level associated with $t = 0.4$ the probability $P(T^+ \geq 0.4)$ is computed.

Step 1. Because $n(1-t) = 10(.6) = 6$, the probabilities f_0 to f_5 need to be computed. The horizontal line with ordinate $1 - t = 0.6$ intersects $F^*(x)$ directly at the top of the jump at $x = 3$, so f_0 equals the ordinate of the

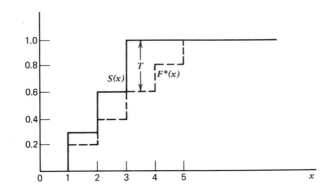

Figure 5. Graphs of $F^*(x)$ and $S(x)$, with T.

horizontal line: $f_0 = 0.6$. For $j = 1$, the horizontal line $1 - t - 1/10 = 0.5$ intersects $F^*(x)$ at a jump, so f_1 equals the height of $F^*(x)$ at the bottom of the jump: $f_1 = 0.4$. Similarly, we find $f_2 = 0.4$, $f_3 = 0.2$, $f_4 = 0.2$, and $f_5 = 0$.

Step 2. The constants e_0 to e_4 are computed from Equation 6.

$$e_0 = 1$$
$$e_1 = 1 - 0.6 = 0.4$$
$$e_2 = 1 - (0.6)^2 - 2(0.4)(0.4) = 0.32$$
$$e_3 = 1 - (0.6)^3 - 3(0.4)^2(0.4) - 3(0.4)(0.32) = 0.208$$
$$e_4 = 1 - (0.6)^4 - 4(0.4)^3(0.4) - 6(0.4)^2(0.32) - 4(0.2)(0.208) = 0.2944$$

Step 3. The one-sided critical level $P(T^+ \geq t)$ is computed from Equation 7.

$$P(T^+ \geq t) = f_0^{10} + \binom{10}{1}f_1^9 e_1 + \binom{10}{2}f_2^8 e_2 + \binom{10}{3}f_3^7 e_3 + \binom{10}{4}f_4^6 e_4$$

(11) $$= .02081$$

Because $F^*(x)$ is symmetric, computation of the other one-sided critical level $P(T^- \geq 0.4)$ is identical with the preceding, so $P(T^- \geq 0.4) = .02081$ and the critical level for the two-sided Kolomogorov test is approximately

(12) $$P(T \geq 0.4) \doteq 2(.02081) = .04162$$

It is interesting to note that this value for the critical level shows that the correct decision is to reject the null hypothesis at $\alpha = 0.05$, while the use of Table A14 leads to the erroneous acceptance of $F^*(x)$ as the true distribution function at the same α level.

COMMENT. One of the most valuable features of the Kolmogorov two-sided test statistic is that its $1 - \alpha$ quantile $w_{1-\alpha}$ may be used to form a confidence band for the true unknown distribution function. Recall that in finding a confidence interval for some unknown parameter, we first drew a random sample and then, from that sample, computed an upper value U and a lower value L that contained the unknown parameter between them with a certain probability $1-\alpha$, called the confidence coefficient. It would be convenient if we could do the same thing to obtain a "confidence band" within which the entire unknown distribution function would lie, with probability $1-\alpha$. Then we could draw a random sample for some population whose distribution function is completely unknown, and we could place some bounds on a graph and make the statement that the unknown distribution function lies entirely within those bounds, with some probability $1-\alpha$ that the statement is correct.

Confidence Band for the Population Distribution Function

DATA. The data consist of a random sample $X_1, X_2, \ldots, X_n$ of size n associated with some unknown distribution function, denoted by $F(x)$.

ASSUMPTIONS

1. The sample is a random sample.
2. For the confidence coefficient to be exact, the random variables should be continuous. If the random variables are discrete, the confidence band is conservative; that is, the true but unknown confidence coefficient is greater than the stated one.

METHOD. Draw a graph of the empirical distribution function $S(x)$ based on the random sample. To form a confidence band with a confidence coefficient $1-\alpha$, find the $1-\alpha$ quantile of the Kolmogorov test statistic from Table A14 for the two-sided test (if a two-sided confidence band is desired) and for the appropriate sample size n. Let $w_{1-\alpha}$ denote this quantile. Draw a graph above $S(x)$ a distance $w_{1-\alpha}$ and call this graph $U(x)$. Draw a second graph a distance $w_{1-\alpha}$ below $S(x)$ and call this second graph $L(x)$. Then $U(x)$ and $L(x)$ form the upper and lower boundaries, respectively, of a $1-\alpha$ confidence band that contains the unknown $F(x)$ completely within its boundaries.

There is no reason for $U(x)$ to be drawn above 1.0 even though $S(x)+w_{1-\alpha}$ might exceed 1.0, because we know that no distribution function ever exceeds 1.0. For the same reason $L(x)$ should not extend below the horizontal axis. The formal mathematical definitions of $U(x)$ and $L(x)$ are as follows.

(13) $\begin{aligned} U(x) &= S(x)+w_{1-\alpha} & \text{if} \quad & S(x)+w_{1-\alpha} \leq 1 \\ U(x) &= 1.0 & \text{if} \quad & S(x)+w_{1-\alpha} > 1 \end{aligned}$

(14) $\begin{aligned} L(x) &= S(x)-w_{1-\alpha} & \text{if} \quad & S(x)-w_{1-\alpha} \geq 0 \\ L(x) &= 0 & \text{if} \quad & S(x)-w_{1-\alpha} < 0 \end{aligned}$

The resulting probability statement is

(15) $P[L(x) \leq F(x) \leq U(x), \text{ for all } x] \geq 1-\alpha$

where the last inequality applies only when the random variables are discrete.

Example 3. Suppose we wish to form a 90% confidence band for an unknown distribution function $F(x)$. A random sample of size 20 is obtained from the population with that distribution function. The results are ordered from smallest to largest for convenience.

$$16.7 \quad 17.4 \quad 18.1 \quad 18.2 \quad 18.8 \quad 19.3 \quad 22.4 \quad 22.4 \quad 24.0 \quad 24.7$$
$$25.9 \quad 27.0 \quad 25.1 \quad 35.8 \quad 36.5 \quad 37.6 \quad 39.8 \quad 42.1 \quad 43.2 \quad 46.2$$

The .90 quantile is found from Table A14 to equal $w_{.90} = 0.265$ for $n = 20$. The confidence band is $S(x) \pm 0.265$ as long as the band is between 0 and 1. Figure 6 shows $S(x)$, $U(x)$, and $L(x)$. The statement "$F(x)$ lies entirely between $U(x)$ and $L(x)$" is true with probability .90.

The derivation of the distribution of the Kolmogorov statistic is complicated and is not presented here. We will mention only some basic papers on the subject.

The asymptotic distribution of the two-sided statistic T was found by Kolmogorov (1933) and was tabulated by Smirnov (1948). The asymptotic distributions of the one-sided statistics T^+ and T^- were obtained by Smirnov (1939). The exact distribution of the test statistics for finite (small) sample sizes was studied by Wald and Wolfowitz (1939) and tabulated by Massey (1950a). The distribution function of T^- for finite sample sizes was derived by Birnbaum and Tingey (1951), and comparisons were made between the exact quantiles obtained from their distribution function and the asymptotic quantiles given by Smirnov (1939, 1948). It was found that use of the asymptotic quantiles leads to a conservative test.

The two-sided Kolmogorov test has the desirable property of being consistent against *all* differences between $F(x)$ and $F^*(x)$, the true and hypothesized

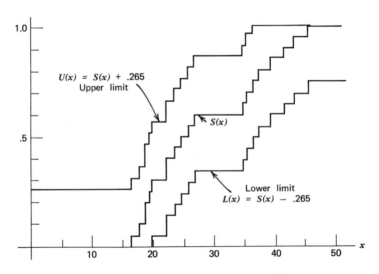

Figure 6. A confidence band for $F(x)$.

distribution functions. However, it is biased for finite sample sizes (Massey, 1950b). A lower bound for the power of the two-sided test is given by Massey (1950b). The greatest lower bound for the power, under a certain class of alternative hypotheses was obtained by Birnbaum (1953), and another greatest lower bound for the power, under a different class of alternative hypotheses, was obtained by Lee (1966).

Lee (1966) also compared the exact power of the Kolmogorov test with a standard parametric test. Some of his findings will now be presented. A random sample of size 5 was considered drawn from a population with the normal distribution with mean μ_1 and variance σ^2. The null hypothesis is that the distribution is normal with mean μ_0, not μ_1, but with the same variance. The power of the Kolmogorov test was obtained for several differences between μ_0 and μ_1, relative to the size of the standard deviation σ, and compared to the "normal test," which is the most powerful parametric test that exists for that situation. Even under these most unfavorable conditions, the power of the Kolmogorov test is not much worse than the normal test.

	Power when $\alpha = .10$		$\alpha = .05$		$\alpha = .01$	
	Kolmogorov Test	*Normal Test*	*Kolmogorov Test*	*Normal Test*	*Kolmogorov Test*	*Normal Test*
$\dfrac{\mu_0 - \mu_1}{\sigma}$	*(Percent)*	*(Percent)*	*(Percent)*	*(Percent)*	*(Percent)*	*(Percent)*
.5	31.65	43.48	25.09	29.91	8.66	11.35
1.0	74.95	82.99	61.57	72.27	33.16	46.41
1.5	94.93	98.09	89.50	95.58	68.19	84.80
2.0	99.50	99.93	97.89	99.77	90.53	98.40

Other power comparisons were made by van der Waerden (1953), Suzuki (1968), Shapiro, Wilk, and Chen (1968), and Knott (1970). Note that if a deviation from the hypothesized variance exists instead of a deviation from the hypothesized mean, as before, the normal test is powerless to detect the difference, and the Kolmogorov test is more powerful than the normal test. Other papers on the Kolmogorov test and similar goodness-of-fit tests are by Finkelstein and Schafer (1971), Maag and Dicaire (1971), Carnal and Riedwyl (1972), and Stephens (1974). Barr and Davidson (1973) and Pettitt and Stephens (1976) present modifications for censored data, while Barr and Shudde (1973) discuss a modification for observations on a circle. Govindarajulu and Klotz (1973) present a note on the asymptotic distribution. Estimation and testing symmetric distributions is the topic of papers by Schuster and Narvarte (1973), Schuster (1973), and Srinivasan and Godio (1974).

The modification for discrete distributions was developed independently by Conover (1972) and Coberly and Lewis (1973). Further analysis of this procedure appears in papers by Horn and Pyne (1976), Horn (1977), Bartels, Horn, Liebetrau, and Harris (1977), and Pettitt and Stephens (1977), who also present some tables. Maag, Streit, and Drouilly (1973) discuss goodness of-fit-tests for grouped data. Wood and Altavela (1978) suggest using simulation techniques for large samples.

Another goodness-of-fit test is the Cramér–von Mises test, developed by Cramér (1928), von Mises (1931), and Smirnov (1936). Although it has more intuitive appeal than the Kolmogorov test to many people, there is not sufficient difference between the two tests to warrant its presentation here. The interested reader may find the asymptotic distribution of the Cramér–von Mises test given by Anderson and Darling (1952) and exact tables for finite sample sizes given by Stephens and Maag (1968). Earlier studies on this test and the Kolmogorov test are by Stephens (1964, 1965a), Tiku (1965), Suzuki (1967), Cronholm (1968), and Noé and Vandewiele (1968). The effect of discreteness (ties) on the two tests is discussed by Walsh (1960, 1963). Bias and power of Cramér–von Mises test are examined by Thompson (1966). Relative efficiency of the Kolmogorov test is studied in Gelzer and Pyke (1965), Quade (1965), and Abrahamson (1967).

Goodness of fit for a sample density is discussed by Woodroofe (1967), for a circle by Stephens (1969), and in general by Riedwyl (1967). A different type of confidence interval for the distribution function is introduced by Durbin (1968).

EXERCISES

1. Five fourth-grade children were selected at random from the entire class and timed in a short race. The times in seconds were 6.3, 4.2, 4.7, 6.0, and 5.7. Give a 90% confidence band, either graphically or in tabled form, for the distribution function of times for all fourth-grade children in the class.

2. As a rural grocery store receives eggs from the neighboring farmers it "candles" the eggs to detect any eggs that are not fresh. Eight crates of eggs, 144 eggs per crate, were candled with the following numbers of eggs rejected from each crate: 4, 0, 2, 0, 2, 0, 2, 0. Present a 95% confidence band, either graphically or in tabled form, for the distribution function of the number of rejected eggs for the population of all crates received.

3. For the data in Exercise 1, test the hypothesis that the distribution of times is uniform on the interval from 4 to 8 seconds. Note that such a distribution is given by

$$
\begin{aligned}
F^*(x) &= 0 && \text{for} && x < 4 \\
&= (x-4)/4 && \text{for} && 4 \le x < 8 \\
&= 1 && \text{for} && 8 \le x
\end{aligned}
$$

4. Previous records have indicated that the number of rejected eggs per crate follows the Poisson distribution with mean 1.5. For the data in Exercise 2 test the hypothesis that these eight crates came from the same distribution function. Note that the Poisson distribution with mean 1.5 has the following probabilities: $P(0) = .223$, $P(1) = .335$, $P(2) = .251$, $P(3) = .126$, $P(4) = .047$, $P(5) = .014$, and $P(6) = .004$.

PROBLEM

1. Show that the confidence band given by Equation 15 is valid. That is, show that if $w_{1-\alpha}$ is a $1-\alpha$ quantile of the Kolmogorov statistic, it follows that Equation 15 is also true.

6.2. GOODNESS-OF-FIT TESTS FOR FAMILIES OF DISTRIBUTIONS

The Kolmogorov goodness-of-fit test presented in Section 6.1 is a good test to use to see if a random sample agrees with some specified distribution function. The Kolmogorov test is intended for use only when the hypothesized distribution function is completely specified, that is, when there are no unknown parameters that must be estimated from the sample. Otherwise the test becomes conservative. The chi-square goodness-of-fit test is flexible enough to allow for some parameters to be estimated from the data. One degree of freedom is simply subtracted for each parameter estimated in the "minimum chi-square" manner described earlier. However, the chi-square test requires that the data be grouped, and such a grouping of data is usually arbitrary. Also, the distribution of the test statistic is known only approximately, and sometimes the power of the chi-square test is not very good. For these reasons other goodness-of-fit tests are sought, especially for frequently tested distributions.

The Kolmogorov test has been modified to allow it to be used in several situations where parameters are estimated from the data. Actually, the test statistic remains unchanged, but different tables of critical values are used. These tables are no longer the same for all distributions; they change from one hypothesized distribution to another. The test is still a nonparametric test because the validity of the test (the α level) does not depend on untested assumptions regarding the population distribution; instead, the population distributional form is the hypothesis being tested.

The first such modification is the Kolmogorov test as modified to test the composite hypothesis of normality. That is, the null hypothesis states that the population is one of the family of normal distributions without specifying the mean or the variance of the normal distribution. This test was first presented by Lilliefors (1967). One interesting feature of this test is the manner in which the computer was used to generate random numbers in order to obtain accurate estimates of the true quantiles of the exact distribution of the test statistic.

The Lilliefors Test for Normality

DATA. The data consist of a random sample $X_1, X_2, \ldots, X_n$ of size n associated with some unknown distribution function, denoted by $F(x)$. Compute the sample mean

(1)
$$\bar{X} = \frac{1}{n} \sum_{i=1}^{n} X_i$$

for use as an estimate of μ and compute

(2)
$$s = \sqrt{\frac{1}{n-1} \sum_{i=1}^{n} (X_i - \bar{X})^2}$$

as an estimate of σ. Then compute the "normalized" sample values Z_i, defined by

(3)
$$Z_i = \frac{X_i - \bar{X}}{s} \qquad i = 1, 2, \ldots, n$$

The test statistic is computed from the Z_is instead of from the original random sample.

ASSUMPTIONS

1. The sample is a random sample.

HYPOTHESES

H_0: The random sample has the normal distribution, with unspecified mean and variance
H_1: The distribution function of the X_is is nonnormal

TEST STATISTIC. Ordinarily the test statistic is the usual two-sided Kolmogorov test statistic, defined as the maximum vertical distance between the empirical distribution function of the X_is and the normal distribution function with mean $\bar{X}$ and standard deviation s, as given by Equations 1 and 2. However, the following method of computing the test statistic is slightly easier and is equivalent to the method indicated.

Draw a graph of the standard normal distribution function, and call it $F^*(x)$. Table A1 may be of assistance. Also draw a graph of the empirical distribution function of the normalized sample, the Z_is defined by Equation 3, using the same set of coordinates as just used for $F^*(x)$. Find the maximum vertical distance between the two graphs, $F^*(x)$ and the empirical distribution function, which we will call $S(x)$. This distance is the test statistic. That is, the Lilliefors test statistic T_1 is defined by

(4)
$$T_1 = \sup_x |F^*(x) - S(x)|$$

The difference between T_1 and the Kolmogorov test statistic is that the empirical distribution function $S(x)$ in Equation 4 was obtained from the normalized sample, while $S(x)$ in the Kolmogorov test was based on the original unadjusted observations.

DECISION RULE. Reject H_0 at the approximate level of significance α if T_1 exceeds the $1 - \alpha$ quantile as given in Table A15.

   Example 1. The same data used to illustrate the chi-square test for normality in Example 4.5.3 will be used to illustrate the Lilliefors test.

Fifty two-digit numbers were drawn at random from a telephone book. Although the random variable sampled is clearly discrete, we may still justify testing for normality if we realize that acceptance of the null hypothesis of normality does not imply that the random variable has the normal distribution and is therefore continuous, but merely indicates that the difference between the normal distribution function and the true distribution function is sufficiently insignificant so as to remain undetected.

The numbers X_i are arranged from smallest to largest and converted to Z_i by subtracting $\bar{X} = 55.04$ and dividing by $s = 19.00$, as computed from Equations 1 and 2.

X_i	Z_i	X_i	Z_i	X_i	Z_i	X_i	Z_i	X_i	Z_i
23	−1.69	36	−1.00	54	−0.05	61	0.31	73	0.95
23	−1.69	37	−0.95	54	−0.05	61	0.31	73	0.95
24	−1.63	40	−0.79	56	0.05	62	0.37	74	1.00
27	−1.48	42	−0.69	57	0.10	63	0.42	75	1.05
29	−1.37	43	−0.63	57	0.10	64	0.47	77	1.16
31	−1.27	43	−0.63	58	0.16	65	0.52	81	1.37
32	−1.21	44	−0.58	58	0.16	66	0.58	87	1.68
33	−1.16	45	−0.53	58	0.16	68	0.68	89	1.79
33	−1.16	48	−0.37	58	0.16	68	0.68	93	2.00
35	−1.05	48	−0.37	59	0.21	70	0.79	97	2.21

The null hypothesis of normality is tested with the Lilliefors test statistic

$$T_1 = \sup_x |F^*(x) - S(x)|$$

where $F^*(x)$ is the standard normal distribution function and $S(x)$ is the empirical distribution function of the Z_is. Figure 7 presents the graphs of

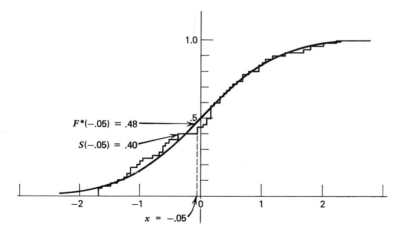

Figure 7. Graphs of $F^*(x)$ and $S(x)$ showing the maximum distance between them.

$F^*(x)$ and $S(x)$. The maximum vertical distance between $F^*(x)$ and $S(x)$ is seen from Figure 7 to occur just to the left of $x = -0.05$, where $S(x) = 0.40$, $F^*(x) = 0.48$, and so $T = 0.08$. The vertical distance between the two curves equals 0.08 at other points, too, such as at $x = +0.05$ and $x = 0.10$. But at no point does the distance separating the two curves exceed 0.08.

The Lilliefors test calls for rejection of H_0 at $\alpha = .05$ if T_1 exceeds its .95 quantile, which is given by Table A15 as

$$w_{.95} = \frac{.886}{\sqrt{n}} = \frac{.886}{\sqrt{50}} = 0.125$$

Because $T_1 = 0.08$ and is less than 0.125, the null hypothesis is accepted. In fact, the null hypothesis would still be accepted at $\alpha = 0.20$, because the .80 quantile is found to equal 0.104. Because Table A15 does not present smaller quantiles, we conclude that the critical level is some value greater than .20. Recall that the chi-square test resulted in about the same conclusion.

Acceptance of the null hypothesis does not mean that the parent population is normal, but it does mean that the normal distribution does not seem to be an unreasonable approximation to the true unknown distribution; therefore either nonparametric methods or the parametric statistical procedures that assume a normal parent distribution may be appropriate for further testing with these data.

☐ *Theory.* One of the principal reasons for presenting the Lilliefors test is to show how the quantiles in Table A15 were obtained. The problem of finding the distribution of T_1 so that the Kolmogorov test could be used to test the composite hypothesis of normality with unspecified mean and variance had been too difficult to solve analytically. Therefore Lilliefors used a high-speed computer and random numbers to obtain an approximate solution. This same technique, described next, may be used to obtain an approximate solution to almost any problem in statistical inference.

Recall that to develop a statistical hypothesis test one must first invent a test statistic that acts as a reasonably sensitive indicator, indicating whether the null hypothesis is true or false. The statistic T_1 satisfies this requirement. Then one must select a certain region of values, corresponding to the critical region, that is unlikely to occur if H_0 is true but is more likely to occur if H_0 is false. Large values of T_1 meet this requirement. Then the difficulty comes in trying to find α, the probability of getting a point in the critical region when H_0 is true. To do this Lilliefors generated random normal deviates on a high-speed computer. Random normal deviates, as mentioned in Section 5.10, are numbers that seem to be observations on independent standard normal random variables. These numbers were grouped into samples of size n for various values of n. For illustration let us say $n = 8$. A simulated sample of size 8 from a standard normal distribution, so that H_0 is true, was obtained from the computer. The sample mean $\bar{X}$ was computed and subtracted, and the result was divided by s as computed by Equation 2 for that sample to obtain the Z_i values. The empirical distribution function based on those Z_is was compared with the standard normal distribution function, and the maximum vertical difference T_1 was written down. The process was repeated with another set of eight computer-generated numbers to obtain another observed value of T_1. In all, over 1000 samples of size 8 were obtained and over 1000 values of T_1 were computed under the condition that H_0 is true. The empirical distribution function based on those 1000 or more values of T_1 was then used as an approximation to the true but unknown distribution function of T_1. From that empirical distribution function the selected quantiles given in Table A15 for $n = 8$ were obtained. Now a critical region of approximate size α may be specified.

The same procedure was repeated for other sample sizes ranging from $n = 4$ to $n = 30$. To obtain the approximation suggested by Lilliefors for n greater than 30, samples of size 40 were obtained, the quantiles were determined in the manner described, and the quantiles were multiplied by $\sqrt{40}$ and given in the table. This procedure is based on the unproved conjecture that T_1 approaches its asymptotic distribution in much the same way that the Kolmogorov statistic is known to approach its limiting asymptotic distribution, as a function of $\sqrt{n}$. The conjecture seems reasonable in view of the approximate quantiles obtained by Lilliefors in Table A15 for various values of n.

Lilliefors (1967) also compared the power of his test with the power of the chi-square test in several nonnormal situations and found his test to be more powerful in the situations reported.

A parametric confidence band for normal distributions was derived by Srinivasan and Wharton (1973). Other related papers are by Kanofsky and Srinivasan (1972) and Dyer (1974). A general discussion of simulation and presimulation days is found in Teichroew (1965).

A second modification of the Kolmogorov test was presented by Lilliefors in 1969. It tests the hypothesis that the parent distribution function is the exponential distribution $F(x) = 1 - e^{-x/t}$, $0 < x$, where t is an unspecified parameter that must be estimated from the data ($e = 2.718 \ldots$ is a well-known constant). Although Lilliefors obtained approximate critical values using random numbers as described, the exact distribution of the test statistic was subsequently obtained by Durbin (1975) and Margolin and Maurer (1976) using methods that are beyond the scope of this book and therefore not presented.

The exponential distribution is used to describe the length of time between consecutive "events," when the events occur randomly in time, according to a popular theory. Thus a test for the exponential ditribution such as this one is actually used primarily as a test for randomness.

The Lilliefors Test for the Exponential Distribution

DATA. The data consist of a random sample $X_1, X_2, \ldots, X_n$ of size n associated with some unknown distribution function, denoted by $F(x)$. Compute the sample mean for use as an estimate of the unknown parameter. For each X_i, compute Z_i, defined by

$$(5) \qquad\qquad Z_i = X_i / \bar{X}$$

for use in computing the test statistic.

ASSUMPTIONS

1. The sample is a random sample.

HYPOTHESES

H_0: The random sample has the exponential distribution

$$(6) \qquad\qquad F(x) = \begin{cases} 1 - e^{-x/t}, & x > 0 \\ 0, & x < 0 \end{cases}$$

where t is an unknown parameter.
H_1: The distribution of X is not exponential.

TEST STATISTIC. First, the empirical distribution function $S(x)$ based on $Z_1, \ldots, Z_n$ is plotted on a graph. On the same graph the function $F^*(x) = 1 - e^{-x}$ is plotted for $x > 0$; actually, only values at n points need to be determined, the points being at $x = Z_1$, $x = Z_2$, and so on. Tables are available for evaluating e^{-x}; calculators that have this function may also be used. The maximum vertical distance between the two functions

$$(7) \qquad\qquad T_2 = \sup_x |F^*(x) - S(x)|$$

is the test statistic.

Although this is only the two-sided version of the test, one-sided versions are presented by Durbin (1975) along with tables.

DECISION RULE. Reject H_0 at the level of significance α if T_2 exceeds the $1 - \alpha$ quantile as given in Table A16.

Example 2. The placement of long-distance telephone calls through a certain switchboard is believed to be a random process, with times between calls having an exponential distribution. The first 10 calls after 1 P.M. on a certain Monday occurred at 1:06, 1:08, 1:16, 1:22, 1:23, 1:34, 1:44, 1:47, 1:51, and 1:57. The successive times between calls, counting the first time from 1:00 to 1:06, are (in minutes) 6, 2, 8, 6, 1, 11, 10, 3, 4, and 6, with sample mean $\bar{X} = 5.7$. The resulting values of Z_i, $1 - e^{-Z_i}$, and the differences between $S(x)$ and $F^*(x)$ on both sides of each jump in $S(x)$ are given as follows. Note that the Xs are listed from smallest to largest for convenience.

i	X_i	$Z_i = X_i/\bar{X}$	$1 - e^{-Z_i}$	$i/10 - 1 + e^{-Z_i}$	$1 - e^{-Z_i} - (i-1)/10$
1	1	0.1754	.1609	−.0609	.1609
2	2	0.3508	.2959	−.0959	.1959
3	3	0.5263	.4092	−.2092	.2092
4	4	0.7018	.5043	−.1043	.2043
5	6	1.0526	.6510	−.1510	.2510[b]
6	6	1.0526	.6510	−.0510	.1510
7	6	1.0526	.6510	.0490	.0510
8	8	1.4035	.7543	.0457	.0543
9	10	1.7544	.8270	.0730	.0270
10	11	1.9298	.8548	.1452[a]	−.0452

[a] Large difference $S(x) - F^*(x)$.
[b] Largest difference $F^*(x) - S(x)$.

The largest absolute deviation between $S(x)$ and $F^*(x)$ is seen to equal .2510. The null hypothesis of an exponential distribution may be rejected at $\alpha = .05$ only if T_2 exceeds .3244 (from Table A16, $n = 10$, $1 - \alpha = .95$). Since $T_2 = .2510$, the null hypothesis is accepted. The critical level is obtained by interpolation in Table A16: $\hat{\alpha} = .25$. The times for the long-distance phone calls could be following a random process.

The Kolmogorov test has been extended to the gamma distribution when parameters must be estimated by Lilliefors (1973) and Schneider and Clickner (1976). A similar version of the Cramér–von Mises test is presented by Pettitt (1978). Other tests of a similar type are discussed by Green and Hegazy (1976).

We conclude this section by presenting a well-known goodness-of-fit test for normality that may be used instead of the Lilliefors test if desired. Some empirical studies indicate that this test has good power in many situations when compared with many other tests of the composite hypothesis of normality, including the Lilliefors test and the chi-square test (Shapiro, Wilk, and Chen, 1960; La Brecque, 1977). Although this test is not a Kolmogorov-type test, it is included here because of its usefulness.

The Shapiro–Wilk Test for Normality

DATA. The data consist of a random sample $X_1, X_2, \ldots, X_n$ of size n associated with some unknown distribution function $F(x)$.

ASSUMPTIONS

1. The sample is a random sample.

HYPOTHESES

H_0: $F(x)$ is a normal distribution function with unspecified mean and variance
H_1: $F(x)$ is nonnormal

TEST STATISTIC. First compute the denominator D of the test statistic

$$(8) \qquad D = \sum_{i=1}^{n} (X_i - \bar{X})^2$$

where $\bar{X}$ is the sample mean. Then order the sample from smallest to largest,

$$X^{(1)} \le X^{(2)} \le \cdots \le X^{(n)}$$

and let $X^{(i)}$ denote the ith order statistic. From Table A17, for the observed sample size n, obtain the coefficients $a_1, a_2, \ldots, a_k$ where k is approximately $n/2$.

The test statistic T_3 is given by

$$(9) \qquad T_3 = \frac{1}{D} \left[\sum_{i=1}^{k} a_i (X^{(n-i+1)} - X^{(i)}) \right]^2$$

Note that this test statistic is often denoted by W, and the test is often called the W test.

DECISION RULE. Reject H_0 at the level of significance α if T_3 is less than the α quantile as given by Table A18. If a more precise critical level for an observed value of T_3 is desired, the instructions in Table A19 allow T_3 to be converted to an approximately normal random variable, which may then be compared with the normal distribution in Table A1 to obtain $\hat{\alpha}$.

COMMENT. Although existing tables allow the Shapiro–Wilk test to be used only if $n \le 50$, D'Agostino (1971) presents a test that may be used for n larger than 50, and Shapiro and Francia (1972) suggest an approximate test for n greater than 50 that is similar to the Shapiro–Wilk test.

Example 3. The 50 two-digit numbers in Example 4.5.3 were drawn from a telephone book. The chi-square goodness-of-fit test accepted the hypothesis of normality with $\hat{\alpha}$ well above .25. The Lilliefors test accepted the same hypothesis in Example 1 with $\hat{\alpha}$ greater than .20. The same data will be analyzed using the Shapiro–Wilk test.

The coefficients from Table A17 and the order statistics $X^{(n-i+1)} - X^{(i)}$ are given next.

i	a_i	$X^{(n-i+1)} - X^{(i)}$	i	a_i	$X^{(n-i+1)} - X^{(i)}$
1	.3751	$97 - 23$	14	.0846	$66 - 42$
2	.2574	$93 - 23$	15	.0764	$65 - 43$
3	.2260	$98 - 24$	16	.0685	$64 - 43$
4	.2032	$87 - 27$	17	.0608	$63 - 44$
5	.1847	$81 - 29$	18	.0532	$62 - 45$
6	.1691	$77 - 31$	19	.0459	$61 - 48$
7	.1554	$75 - 32$	20	.0386	$61 - 48$
8	.1430	$74 - 33$	21	.0314	$59 - 54$
9	.1317	$73 - 33$	22	.0244	$58 - 54$
10	.1212	$73 - 35$	23	.0174	$58 - 56$
11	.1113	$70 - 36$	24	.0104	$58 - 57$
12	.1020	$68 - 37$	25	.0035	$58 - 57$
13	.0932	$68 - 40$			

The numerator of the test statistic becomes

$$\left[\sum_{i=1}^{k} a_i (X^{(n-i+1)} - X^{(i)})\right]^2 = [(.3751)(97-23) + \cdots + (.0035)(58-57)]^2$$

$$= [130.63]^2 = 17,064$$

and the denominator is given by

$$D = \sum_{i=1}^{n} (X_i - \bar{X})^2 = 17,698$$

so the test statistic becomes

$$T_3 = \frac{17,064}{17,698} = .9642$$

which lies somewhere between the .10 and the .50 quantiles of the distribution. Interpolation in Table A18 gives $\hat{\alpha} = .29$ approximately.

In order to find a more precise value for $\hat{\alpha}$, the coefficients from Table A19 are obtained for $n = 50$; $b_{50} = -7.677$, $c_{50} = 2.212$, and $d_{50} = .1436$. The observed value of T_3 is substituted into the formula

$$G = b_{50} + c_{50} \ln \left(\frac{T_3 - d_{50}}{1 - T_3}\right)$$

$$= -7.677 + (2.212) \ln \left(\frac{.9642 - .1463}{1 - .9642}\right)$$

$$= -.7488$$

which corresponds to $\hat{\alpha} = .227$ from Table A1. This is a more precise value of the critical level than the one we obtained by interpolation earlier.

The theory behind the Shapiro–Wilk test is too lengthy to present here, but the interested reader is referred to the original papers by Shapiro and Wilk (1965, 1968). Some efforts to extend existing tables (Stephens, 1975) apparently have not yet resulted in extended tables as far as we know. Other goodness-of-fit tests for the same composite hypothesis of normality have been offered by Hartley and Pfaffenberger (1972), Bowman and Shenton (1975), and Pearson, D'Agostino, and Bowman (1977).

One useful feature of the Shapiro–Wilk test is that several independent goodness-of-fit tests may be combined into one overall test of normality. This is convenient when several small samples from possibly different populations are insufficient by themselves to reject the hypothesis of normality, but their combined evidence is enough to disprove normality. The technique is illustrated in the following example.

Example 4. When an offshore lease is made available for bids, several oil companies usually submit bids for the right to drill for oil in that area. The distribution of these bids is often assumed to follow the "lognormal" distribution; that is, the logarithm of the bids is assumed to follow the normal distribution. However, the means and variances may vary from lease to lease. Also, the number of bids on any one lease is usually too small to be able to tell whether the normality assumption on the logarithms of the bids is reasonable or not.

To test the hypothesis

H_0: The bids are lognormally distributed

against the alternative that they are not lognormal, the bids on 16 different leases are observed. The Shapiro–Wilk test is conducted on the logarithms of the bids on each lease separately, with the result that the null hypothesis is rejected on 4 of the 16 leases at $\alpha = .05$. However, some of the leases show good agreement with the null hypothesis, with critical levels well above .50. To combine the results from the 16 tests, the following steps are followed.

1. Each value of T_3 is converted to values of G, as described in Table A19.
2. All $n = 16$ values of G are added together.
3. The result is divided by $\sqrt{n}$ to get Z, which is approximately standard normal under the null hypothesis.
4. If Z is less than the α quantile from Table A1, the null hypothesis is rejected at the level α.

For these leases the calculations are as follows.

Lease Number	Number of Bids	T_3	G
1	14	.9243	−.6550
2	14	.9757	1.3559
3	14	.9717	1.0939
4	14	.8772	−1.5848
5	14	.9537	.2345
6	15	.9135	−1.0093
7	15	.8629[a]	−1.9321
8	15	.8786[a]	−1.6806
9	15	.8515[a]	−2.1011
10	15	.9226	−.7966
11	15	.9581	.3354
12	15	.9625	.5344
13	16	.9178	−1.0151
14	16	.8596[a]	−2.1011
15	15	.9603	.4323
16	16	.9669	.6795

Total   −8.2099

[a] Significant at $\alpha = .05$.

$$Z = \frac{-8.2099}{\sqrt{16}} = -2.0525$$

This value of Z is smaller than -1.6449 from Table A1, so H_0 is rejected at $\alpha = .05$. The critical level is seen from Table A1 to equal .020. The assumption of lognormally distributed bids does not seem to be justified.

We would be remiss if we did not point out that almost any goodness-of-fit test will result in rejection of the null hypothesis if the number of observations is very large. In other words, real data never really are distributed according to any distribution known to man. However, these known distributions are often "close enough" to the data for some reasonably accurate results to be obtained by assuming that the hypothesized distribution is the real one. A goodness-of-fit test is one way of ascertaining whether or not the agreement is close enough.

EXERCISES

1. The return on investment for 12 months on 20 randomly selected stocks is as follows.

9.1	5.0	7.3	7.4	5.5
8.6	7.0	4.3	4.7	8.0
4.0	8.5	6.4	6.1	5.8
9.5	5.2	6.7	8.3	9.2

   Test the composite null hypothesis of normality using the Lilliefors test.

2. Fifteen entering freshmen had achievement scores as follows.

481	620	642	515	740
562	395	615	596	618
525	584	540	580	598

   Test for normality using the Lilliefors test.

3. Test the hypothesis of Exercise 1 using the Shapiro–Wilk test.
4. Test the hypothesis of Exercise 2 using the Shapiro–Wilk test.
5. A certain store manager wanted to test the hypothesis that customers arrived randomly at her store, so she recorded the times between successive arrivals of customers one morning. These times (in minutes) were as follows.

3.6	6.2	12.7
14.2	38.0	3.8
10.8	6.1	10.1
22.1	4.2	4.6
1.4	3.3	8.2

   Test the null hypothesis that these interarrival times follow an exponential distribution.

6. Twenty accidents occurred along a particular stretch of interstate highway one month. The nineteen distances between accidents, in miles, are as follows.

0.3	6.1	4.3	3.3	1.9
4.8	0.3	1.2	0.8	10.3
1.2	0.1	10.0	1.6	27.6
12.0	14.2	19.7	15.5	

   Do the accidents appear to be distributed at random along the highway?

7. It is sometimes assumed that stream flow data (the amount of water flowing through a particular stream or river) are lognormally distributed. In order to test

this assumption, data were collected on eight streams and rivers of various sizes. The data consisted of stream flow (cubic feet per second) measurements taken once a week for various numbers of weeks. The logarithms of the data were tested for normality using the Shapiro–Wilk test, with the following results.

Stream Number	Weeks of Record	Value of T_3
1	8	.972
2	10	.858
3	6	.875
4	14	.840
5	9	.966
6	10	.924
7	14	.881
8	12	.868

Do the combined results indicate that stream flow data tend to follow a lognormal distribution?

8. The total yearly rainfall is sometimes assumed to follow a normal distribution. Ten cities across the United States were selected to test this assumption. Annual rainfall records were analyzed using the Shapiro–Wilk test, with the following results.

City	Years of Record	Value of T_3
1	18	.875
2	34	.874
3	26	.948
4	43	.980
5	40	.937
6	29	.915
7	35	.915
8	38	.890
9	42	.963
10	47	.941

Do the combined results indicate that annual rainfall follows a normal distribution?

6.3. TESTS ON TWO INDEPENDENT SAMPLES

The tests presented in this section are useful in situations where two samples are drawn, one from each of two possibly different populations, and the experimenter wishes to determine whether the two distribution functions associated with the two populations are identical or not. While other tests such as the median test, the Mann–Whitney test, or the parametric t test may also be appropriate, they are sensitive to differences between the two means or medians, but they may not detect differences of other types, such as differences in variances. One of the advantages of the two two-sided tests presented in this section is that both tests are consistent against all types of differences that may exist between the two distribution functions.

The first test presented is the Smirnov test (Smirnov, 1939). It is a two-sample version of the Kolmogorov test presented in Section 6.1 and is sometimes called the Kolmogorov–Smirnov two-sample test, while the Kolmogorov test is sometimes called the Kolmogorov–Smirnov one-sample test. The Smirnov test is presented in the one-sided and two-sided versions. Another two-sided test, the Cramér–von Mises test for two samples, is also presented. It is slightly more difficult to compute than the Smirnov test, but it appeals to some people because it seems to make more effective use of the data. Actually, there is probably little difference in power between the two tests.

The Smirnov Test

DATA. The data consist of two independent random samples, one of size n, $X_1, X_2, \ldots, X_n$, and the other of size m, $Y_1, Y_2, \ldots, Y_m$. Let $F(x)$ and $G(x)$ represent their respective, unknown, distribution functions.

ASSUMPTIONS

1. The samples are random samples.
2. The two samples are mutually independent.
3. The measurement scale is at least ordinal.
4. For this test to be exact the random variables are assumed to be continuous.

If the random variables are discrete, the test is still valid but becomes conservative (Noether, 1967a).

HYPOTHESES

A. (Two-Sided Test)

$$H_0: F(x) = G(x) \quad \text{for all } x \text{ from } -\infty \text{ to } +\infty$$

$$H_1: F(x) \neq G(x) \quad \text{for at least one value of } x$$

B. (One-Sided Test)

$$H_0: F(x) \leq G(x) \quad \text{for all } x \text{ from } -\infty \text{ to } +\infty$$

$$H_1: F(x) > G(x) \quad \text{for at least one value of } x$$

This alternative hypothesis is sometimes stated as, "The Xs tend to be *smaller* than the Ys," which is a more general form of location alternatives than the statement that the Xs and Ys differ only by a location parameter (means or medians).

C. (One-Sided Test)

$$H_0: F(x) \geq G(x) \qquad \text{for all } x \text{ from } -\infty \text{ to } +\infty$$

$$H_1: F(x) < G(x) \qquad \text{for at least one value of } x$$

This is the one-sided test to use if it is suspected that the Xs might be shifted to the right (i.e., larger) of the Ys.

TEST STATISTIC. Let $S_1(x)$ be the empirical distribution function based on the random sample $X_1, X_2, \ldots, X_n$, and let $S_2(x)$ be the empirical distribution function based on the other random sample $Y_1, Y_2, \ldots, Y_m$. The test statistic is defined differently for the three different sets of hypotheses.

A. (Two-Sided Test) Define the test statistic T_1 as the greatest vertical distance between the two empirical distribution functions.

(1) $$T_1 = \sup_x |S_1(x) - S_2(x)|$$

B. (One-Sided Test) Denote the test statistic by T_1^+ and let it equal the greatest vertical distance attained by $S_1(x)$ above $S_2(x)$.

(2) $$T_1^+ = \sup_x [S_1(x) - S_2(x)]$$

C. (One-Sided Test) For the one-sided hypotheses in C above, let the test statistic, denoted by T_1^-, be the greatest vertical distance attained by $S_2(x)$ above $S_1(x)$.

(3) $$T_1^- = \sup_x [S_2(x) - S_1(x)]$$

DECISION RULE. Reject H_0 at the level of significance α if the appropriate test statistic T_1, T_1^+, or T_1^-, as the case may be, exceeds its $1 - \alpha$ quantile as given by Table A20 if $n = m$ and by Table A21 if $n \neq m$. For the one-sided tests those tables give the .90, .95, .975, .99, and .995 quantiles. For the two-sided test the .80, .90, .95, .98, and .99 quantiles are furnished. The large sample approximations given at the end of the tables may be used for the sample sizes not covered by the tables.

Example 1. A random sample of size 9, $X_1, \ldots, X_9$ is obtained from one population, and a random sample of size 15, $Y_1, \ldots, Y_{15}$ is obtained from a second population. The null hypothesis is that the two populations have identical distribution functions. If the respective distribution functions are denoted by $F(x)$ and $G(x)$, then the null hypothesis may be written as

$$H_0: F(x) = G(x) \qquad \text{for all } x \text{ from } -\infty \text{ to } +\infty$$

The alternative hypothesis may be stated as

$$H_1: F(x) \neq G(x) \qquad \text{for at least one value of } x$$

The two samples are ordered from smallest to largest for convenience, and their values, along with other pertinent information about their empirical distribution functions, are given next.

X_i	Y_i	$S_1(x) - S_2(x)$	X_i	Y_i	$S_1(x) - S_2(x)$
	5.2	$0 - 1/15 = -1/15$		9.8	$5/9 - 8/15 = 1/45$
	5.7	$0 - 2/15 = -2/15$	9.9		$6/9 - 8/15 = 2/15$
	5.9	$0 - 3/15 = -1/5$	10.1		$7/9 - 8/15 = 11/45$
	6.5	$0 - 4/15 = -4/15$	10.6		$8/9 - 8/15 = 16/45$
	6.8	$0 - 5/15 = -1/3$		10.8	$8/9 - 9/15 = 13/45$
7.6		$1/9 - 5/15 = -2/9$	11.2		$1 - 9/15 = 2/5$
	8.2	$1/9 - 6/15 = -13/45$	11.3		$1 - 10/15 = 1/3$
8.4		$2/9 - 6/15 = -8/45$	11.5		$1 - 11/15 = 4/15$
8.6		$3/9 - 6/15 = -1/15$	12.3		$1 - 12/15 = 1/5$
8.7		$4/9 - 6/15 = 2/45$	12.5		$1 - 13/15 = 2/15$
	9.1	$4/9 - 7/15 = -1/45$	13.4		$1 - 14/15 = 1/15$
9.3		$5/9 - 7/15 = 4/45$	14.6		$1 - 1 = 0$

The test statistic for the two-sided test is given by Equation 1 as

$$T_1 = \sup_x |S_1(x) - S_2(x)|$$
$$= \tfrac{2}{5} = .400$$

the largest absolute difference between $S_1(x)$ and $S_2(x)$, which happens to occur between $x = 11.2$ and $x = 11.3$. The value of .400 for T_1 could also have been determined graphically by drawing graphs of $S_1(x)$ and $S_2(x)$ on the same coordinate axes. From the graphs one can easily see that the difference $S_1(x) - S_2(x)$ changes only at those observed values $x = X_i$ or $x = Y_j$, and that is why it is sufficient to compute $S_1(x) - S_2(x)$ only at the observed sample values, as done here.

From Table A21 we see that the .95 quantile of T_1, for the two-sided test and for $n = 9 = N_1$ and $m = 15 = N_2$, is given as $w_{.95} = 8/15$. For these data T_1 equals 2/5 or 6/15. Therefore H_0 is accepted at the .05 level. From the table, the critical level $\hat{\alpha}$ may be estimated as slightly larger than .20.

For the sake of comparison, the approximate .95 quantile based on the asymptotic distribution is found to be

$$w_{.95} \cong 1.36\sqrt{\frac{m+n}{mn}} = .573$$

which is slightly larger than the exact value of $8/15 = .533$. This illustrates the tendency of the asymptotic approximation to furnish a conservative test.

Note that many of the calculations performed in this example could have been eliminated because, either by an inspection of the data or a preliminary sketch of $S_1(x)$ and $S_2(x)$, many of the values of X_i and Y_j may be seen to be unlikely candidates for yielding the maximum value of $|S_1(x) - S_2(x)|$ and therefore may be ignored in favor of the more likely values of X_i and Y_j.

If a one-sided test had been appropriate instead of the two-sided test, the statistics

$$T_1^+ = \sup_x [S_1(x) - S_2(x)] = \tfrac{2}{5} = .400$$

for the set B of hypotheses, and

$$T_1^- = \sup_x [S_2(x) - S_1(x)] = \tfrac{1}{3} = .333$$

for the set C of hypotheses are easily determined from the preceding table of data. The critical levels for both of the one sided tests are seen from Table A21 to be greater than .10.

☐ *Theory.* Although it may not be apparent at first, the statistics T_1, T_1^+, and T_1^- depend only on the order of the Xs and Ys in the ordered combined sample of Xs and Ys and do not require actual knowledge of the numerical values of the observations. To illustrate this, suppose there are 3Xs and 2Ys. There are $\binom{5}{2} = 10$ different ordered arrangements of the combined sample. These arrangements are given next, along with the values of T_1, T_1^+, and T_1^- for each ordered arrangement.

Arrangement	T_1	T_1^+	T_1^-	Arrangement	T_1	T_1^+	T_1^-
$X<X<X<Y<Y$	1	1	0	$X<Y<X<Y<X$	$\frac{1}{3}$	$\frac{1}{3}$	$\frac{1}{3}$
$X<X<Y<X<Y$	$\frac{2}{3}$	$\frac{2}{3}$	0	$Y<X<X<Y<X$	$\frac{1}{2}$	$\frac{1}{6}$	$\frac{1}{2}$
$X<Y<X<X<Y$	$\frac{1}{2}$	$\frac{1}{2}$	$\frac{1}{6}$	$X<Y<Y<X<X$	$\frac{2}{3}$	$\frac{1}{3}$	$\frac{2}{3}$
$Y<X<X<X<Y$	$\frac{1}{2}$	$\frac{1}{2}$	$\frac{1}{2}$	$Y<X<Y<X<X$	$\frac{2}{3}$	0	$\frac{2}{3}$
$X<X<Y<Y<X$	$\frac{2}{3}$	$\frac{2}{3}$	$\frac{1}{3}$	$Y<Y<X<X<X$	1·	0	1

If the null hypothesis in the two-sided test is true, the two distribution functions are equal and each ordered arrangement is equally likely under the assumption of continuous random variables. This same point was discussed more thoroughly in connection with the Mann–Whitney test in Section 5.1. Therefore, in the two-sided test, the probability associated with each ordered arrangement is given by

$$(4) \qquad \text{probability} = \frac{1}{\binom{m+n}{n}} = \frac{1}{\binom{5}{3}} = \frac{1}{10}$$

and from this the following probability distributions can be deduced.

$$P(T_1 = \tfrac{1}{3}) = \tfrac{1}{10} \qquad P(T_1^+ = 0) = \tfrac{1}{5} \qquad P(T_1^- = 0) = \tfrac{1}{5}$$
$$P(T_1 = \tfrac{1}{2}) = \tfrac{3}{10} \qquad P(T_1^+ = \tfrac{1}{6}) = \tfrac{1}{10} \qquad P(T_1^- = \tfrac{1}{6}) = \tfrac{1}{10}$$
$$P(T_1 = \tfrac{2}{3}) = \tfrac{2}{5} \qquad P(T_1^+ = \tfrac{1}{3}) = \tfrac{1}{5} \qquad P(T_1^- = \tfrac{1}{3}) = \tfrac{1}{5}$$
$$P(T_1 = 1) = \tfrac{1}{5} \qquad P(T_1^+ = \tfrac{1}{2}) = \tfrac{1}{5} \qquad P(T_1^- = \tfrac{1}{2}) = \tfrac{1}{5}$$
$$\qquad\qquad\qquad\qquad P(T_1^+ = \tfrac{2}{3}) = \tfrac{1}{5} \qquad P(T_1^- = \tfrac{2}{3}) = \tfrac{1}{5}$$
$$\qquad\qquad\qquad\qquad P(T_1^+ = 1) = \tfrac{1}{10} \qquad P(T_1^- = 1) = \tfrac{1}{10}$$

It is no coincidence that the distributions of T_1^+ and T_1^- are identical with each other for $n = 3$ and $m = 2$. They are identical with each other for all choices of n and m. However, the space-saving technique used in Tables A20 and A21 of stating that the $1 - \alpha$ quantile of T_1 in the two-sided test equals the $1 - \alpha/2$ quantile of T_1^+ in the one-sided test is a valid technique only if α is small. Notice, for example, in the preceding illustration that $P(T_1 \geq 1)$ equals twice $P(T_1^+ \geq 1)$, and $P(T_1 \geq 2/3)$ equals twice $P(T_1^+ \geq 2/3)$, but $P(T_1 \geq 1/2)$ does not equal twice $P(T_1^+ \geq 1/2)$.

The null distribution (i.e., the distribution when H_0 is true) in the one-sided tests is also found in the manner just described because, under the one-sided null hypotheses, the size of the critical region is a maximum when $F(x)$ is identical with $G(x)$. If the two samples are of equal size, it is not necessary to use this method to find the upper quantiles, because the distribution functions for T_1, T_1^+, and T_1^- were derived as a function of the sample size n by Gnedenko and Korolyuk (1951). The derivation of these distribution functions is interesting, and it is within the presumed mathematical grasp of the reader, but its length precludes its presentation here. The reader is referred to Fisz (1963) for a readable presentation of the derivation.

For samples of unequal size the method of finding quantiles is essentially as illustrated. However, many refinements using path-counting methods have simplified the bookkeeping enough so that extensive tables exist (Harter and Owen, 1970). See Steck (1969) for a general discussion of the Smirnov test. Kim (1969) gives some closer approximations to the exact quantiles when exact tables are not available.

A modification of the Smirnov test was suggested by Tsao (1954) so that the test may be applied to truncated samples. That is, perhaps only the Xs and Ys less than $X^{(r)}$ are observed, as sometimes happens in life-testing experiments. The Smirnov test may then be applied to the truncated samples with the aid of tables derived recursively by Tsao (1954). The distribution functions of Tsao's statistics were derived analytically by Conover (1967a). Extensions of the Smirnov test to three or more samples are presented in the next section.

The next test is the Cramér–von Mises goodness-of-fit test. This test is two-sided only and involves slightly more calculations than the Smirnov test does.

The Cramér–von Mises Two-Sample Test

DATA. The data consist of two independent random samples, $X_1, \ldots, X_n$ and $Y_1, \ldots, Y_m$, with unknown distribution functions $F(x)$ and $G(x)$, respectively.

ASSUMPTIONS

1. The samples are random samples, independent of each other.
2. The measurement scale is at least ordinal.

3. The random variables are continuous. If the random variables are actually discrete the test is likely to be conservative.

HYPOTHESES

$$H_0: F(x) = G(x) \qquad \text{for all } x \text{ from } -\infty \text{ to } +\infty$$

$$H_1: F(x) \neq G(x) \qquad \text{for at least one value of } x$$

TEST STATISTIC. Let $S_1(x)$ and $S_2(x)$ be the empirical distribution functions of the two samples. The test statistic T_2 is defined as

$$(5) \qquad T_2 = \frac{mn}{(m+n)^2} \sum_{\substack{x=X_i \\ x=Y_j}} [S_1(x) - S_2(x)]^2$$

where the squared difference in the summation is computed at each X_i and at each Y_j. Perhaps it is clearer to write the test statistic as

$$(6) \qquad T_2 = \frac{mn}{(m+n)^2} \left\{ \sum_{i=1}^{n} [S_1(X_i) - S_2(X_i)]^2 + \sum_{j=1}^{m} [S_1(Y_j) - S_2(Y_j)]^2 \right\}$$

An equivalent form for T_2 may be obtained by letting $R(X^{(i)})$ and $R(Y^{(j)})$ be the ranks, in the combined ordered sample, of the ith smallest of the Xs denoted by $X^{(i)}$ and the jth smallest of the Ys denoted by $Y^{(j)}$, respectively. Then we can write Equation 6, if there are no ties, as

$$(7) \qquad T_2 = \frac{mn}{(m+n)^2} \left\{ \sum_{i=1}^{n} \left[\frac{R(X^{(i)})}{m} - i \frac{n+m}{nm} \right]^2 + \sum_{j=1}^{m} \left[\frac{R(Y^{(j)})}{n} - j \frac{n+m}{nm} \right]^2 \right\}$$

If $n = m$, Equation 7 reduced to

$$(8) \qquad T_2 = \frac{1}{4n^2} \left\{ \sum_{i=1}^{n} [R(X^{(i)}) - 2i]^2 + \sum_{j=1}^{m} [R(Y^{(j)}) - 2j]^2 \right\}$$

DECISION RULE. Reject H_0 at the approximate level α if T_2 exceeds the $1 - \alpha$ quantile $w_{1-\alpha}$, as shown. These quantiles are approximations based on the asymptotic distribution, valid for large m and n, but they are considered fairly accurate even if the sample sizes are small (Burr, 1964).

$w_{.10} = 0.046$	$w_{.50} = 0.119$	$w_{.90} = 0.347$
$w_{.20} = 0.062$	$w_{.60} = 0.147$	$w_{.95} = 0.461$
$w_{.30} = 0.079$	$w_{.70} = 0.184$	$w_{.99} = 0.743$
$w_{.40} = 0.097$	$w_{.80} = 0.241$	$w_{.999} = 1.168$

These values were taken from Anderson and Darling (1952). Exact quantiles for $n + m \leq 17$ are given by Burr (1964).

Example 2. From the data of Example 1 the test statistic T_2 may be computed by first finding

$$\sum_{i=1}^{9} [S_1(X_i) - S_2(X_i)]^2 = 0.459$$

and

$$\sum_{j=1}^{15}[S_1(Y_j) - S_2(Y_j)]^2 = 0.657$$

Then, from Equation 6, we have

$$T_2 = \frac{mn}{(m+n)^2}\left\{\sum_{i=1}^{n}[S_1(X_i) - S_2(X_i)]^2 + \sum_{j=1}^{m}[S_1(Y_j) - S_2(Y_j)]^2\right\}$$

$$= \frac{(15)(9)}{(24)^2}(.459 + .657)$$

$$= 0.262$$

The null hypothesis of identical distribution functions is accepted at $\alpha = .05$, because $T_2 = 0.262$ is less than $w_{.95} = 0.461$, as just given. The critical level is seen to be about $\hat{\alpha} = 0.18$. This is slightly smaller than the critical level for the Smirnov test on the same data.

☐ *Theory*. The exact distribution of the Cramér–von Mises two-sample test statistic may be obtained in the same way as with the Smirnov test statistic. The different ordered arrangements of the combined sample are equally likely under the null hypothesis, and the statistic T_2 may be computed from the ordered combined sample. Exact quantiles of T_2 were obtained by Anderson (1962) and Burr (1963, 1964) for small samples in essentially ☐ the manner just described, with some computational shortcuts.

The statistic T_2 was apparently introduced by Fisz (1960). He credits the statistic to Lehmann (1951) and the asymptotic distribution of the statistic to Rosenblatt (1952), but the statistic studied by Lehmann and Rosenblatt is

$$(9) \quad T_3 = \frac{m}{2(m+n)}\sum_{i=1}^{n}\left[\frac{R(X^{(i)})}{m} - i\frac{m+n}{mn}\right]^2 + \frac{n}{2(m+n)}\sum_{j=1}^{m}\left[\frac{R(Y^{(j)})}{n} - j\frac{m+n}{mn}\right]^2$$

which differs from Fisz's statistic T_2 unless $m = n$. Fisz showed that T_2 has the same asymptotic distribution as T_3, which was shown by Rosenblatt to be the same as the asymptotic distribution of the Cramér–von Mises goodness-of-fit statistic. Therefore the asymptotic distribution of T_2 was actually obtained by Anderson and Darling (1952) in their paper on the Cramér–von Mises goodness-of-fit statistic. That is why T_2 is called the Cramér–von Mises two-sample test statistic, even though neither Cramér nor von Mises is credited with its invention.

For a two-sample test designed for points on a circle, see Stephens (1965b), Maag (1966), and Maag and Stephens (1968). A multivariate Smirnov test is described by Bickel (1969). Fine (1966) is concerned with the Cramér–von Mises statistic, while Csörgö (1965) and Percus and Percus (1970) work with variations of the Smirnov test. Papers on the asymptotic efficiency of the Smirnov test are by Capon (1965), Ramachandramurty (1966), Andel (1967), and Klotz (1967).

Adaptations of these tests to test for symmetry are presented by Rothman and Woodroofe (1972) and Rao, Schuster and Littell (1975). Gail and Green (1976a) present more extensive tables for the one-sided test and discuss an interesting use of the test in another paper (1976b). More theoretical discussions are given by Takacs (1971) and Kalish and Mikulski (1971).

EXERCISES

1. Test the null hypothesis $F(x) \leq G(x)$, where the observations from $F(x)$ are 0.6, 0.8, 0.8, 1.2, and 1.4 and the observations from $G(x)$ are 1.3, 1.3, 1.8, 2.4, and 2.9.

2. A random sample of five sixth-grade boys in one section of town were given a literacy test with the following results; 82, 74, 87, 86, 75. A random sample of eight sixth-grade boys from a different section of town were given the same literacy test with these scores resulting: 88, 77, 91, 88, 94, 93, 83, 94. Is there a difference in literacy, as measured by this test, in the two populations of sixth-grade boys? (Use the Smirnov test).

3. Use the Cramér–von Mises test on the data in Exercise 2 and compare results with the Smirnov test.

PROBLEMS

1. Find the .80, .90, and .95 quantiles of T_1 for $n = 3$ and $m = 2$ from the exact distribution obtained in the text. Compare these with the quantiles in Table A21 and explain any differences.

2. Find the exact distribution of T_1, T_1^+, T_1^-, and T_2 when $n = 3$ and $m = 3$.

3. Compare the exact .95 quantiles of T_1 with the approximation based on the asymptotic distribution, for $n = m = 30$ and $n = m = 10$.

6.4. TESTS ON SEVERAL INDEPENDENT SAMPLES

The tests presented in this section are multisample analogues of the Smirnov test for two samples. These tests may be applied to several independent samples and, as such, may be compared with the Kruskal–Wallis test and the normal scores test for several independent samples. The Kruskal–Wallis and normal scores tests are intended to be sensitive to differences in means or medians of the various populations, but they may be insensitive to other differences, in particular differences in variances. In fact, a wide disparity of variances may tend to hide the differences in means. These Smirnov-type tests are not as sensitive to differences only in means, but they are consistent against a wider variety of differences and, therefore, are often more powerful than the Kruskal–Wallis and normal scores tests if the differences in means are accompanied by differences in variances and other differences, as is often the case. A major drawback of these tests is that they may be applied only to samples of equal sizes, because the tables for the case of unequal sample sizes have not been developed.

The first test presented is a direct analogue of the two-sided Smirnov test. The test statistic and the method of obtaining its distribution may be used for any number of samples and any sample sizes, whether equal or unequal. But the distribution of the test statistic has been obtained only for the case of three samples of equal size, so from a practical standpoint the test is only a three-sample test at present. Two other tests will also be presented because the distribution of the test statistics in those tests has been obtained for up to 10 samples. Their test statistics are also of the Smirnov type, but the tests are not consistent against all alternatives, as is the first test proposed by Birnbaum and Hall (1960).

The Birnbaum–Hall Test

DATA. The data consist of three independent random samples, each of size n. Denote the empirical distribution functions of the three samples by $S_1(x)$, $S_2(x)$, and $S_3(x)$ and denote their unknown distribution functions by $F_1(x)$, $F_2(x)$, and $F_3(x)$, respectively.

ASSUMPTIONS

1. The samples are random samples, mutually independent of each other.
2. The measurement scale is at least ordinal.
3. In order for the test to be exact the random variables need to be continuous. Otherwise the test is conservative.

HYPOTHESES

H_0: $F_1(x)$, $F_2(x)$, and $F_3(x)$ are identical with each other

H_1: At least two of the distribution functions are different from each other

TEST STATISTIC. Consider the largest vertical distance between $S_1(x)$ and $S_2(x)$, between $S_2(x)$ and $S_3(x)$, and between $S_1(x)$ and $S_3(x)$, without regard to sign. The Birnbaum–Hall test statistic T_1 equals the largest of these distances. Mathematically T_1 may be stated as

(1) $$T_1 = \sup_{x,i,j} |S_i(x) - S_j(x)|$$

which is read, "T_1 equals the supremum, over all x, and all i and j (from 1 to 3), of the absolute difference between the empirical distribution functions $S_i(x)$ and $S_j(x)$."

DECISION RULE. Reject H_0 at the level of significance α if T_1 exceeds the $1-\alpha$ quantile $w_{1-\alpha}$ as given by Table A22. Otherwise accept the null hypothesis that the three distribution functions are identical.

Example 1. Twelve volunteers were assigned to each of three weight-reducing plans. The assignment of the volunteers to the plans was at random, and it was assumed that the 36 volunteers in all would resemble a random sample of people who might try a weight-reducing program. The null hypothesis is that there is no difference in the probability distributions of the amount of weight lost under the three programs, and the alternative is that there is a difference. The results are given as the number of pounds lost by each person

Plan A		Plan B		Plan C	
2	17	17	5	29	5
12	4	15	6	3	25
5	25	3	19	25	32
4	6	19	4	28	24
26	21	5	9	11	36
8	6	14	7	7	20

The empirical distribution functions appear in Figure 8. The greatest vertical distance between any two empirical distribution functions is seen from Figure 8 to occur at $x = 19$, between $S_2(x)$ and $S_3(x)$, plans B and C. This distance is 8/12. The critical region of size $\alpha = .05$ corresponds to all values of T_1 greater than 7/12, the .95 quantile as obtained from Table A22 for $n = 12$. Therefore the null hypothesis is rejected, and we may conclude that the different weight-reducing plans do in fact differ with regard to the probability distribution of the number of pounds lost. The critical level is estimated from Table A22 as slightly less than .05. The exact critical level may be obtained from the more extensive tables furnished by Birnbaum and Hall (1960), and equals .022.

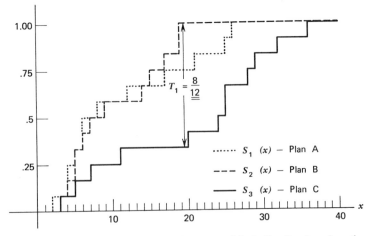

Figure 8. Graphs showing the three empirical distribution functions and the Birmbaum-Hall test statistic T_1.

☐ *Theory.* The exact distribution of the Birnbaum–Hall statistic is found in the same way that the exact distribution functions of the Smirnov statistic and the Kruskal–Wallis statistic were found. That is, under the null hypothesis of identical distributions, each arrangement of the combined ordered sample is equally likely. The statistic T_1 is computed for each arrangement, and the distribution of T_1 is obtained. The procedure is simplified to a great extent by the difference equations used by Birnbaum and Hall (1960). The asymptotic quantiles are based on an unproved conjecture by the author, which states that the large sample distribution of T_1 is likely to be very similar to the large sample distribution of the one-sided statistic T_2 for seven samples, to be introduced next, because of similarities in the
☐ structure of their critical regions.

The following extension of the one-sided Smirnov statistic to the multisample case is appropriate if the alternative hypothesis of interest is one that not only says that differences exist, but states in which direction differences will exist, if indeed there are differences. For example, if the populations are identical except for one variable, such as dosage level on populations of plants or animals, or age levels in populations of humans, then the experimenter can often state that if differences exist among populations, those differences will be exhibited as a tendency for observations to be larger in older populations, or smaller as the dosage level increases, and so on. These are the one-sided alternatives that make this one-sided extension of the Smirnov test appropriate.

The One-Sided *k*-Sample Smirnov Test

DATA. The data consist of k random samples of equal size n. Let their respective empirical distribution functions be denoted by $S_1(x)$, $S_2(x), \ldots, S_k(x)$ and let their respective, unknown, distribution functions be denoted by $F_1(x), F_2(x), \ldots, F_k(x)$.

ASSUMPTIONS

1. The samples are random samples, mutually independent of each other.
2. The measurement scale is at least ordinal.
3. In order for the test to be exact the random variables need to be continuous. Otherwise the test is likely to be conservative.

HYPOTHESES

$$H_0: F_1(x) \leq F_2(x) \leq \cdots \leq F_k(x) \qquad \text{for all } x$$
$$H_1: F_i(x) > F_j(x) \qquad \text{for some } i < j, \text{ and for some } x$$

These hypotheses are used when the alternative hypothesis is that the ith sample tends to have smaller values than the jth sample, for some i less than j. The null hypothesis is usually interpreted as, "All samples come from identical populations," because this one-sided test is usually appropriate if, for some physical reasons, differences between populations will only occur in the direction indicated by H_1.

TEST STATISTIC. The test statistic T_2 is defined as the largest vertical distance achieved by $S_i(x)$ above $S_{i+1}(x)$, where the adjacent samples are being compared as i ranges from 1 to $k-1$. This may be stated mathematically as

$$(2) \qquad T_2 = \sup_{x, i < k} [S_i(x) - S_{i+1}(x)]$$

which is read, "T_2 equals the supremum, over all x, and over all i less than k (the number of samples) of the difference $S_i(x)$ minus $S_{i+1}(x)$." To evaluate T_2, one computes the usual one-sided Smirnov statistic, defined in the previous section, first for samples 1 and 2, then for samples 2 and 3, and so on to samples $k-1$ and k, and lets T_2 equal the largest of these.

DECISION RULE. Reject H_0 at the level α if T_2 exceeds the $1-\alpha$ quantile as given in Table A23. Table A23 is entered with k, the number of samples, and n, the size of each of the k samples. The entry in the column under $p = 1 - \alpha$ is divided by n to give the $1-\alpha$ quantile. The approximation based on the asymptotic distribution, given at the bottom of each column, requires division by $\sqrt{n}$ as indicated.

Example 2. As the human eye ages, it loses its ability to focus on objects close to the eye. This is a well-recognized characteristic of people over 40 years old. In order to see if people in the 15- to 30-year-old range also exhibit this loss of ability to focus on nearby objects as they get older, eight people were selected from each of four age groups; about 15 years old, about 20, about 25, and about 30 years old. It was assumed that these people would behave as a random sample from their age group populations would, with regard to the characteristic being measured. Each person held a printed paper in front of their right eye, with the left eye covered. The paper was moved closer to the eye until the person declared that the print began to look fuzzy. The closest distance at which the print was still sharp was mesured once for each person.

The null hypothesis was that the distance measured was identically distributed for all populations. The alternative hypothesis was that the older groups tended to furnish greater distances measured. The samples were numbered from 1 to 4 in order of age group.

$$H_0: F_1(x) = F_2(x) = F_3(x) = F_4(x) \qquad \text{for all } x$$

$$H_1: F_i(x) > F_j(x) \qquad \text{for some } x \text{ and some } i < j$$

We are assuming that ability to focus on close objects does not improve with age and, therefore, we are able to state the null hypothesis in this slightly simpler form.

The distances, measured in inches, are given next. The samples are ordered within themselves for convenience.

1. 15 *years old*		2. 20 *years old*		3. 25 *years old*		4. 30 *years old*	
4.6	6.3	4.7	6.4	5.6	6.8	6.0	8.6
4.9	6.8	5.0	6.6	5.9	7.4	6.8	8.9
5.0	7.4	5.1	7.1	6.6	8.3	8.1	9.8
5.7	7.9	5.8	8.3	6.7	9.6	8.4	11.5

The largest differences between $S_i(x)$ above $S_{i+1}(x)$ occur at the jump points of $S_i(x)$, which are the values in the ith random sample. Therefore the differences $S_i(x) - S_{i+1}(x)$ need to be computed only at the $n = 8$ numbers in the ith sample.

$$S_1(x) - S_2(x)$$

$$\tfrac{1}{8} - 0 = \tfrac{1}{8}$$
$$\tfrac{2}{8} - \tfrac{1}{8} = \tfrac{1}{8}$$
$$\tfrac{3}{8} - \tfrac{2}{8} = \tfrac{1}{8}$$
$$\tfrac{4}{8} - \tfrac{3}{8} = \tfrac{1}{8}$$
$$\tfrac{5}{8} - \tfrac{4}{8} = \tfrac{1}{8}$$
$$\tfrac{6}{8} - \tfrac{6}{8} = 0$$
$$\tfrac{7}{8} - \tfrac{7}{8} = 0$$
$$\tfrac{8}{8} - \tfrac{7}{8} = \tfrac{1}{8}$$

$$\sup_x [S_1(x) - S_2(x)]$$

$$= \tfrac{1}{8}$$

$$S_2(x) - S_3(x)$$

$$\tfrac{1}{8} - 0 = \tfrac{1}{8}$$
$$\tfrac{2}{8} - 0 = \tfrac{2}{8}$$
$$\tfrac{3}{8} - 0 = \tfrac{3}{8}$$
$$\tfrac{4}{8} - \tfrac{1}{8} = \tfrac{3}{8}$$
$$\tfrac{5}{8} - \tfrac{2}{8} = \tfrac{3}{8}$$
$$\tfrac{6}{8} - \tfrac{3}{8} = \tfrac{3}{8}$$
$$\tfrac{7}{8} - \tfrac{5}{8} = \tfrac{2}{8}$$
$$\tfrac{8}{8} - \tfrac{7}{8} = \tfrac{1}{8}$$

$$\sup_x [S_2(x) - S_3(x)]$$

$$= \tfrac{3}{8}$$

$$S_3(x) - S_4(x)$$

$$\tfrac{1}{8} - 0 = \tfrac{1}{8}$$
$$\tfrac{2}{8} - 0 = \tfrac{2}{8}$$
$$\tfrac{3}{8} - \tfrac{1}{8} = \tfrac{2}{8}$$
$$\tfrac{4}{8} - \tfrac{1}{8} = \tfrac{3}{8}$$
$$\tfrac{5}{8} - \tfrac{2}{8} = \tfrac{3}{8}$$
$$\tfrac{6}{8} - \tfrac{2}{8} = \tfrac{4}{8}$$
$$\tfrac{7}{8} - \tfrac{3}{8} = \tfrac{4}{8}$$
$$\tfrac{8}{8} - \tfrac{6}{8} = \tfrac{2}{8}$$

$$\sup_x [S_3(x) - S_4(x)]$$

$$= \tfrac{4}{8}$$

The test statistic $T_2 = 4/8$, the largest of the largest differences given at the bottom of the columns. The critical region corresponds to values of T_2 greater than $w_{.95}$, at $\alpha = .05$, where $w_{.95}$ is given by Table A23 as

$$w_{.95} = \frac{5}{n} = \frac{5}{8}$$

for $k = 4$ samples and $n = 8$ observations per sample. Because T_2 does not exceed $5/8$, the null hypothesis is accepted. The critical level is seen from Table A23 to be somewhat greater than .10. An examination of the data reveals a slight tendency of the observations to increase in the direction predicted by the alternative hypothesis, but the difference, if it is real, is too slight to be detected with such a small sample size.

Theory. As with the Birnbaum–Hall test, the exact distribution of the one-sided k-sample Smirnov test statistic may be obtained by considering each ordered arrangement of the combined k samples to be equally likely, evaluating the test statistic for each arrangement, and then tabulating the distribution function of the test statistic. However, this tedious task is not necessary, because the distribution function of T_2 was derived by Conover (1967b) as a mathematical function of k and n. The asymptotic distribution was also derived, and these distribution functions were used to obtain

Table A23. The derivations of these distribution functions are beyond the mathematical level of this book, and are therefore omitted. Critical values of T_2 are also given by Wolf and Naus (1973).

Another k-sample Smirnov test was suggested by Conover (1965). This differs from the previous test in that it is a two-sided test. It differs from the Birnbaum–Hall test in that it may be applied to more than three samples because of the more extensive tables available. This is a two-sided test for use with k independent samples, necessarily of the same size, where the alternative of interest indicates that some populations may tend to yield larger values than other populations. It is particularly suited to biological and agricultural type data, or other types of data that are bounded below by some number such as zero but not bounded above by any number, where larger means in a population are usually accompanied by other differences, such as larger variances.

The Two-Sided k-Sample Smirnov Test

DATA. The data consist of k random samples of equal size n. Denote the respective unknown distribution functions by $F_1(x), F_2(x), \ldots, F_k(x)$.

ASSUMPTIONS

1. The samples are random samples, mutually independent of each other.
2. The measurement scale is at least ordinal.
3. In order for the test to be exact the random variables need to be continuous. Otherwise the test is likely to be conservative.

HYPOTHESES

$$H_0: F_1(x) = F_2(x) = \cdots = F_k(x) \qquad \text{for all } x$$

(The population distribution functions are identical.)

$$H_1: F_i(x) \neq F_j(x) \qquad \text{for some } i, j \text{ and } x$$

(The population distribution functions are not identical.)

TEST STATISTIC. The test statistic is evaluated by comparing the "largest" sample with the "smallest" sample. That is, find the largest observation in each sample and denote these by $Z_1, Z_2, \ldots, Z_k$. The sample with the *largest* Z_i (i.e., the largest observation of all) is called the "sample of rank k," or the largest sample, and its empirical distribution function is denoted by $S^{(k)}(x)$. The sample with the *smallest* Z_i is called the "sample of rank 1," or the smallest sample, and its empirical distribution function is denoted by $S^{(1)}(x)$. The test statistic T_3 is defined as the greatest vertical distance attained by $S^{(1)}(x)$ above $S^{(k)}(x)$. Mathematically this may be stated as

$$(3) \qquad T_3 = \sup_x [S^{(1)}(x) - S^{(k)}(x)]$$

DECISION RULE. Reject H_0 at the level α if T_3 exceeds the $1-\alpha$ quantile as given by Table A24. To obtain the $1-\alpha$ quantile $w_{1-\alpha}$ from Table A24, enter the row corresponding to the correct sample size n, and the column corresponding to $p = 1-\alpha$. Then choose the table entry listed with the correct number of samples k. This table entry is then divided by n to obtain the desired quantile. The approximate quantiles for the asymptotic distribution require only division by $\sqrt{n}$ as indicated and do not depend on k.

Example 3. For this example we will use the same data as given in Example 1, where 12 volunteers were assigned to each of three weight-reducing plans. The null hypothesis is that there is no difference among plans, and the alternative is that some difference exists. The data were as follows.

Plan A		Plan B		Plan C	
2	17	17	5	29	5
12	4	15	6	3	25
5	25	3	19	25	32
4	6	19	4	28	24
26	21	5	9	11	36
8	6	14	7	7	20

The largest observation for each sample is underlined.

$$Z_1 = 26$$
$$Z_2 = 19$$
$$Z_3 = 36$$

The largest Z_i is 36, so plan C is called the sample of rank 3. The smallest Z_i is 19, so plan B is called the sample of rank 1. The test statistic is computed from those two samples. The difference $S^{(1)}(x) - S^{(k)}(x)$ needs to be computed only at the numbers listed in the sample of rank 1, because these are where the jumps of $S^{(1)}(x)$ occur and, therefore, these are where $S^{(1)}(x)$ will achieve its largest value above $S^{(k)}(x)$. The samples are ordered for convenience.

Plan B	Plan C	$S^{(1)}(x) - S^{(k)}(x)$	Plan B	Plan C	$S^{(1)}(x) - S^{(k)}(x)$
3	3	$\frac{1}{12} - \frac{1}{12} = 0$	9	25	$\frac{7}{12} - \frac{3}{12} = \frac{4}{12}$
4	5	$\frac{2}{12} - \frac{1}{12} = \frac{1}{12}$	14	25	$\frac{8}{12} - \frac{4}{12} = \frac{4}{12}$
5	7		15	28	$\frac{9}{12} - \frac{4}{12} = \frac{5}{12}$
5	11	$\frac{4}{12} - \frac{2}{12} = \frac{2}{12}$	17	29	$\frac{10}{12} - \frac{4}{12} = \frac{6}{12}$
6	20	$\frac{5}{12} - \frac{2}{12} = \frac{3}{12}$	19	32	
7	24	$\frac{6}{12} - \frac{3}{12} = \frac{3}{12}$	19	36	$\frac{12}{12} - \frac{4}{12} = \frac{8}{12}$

The test statistic T_3, defined by Equation 3 as

$$T_3 = \sup_x [S^{(1)}(x) - S^{(k)}(x)]$$

equals 8/12 for these data. The critical region of size $\alpha = .05$ corresponds to values of T_3 greater than $w_{.95}$, which is given by Table A24 for $n = 12$ and $k = 3$ as

$$w_{.95} = \frac{6}{n} = \frac{6}{12}$$

Because T_3 exceeds 6/12, the null hypothesis is rejected. In fact, the null hypothesis may be rejected at a level of significance as small as .01, so the critical level $\hat{\alpha}$ is about .01.

☐ *Theory.* As with all the tests of this section the exact distribution of the test statistic may be obtained by considering each ordered arrangement of the combined sample as equally likely when the null hypothesis is true. However, this tedious procedure is not necessary, because the distribution function of T_3 is given as a mathematical function of n and k by Conover (1965). This simplifies tabulation, so that tables of quantiles may be easily obtained on a computer. The asymptotic distribution of T_3 is the same as that of the Smirnov one-sided test statistic for two samples, defined in the previous section and also defined as a special case of the previous test in
☐ this section.

The tests of this section were restricted to samples of equal size, and the Birnbaum–Hall test was further restricted to three samples. Actually, any of these tests could be applied to any number of samples, and the samples could be of differing sizes if tables of the distributions of the test statistics were available. Theoretically, these tables are possible and may be obtained by the enumeration method described following each example. From a practical standpoint, however, this enumeration method of considering all ordered arrangements of the combined sample is too exhaustive even for computers. At least that has been the feeling so far, except for the case of three samples of equal sizes. Only when the mathematical formulas for the distribution functions are known have the distributions been obtained for more than three samples, and those formulas are known only when the sample sizes are equal.

EXERCISES

1. Do the following data indicate any difference due to gender in the lengths of Latin words? The observations represent the number of letters in Latin words selected at random from among those of the three genders, masculine, feminine, and neuter.

Masculine		Feminine		Neuter	
5	7	4	6	7	8
7	5	8	3	10	7
6	9	5	6	7	12

Use both tests of this section and compare results.

2. In order to see whether a longer time lapse between the last day of class and the time of the final exam tends to improve student performance on the final exam, a class of 48 students was divided at random into four groups of 12 students each. Group 1 took the final exam 2 days after the last class period. Group 2 took the final exam 4 days after the last class period. Group 3 was given 6 days and Group 4 8 days. All groups were given comparable exams under otherwise comparable conditions. The final exam scores are as follows.

Group 1			Group 2			Group 3			Group 4		
48	71	80	42	70	77	38	73	83	49	77	84
61	74	82	48	71	81	58	74	87	58	79	93
67	75	87	62	73	89	70	75	90	73	80	94
68	79	89	67	75	92	71	79	94	74	84	97

Does the increased time lapse tend to improve test performance?

3. The amount of iron present in the livers of white rats is measured after the animals had been fed one of five diets for a prescribed length of time. There were 10 animals randomly assigned to each of the five diets.

Diet A	Diet B	Diet C	Diet D	Diet E
2.23	5.59	4.50	1.35	1.40
1.14	0.96	3.92	1.06	1.51
2.63	6.96	10.33	0.74	2.49
1.00	1.23	8.23	0.96	1.74
1.35	1.61	2.07	1.16	1.59
2.01	2.94	4.90	2.08	1.36
1.64	1.96	6.84	0.69	3.00
1.13	3.68	6.42	0.68	4.81
1.01	1.54	3.72	0.84	5.21
1.70	2.59	6.00	1.34	5.12

Do the different diets appear to affect the amount of iron present in the livers?

6.5. REVIEW PROBLEMS FOR CHAPTERS 1 TO 6

1. A test is conducted to determine the breaking point for a particular type of rope. Ten pieces of rope were obtained. Their breaking points were recorded as follows (in pounds): 780, 620, 910, 900, 730, 700, 630, 690, 730, and 840.
   (a) Sketch the empirical distribution function.
   (b) Sketch a 90% confidence band for the population distribution function.
   (c) Find an approximate 90% confidence interval for the population median.

2. A certain town consists of five wards. Ten houses are selected at random from each ward and given a score from 0 to 100, depending on the level of deterioration of the house and yard (0 = no deterioration, 100 = no redeeming social value). These are the results.

House	Ward 1	Ward 2	Ward 3	Ward 4	Ward 5
1	08	74	92	03	37
2	45	42	79	09	28
3	43	77	99	22	42
4	64	09	38	06	44
5	03	32	31	26	01
6	85	66	83	20	32
7	74	16	27	56	65
8	48	45	76	20	02
9	19	15	82	04	80
10	57	24	37	29	93

Make a list of all of the nonparametric tests that you could use to analyze these data to detect differences from ward to ward. Carefully state the advantages and disadvantages of each test. Select the test that you feel is the best one to use and test the hypothesis of no differences from ward to ward.

3. Gwen and Rich were teaching different sections of the same course. When the course was completed, they compared the grades they gave to see whether their distributions of grades were essentially the same.

	A	B	C	D & F
Gwen	14	28	17	3
Rich	6	22	23	7

Use a chi-square test to see whether the difference is significant.

4. Use a Kruskal–Wallis test for the data in Problem 3 and compare results. Under what circumstances would you recommend a chi-square test, and when would you recommend the Kruskal–Wallis test?

5. (a) Answer True or False.
   (1) If two events are mutually exclusive, they are also independent.
   (2) The power of a consistent test approaches α as the sample size gets large.
   (3) The A.R.E. is a good approximation to use for the small sample relative efficiency.
   (4) When sampling from the double exponential distribution, the sign test is more powerful than the Wilcoxon signed ranks test.
   (5) The exact distribution of a rank statistic under H_0 can always be found by simple randomization methods.
   (6) If extensive ties are present, the median test should be used rather than the Kruskal–Wallis test.
(b) Fill in the Blanks with words.
   (1) _____ is the size of the critical region.
   (2) _____ is a subset of the sample space.

(3) _____ is the set of all possible outcomes of an experiment.

(4) If the experiment yields an outcome in the_____, the null hypothesis is rejected.

(5) The smallest level of significance at which the null hypothesis may be rejected is called the_____ .

(6) The probability of rejecting a false null hypothesis is called_____.

6. A biased coin is tossed six times to test $H_0 : P(H) = 1/3$ against $H_1 : P(H) \neq 1/3$. If the outcome is "all tails" or "more than four heads" the null hypothesis is rejected.

(a) Is H_0 simple or composite?

(b) Is H_1 simple or composite?

(c) List the points in the critical region.

(d) What is the value of α?

(e) What is the equation of the power function?

7. An economist has been computing a monthly "prosperity index." For the last 24 months the values obtained are 123.6, 121.0, 124.1, 123.4, 125.7, 129.0, 126.8, 127.1, 127.3, 126.7, 124.8, 125.9, 124.7, 125.9, 125.6, 126.0, 125.7, 127.3, 127.7, 129.0, 128.2, 127.9, 127.8, and 127.1. Do these figures indicate a trend in the prosperity index?

8. The "maximum annual river stages" reported at a certain point each year for 16 years were (in feet): 7.4, 7.8, 6.9, 8.1, 8.0, 7.1, 7.4, 6.8, 6.9, 7.3, 7.6, 8.0, 8.3, 7.5, 7.8, and 7.1. Test the null hypothesis that the median maximum annual river stage is no greater than 8.0 feet. Find a 90% confidence interval for the median maximum annual river stage.

9. Sixty rolls of a die resulted in the following occurrences.

Number of Spots Showing	1	2	3	4	5	6
Number of Times Occurred	12	10	14	8	9	7

Test the hypothesis that the die is balanced, that is, that each number of spots has an equal probability of occurring.

10. Ninety graduating seniors, including 30 from Arts and Sciences, 30 from Engineering, and 30 from Agriculture, were selected at random from among those who had accepted salaried positions. Half of the 90 seniors accepted positions paying more than $1000 per month; the other half to receive less than $1000 per month. Of those to receive more than $1000 per month, 9 were in Agriculture, 17 were in Arts and Sciences, and (therefore) 19 were in Engineering. Test the hypothesis that the median salary is the same for seniors in all three categories.

11. One hundred people were asked to taste four new brands of cough syrup and state which new brands tasted better to them than the present formula and which brands did not. As indicated in the following, 15 subjects preferred the new taste to the old for all four brands, 3 subjects preferred brands A, B, and C over the old brand but did not prefer brand D over the present formula, and so on. Test the null hypothesis that there is no significant difference in preferences among the four new brands of cough syrup.

	Brand			Number of Subjects
A	B	C	D	with this Response
1	1	1	1	15
1	1	1	0	3
1	1	0	1	3
1	0	1	1	6
0	1	1	1	21
1	1	0	0	1
1	0	1	0	1
0	1	1	0	1
1	0	0	1	2
0	1	0	1	2
0	0	1	1	19
1	0	0	0	3
0	1	0	0	3
0	0	1	0	2
0	0	0	1	13
0	0	0	0	5
				___
				100

12. The following data presents survival time in days of four samples of hypophysec-
tomized rats without and with different dosages of adrenal cortical hormone.

A (None)	B	C	D
3	2	4	13
2	1	4	4
2	2	3	6
3	6	4	8
5	14	6	19
4	7	5	19
2	15	4	12
2	2	3	1
3	1	4	4
	4	5	4
			1
			12

Test the null hypothesis of no differences in treatment effects.

13. These data represent the yields of four varieties of wheat grown in 13 different
locations.

		Variety		
Location	A	B	C	D
1	43.60	24.05	19.47	19.41
2	40.40	21.76	16.61	23.84
3	18.08	14.19	16.69	16.08

	Variety			
Location	A	B	C	D
4	19.57	18.61	17.78	18.29
5	45.20	29.33	20.19	30.08
6	25.87	25.60	23.31	27.04
7	55.20	38.77	21.15	39.95
8	55.32	34.19	18.56	25.12
9	19.79	21.65	23.31	22.45
10	46.24	31.52	22.48	29.28
11	14.88	15.68	19.79	22.56
12	7.52	4.69	20.53	22.08
13	41.17	32.59	29.25	43.95

Test the hypothesis of no difference in yields of the different varieties.

14. The following data represent days to death of 180 mice inoculated with three strains of typhoid organisms.

	Days to Death												
Strain	2	3	4	5	6	7	8	9	10	11	12	13	14
9D	10	8	18	16	3	4	1						
11C	1	3	3	6	6	14	11	4	6	2	3	1	
DSC1	1	2	1	3	8	11	10	7	7	3	4	2	1

For example, 10 mice inoculated with 9D died on the second day. Does there seem to be a significant difference in the reaction time to the various strains?

15. In Problem 14, is it reasonable to assume that the time till death for mice inoculated with strain 9D follows a normal distribution?

16. The pain threshold values for 12 males and 12 females were as follows.

Males	Females
8.5	6.4
7.9	7.8
6.7	7.1
7.4	8.0
7.5	6.6
8.6	7.3
8.0	8.1
8.1	7.4
7.2	8.3
8.0	8.9
7.8	7.8
7.8	7.7

Test the hypothesis of equal means. Test the hypothesis of equal variances.

17. The following data represent the 1951 and 1952 net earnings of common stocks in 20 representative corporations.

1951	1952	1951	1952
$1.68	$1.71	$4.64	$4.79
1.72	2.17	4.76	4.33
2.50	2.25	5.35	6.05
2.90	2.43	5.81	7.09
3.11	2.32	6.11	6.38
3.35	3.15	6.35	6.00
3.80	3.30	6.69	6.01
3.85	5.52	8.41	7.41
3.89	3.32	8.83	9.33
4.36	3.76	8.97	9.25

Does there seem to be a significant trend in earnings from 1951 to 1952?

18. As an experiment in a sophomore statistics class, 10 students volunteered to take two tests. The first test was a basic mathematics test, and the second was a test on current events. The results were as follows.

Students	Math Scores	Current Events
1	37	23
2	44	34
3	55	59
4	70	25
5	26	16
6	39	12
7	26	16
8	30	25
9	85	60
10	83	69

Does there seem to be a significant correlation between the scores on the two tests?

19. Four different types of automobile tires, 10 tires of each type, were tested under laboratory conditions for a given length of time. The average tread depth of each tire (in centimeters) was measured at the end of the experiment, with the following results.

Tire	Type 1	Type 2	Type 3	Type 4
1	0.34	0.18	0.40	0.33
2	0.31	0.31	0.21	0.29
3	0.08	0.16	0.27	0.13
4	0.26	0.00	0.38	0.24
5	0.29	0.07	0.00	0.10
6	0.00	0.12	0.08	0.45
7	0.09	0.00	0.19	0.37
8	0.14	0.00	0.36	0.19
9	0.26	0.04	0.34	0.53
10	0.19	0.09	0.44	0.56

The primary interest was in choosing the tire (or tires) that tended to show less wear. Which tire (or tires), if any, should be chosen?

20. The following data represent cancer patients diagnosed as terminally ill and undergoing a new treatment.

Case	Primary Tumor Type	Sex	Age	Survival Time (days)
1	Stomach	F	61	121
2	Stomach	M	69	12
3	Stomach	F	62	9
4	Stomach	F	66	18
5	Stomach	M	42	258
6	Stomach	M	79	43
7	Stomach	M	76	142
8	Stomach	M	54	36
9	Stomach	M	62	149
10	Stomach	F	69	182
11	Stomach	M	45	82
12	Stomach	M	57	64
13	Bronchus	M	74	39
14	Bronchus	M	74	427
15	Bronchus	M	66	17
16	Bronchus	M	52	460
17	Bronchus	F	48	90
18	Bronchus	F	64	187
19	Bronchus	M	70	58
20	Brohchus	M	78	52
21	Bronchus	M	71	100
22	Bronchus	M	70	200
23	Bronchus	M	39	42
24	Bronchus	M	70	167
25	Bronchus	M	70	33
26	Esophagus	M	72	50
27	Esophagus	F	80	43
28	Colon	F	76	57
29	Colon	F	58	32
30	Colon	M	49	201
31	Colon	M	69	1267
32	Colon	F	70	144
33	Colon	F	68	170
34	Colon	M	50	428
35	Colon	F	74	157
36	Colon	M	66	58
37	Colon	F	76	123
38	Colon	F	56	861
39	Rectum	F	56	62
40	Rectum	F	75	223
41	Rectum	M	56	18
42	Rectum	F	57	223

Case	Primary Tumor Type	Sex	Age	Survival Time (days)
43	Rectum	M	68	140
44	Rectum	M	64	198
45	Rectum	M	59	759
46	Ovary	F	49	226
47	Ovary	F	68	33
48	Ovary	F	49	183
49	Ovary	F	67	240
50	Ovary	F	56	123
51	Breast	F	56	4
52	Breast	F	57	22
53	Breast	F	53	576
54	Breast	F	66	342
55	Breast	F	68	567
56	Breast	F	53	86
57	Breast	F	75	590
58	Breast	F	74	8
59	Breast	F	59	35
60	Breast	F	50	1644
61	Breast	F	53	173
62	Bladder	M	93	241
63	Bladder	F	70	253
64	Bladder	F	73	110
65	Bladder	F	77	34
66	Bladder	M	44	34
67	Bladder	M	62	669
68	Bladder	M	69	30
69	Gallbladder	F	71	22
70	Gallbladder	M	67	209
71	Kidney	F	71	176
72	Kidney	F	63	89
73	Kidney	F	51	147
74	Kidney	M	53	58
75	Kidney	M	55	659
76	Kidney	M	73	293
77	Kidney	M	45	3
78	Kidney	M	69	24
79	Kidney	M	74	1554
80	Lymphoma	M	40	1016
81	Lymphoma	M	65	82
82	Prostrate	M	47	166
83	Uterus	F	56	68
84	Chondrosarcoma	M	63	9
85	"Brain"	M	49	37
86	Pancreas	M	77	317
87	Pancreas	M	67	21
88	Pancreas	F	60	16
89	Fibrosarcoma	F	54	22

Case	Primary Tumor Type	Sex	Age	Survival Time (days)
90	Testicle	M	42	15
91	Pseudomyxoma	M	47	132
92	Carcinoid	F	68	162
93	Leiomyosarcoma	F	32	453
94	Leukemia	F	59	430
95	Stomach	M	55	27
96	Ovary	F	51	82
97	Bronchus	M	69	31
98	Bronchus	F	67	138
99	Colon	M	77	15
100	Colon	M	38	152

Does there seem to be a significant difference in the distribution of ages of female patients with sex-related cancer (ovary, breast, uterua) versus female patients with other types of cancer?

21. For the data in Problem 20, does age seem to be monotonically related to survival time?

22. For the data in Problem 20, does the survival time for stomach cancer patients seem to equal the survival time for patients with cancer of the bronchus? Discuss the concept of the sampled population versus the target population as it pertains to this problem.

23. Consider the stomach, bronchus, colon, rectum, and bladder patients as one group for the data in Problem 20. Does age seem to be correlated with sex in this group?

24. Ignore those types of cancer that are exclusively male or female in Problem 20. Does the male–female distribution seem to be proportionately the same for the different cancer types?

25. The ages and blood pressures of 15 women are recorded as follows.

Age	Blood Pressure	Age	Blood Pressure
48	144	54	151
60	168	56	152
35	135	31	141
38	125	24	144
55	159	77	170
51	148	63	157
49	128	67	162
38	134		

Does there seem to be a significant monotonic relation between age and blood pressure?

26. Using the data in Problem 25 as a random sample, predict the mean blood pressure of 50-year-old women.

References

Abrahamson, I.G. (1967). Exact Bahadur efficiencies for the Kolmogorov–Smirnov and Kuiper one- and two-sample statistics. *The Annals of Mathematical Statistics, 38,* 1475–1490 (6.1).

Adichie, J.N. (1967a). Asymptotic efficiency of a class of nonparametric tests for regression parameters. *The Annals of Mathematical Statistics, 38,* 884–893 (5.4).

Adichie, J.N. (1976b). Estimates of regression parameters based on rank tests. *The Annals of Mathematical Statistics, 38,* 894–904 (5.4).

Adichie, J.N. (1974). Rank score comparison of several regression parameters. *The Annals of Statistics, 2,* 396–402 (5.5).

Adichie, J.N. (1975). On the use of ranks for testing the coincidence of several regression lines. *The Annals of Statistics, 3,* 521–527 (5.5).

Agresti, A. (1977). Considerations in measuring partial association for ordinal categorical data. *Journal of the American Statistical Association, 72,* 37–45 (5.4).

Aitkin, M.A., and Hume, M.W. (1965). Correlation in a singly truncated bivariate normal distribution. II. Rank correlation. *Biometrika, 52,* 639–643 (5.4).

Aitchison, J., and Aitken, C.G.G. (1976). Multivariate binary discrimination by the kernel method. *Biometrika, 63,* 413–420 (2.4).

Allan, G.J.B. (1976). Ordinal-scaled variables and multivariate analysis: Comment on Hawkes. *American Journal of Sociology, 81,* 1498–1500 (5.12).

Alling, D.W. (1963). Early decision in the Wilcoxon two-sample test. *Journal of the American Statistical Association, 58,* 713–720 (5.1).

Altham, P.M.E. (1971). The analysis of matched proportions. *Biometrika, 58,* 561–566 (3.5).

Andel, J. (1967). Local asymptotic power and efficiency of tests of Kolmogorov–Smirnov type. *The Annals of Mathematical Statistics, 38,* 1705–1725 (6.3).

Anderson T.W. (1962). On the distribution of the two-sample Cramer-von Mises criterion. *The Annals of Mathematical Statistics, 33,* 1148–1159. (6.3).

Anderson, T.W., and Burstein, H. (1967). Approximating the upper binomial confidence limit. *Journal of the American Statistical Association, 62,* 857–861 (3.1).

Anderson, T.W., and Burstein H. (1968). Approximating the lower binomial confidence limit. *Journal of the American Statistical Association, 63,* 1413–1415 (corrections appear in Vol. 64, p. 669) (3.1).

Anderson, T.W., and Darling, D.A. (1952). Asymptotic theory of certain "goodness of fit" criteria based on stochastic processes. *The Annals of Mathematical Statistics, 23,* 193–212 (6.1, 6.3).

Ansari, A.R., and Bradley, R.A. (1960). Rank-sum tests for dispersion. *The Annals of Mathematical Statistics, 31,* 1174–1189 (5.3).

Arbuthnott, J. (1710). An argument for divine providence, taken from the constant regularity observed in the births of both sexes. *Philosophical Transactions, 27,* 186–190 (3.4).

Arnold, H.J. (1965). Small sample power for the one sample Wilcoxon test for non-normal shift alternatives. *The Annals of Mathematical Statistics, 36,* 1767–1778 (5.7).

Barlow, R.E., and Gupta, S.S. (1966). Distribution-free life test sampling plans. *Technometrics, 8,* 591–614 (3.2).

Barr, D.R., and Shudde, R.H. (1973). A note on Kuiper's V_n statistic. *Biometrika, 60,* 663–664 (6.1).

Barr, D.R., and Davidson, T. (1973). A Kolmogorov–Smirnov test for censored samples. *Technometrics, 15,* 739–757 (6.1).

Bartels, R.H., Horn, S.D., Liebetrau, A.M., and Harris, W.L. (1977). A computational investigation of Conover's Kolmogorov–Smirnov test for discrete distributions. Department of Mathematical Sciences, Johns Hopkins University, Technical Report No. 260 (6.1).

Basu, A.P. (1967a). On the large sample properties of a generalized Wilcoxon–Mann–Whitney statistic. *The Annals of Mathematical Statistics, 38,* 905–915 (5.1).

Basu, A.P. (1967b). On two k-sample rank tests for censored data. *The Annals of Mathematical Statistics, 38,* 1520–1535 (5.2).

Basu, A.P. (1968). On a generalized Savage statistic with applications to life testing. *The Annals of Mathematical Statistics, 39,* 1591–1604 (5.1).

Basu, A.P., and Woodworth, G. (1967). A note on nonparametric tests for scale. *The Annals of Mathematical Statistics, 38,* 274–277 (5.3).

Batschelet, E. (1965). *Statistical Methods for the Analysis of Problems in Animal Orientation and Certain Biological Rhythms.* The American Institute of Biological Sciences, Washington, D.C. (3.4, 5.1, 5.7).

Bauer, D.F. (1972). Constructing confidence sets using rank statistics. *Journal of the American Statistical Association, 67,* 687–690 (5.1, 5.3).

Behnen, K. (1976). Asymptotic comparison of rank tests for the regression problem when ties are present. *The Annals of Statistics, 4,* 157–174 (5.5).

Bell, C.B. (1964). Some basic theorems of distribution-free statistics. *The Annals of Mathematical Statistics, 35,* 150–156 (2.5).

Bell, C.B., and Doksum, K.A. (1965). Some new distribution-free statistics. *The Annals of Mathematical Statistics, 36,* 203–214 (5.10).

Bell, C.B., and Doksum, K.A. (1967). Distribution-free tests of independence. *The Annals of Mathematical Statistics, 38,* 429–446 (5.4).

Bell, C.B., and Haller, H.S. (1969). Bivariate symmetry tests: Parametric and nonparametric. *The Annals of Mathematical Statistics, 40,* 259–269 (5.7).

Benard, A., and van Elteren, P. (1953). A generalization of the method of *m* rankings. *Proceedings Koninklijke Nederlandse Akademie van Wetenschappen* (*A*), *56* (Indagationes Mathematicae 15), 358–369 (5.9).

Bennett, B.M. (1965). On multivariate signed rank tests. *Annals of the Institute of Statistical Mathematics, 17,* 55–61 (5.7).

Bennett, B.M., and Nakamura, E. (1963). Tables for testing significance in a 2×3 contingency table. *Technometrics, 5,* 501–511 (4.2).

Bennett, B.M., and Nakamura, E. (1964). The power function of the exact test for the 2×3 contingency table. *Technometrics, 6,* 439–458 (4.2).

Bennett, B.M., and Underwood, R.E. (1970). On McNemar's test for the 2×2 table and its power function. *Biometrics, 26,* 339–343 (3.5).

Beran, R.J. (1969). The derivation of nonparametric two-sample tests from tests for uniformity of a circular distrtribution. *Biometrika, 56,* 561–570 (5.1).

Beran, R.J. (1977). Robust location estimates. *The Annals of Statistics, 5,* 431–444 (5.7).

Berger, A., and Gold, R.Z. (1973). Note on Cochran's Q-test for the comparison of correlated proportions. *Journal of the American Statistical Association, 68,* 989–993 (4.6).

Best, D.J. (1973). Extended tables for Kendall's tau. *Biometrika, 60,* 429–430 (5.4).

Best, D.J. (1974). Tables for Kendall's tau and an examination of the normal approximation. Division of Mathematical Statistics Technical Paper No. 39, Commonwealth Scientific and Industrial Research Organization, Australia (5.4, Appendix).

Bhapkar, V.P., and Deshpande, J.V. (1968). Some nonparametric tests for multisample problems. *Technometrics, 10,* 578–585 (5.2).

Bhapkar, V.P., and Koch, G.G. (1968). Hypotheses of "no interaction" in multidimensional contingency tables. *Technometrics, 10,* 107–124 (4.2).

Bhapkar, V.P., and Patterson, K.W. (1977). On some nonparametric tests for profile analysis of several multivariate samples. *Journal of Multivariate Analysis, 7,* 265–277 (2.5).

Bhapkar, V.P., and Somes, G.W. (1977). Distribution of Q when testing equality of matched proportions. *Journal of the American Statistical Association, 72,* 658–661 (4.6).

Bhattacharyya, G.K. (1967). Asympototic efficiency of multivariate normal score test. *The Annals of Mathematical Statistics, 39,* 1731–1743 (5.1).

Bhattacharyya, G.K., and Johnson, R.A. (1968). Nonparametric tests for shift at unkown time point. *The Annals of Mathematical Statistics, 39,* 1731–1743 (5.1).

Bhattacharyya, G.K., Johnson, R.A., and Neave, H.R. (1971). A comparative power study of the bivariate rank sum test and T^2. *Technometrics, 13,* 191–198 (5.7).

Bhattacharyya, H.T. (1977). Nonparametric estimation of ratio of scale parameters. *Journal of the American Statistical Association, 72,* 459–463 (5.3).

Bickel, P.J. (1969). A distribution free version of the Smirnov two sample test in the *p*-variate case. *The Annals of Mathematical Statistics, 40,* 1–23 (6.3).

Bickel, P.J., and Lehmann, E.L. (1975). Descriptive statistics for nonparametric models. *The Annals of Statistics, 3,* 1038–1069 (5.1, 5.7).

Birnbaum, Z.W. (1953). On the power of a one-sided test of fit for continuous probability functions. *The Annals of Mathematical Statistics, 24,* 484–489 (6.1).

Birnbaum, Z.W. (1962). *Introduction to Probability and Mathematical Statistics.* Harper, New York (4.5).

Birnbaum, Z.W., and Hall, R.A. (1960). Small sample distributions for multisample statistics of the Smirnov type. *The Annals of Mathematical Statistics, 31,* 710–720 (6.4, Appendix).

Birnbaum, Z.W, and Tingey, F.H. (1951). One-sided confidence contours for probability distribution functions. *The Annals of Mathematical Statistics, 22,* 592–596 (6.1).

Birnbaum, Z.W., and Zuckerman, H.S. (1949). A graphical determination of sample size for Wilks' tolerance limits. *The Annals of Mathematical Statistics, 20,* 313–317 (3.3).

Bishop, Y.M.M. (1969). Full contingency tables, logits, and split contingency tables. *Biometrics, 25,* 383–400 (4.7).

Bishop, Y.M.M. (1971). Effects of collapsing multidimensional contingency tables. *Biometrics, 27,* 545–562 (4.7).

Bishop, Y.M.M., Fienberg, S.E., and Holland, P.W. (1975). *Discrete Multivariate Analysis: Theory and Practice.* The MIT Press, Cambridge, Mass. (4.7).

Blomqvist, N. (1951). Some tests based on dichotomization. *The Annals of Mathematical Statistics, 22,* 362–371 (4.6).

Blum, J.R. and Fattu, N.A. (1954). Nonparametric methods. *Review of Educational Research,* 24:5, 467–487 (2.5).

Bohrer, R. (1968). A note on tolerance limits with type I censoring. *Technometrics, 10,* 392 (3.3).

Bowden, D.C. (1968). Query: Tolerance interval in regression. *Technometrics, 10,* 207–210 (3.3).

Bowman, K.O., and Shenton, L.R. (1975). Omnibus test contours for departures from normality based on $\sqrt{b_1}$ and b_2. *Biometrika, 62,* 243–250 (6.2).

Boyd, W.C. (1965). A nomogram for chi-square. *Journal of the American Statistical Association, 60,* 344–346 (1.5).

Boyett, J.M., and Shuster, J.J. (1977). Nonparametric one-sided tests in multivariate analysis with medical applications. *Journal of the American Statistical Association, 72,* 665–668 (5.11).

Bradley, J.V. (1968). *Distribution-Free Statistical Tests.* Prentice-Hall, Englewood Cliffs, N.J. (5.10).

Bradley, R.A., Martin, D.C., and Wilcoxon, F. (1965). Sequential rank tests I. Monte Carlo studies of the two-sample procedure. *Technometrics, 7,* 463–483 (5.1).

Bradley, R.A., Merchant, S.D., and Wilcoxon, F. (1966). Sequential rank tests II. Modified two-sample procedures. *Technometrics, 8,* 615–624 (5.1).

Bradley, R.A., Patel, K.M., and Wackerly, D.D. (1971). Approximate small sample distributions for multivariate two-sample nonparametric tests. *Biometrics, 27,* 515–530 (5.10).

Breslow, N. (1970). A generalized Kruskal–Wallis test for comparing k samples and subject to unequal patterns of censorship. *Biometrika, 57,* 579–594 (5.2).

Broffitt, J.D., Randles, R.H., and Hogg, R.V. (1976). Distribution-free partial discriminant analysis. *Journal of the American Statistical Association, 71,* 934–939 (2.5).

Brunden, M.N. (1972). The analysis of non-independent 2×2 tables from $2 \times C$ tables using rank sums. *Biometrics, 28,* 603–607 (5.2).

Buckle, N., Kraft, C.H., and van Eeden, C. (1969). An approximation to the Wilcoxon–Mann–Whitney distribution. *Journal of the American Statistical Association, 64,* 591–599 (5.1).

Burr, E.J. (1963). Distribution of the two-sample Cramér–von Mises criterion for small equal samples. *The Annals of Mathematical Statistics, 34,* 95–101 (6.3).

Burr, E.J. (1964). Small-sample distributions of the two-sample Cramér–von Mises' W^2 and Watson's U^2. *The Annals of Mathematical Statistics, 35,* 1091–1098 (6.3).

Capon, J. (1965). On the asymptotic efficiency of the Kolomogorov–Smirnov test. *Journal of the American Statistical Association, 60,* 843–853 (6.3).

Carnal, H., and Riedwyl, H. (1972). On a one-sample distribution-free test statistic V. *Biometrika, 59,* 465–467 (6.1).

Casady, R.J., and Cryer, J.D. (1976). Monotone percentile regression. *The Annals of Statistics, 4,* 532–541 (5.6).

Chacko, V.J. (1966). Modified chi-square test for ordered alternatives. *Sankhya (B), 28,* 185–190 (4.2).

Chanda, K.C. (1963). On the efficiency of two-sample Mann–Whitney test for discrete populations. *The Annals of Mathematical Statistics, 34,* 612–617 (5.1).

Chapman, D.G., and Meng, R.C. (1966). The power of chi-square tests for contingency tables. *Journal of the American Statistical Association, 61,* 965–975 (4.2).

Chase, G.R. (1972). On the chi-square test when the parameters are estimated independently of the sample. *Journal of the American Statistical Association, 67,* 609–611 (4.5).

Chatterjee, S.K. (1966). A bivariate sign test for location. *The Annals of Mathematical Statistics, 37,* 1771–1782 (3.5).

Chen, T., and Fienberg, S. E. (1974). Two-dimensional contingency tables with both completely and partially cross-classified data. *Biometrics, 30,* 629–642 (4.2).

Chen, T., and Fienberg, S.E. (1976). The analysis of contingency tables with incompletely classified data. *Biometrics, 32,* 133–144 (4.7).

Chernoff, H. (1967). Query: Degrees of freedom for chi-square. *Technometrics, 9,* 489–490 (4.5).

Chernoff, H., and Lehmann, E.L. (1954). The use of maximum likelihood estimates in χ^2 tests for goodness of fit. *The Annals of Mathematical Statistics, 25,* 579–686 (4.5).

Chiacchierini, R.P., and Arnold, J.C. (1977). A two-sample test for independence in 2×2 contingency tables with both margins subject to misclassification. *Journal of the American Statistical Association, 72,* 170–174 (4.1).

Chmiel, J.J. (1976). Some properties of Spearman-type estimators of the variance and percentiles in bioassay. *Biometrika, 63,* 621–626 (2.5).

Choi, S.C. (1973). On nonparametric sequential tests for independence. *Technometrics, 15,* 625–629 (5.4).

Chow, W.K., and Hodges, J.L., Jr. (1965). An approximation for the distribution of the Wilcoxon one-sample statistic. *Journal of the American Statistical Association, 70,* 648–655 (5.7).

Chung, J.H., and Fraser, D.A.S. (1958). Randomization test for a multivariate two-sample problem. *Journal of the American Statistical Association, 53*, 729–735 (5.11).

Claringbold, P.J. (1961). The use of orthogonal polynomials in the partition of chi-square. *The Australian Journal of Statistics, 3*, 48–63 (4.2).

Claypool, P.L. (1970). Linear interpolation within McCormack's table of the Wilcoxon signed rank statistic. *Journal of the American Statistical Association, 65*, 974–975 (5.7).

Clayton, D.G. (1974). Some odds ratio statistics for the analysis of ordered categorical data. *Biometrika, 61*, 525–531 (4.2).

Cleroux, R. (1969). First and second moments of the randomization test in two-associate PBIB designs. *Journal of the American Statistical Association, 64*, 1424–1433 (5.11).

Clopper, C.J., and Pearson, E.S. (1934). The use of confidence or fiducial limits illustrated in the case of the binomial. *Biometrika, 26*, 404–413 (3.1).

Coberley, W.A., and Lewis, T.O. (1973). A note on a one-sided Kolmogorov–Smirnov test of fit for discrete distribution functions. *Annals of the Institute of Statistical Mathematics, 24*, 183–187 (6.1).

Cochran, W.G. (1937). The efficiencies of the binomial series tests of significance of a mean and of a correlation coefficient. *Journal of the Royal Statistical Society, 100*, 69–73 (3.5).

Cochran, W.G. (1950). The comparison of percentages in matched samples. *Biometrika, 37*, 256–266 (4.6).

Cochran, W.G. (1952). The χ^2 test of goodness of fit. *Annals of Mathematical Statistics, 23*, 315–345 (4.2, 4.5).

Cochran, W.G. (1963). *Sampling Techniques*, 2nd ed. J. Wiley, New York (2.1).

Cohen, A., and Sackrowitz, H.B. (1975). Unibiasedness of the chi-square, likelihood ratio, and other goodness of fit tests for the equal cell case. *The Annals of Statistics, 3*, 959–964 (4.5).

Conover, W.J. (1965). Several k-sample Kolmogorov–Smirnov tests. *The Annals of Mathematical Statistics, 36*, 1019–1026 (6.4).

Conover, W.J. (1967a). The distribution functions of Tsao's truncated Smirnov statistics. *The Annals of Mathematical Statistics, 38*, 1208–1215 (6.3).

Conover, W.J. (1967b). A k-sample extension of the one-sided two-sample Smirnov test statistic. *The Annals of Mathematical Statistics, 38*, 1726–1730 (6.4).

Conover, W.J. (1972). A Kolmogorov goodness-of-fit test for discontinuous distributions. *Journal of the American Statistical Association, 67*, 591–596 (6.1).

Conover, W.J. (1973a). Rank tests for one sample, two samples and k samples without the assumption of a continuous distribution function. *The Annals of Statistics, 1*, 1105–1125 (5.1, 5.2, 5.7).

Conover, W.J. (1973b). On methods of handling ties in the Wilcoxon signed-rank test. *Journal of the American Statistical Association, 68*, 985–988 (5.7).

Conover, W.J. (1974). Some reasons for not using the Yates continuity correction on 2×2 contingency tables. *Journal of the American Statistical Association, 69*, 374–376 (4.1).

Conover, W.J., and Iman, R.L. (1976). On some alternative procedures using ranks for the analysis of experimental designs. *Communications in Statistics—Theory and Methods, A5,* 1349–1368 (5.12).

Conover, W.J., and Iman, R.L. (1978). Some exact tables for the squared ranks test. *Communications in Statistics, B7,* 491–513 (5.3).

Conover, W.J., and Iman, R.L. (1978). The rank transformation as a method of discrimination with some examples. Technical Report SAND78-0583, Sandia Laboratories, Albuquerque (5.12).

Conover, W.J., and Iman, R.L. (1979). On multiple comparison procedures. Technical Report LA-7677-MS, Los Alamos Scientific Laboratory (5.2).

Conover, W.J., and Iman, R.L. (1980). The rank transformation as a method of discrimination with some examples. *Communications in Statistics, A9,* (in press) (2.5, 5.12).

Conover, W.J., and Kemp, K.E. (1976). Comparisons of the asymptotic efficiencies of two sample tests for discrete distributions. *Communications in Statistics—Theory and Methods, A5,* 1–15 (5.1).

Conover, W.J., Wehmanen, O., and Ramsey, F.L. (1978). A note on the small sample power functions for nonparametric tests of location in the double exponential family. *Journal of the American Statistical Association, 73,* 188–190 (5.1).

Cox, D.R., and Stuart, A. (1955). Some quick tests for trend in location and dispersion. *Biometrika, 42,* 80–95 (3.5).

Cramér, H. (1928). On the composition of elementary errors. *Skandinavisk Aktuarietidskrift, 11,* 13–74, 141–180 (6.1).

Cramér, H. (1946). *Mathematical Methods of Statistics.* Princeton University Press, Princeton, N.J. (4.1, 4.2, 4.4, 4.5).

Cronholm, J.N. (1968). Two tables connected with goodness-of-fit tests for equiprobable alternatives. *Biometrika, 55,* 441 (6.1).

Crouse, C.F. (1966). Distribution-free tests based on the sample distribution function. *Biometrika, 53,* 99–108 (5.2).

Crouse, C.F. (1967). A class of distribution-free analysis of variance tests. *South African Statistics Journal, 1,* 75-80 (5.12).

Crouse, C.F. (1968). A distribution-free method of analyzing a 2^m factorial experiment. *South African Statistics Journal, 2,* 101–108 (5.12).

Crowley, J., and Breslow N. (1975). Remarks on the conservatism of $\sum(0-E)^2/E$ in survival data. *Biometrics, 31,* 957–961 (4.2).

Cryer, J.D., Robertson, T., Wright, F.T., and Casady, R.J. (1972). Monotone median regression. *The Annals of Mathematical Statistics, 43,* 1459–1469 (5.6).

Csörgo, M. (1965). Some Smirnov type theorems of probability. *The Annals of Mathematical Statistics, 36,* 1113–1119 (6.3).

Cureton, E.E. (1967). The normal approximation to the signed rank sampling distribution when zero differences are present. *Journal of the American Statistical Association, 62,* 1068–1069 (5.7).

D'Agostino, R.B. (1971). An omnibus test of normality for moderate and large size samples. *Biometrika, 58,* 341–348 (6.2).

Dahiya, R.C. (1971). On the Pearson chi-squared goodness-of-fit test statistic. *Biometrika, 58,* 685–686 (4.5).

Dahiya, R.C., and Gurland, J. (1972). Pearson chi-squared test of fit with random intervals. *Biometrika, 59*, 147–153 (4.5).

Dahiya, R.C., and Gurland, J. (1973). How many classes in the Pearson chi-square test? *Journal of the American Statistical Association, 68*, 707–712 (4.5).

Daniels, H.E. (1950). Rank correlation and population models. *Journal of the Royal Statistical Society (B), 12*, 171–181 (5.4).

Danziger, L., and Davis, S.A. (1964). Tables of distribution-free tolerance limits. *The Annals of Mathematical Statistics, 35*, 1361–1365 (3.3).

Darroch, J.N. (1974). Multiplicative and additive interaction in contingency tables. *Biometrika, 61*, 207–214 (4.2).

Davis, J.A. (1967). A partial coefficient for Goodman and Kruskal's gamma. *Journal of the American Statistical Association, 62*, 189–193 (4.4).

Dempster, A.P., and Schatzoff, M. (1965). Expected significance level as a sensitivity index for test statistics. *Journal of the American Statistical Association, 60*, 420–436 (2.3).

Deshpande, J.V. (1970). A class of multisample distribution-free tests. *The Annals of Mathematical statistics, 41*, 227–236 (5.2).

Diamond, E.L. (1963). The limiting power of categorical data chi-square tests analogous to normal analysis of variance. *The Annals of Mathematical Statistics, 34*, 1432–1441 (4.2).

Dixon, W.J. (1953). Power functions of the sign test and power efficiency for normal alternatives. *The Annals of Mathematical Statistics, 24*, 467–473 (2.4, 3.4).

Doksum, K.A. (1967). Robust procedures for some linear models with one observation per cell. *The Annals of Mathematical Statistics, 38*, 878–883 (5.8).

Doksum, K.A., and Sievers, G.L. (1976). Plotting with confidence: Graphical comparisons of two populations. *Biometrika, 63*, 421–434 (5.1).

Dunn, O.J. (1964). Multiple comparisons using rank sums. *Technometrics, 6*, 241–252 (5.8).

Duran, B.S. (1976). A survey of nonparametric tests for scale. *Communications in Statistics—Theory and Methods, A5*, 1287–1312 (5.3).

Duran, B.S., and Mielke, P.W., Jr. (1968). Robustness of the sum of squared ranks test. *Journal of the American Statistical Association, 63*, 338–344 (5.3).

Duran, B.S., Tsai, W.S., and Lewis, T.O. (1976). A class of location-scale nonparametric tests. *Biometrika, 63*, 173–176 (5.3).

Durbin, J. (1951). Incomplete blocks in ranking experiments. *British Journal of Psychology (Statistical Section), 4*, 85–90 (5.9).

Durbin, J. (1961). Some methods of constructing exact tests. *Biometrika, 48*, 41–55 (5.10).

Durbin, J. (1968). The probability that the sample distribution function lies between two parallel straight lines. *The Annals of Mathematical Statistics, 39*, 398–411 (6.1).

Durbin, J. (1975). Kolmogorov–Smirnov tests when parameters are estimated with applications to tests of exponentiality and tests on spacings. *Biometrika, 62*, 5–22 (6.2, Appendix).

Dyer, A.R. (1974). Comparison of tests for normality with a cautionary note. *Biometrika, 61*, 185–189 (6.2).

Ehrenberg, A.S.C. (1951). Note on normal transformations of ranks. *British Journal of Psychology (Statistical Section)*, *4*, 133–134 (5.10).

Elston, R.C. (1970). A new test of association for continuous variables. *Biometrics*, *26*, 305–314 (4.2).

Erdos, P., and Renyi, A. (1959). On the central limit theorem for samples from a finite population. *Matem. Kutato Intezet Kolzem.*, *4*, 49 (1.5).

Evans, L.S. (1973). A mechanical interpretation of the coefficient of rank correlation and other analogies. *The American Statistician*, *27*, (2), 79–81 (5.4).

Federer, W.T. (1963). *Experimental Design*. Macmillan, New York (5.9).

Feller, W. (1968). *An Introduction to Probability Theory and Its Applications*, Vol. I, 3rd ed. J. Wiley, New York (1.1).

Festinger, L. (1946). The significance of difference between means without reference to the frequency distribution function. *Psychometrika*, *11*, 97–105 (5.1).

Fienberg, S.E. (1970). Quasi-independence and maximum likelihood estimation in incomplete contingency tables. *Journal of the American Statistical Association*, *65*, 1610–1616 (4.7).

Fienberg, S.E. (1972). The analysis of incomplete multiway contingency tables. *Biometrics*, *28*, 177–202 (4.7).

Fienberg, S.E. (1970). An iterative procedure for estimation in contingency tables. *The Annals of Mathematical Statistics*, *41*, 907–917 (correction appears in Vol. 42, p. 1778) (4.2).

Fienberg, S.E. (1977). *The Analysis of Cross-Classified Categorical Data*. The MIT Press, Cambridge, Mass. (4.7).

Fienberg, S.E., and Gilbert, J.P. (1970). The geometry of a two by two contingency table. *Journal of the American Statistical Association*, *65*, 694–701 (4.1).

Fienberg, S.E., and Larntz, K. (1976). Log linear representation for paired and multiple comparison models. *Biometrika*, *63*, 245–254 (4.7).

Fine, T. (1966). On the Hodges and Lehmann shift estimator in the two sample problem. *The Annals of Mathematical Statistics*, *37*, 1814–1818 (6.3).

Finkelstein, J.M., and Schafer, R.E. (1971). Improved goodness-of-fit tests. *Biometrika*, *58*, 641–646 (6.1).

Finney, D.J. (1948). The Fisher–Yates test of significance in 2×2 contingency tables. *Biometrika*, *35*, 145–156 (4.2).

Fisher, R.A. (1935). *The Design of Experiments*. Oliver & Boyd, Edinburgh–London (7th ed., 1960) (5.11).

Fisher, R.A., and Yates, F. (1957). *Statistical Tables for Biological, Agricultural, and Medical Research*, 5th ed. Oliver & Boyd, Edinburgh (5.10).

Fisz, M. (1960). On a result by M. Rosenblatt concerning the von Mises–Smirnov test. *The Annals of Mathematical Statistics*, *31*, 427–429 (6.3).

Fisz, M. (1963). *Probability Theory and Mathematical Statistics*, 3rd ed. J. Wiley, New York (1.5, 6.3).

Fleiss, J.L. (1965). A note on Cochran's Q test. *Biometrics*, *21*, 1008–1010 (4.6).

Fleiss, J.L. (1973). *Statistical Methods for Rates and Proportions*. J. Wiley, New York (4.1).

Fligner, M.A., Hogg, R.V., and Killeen, T.J. (1976). Some distribution-free rank-like statistics having the Mann–Whitney–Wilcoxon null distribution. *Communications in Statistics—Theory and Methods, A5,* 373–376 (5.1).

Fligner, M.A., and Killeen, T.J. (1976). Distribution-free two-sample tests for scale. *Journal of the American Statistical Association, 71,* 210–213 (5.3).

Fraser, D.A.S. (1957). *Nonparametric Methods in Statistics.* J. Wiley, New York (5.10).

Freund, J.E. (1962). *Mathematical Statistics.* Prentice-Hall, Englewood Cliffs, N.J. (1.5).

Freund, J.E., and Ansari, A.R. (1957). *Two-way rank sum test for variances.* Technical Report No. 34, Virginia Polytechnic Institute, Blacksburg (5.1).

Friedman, M. (1937). The use of ranks to avoid the assumption of normality implicit in the analysis of variance. *Journal of the American Statistical Association, 32,* 675–701 (5.8).

Gabriel, K.R. (1966). Simultaneous test procedures for multiple comparisons on categorical data. *Journal of the American Statistical Association, 61,* 1081–1096 (4.3).

Gabriel, K.R., and Lachenbruch, P.A. (1969). Non-parametric ANOVA in small samples: A Monte Carlo study of the adequacy of the asymptotic approximation. *Biometrics, 25,* 593–596 (5.2).

Gail, M.H., and Gart, J.J. (1973). The determination of sample sizes for use with the exact conditional test in 2×2 comparative trials. *Biometrics, 29,* 441–448 (4.1).

Gail, M.H., and Green, S.B. (1976a). Critical values for the one-sided two-sample Kolmogorov–Smirnov statistic. *Journal of the American Statistical Association, 71,* 757–760 (6.3).

Gail, M.H., and Green, S.B. (1976b). A generalization of the one-sided two-sample Kolmogorov–Smirnov statistic for evaluating diagnostic tests. *Biometrics, 32,* 561–570 (6.3).

Garside, G.R., and Mack, C. (1976). Actual type 1 error probabilities for various tests in the homogeneity case of the 2×2 contingency table. *The American Statistician, 30* (1), 18–21 (4.1).

Gart, J.J. (1966). Alternative analyses of contingency tables. *Journal of the Royal Statistical Society (B), 28,* 164–179 (4.2).

Gart, J.J. (1972). Interaction tests for $2 \times s \times t$ contingency tables. *Biometrika, 59,* 309–316 (4.7).

Gastwirth, J.L. (1965a). Asymptotically most powerful rank tests for the two-sample problem with censored data. *The Annals of Mathematical Statistics, 36,* 1243–1248 (5.1).

Gastwirth, J.L. (1965b). Percentile modifications of two sample rank tests. *Journal of the American Statistical Association, 60,* 1127–1141 (5.1).

Geertsema, J.C. (1970). Sequential confidence intervals based on rank tests. *The Annals of Mathematical Statistics, 41,* 1016–1026 (5.7).

Gehan, E.A. (1965a). A generalized Wilcoxon test for comparing arbitrarily singly censored samples. *Biometrika, 52,* 203–224 (5.1).

Gehan, E.A. (1965b). A generalized two-sample Wilcoxon test for doubly censored data. *Biometrika, 52,* 650–653 (5.1).

Gehan, E.A., and Thomas, D.G. (1969). The performance of some two-sample tests in small samples with and without censoring. *Biometrika, 56,* 127–132 (5.1).

Gelzer, J., and Pyke, R. (1965). The asymptotic relative efficiency of goodness-of-fit tests against scalar alternatives. *Journal of the American Statistical Association, 60,* 410–419 (6.1).

Gerig, T.M. (1969). A multivariate extension of Friedman's χ_r^2 test. *Journal of the American Statistical Association, 64,* 1595–1608 (5.8).

Gerig, T.M. (1975). A multivariate extension of Friedman's χ_r^2 test with random covariates. *Journal of the American Statistical Association, 70,* 443–447 (5.8).

Gessaman, M.P., and Gessaman, P.H. (1972). A comparison of some multivariate discrimination procedures. *Journal of the American Statistical Association, 67* (338), 468–472 (2.5).

Ghosh, M., Grizzle, J.E., and Sen, P.K. (1973). Nonparametric methods in longitudinal studies. *Journal of the American Statistical Association. 68,* 29–36 (2.5).

Gibbons, J.D. (1964). Effect of non-normality on the power function of the sign test. *Journal of the American Statistical Association, 59,* 142–148 (3.4).

Gibbons, J.D. (1967). Correlation coefficients between nonparametric tests for location and scale. *Annals of the Institute of Statistical Mathematics, 19,* 519–526 (5.3).

Gibbons, J.D., and Gastwirth, J.L. (1970). Properties of the percentile modified rank tests. *Annals of the Institute of Statistical Mathematics,* Supplement 6, 95–114 (5.1).

Gilbert, R.O. (1972). A Monte Carlo study of analysis of variance and competing rank tests for Scheffe's mixed model. *Journal of the American Statistical Association, 67,* 71–75 (5.8).

Glasser, G.J., and Winter, R.F. (1961). Critical values of the coefficient of rank correlation for testing the hypothesis of independence. *Biometrika, 48,* 444–448 (Appendix).

Gnedenko, B.V., and Korolyuk, V.S. (1951). On the maximum discrepancy between two empirical distributions (Russian). *Doklady Akademii Nauk SSSR* (N.S.), *80,* 525–528. English translation in *IMS and American Mathematical Society* (1961) (6.3).

Gokhale, D.V. (1968). On asymptotic relative efficiencies of a class of rank tests for independence of two variables. *Annals of the Institute of Statistical Mathematics, 20,* 255–261 (5.4, 5.10).

Goodman, L.A. (1964). Simple methods for analyzing three-factor interaction in contingency tables. *Journal of the American Statistical Association, 59,* 319–352 (4.2).

Goodman, L.A. (1965). On simultaneous confidence intervals for multinomial proportions. *Technometrics, 7,* 247–254 (3.1).

Goodman, L.A. (1968). The analysis of cross-classified data: Independence, quasi-independence, and interactions in contingency tables with or without missing entries. *Journal of the American Statistical Association, 63,* 1091–1113 (4.2).

Goodman, L.A. (1970). The multivariate analysis of qualitative data: Interactions among multiple classifications. *Journal of the American Statistical Association, 65,* 226–256 (4.2).

Goodman, L.A. (1971). Partitioning of chi-square, analysis of marginal contingency tables, and estimation of expected frequencies in multidimensional contingency tables. *Journal of the American Statistical Association, 66*, 339–344 (4.2).

Goodman, L.A. (1971). The analysis of multidimensional contingency tables: Stepwise procedures and direct estimation methods for building models for multiple classifications. *Technometrics, 13*, 33–62 (4.7).

Goodman, L.A., and Kruskal, W.H. (1954). Measures of association for cross-classifications. *Journal of the American Statistical Association, 49*, 732–764 (correction appears in Vol. 52, p. 578) (4.4).

Goodman, L.A., and Kruskal, W.H. (1959). Measures of association for cross-classifications. II: Further discussion and references. *Journal of the American Statistical Association, 54*, 123–163 (4.4).

Goodman, L.A., and Kruskal, W.H. (1963). Measures of association for cross-classifications. III: Approximate sample theory. *Journal of the American Statistical Association, 58*, 310–364 (4.4).

Goodman, L.A., and Madansky, A. (1962). Parameter-free and nonparametric tolerance limits: The exponential case. *Technometrics, 4*, 75–95 (3.3).

Govindarajulu, Z. (1968). Distribution-free confidence bounds for $P(X < Y)$. *Annals of the Institute of Statistical Mathematics, 20*, 229–238 (5.1).

Govindarajulu, Z. (1976). A brief survey of nonparametric statistics. *Communications in Statistics—Theory and Methods, A5*, 429–453 (2.5).

Govindarajulu, Z., and Klotz, J.H. (1973). A note on the asymptotic distribution of the one-sample Kolmogorov–Smirnov statistic. *The American Statistician, 27* (4), 164–165 (6.1).

Govindarajulu, Z., and Leslie, R.T. (1972). Annotated bibliography on robustness studies of statistical procedures. Department of Health, Education, and Welfare Publication No. (HSM) 72–1051 (2.5).

Green, J.R., and Hegazy, Y.A.S. (1976). Powerful modified-EDF goodness-of-fit tests. *Journal of the American Statistical Association, 71*, 204–209 (6.2).

Gregory, G. (1961). Contingency tables with a dependent classification. *The Australian Journal of Statistics, 3*, 42–47 (4.2).

Grizzle, J.E. (1967). Continuity correction in the χ^2-test for 2×2 tables. *The American Statistician, 21* (4), 28–32 (4.1).

Grizzle, J.E., Starmer, C.F., and Koch, G.G. (1969). Analysis of categorical data by linear models. *Biometrics, 25*, 489–504 (4.7).

Grizzle, J.E., and Williams, O.D. (1972). Log linear models and tests of independence for contingency tables. *Biometrics, 28*, 137–156 (4.7).

Groeneveld, R.A. (1972). Asymptotically optimal group rank tests for location. *Journal of the American Statistical Association, 67*, 847–849 (5.7).

Haberman, S.J. (1973). Log-linear models for frequency data: Sufficient statistics and likelihood equations. *The Annals of Statistics, 1*, 617–632 (4.7).

Haga, T. (1960). A two-sample rank test on location. *Annals of the Institute of Statistical Mathematics, 11*, 211–219 (5.1).

Hajek, J., and Sidak, Z. (1967). *Theory of Rank Tests*. Academic Press, New York (5.1, 5.10).

Halperin, M., Ware, J.H., Byar, D.P., Mantel, N., Brown, C.C., Koziol, J., Gail, M., and

Greer, S.B. (1977). Testing for interaction in an $I \times J \times K$ contingency table. *Biometrika*, *64*, 271–275 (4.2).

Hamilton, B.L. (1976). A Monte Carlo test of the robustness of parametric and nonparametric analysis of covariance against unequal regression slopes. *Journal of the American Statistical Association*, *71*, 864–869 (5.12).

Hanson, D.L., and Owen, D.B. (1963). Distribution-free tolerance limits, elimination of the requirement that cumulative distribution functions be continuous. *Technometrics*, *5*, 518–522 (3.3).

Harkness, W.L., and Katz, L. (1964). Comparison of the power functions for the test of independence in 2×2 contingency tables. *The Annals of Mathematical Statistics*, *35*, 1115–1127 (4.1).

Harter, H.L. (1964). A new table of percentage points of the chi-square distribution. *Biometrika*, *51*, 231–240 (1.5).

Harter, H.L., and Owen, D.B. (1970). *Selected Tables in Mathematical Statistics*, Vol. 1. Markham, Chicago (5.7, Appendix).

Harter, H.L., and Owen, D.B. (1975). *Selected Tables in Mathematical Statistics*, Vol. 3. American Mathematical Society, Providence (Appendix).

Hartley, H.O., and Pfaffenberger, R.C. (1972). Quadratic forms in order statistics used as goodness-of-fit criteria. *Biometrika*, *59*, 605–611 (6.2).

Haynam, G.E., and Govindarajulu, Z. (1966). Exact power of the Mann-Whitney test for exponential and rectangular alternatives. *The Annals of Mathematical Statistics*, *37*, 945–953 (5.1).

Haynam, G.E., and Leone, F.C. (1965). Analysis of categorical data. *Biometrika*, *52*, 654–660 (4.2).

Healy, M.J.R. (1969). Exact tests of significance in contingency tables. *Technometrics*, *11*, 393–395 (4.2).

Hemelrijk, J. (1952). A theorem on the sign test when ties are present. *Proceedings Koninklijke Nederlandse Akademie van Wetenschappen* (A), *55*, 322–326 (3.4).

Hettmansperger, T.P. (1968). On the trimmed Mann–Whitney statistic. *The Annals of Mathematical Statistics*, *39*, 1610–1614 (5.1).

Hettmansperger, T.P., and Malin, J.S. (1975). A modified Mood's test for location with no shape assumptions on the underlying distributions. *Biometrika*, *62*, 527–529 (5.1).

Hettmansperger, T.P., and McKean, J.W. (1977). A robust alternative based on ranks to least squares in analyzing linear models. *Technometrics*, *19*, 275–284 (5.5).

Hettmansperger, T.P., and McKean, J.W. (1978). Statistical inference based on ranks. *Psychometrika*, *43*, 69–79 (5.12).

Hewett, J.E., and Tsutakawa, R.K. (1972). Two-stage chi-square goodness-of-fit test. *Journal of the American Statistical Association*, *67*, 395–401 (4.5).

Hocking, R.R., and Oxspring, H.H. (1974). The analysis of partially categorized contingency data. *Biometrics*, *30*, 469–484 (4.2).

Hodges, J.L., Jr. and Lehmann, E. (1956). The efficiency of some nonparametric competitors of the *t*-test. *The Annals of Mathematical Statistics*, *27*, 324–335 (3.4, 5.1).

Hodges, J.L., Jr. and Lehmann, E.L. (1963). Estimates of location based on rank tests. *The Annals of Mathematical Statistics*, *34*, 598–611 (5.1).

Hoeffding, W. (1952). The large-sample power of tests based on permutations of observations. *The Annals of Mathematical Statistics, 23,* 169–192 (5.11).

Hoeffding, W. (1965). Asymptotically optimal tests for multinomial distributions. *The Annals of Mathematical Statistics, 36,* 369–400 (4.2).

Hogg, R.V. (1974). Adaptive robust procedures. *Journal of the American Statistical Association, 69,* 909–923 (2.5).

Hogg, R.V. (1975). Estimates of percentile regression lines using salary data. *Journal of the American Statistical Association, 70,* 56–59 (5.6).

Hogg, R.V. (1976). A new dimension to nonparametric tests. *Communications in Statistics—Theory and Methods, A5,* 1313–1325 (5.10).

Hogg, R.V. (ed.) (1977). Robustness; A special issue of *Communications in Statistics—Theory and Methods, A6,* 789–894 (5.12).

Hollander, M. (1963). A nonparametric test for the two-sample problem. *Psychometrika, 28,* 395–403 (5.3).

Hollander, M. (1967a). Asymptotic efficiency of two nonparametric competitors of Wilcoxon's two sample test. *Journal of the American Statistical Association, 62,* 939–949 (5.1).

Hollander, M. (1967b). Rank tests for randomized blocks when the alternatives have an a priori ordering. *The Annals of Mathematical Statistics, 38,* 867–877 (5.8).

Hollander, M. (1968). Certain uncorrelated nonparametric test statistics. *Journal of the American Statistical Association, 63,* 707–714 (5.3).

Hollander, M. (1970). A distribution-free test for parallelism. *Journal of the American Statistical Association, 65,* 387–394 (5.7).

Hollander, M. (1971). A nonparametric test for bivariate symmetry. *Biometrika, 58,* 203–212 (5.7).

Hollander, M., Pledger, G., and Lin, P. (1974). Robustness of the Wilcoxon test to a certain dependency between samples. *The Annals of Statistics, 1,* 177–181 (5.1).

Hollander, M., and Wolfe, D.A. (1973). *Nonparametric Statistical Methods.* J. Wiley, New York (5.5).

Holst, Lars. (1972). Asymptotic normality and efficiency for certain goodness-of-fit tests. *Biometrika, 59,* 137–145 (4.5).

Horn, S.D. (1977). Goodness-of-fit tests for discrete data: A review and an application to a health impairment scale. *Biometrics, 33,* 237–248 (4.5, 6.1).

Horn, S.D., and Pyne, D. (1976). Comparison of exact and approximate goodness-of-fit tests for discrete data. Department of Mathematical Sciences, Johns Hopkins University, Technical Report No. 257 (6.1).

Hotelling, H., and Pabst, M.R. (1936). Rank correlation and tests of significance involving no assumption of normality. *The Annals of Mathematical Statistics, 7,* 29–43 (5.4).

Høyland, A. (1965). Robustness of the Hodges-Lehmann estimates for shift. *The Annals of Mathematical Statistics, 36,* 174–197 (5.1).

Høyland, A. (1968). Robustness of the Wilcoxon estimate of location against a certain dependence. *The Annals of Mathematical Statistics, 39,* 1196–1201 (5.7).

Huber, P.J. (1972). Robust statistics: A review. *The Annals of Mathematical Statistics, 43,* 1041–1067 (5.12).

Huber, P.J. (1973). Robust regression: Asymptotics, conjectures and Monte Carlo. *The Annals of Statistics, 1,* 799–821 (5.5).

Hudimoto, H. (1959). On a two-sample nonparametric test in the case that ties are present. *Annals of the Institute of Statistical Mathematics, 11,* 113–120 (5.3).

Hwang, T.Y., and Klotz, J.H. (1975). Bahadur efficiency of linear rank statistics for scale alternatives, *The Annals of Statistics, 3,* 947–954 (5.3).

Iman, R.L. (1970). Use of summation operators for the derivation of common formulae. *Mathematics Teacher, 43* (4), 296–297 (1.6).

Iman, R.L. (1974a). Use of a *t*-statistic as an approximation to the exact distribution of the Wilcoxon signed ranks test statistic. *Communications in Statistics, 3,* 795–806 (5.7).

Iman, R.L. (1974b). A power study of a rank transform for the two-way classification model when interaction may be present. *The Canadian Journal of Statistics Section C: Applications, 2,* 227–239 (5.12).

Iman, R.L. (1976). An approximation to the exact distribution of the Wilcoxon–Mann–Whitney rank sum statistic. *Communications in Statistics—Theory and Methods, A5,* 587–598 (5.1).

Iman, R.L., and Conover, W.J. (1978). Approximations of the critical region for Spearman's rho with and without ties present. *Communications in Statistics—Series B, Computations and Simulation, 7,* 269–282 (5.4).

Iman, R.L., and Conover, W.J. (1979). The use of the rank transform in regression. *Technometrics, 21,* 499–509 (5.6).

Iman, R.L., and Davenport, J.M. (1976). New approximations to the exact distribution of the Kruskal-Wallis test statistic. *Communications in Statistics—Theory and Methods, A5,* 1335–1348 (5.2).

Iman, R.L., and Davenport, J.M. (1980). Approximations of the critical region of the Friedman statistic. *Communications in Statistics, A9(6)* (in press) (5.8).

Iman, R.L., Quade, D., and Alexander, D.A. (1975). Exact probability levels for the Kruskal-Wallis test. *Selected Tables in Mathematical Statistics, 3,* 329–384 (5.2, Appendix).

Ireland, C.T., Ku, H.H., and Kullback, S. (1969). Symmetry, and marginal homogeneity of an $r \times r$ contingency table. *Journal of the American Statistical Association, 64,* 1323–1341 (4.2).

Ireland, C.T., and Kullback, S. (1968). Contingency tables with given marginals. *Biometrika, 55,* 179–188 (4.2).

Ishii, G. (1960). Intraclass contingency tables. *Annals of the Institute of Statistical Mathematics, 12,* 161–207 (corrections appear in Vol. 12, p. 279) (4.2).

Ives, K.H., and Gibbons, J.D. (1967). A correlation measure for nominal data. *The American Statistician, 21* (5), 16–17 (4.4).

Jacobson, J.E. (1963). The Wilcoxon two-sample statistic: Tables and bibliography. *Journal of the American Statistical Association, 58,* 1086–1103 (5.1).

Jaeckel, L.A. (1972). Estimating regression coefficients by minimizing the dispersion of the residuals. *The Annals of Mathematical Statistics, 43,* 1449–1458 (5.5).

Jogdeo, K. (1966). On randomized rank score procedures of Bell and Doksum. *The Annals of Mathematical Statistics, 37,* 1697–1703 (5.10).

Johns, M.V., Jr. (1974). Nonparametric estimation of location. *Journal of the American Statistical Association, 69,* 453–460 (5.7).

Johnson, R.A., and Mehrotra, K.G. (1972). Locally most powerful rank tests for the two-sample problem with censored data. *The Annals of Mathematical Statistics, 43,* 823–831 (5.10).

Jureckova, J. (1971). Nonparametric estimate of regression coefficients. *The Annals of Mathematical Statistics, 42,* 1328–1338 (5.5).

Jureckova, J. (1977). Asymptotic relations of M-estimates and R-estimates in linear regression model. *The Annals of Statistics, 5,* 464–472 (5.5).

Kalbfleish, J.D. (1974). Some efficiency calculations for survival distributions. *Biometrika, 61,* 31–38 (5.5).

Kalish, G., and Mikulski, P.W. (1971). The asymptotic behavior of the Smirnov test compared to standard "optimal procedures." *The Annals of Mathematical Statistics, 42,* 1742–1747 (6.3).

Kanofsky, P., and Srinivasan, R. (1972). An approach to the construction of parametric confidence bands on cumulative distribution functions. *Biometrika, 59,* 623–631 (6.2).

Kempthorne, O. (1955). The randomization theory of experimental inference. *Journal of the American Statistical Association, 50,* 946–967 (5.11).

Kempthorne, O., and Doerfler, T.E. (1969). The behavior of some significance tests under experimental randomization. *Biometrika, 56,* 231–248 (5.11).

Kendall, M.G. (1938). A new measure of rank correlation. *Biometrika, 30,* 81–93 (5.4).

Kendall, M.G. (1942). Partial rank correlation. *Biometrika, 32,* 277–283 (5.4).

Kendall, M.G. (1955). *Rank Correlation Methods,* 2nd ed. Hafner, New York (5.4).

Kendall, M.G., and Bavington-Smith, B. (1939). The problem of m rankings. *The Annals of Mathematical Statistics, 10,* 275–287 (5.8).

Kendall, M.G., and Sundrum, R.M. (1953). Distribution-free methods and order properties. *Review of the International Statistical Institute, 21* (3), 124–134 (2.5).

Kim Jae-On (1975). Multivariate analysis of ordinal variables. *American Journal of Sociology, 81,* 261–298 (5.12).

Kim, P.J. (1969). On the exact and approximate sampling distribution of the two sample Kolmogorov–Smirnov criterion $D_{mn}m \le n$. *Journal of the American Statistical Association, 64,* 1625–1635 (6.3).

Klotz, J. (1962). Nonparametric tests for scale. *The Annals of Mathematical Statistics, 33,* 498–512 (5.10).

Klotz, J. (1963). Small sample power and efficiency for the one sample Wilcoxon and normal scores tests. *The Annals of Mathematical Statistics, 34,* 624–632 (5.7).

Klotz, J. (1965). Alternative efficiencies for signed rank tests. *The Annals of Mathematical Statistics, 36,* 1759–1766 (5.7).

Klotz, J. (1966). The Wilcoxon, ties, and the computer. *Journal of the American Statistical Association, 61,* 772–787 (5.1).

Klotz, J. (1967). Asymptotic efficiency of the two sample Kolmorgorov–Smirnov test. *Journal of the American Statistical Association, 62,* 932–938 (6.3).

Klotz, J., and Teng, J. (1977). One-way layout for counts and the exact enumeration of the Kruskal–Wallis H distribution with ties. *Journal of the American Statistical Association, 72,* 165–169 (5.2).

Knight, W.R. (1966). A computer method for calculating Kendall's tau with ungrouped data. *Journal of the American Statistical Association, 61*, 436–439 (5.4).

Knoke, J.D. (1976). Multiple comparisons with dichotomous data. *Journal of the American Statistical Association, 71*, 849–853 (4.3).

Knott, M. (1970). The small sample power of one-sided Kolmogorov tests for a shift in location of the normal distribution. *Journal of the American Statistical Association, 65*, 1384–1391 (6.1).

Koch, G.G. (1970). The use of non-parametric methods in the statistical analysis of a complex split plot experiment. *Biometrics, 26*, 105–128 (5.8).

Koch, G.G., Imrey, P.B., and Reinfurt, D.W. (1972). Linear model analysis of categorical data with incomplete response vectors. *Biometrics, 28*, 663–692 (4.7).

Koch, G.G., Johnson, W.D., and Tolley, H.D. (1972). A linear models approach to the analysis of survival and extent of disease in multidimensional contingency tables, *Journal of the American Statistical Association, 67*, 783–796 (4.2).

Koch, G.G., and Reinfurt, D.W. (1971). The analysis of categorical data from mixed models. *Biometrics, 27*, 157–175 (4.7).

Kolmogorov, A.N. (1933). Sulla determinazione empirica di una legge di distribuzione. *Giornale dell' Istituto Italiano degli Attuari, 4*, 83–91 (6.1).

Konijn, H.S. (1961). Non-parametric, robust and short-cut methods in regression and structural analysis. *The Australian Journal of Statistics, 3*, 77–86 (5.4).

Kraft, C.H., and van Eeden, C. (1972). Asymptotic efficiencies of quick methods of computing efficient estimates based on ranks. *Journal of the American Statistical Association, 67*, 199–202 (5.1, 5.7).

Krewski, D. (1976). Distribution-free confidence intervals for quantile intervals. *Journal of the American Statistical Association, 71*, 420–422 (3.2).

Kruskal, W.H. (1952). A nonparametric test for the several sample problem. *The Annals of Mathematical Statistics, 23*, 525–540 (5.2).

Kruskal, W.H. (1958). Ordinal measures of association. *Journal of the American Statistical Association, 53*, 814–861 (5.4).

Kruskal, W.H., and Wallis, W.A. (1952). Use of ranks on one-criterion variance analysis. *Journal of the American Statistical Association, 47*, 583–621 (corrections appear in Vol. 48, pp. 907–911) (5.2).

Ku, H.H. (1963). A note on contingency tables involving zero frequencies and the 2I test. *Technometrics, 5*, 398–400 (4.2).

Ku, H.H., and Kullback, S. (1974). Loglinear models in contingency table analysis. *The American Statistician, 28* (4), 115–122 (4.7).

Ku, H.H., Varner, R.N., and Kullback, S. (1971). On the analysis of multidimensional contingency tables. *Journal of the American Statistical Association, 66*, 55–64 (4.2).

Kullback, S. (1971). Marginal homogeneity of multidimensional contingency tables. *The Annals of Mathematical Statistics, 42*, 594–606 (4.2).

Kullback, S., Kupperman, M., and Ku, H.H. (1962). Tests for contingency tables and Markov chains. *Technometrics, 4*, 573–608 (4.2).

Labovitz, S. (1970). The assignment of numbers to rank order categories. *American Sociological Review, 35*, 515–524 (5.12).

La Brecque, J. (1977). Goodness-of-fit tests based on non-linearity in probability plots. *Technometrics, 19*, 293–306 (6.2).

Lapunov, A.M. (1901). Nouvelle forme du theoreme sur la limite de probabilite. *Mem. Acad. Sci. St. Petersburg, 12* (5), (1.5).

Laubscher, N.F., and Odeh, R.E. (1976). A confidence interval for the scale parameter based on Sukhatme's two-sample statistic. *Communications in Statistics—Theory and Methods, A5,* 1393–1407 (5.3).

Laubscher, N.F., Steffens, F.E., and DeLange, E.M. (1968). Exact critical values for Mood's distribution-free test statistic for dispersion and its normal approximation. *Technometrics, 10,* 497–508 (5.3).

Lawler, K.R. (1978). A distribution-free generalization of the Wilcoxon signed-ranks test for the two-way layout. Master's Thesis, Texas Tech University, Lubbock.

Lee, S.K. (1978). An example for teaching some basic concepts in multidimensional contingency table analysis. *The American Statistician, 32* (2), 69–70 (4.7).

Lee, S.W. (1966). The power of the one-sided and one-sample Kolmogorov–Smirnov test. Unpublished master's report, Kansas State University (6.1).

Lehmann, E.L. (1951). Consistency and unbiasedness of certain nonparametric tests. *The Annals of Mathematical Statistics, 22,* 165–179 (6.3).

Lehmann, E.L. (1959). *Testing Statistical Hypotheses.* J. Wiley, New York (2.4).

Lehmann, E.L. (1966). Some concepts of dependence. *The Annals of Mathematical Statistics, 37,* 1137–1153 (5.4).

Lehmann, E.L. (1975). *Nonparametrics: Statistical Methods Based on Ranks.* Holden–Day, San Francisco (5.5, 5.8, 5.10).

Lehmann, E.L., and Stein, C. (1949). On the theory of some nonparametric hypotheses. *The Annals of Mathematical Statistics, 20,* 28–45 (5.11).

Lemmer, H.H., and Stoker, D.J. (1967). A distribution-free analysis of variance for the two-way classification. *South African Statistics Journal, 1,* 67–74 (5.12).

Lemmer, H.H., Stoker, D.J., and Reinach, S.G. (1968). A distribution-free analysis of variance technique for block designs. *South African Journal of Statistics, 2,* 9–32 (5.8).

Lepage, Y. (1971). A combination of Wilcoxon's and Ansari–Bradley's statistics. *Biometrika, 58,* 213–217 (5.3).

Lepage, Y. (1973). A table for a combined Wilcoxon Ansari–Bradley statistic. *Biometrika, 60,* 113–116 (5.3).

Lepage, Y. (1977). A class of nonparametric tests for location and scale parameters. *Communications in Statistics-Theory and Methods, A6,* 649–659 (5.3).

Levy, P. (1925). *Calcul des Probabilities.* Gauthiers–Villars, Paris (1.5).

Lewis, T.G. (1975). *Distribution Sampling for Computer Simulation.* D.C. Heath, Lexington, Mass. (5.10).

Li, L., and Schucany, W.R. (1975). Some properties of a test for concordance of two groups of rankings. *Biometrika, 62,* 417–423 (5.8).

Light, R.J., and Margolin, B.H. (1971). An analysis of variance for categorical data. *Journal of the American Statistical Association, 66,* 534–544 (4.2).

Lilliefors, H.W. (1967). On the Kolmogorov–Smirnov test for normality with mean and variance unknown. *Journal of the American Statistical Association, 62,* 399–402 (6.2, Appendix).

Lilliefors, H.W. (1969). On the Kolmogorov–Smirnov test for the exponential distribution with mean unknown. *Journal of the American Statistical Association, 64,* 387–389 (6.2).

Lilliefors, H.W. (1973). The Kolmogorov–Smirnov and other distance tests for the gamma distribution and for the extreme-value distribution when parameters must be estimated. Department of Statistics, George Washington University, unpublished manuscript (6.2).

Lindeberg, J.W. (1922). Eine neue, Herleitung des Exponential-gesetzes. *Mathematische Zeitschrift, 15*, 221 (1.5).

Maag, U.R. (1966). A k-sample analogue of Watson's U^2 statististic. *Biometrika, 53,* 579–584 (6.3).

Maag, U.R., and Dicaire, G. (1971). On Kolmogorov–Smirnov type one-sample statistics. *Biometrika, 58,* 653–656 (6.1).

Maag, U.R., and Stephens, M.A. (1968). The V_{NM} two-sample test. *The Annals of Mathematical Statistics, 39,* 923–935 (6.3).

Maag, U.R., Streit, F., and Drouilly, P.A. (1973). Goodness-of-fit tests for grouped data. *Journal of the American Statistical Association, 68,* 462–465 (6.1).

Macdonald, P. (1971). The analysis of a 2^n experiment by means of ranks. *Journal of the Royal Statistical Society (Series C) Applied Statistics, 20,* 259–275 (5.12).

Mack, C. (1969). Query: Tolerance limits for a binomial distribution. *Technometrics, 11,* 201 (3.3).

MacKinnon, W. J. (1964). Table for both the sign test and distribution-free confidence intervals of the median for sample sizes to 1,000. *Journal of the American Statistical Association, 59,* 935–959 (3.4).

Mann, H.B., and Whitney, D.R. (1947). On a test of whether one of two random variables is stochastically larger than the other. *The Annals of Mathematical Statistics, 18,* 50–60 (5.1).

Mansfield, E. (1962). Power functions for Cox's test of randomness against trend. *Technometrics, 4,* 430–432 (3.5).

Mantel, N., and Fleiss, J.L. (1975). The equivalence of the generalized McNemar tests for marginal homogeneity in 2^3 and 3^2 tables. *Biometrics, 31,* 727–729 (3.5).

Mantel, N., and Greenhouse, S.W. (1968). What is the continuity correction? *The American Statistician, 22*(5), 27–30 (4.1).

Mantel, N., and Haenszel, W. (1959). Statistical aspects of the analysis of data from retrospective studies of disease. *Journal of the National Cancer Institute, 22,* 719–748 (4.2).

Mardia, K.V. (1967a). A non-parametric test for the bivariate two-sample location problem. *Journal of the Royal Statistical Society (B), 29,* 320–342 (5.1).

Mardia, K. V. (1967b). Some contributions to contingency-type bivariate distributions. *Biometrika, 54,* 235–249 (4.2).

Mardia, K.V. (1968). Small sample power of a nonparametric test for the bivariate two-sample location problem in the normal case. *Journal of the Royal Statistical Society (B), 30,* 83–92 (5.1).

Margolin, B.H., and Light, R.J. (1974). An analysis of variance for categorical data, II: Small sample comparisons with chi-square and other competitors. *Journal of the American Statistical Association, 69,* 755–764 (4.2).

Margolin, B.H., and Maurer, W. (1976). Tests of the Kolmogorov–Smirnov type for exponential data with unknown scale, and related problems. *Biometrika, 63,* 149–160 (6.2).

Maritz, J.S., Wu, M, and Staudte, R.G., Jr. (1977). A location estimator based on a U-statistic. *The Annals of Statistics, 5,* 779–786 (5.7).

Marsaglia, G. (1968). Query: Psuedo random normal numbers. *Technometrics, 10,* 401–402 (5.10).

Massey, F.J. (1950a). A note on the estimation of a distribution function by confidence limits. *The Annals of Mathematical Statistics, 21,* 116–119 (6.1).

Massey, F.J. (1950b). A note on the power of a nonparametric test. *The Annals of Mathematical Statistics, 21,* 440–443 (Corrections appear in Vol. 23, pp. 637–638) (6.1).

Massey, F.J. (1952). Distribution table for the deviation between two sample cumulatives. *The Annals of Mathematical Statistics, 23,* 435–441. Corrections appear in Davis, L.S. (1958), *Mathematical Tables and Other Aids to Computation, 12,* 262–263 (Appendix).

Matthes, T.K., and Truax, D.R. (1965). Optimal invariant rank tests for the k-sample problem. *The Annals of Mathematical Statistics, 36,* 1207–1222 (5.2).

Maxwell, E.A. (1961). *Analyzing Qualitative Data.* Wiley, New York (4.2).

Maxwell, E.A. (1976). Analysis of contingency tables and further reasons for not using Yates correction in 2×2 tables. *The Canadian Journal of Statistics Section C: Applications, 4,* 277–290 (4.1).

McCarthy, P.J. (1965). Stratified sampling and distribution-free confidence intervals for a median. *Journal of the American Statistical Association, 60,* 772–783 (5.7).

McCornack, R.L. (1965). Extended tables of the Wilcoxon matched pairs signed rank statistics. *Journal of the American Statistical Association, 60,* 864–871 (5.7).

McDonald, B.J., and Thompson, W.A. (1967). Rank sum multiple comparisons in one- and two-way classifications, *Biometrika, 54,* 487–498 (5.2, 5.8).

McDonald, L.L., Davis, B.M., and Milliken, G.A. (1977). A nonrandomized unconditional test for comparing two proportions in 2×2 contingency tables. *Technometrics, 19,* 145–158 (4.1).

McKean, J.W., and Ryan, T.A., Jr. (1977). An algorithm for obtaining confidence intervals and point estimates based on ranks in the two-sample location problem. *AMC Transactions on Mathematical Software, 3,* 183–185 (5.1).

McKinlay, S.M. (1975). A note on the chi-square test for pair-matched samples. *Biometrics, 31,* 731–735 (3.5).

McNeil, D.R. (1967). Efficiency loss due to grouping in distribution-free tests. *Journal of the American Statistical Association, 62,* 954–965 (5.1).

McNeil, D.R., and Tukey, J.W. (1975). Higher order diagnosis of two-way tables, illustrated on two sets of demographic empirical distributions. *Biometrics, 31,* 487–510 (4.2).

McNemar, Q. (1962). *Psychological Statistics,* 3rd ed. Wiley, New York (4.4).

Meeker, W.Q. (1978). Sequential tests of independence for 2×2 contingency tables. *Biometrika, 65*(1), 85–90 (4.1).

Mehra, K.L., and Sarangi, J. (1967). Asymptotic efficiency of certain rank tests for comparative experiments. *The Annals of Mathematical Statistics, 38,* 90–107 (5.8).

Mielke, P.W., Jr. (1967). Note on some squared rank tests with existing ties. *Technometrics, 9,* 312–314 (5.3).

Mielke, P.W., Jr. (1972). Asymptotic behavior of two-sample tests based on powers of ranks for detecting scale and location alternatives. *Journal of the American Statistical Association. 67*, 850–854 (5.1, 5.3).

Mielke, P.W., Jr., and Siddiqui, M.M. (1965). A combinatorial test for independence of dichotomous responses. *Journal of the American Statistical Association, 60*, 437–441 (4.2).

Mikulski, P.W. (1963). On the efficiency of optimal nonparametric procedures in the two sample case. *The Annals of Mathematical Statistics, 34*, 22–32 (5.1).

Miller, L.H. (1956). Table of percentage points of Kolmogorov statistics. *Journal of the American Statistical Association, 51*, 111–121 (Appendix).

Miller, R.G., Jr. (1966). *Simultaneous Statistical Inference.* McGraw-Hill, New York (5.8).

Miller, R.G., Jr. (1970). A sequential signed rank test. *Journal of the American Statistical Association, 65*, 1554–1561 (5.7).

Miller, R.G. (1973). Nonparametric estimators of the mean tolerance in bioassay. *Biometrika, 60*, 535–542 (2.5).

Milton, R.C. (1964). An extended table of critical values for the Mann–Whitney (Wilcoxon) two-sample statistic. *Journal of the American Statistical Association, 59*, 925–934 (5.1).

Molinari, L. (1977). Distribution of the chi-square test in nonstandard situations. *Biometrika, 64*, 115–121 (4.5).

Mood, A.M. (1954). On the asymptotic efficiency of certain nonparametric two-sample tests. *The Annals of Mathematical Statistics, 25*, 514–522 (5.3).

Moses, L.E. (1952). Nonparametric statistics for psychological research. *Psychological Bulletin, 49*, 122–143 (5.11).

Moses, L.E. (1963). Rank tests of dispersion, *The Annals of Mathematical Statistics, 34*, 973–983 (5.3).

Moses, L.E. (1965). Query: Confidence limits from rank tests, *Technometrics, 7*, 257–260 (5.1, 5.7).

Mosteller, F. (1968). Association and estimation in contingency tables. *Journal of the American Statistical Association, 63*, 1–29 (4.2).

Mote, V.L., and Anderson, R.L. (1965). An investigation of the effect of misclassification on the properties of χ^2-tests in the analysis of categorical data. *Biometrika, 52*, 95–110 (4.2).

Murphy, R.B. (1948). Nonparametric tolerance limits. *The Annals of Mathematical Statistics, 19*, 581–588 (3.3).

Neave, H.R., and Granger, W.J. (1968). A Monte Carlo study comparing various two-sample tests for differences in mean. *Technometrics, 10*, 509–522 (7.1).

Nelson, L.S. (1966). Query: Combining values of observed χ^2's. *Technometrics, 8*, 709 (15, 4.1).

Nemenyi, P. (1969). Variances: An elementary proof and a nearly distribution-free test. *The American Statistician, 23*(5), 35–37 (5.3).

Noe, M., and Vandewiele, G. (1968). The calculation of distributions of Kolmogorov–Smirnov type statistics including a table of significance points for a particular case. *The Annals of Mathematical Statistics, 39*, 233–241 (6.1).

Noether, G.E. (1963). Efficiency of the Wilcoxon two-sample statistic for randomized blocks. *Journal of the American Statistical Association, 58,* 894–898 (5.1).

Noether, G.E. (1976a). *Elements of Nonparametric Statistics.* Wiley, New York (2.4, 3.3, 5.1, 5.7, 5.8, 5.9).

Noether, G.E. (1967b). Wilcoxon confidence intervals for location parameters in the discrete case. *Journal of the American Statistical Association, 62,* 184–188 (5.7).

Noether, G.E. (1973). some simple distribution-free confidence intervals for the center of a symmetric distribution. *Journal of the American Statistical Association, 68,* 716–719 (5.7).

Odeh, R.E. (1967). The distribution of the maximum use of ranks. *Technometrics, 9,* 271–278 (5.2).

Odeh, R.E. (1971). On Jonckheere's k-sample test against ordered alternatives. *Technometrics, 13,* 912–918 (5.2).

Odeh, R.E. (1972). On the power of Jonckheere's k-sample test against ordered alternatives. *Biometrika, 59,* 467–471 (5.2).

Odeh, R.E. (1977). Extended tables of the distribution of Friedman's S-statistic in the two-way layout. *Communications in Statistics-Simulation and Computation, B6,* 29–48 (5.8).

Oden, A., and Wedel, H. (1975). Arguments for Fisher's permutation test. *The Annals of Statistics, 3,* 518–520 (5.11).

Odoroff, C.L. (1970). A comparison of minimum logit chi-square estimation and maximum likelihood estimation in $2 \times 2 \times 2$ and $3 \times 2 \times 2$ contingency tables: Tests for interaction. *Journal of the American Statistical Association, 65,* 1617–1631 (4.7).

Olshen, R.A. (1967). Sign and Wilcoxon tests for linearity. *The Annals of Mathematical Statistics, 38,* 1759–1769 (3.5).

Ott, R.L., and Free, S.M. (1969). A short-cut rule for a one-sided test of hypothesis for qualitative data. *Technometrics, 11,* 197–200 (4.1).

Owen, D.B. (1962). *Handbook of Statistical Tables.* Addison-Wesley, Reading, Mass. (3.3, 5.7, 5.8, 5.10).

Page, E.B. (1963). Ordered hypotheses for multiple treatments: A significance test for linear ranks. *Journal of the American Statistical Association, 58,* 216–230 (5.8).

Pahl, P.J. (1969). On testing for goodness-of-fit of the negative binomial distribution when expectations are small. *Biometrics, 25,* 143–151 (4.5).

Patel, K.D. (1975). Cochran's Q test: Exact distribution. *Journal of the American Statistical Association. 70,* 186–189 (4.6).

Patel, K.M., and Hoel, David G. (1973). A nonparametric test for interaction in factorial experiments. *Journal of the American Statistical Association, 68,* 615–620 (5.8).

Pearson, E.S. (1947). The choice of statistical test illustrated on the interpretation of data classed in a 2×2 table. *Biometrika, 34,* 139–167 (4.1).

Pearson, E.S., D'Agostino, R.B., and Bowman, K.O. (1977). Tests for departure from normality: Comparison of powers. *Biometrika, 64,* 231–246 (6.2).

Pearson, E.S., and Hartley, H. O. (1962). *Biometrika Tables for Statisticians,* Vol. I, 2nd ed. Cambridge University Press, Cambridge, England (5.10).

Pearson, E. S., and Hartley, H.O. (1970). *Biometrika Tables for Statisticians*, Vol. I. 3rd ed. Reprinted with corrections 1976. Cambridge University Press, Cambridge, England (Appendix).

Pearson, E.S., and Hartley, H.O. (1972). *Biometrika Tables for Statisticians*, Vol. II. Reprinted with corrections 1976. Cambridge University Press, Cambridge, England (Appendix).

Pearson, E.S., and Please, N.W. (1975). Relation between the shape of population distribution and the robustness of four simple test statistics. *Biometrika*, *62*, 223–241 (2.5).

Pearson, K. (1900). On the criterion that a given system of deviations from the probable in the case of a correlated system of variables is such that it can reasonably be supposed to have arisen from random sampling. *Philosophical Magazine (5)*, *50*, 157–175 (4.5).

Percus, O.E., and Percus, J.K. (1970). Extended criterion for comparison of empirical distributions. *Journal of Applied Probability*, *7*, 1–20 (6.3).

Pettitt, A.N. (1976). A two-sample Anderson–Darling rank statistic. *Biometrika*, *63*, 161–168 (5.1).

Pettitt, A.N. (1978). Generalized Cramér–von Mises statistics for the gamma distribution. *Biometrika*, *65*(1), 232–235 (6.2).

Pettitt, A.N., and Stephens, M.A. (1976). Modified Cramér-von Mises statistics for censored data. *Biometrika*, *63*, 291–298 (6.1).

Pettitt, A.N., and Stephens, M.A. (1977). The Kolmogorov–Smirnov goodness-of-fit statistic with discrete and grouped data. *Technometrics*, *19*, 205–210 (6.1).

Pirie, W.R. (1974). Comparing rank tests for ordered alternatives in randomized blocks. *The Annals of Statistics*, *2*, 374–382 (correction appears in Vol. 3, p. 796) (5.8).

Pirie, W.R., and Hamdan, M.A. (1972). Some revised continuity corrections for discrete distributions. *Biometrics*, *28*, 693–701 (4.1).

Pirie, W.R., and Hollander, M. (1972). A distribution-free normal scores test for ordered alternatives in the randomized block design. *Journal of the American Statistical Association*, *67*, 855–857 (5.9).

Pitman, E.J.G. (1937/1938). Significance tests which may be applied to samples from any populations, I. *Supplement to the Journal of the Royal Statistical Society*, *4*, (1937), 119–130, II, The correlation coefficient test. *Supplement to the Journal of the Royal Statistical Society*, *4*, (1937), 225–232. III. The analysis of variance test. *Biometrika*, *29* (1938), 322–335 (5.11).

Plackett, R.L. (1964). The continuity correction in 2×2 tables. *Biometrika*, *51*, 327–338 (4.1).

Plackett, R.L. (1965). A class of bivariate distributions. *Journal of the American Statistical Association*, *60*, 516–522 (4.2).

Plackett, R.L. (1977). The marginal totals of a 2×2 table. *Biometrika*, *64*, 37–42 (4.1).

Policello, G.E., II, and Hettmansperger, T.P. (1976). Adaptive robust procedures for the one-sample location problem. *Journal of the American Statistical Association*, *71*, 624–633 (2.5).

Potthoff, R.F. (1963). Use of the Wilcoxon statistic for a generalized Behrens–Fisher problem. *The Annals of Mathematical Statistics*, 1596–1599 (5.1).

Potthoff, R.F. (1974). A non-parametric test of whether two simple regression lines are parallel. *The Annals of Statistics, 2,* 295–310 (5.5).

Pratt, J.W. (1959). Remarks on zeros and ties in the Wilcoxon signed rank procedures. *Journal of the American Statistical Association, 54,* 655–667 (5.7).

Puri, M.L. (1965). On some tests of homogeneity of variances. *Annals of the Institute of Statistical Mathematics, 17,* 323–330 (5.3).

Puri, M.L. (1968). Multi-sample scale problem: Unknown location parameters. *Annals of the Institute of Statistical Mathematics, 20,* 99–106 (5.3).

Puri, M.L., and Puri, P.S. (1969). Multiple decision procedures based on ranks for certain problems in analysis of variance. *The Annals of Mathematical Statistics, 40,* 619–632 (5.2).

Puri, M.L., and Sen, P.K. (1967). On some optimum nonparametric procedures in two-way layouts. *Journal of the American Statistical Association, 62,* 1214–1229 (5.8).

Puri, M.L., and Sen, P.K. (1968). Nonparametric confidence regions for some multivariate location problems. *Journal of the American Statistical Association, 63,* 1373–1378 (5.7).

Puri, M.L., and Sen, P.K. (1969a). Analysis of covariance based on general rank scores. *The Annals of Mathematical Statistics, 40,* 610–618 (5.2).

Puri, M.L., and Sen, P.K., (1969b). On the asymptotic theory of rank order tests for experiments involving paired comparisons. *Annals of the Institute of Statistical Mathematics, 21,* 163–173 (5.9).

Puri, M.L., and Sen, P.K. (1971). *Nonparametric Methods in Multivariate Analysis.* Wiley, New York (2.5).

Putter, J. (1964). The χ^2 goodness-of-fit test for a class of dependent observations. *Biometrika, 51,* 250–252 (4.5).

Quade, D. (1965). On the asymptotic power of the one-sample Kolmogorov–Smirnov tests. *The Annals of Mathematical Statistics, 36,* 1000–1018 (6.1).

Quade, D. (1966). On analysis of variance for the k-sample problem. *The Annals of Mathematical Statistics, 37,* 1747–1785 (5.2).

Quade, D. (1967). Rank analysis of covariance. *Journal of the American Statistical Association, 62,* 1187–1200 (5.2).

Quade, D. (1972). Analyzing randomized blocks by weighted rankings. Technical Report SW 18/72, Stichting Mathematisch Centrum, Amsterdam (5.8).

Quade, D. (1979). Using weighted rankings in the analysis of complete blocks with additive block effects. *Journal of the American Statistical Association, 74,* 680–683 (5.8).

Quade, D., and Salama, I.A. (1975). A note on minimum chi-square statistics in contingency tables. *Biometrics, 31,* 953–956 (4.2).

Quesenberry, C.P., and Gessaman, M.P. (1968). Nonparametric discrimination using tolerance regions. *The Annals of Mathematical Statistics, 39,* 664–673 (3.3).

Quesenberry, C.P., and Hurst, D.C. (1964). Large sample simultaneous confidence intervals for multinomial proportions. *Technometrics, 6,* 191–195 (3.1).

Radhakrishna, S. (1965). Combination of results from several 2×2 contingency tables. *Biometrics, 21,* 86–99 (1.5, 4.1).

Raghavachari, M. (1965a). The two-sample scale problem when locations are unknown. *The Annals of Mathematical Statistics, 36,* 1236–1242 (5.3).

Raghavachari, M. (1965b). On the efficiency of the normal scores test relative to the *F*-test. *The Annals of Mathematical Statistics, 36,* 1306–1307 (5.10).

Rahe, A.J. (1974). Tables of critical values for the Pratt matched pair signed rank statistic. *Journal of the American Statistical Association, 69,* 368–373 (5.7).

Ramachandramurty, P.V. (1966). On the Pitman efficiency of one-sided Kolmogorov and Smirnov tests for normal alternatives. *The Annals of Mathematical Statistics, 37,* 940–944 (6.3).

Ramsey, F.L. (1971). Small sample power functions for nonparametric tests of location in the double exponential family. *Journal of the American Statistical Association, 66,* 149–151 (correction appears in Vol. 72, p. 703) (5.10).

Rand Corporation (1955). *A Million Random Digits with* 100,000 *Normal Deviates.* Free Press, Glencoe, Ill. (5.10).

Randles, R.H., Broffitt, J.D., Ramberg, J.S., and Hogg, R.V. (1978). Discriminant analysis based on ranks. *Journal of the American Statistical Association, 73*(362), 379–384 (2.5).

Rao, P.V., Schuster, Eugene F., and Littell, R.C. (1975). Estimation of shift and center of symmetry based on Kolmogorov–Smirnov statistics. *The Annals of Statistics, 3,* 862–873 (5.1, 5.6, 6.3).

Rao, T.S. (1968). A note on the asymptotic relative efficiencies of Cox and Stuart's tests for testing trend in dispersion of a *p*-dependent time series. *Biometrika, 55,* 381–386 (3.5).

Ray, R.M. (1976). A new $C(\alpha)$ test for 2×2 tables. *Communications in Statistics—Theory and Methods, A5,* 545–563 (4.1).

Read, C.B. (1977). Partitioning chi-square in contingency tables: A teaching approach. *Communications in Statistics—Theory and Methods, A6,* 553–562 (4.7).

Reiss, R.D., and Ruschendorf, L. (1976). On Wilks' distribution-free confidence intervals for quantile intervals. *Journal of the American Statistical Association, 71,* 940–944 (3.2).

Reynolds, Marion R., Jr. (1975). A sequential signed-rank test for symmetry. *The Annals of Statistics, 3,* 382–400 (5.7).

Rhyne, A.L., and Steel, R.G.D. (1965). Tables for a treatments versus control multiple comparisons sign test. *Technometrics, 7,* 293–306 (3.5).

Riedwyl, H. (1967). Goodness of fit. *Journal of the American Statistical Association, 62,* 390–398 (6.1).

Rizvi, M.H., and Sobel, M. (1967). Nonparametric procedures for selecting a subset containing the population with the largest α-quantile. *The Annals of Mathematical Statistics, 36,* 1788–1803 (5.2).

Rizvi, M.H., Sobel, M., and Woodworth, G.G. (1968). Nonparametric ranking procedures for comparison with a control. *The Annals of Mathematical Statistics, 39,* 2075–2093 (5.2).

Robertson, W.H. (1960). Programming Fisher's exact method of comparing two percentages. *Technometrics, 2,* 103–107 (4.2).

Roscoe, J.T., and Byars, J.A. (1971). Sample size restraints commonly imposed on the use of the chi-square statistic. *Journal of the American Statistical Association, 66,* 755–759 (4.2).

Rosenblatt, M. (1952). Limit theorems associated with variants of the von Mises statistic. *The Annals of Mathematical Statistics, 23,* 617–623 (6.3).

Rothman, E.D., and Woodroofe, M. (1972). A Cramér–von Mises type statistic for testing symmetry. *The Annals of Mathematical Statistics, 43,* 2035–2038 (5.7, 6.3).

Ruist, E. (1955). Comparison of tests for nonparametric hypotheses. *Arkiv For Matematik, 3*(2), 131–163 (2.4).

Ruymgaart, F.H. (1973). *Asymptotic Theory of Rank Tests of Independence.* Mathematical Centre Tracts, 43, Mathematisch Centrum, Amsterdam (5.4).

Savage, I.R. (1962). *Bibliography of Nonparametric Statistics.* Harvard University, Cambridge, Mass. (Introduction).

Savage, I.R. (1969). Nonparametric Statistics: A personal review. *Sankhya: The Indian Journal of Statistics. Series A, 31,* 107–144 (2.5).

Saw, J.G. (1966). A nonparametric comparison of two samples, one of which is censored. *Biometrika, 53,* 599–603 (5.1).

Schaafsma, W. (1973). Paired comparisons with order-effects. *The Annals of Statistics, 1,* 1027–1045 (3.5).

Schach, S. (1969a). Nonparametric symmetry tests of circular distributions. *Biometrika, 56,* 571–577 (5.7).

Schach, S. (1969b). On a class of nonparametric two-sample tests for circular distributions. *The Annals of Mathematical Statistics, 40,* 1791–1800 (5.1).

Scheffé, H. (1943). Statistical inference in the nonparametric case. *The Annals of Mathematical Statistics, 14,* 305–332 (5.11).

Scheffé, H., and Tukey, J.W. (1944). A formula for sample sizes for population tolerance limits. *The Annals of Mathematical Statistics, 15,* 217 (3.3).

Scheirer, C.J., Ray, W.S., and Hare, N. (1976). The analysis of ranked data derived from completely randomized factorial designs. *Biometrics, 32,* 429–434 (5.12).

Schneider, B.E., and Clickner, R.P. (1976). On the distribution of the Kolmogorov–Smirnov test statistic for the gamma distribution with unknown parameters. Department of Statistics, Temple University, Mimeo Series No. 36 (6.2).

Schucany, W.R., and Beckett, J., III (1976). Analysis of multiple sets of incomplete rankings. *Communications in Statistics, Theory and Methods, A5,* 1327–1334 (5.8).

Schuster, E.F. (1973). On the goodness-of-fit problem for continuous symmetric distributions. *Journal of the American Statistical Association, 68,* 713–715 (6.1).

Schuster, E.F. (1975). Estimating the distribution function of a symmetric distribution. *Biometrika, 62,* 631–635 (Correction appears in Vol. 63, p. 412) (5.7).

Schuster E.F., and Narvarte, J.A. (1973). A new nonparametric estimator of the center of a symmetric distribution. *The Annals of Statistics, 1,* 1096–1104 (5.7, 6.1).

Sen, P.K. (1962). On studentized non-parametric multi-sample location tests. *Annals of the Institute of Statistical Mathematics, 14,* 119–131 (5.2).

Sen, P.K. (1963). On weighted rank-sum tests for dispersion. *Annals of the Institute of Statistical Mathematics. 15,* 117–135 (5.3).

Sen, P.K. (1966). On nonparametric simultaneous confidence regions and tests for the one criterion analysis of variance problem. *Annals of the Institute of Statistical Mathematics, 18,* 319–336 (5.2).

Sen, P.K. (1967a). A note on the asymptotic efficiency of Friedman's X_r^2 test. *Biometrika, 54,* 677–679 (5.8).

Sen, P.K. (1967b). On some multisample permutation tests based on a class of U-statistics. *Journal of the American Statistical Association, 62,* 1201–1213 (5.11).

Sen, P.K. (1968a). Estimates of the regression coefficient based on Kendall's tau. *Journal of the American Statistical Association, 63,* 1379–1389 (5.4, 5.5).

Sen, P.K. (1968b). On a class of aligned rank order tests in two-way layouts. *The Annals of Mathematical Statistics, 39,* 1115–1124 (5.8).

Sen, P.K. (1972). On a class of aligned rank order tests for the identity of the intercepts of several regression lines. *The Annals of Mathematical Statistics, 43,* 2004–2012 (5.5).

Sen, P.K., and Ghosh, M. (1974). Sequential rank tests for location. *The Annals of Statistics, 2,* 540–552 (5.1, 5.7).

Sen, P.K., and Govindarajulu, Z. (1966). On a class of c-sample weighted rank-sum tests for location and scale. *Annals of the Institute of Statistical Mathematics, 18,* 87–105 (5.2).

Sen, P.K., and Puri, M.L. (1967). On the theory of rank order tests for location in the multivariate one sample problem. *The Annals of Mathematical Statistics, 38,* 1216–1228 (5.7).

Serfling, R.J. (1968). The Wilcoxon two-sample statistic on strongly mixing processes. *The Annals of Mathematical Statistics, 39,* 1202–1209 (5.1).

Shapiro, S.S., and Francia, R.S. (1972). An approximate analysis of variance test for normality. *Journal of the American Statistical Association, 67,* 215–216 (6.2).

Shapiro, S.S., and Wilk, M.B. (1965). An analysis of variance test for normality (complete samples). *Biometrika, 52,* 591–611 (6.2).

Shapiro, S.S., and Wilk, M.B. (1968). Approximations for the null distribution of the W statistic. *Technometrics, 10,* 861–866 (6.2).

Shapiro, S.S., Wilk, M.B., and Chen, Mrs. H.J. (1968). A comparative study of various tests for normality. *Journal of the American Statistical Association, 63,* 1343–1372 (6.1, 6.2).

Sherman, E. (1965). A note on multiple comparisons using rank sums. *Technometrics, 7,* 255–256 (5.2).

Shirahata, S. (1975). Locally most powerful rank tests for independence with censored data. *The Annals of Statistics, 3,* 241–245 (5.4).

Shirahata, S. (1976). Tests for independence in infinite contingency tables. *The Annals of Statistics, 4,* 542–553 (5.4).

Shirley, E. (1977). A non-parametric equivalent of Williams' test for contrasting increasing dose levels of a treatment. *Biometrics, 33,* 386–389 (5.12).

Shorack, G.R. (1967). Testing against ordered alternatives in model I analysis of variance: Normal theory and nonparametric. *The Annals of Mathematical Statistics, 38,* 1740–1752 (5.2).

Shorack, G.R. (1969). Testing and estimating ratios of scale parameters. *Journal of the American Statistical Association, 64,* 999–1013 (5.3).

Shorack, R.A. (1967). Tables of the distribution of the Mann–Whitney–Wilcoxon U-statistic under Lehmann alternatives. *Technometrics, 9,* 666–678 (5.1).

Shorack, R.A. (1968). Recursive generation of the distribution of several nonparametric test statistics under censoring. *Journal of the American Statistical Association, 63,* 353–366 (5.1).

Shuster, J.J., and Downing, D.J. (1976). Two-way contingency tables for complex sampling schemes. *Biometrika, 63,* 271–276 (4.2).

Siegel, S. (1956). *Nonparametric Statistics for the Behavioral Sciences.* McGraw-Hill, New York (4.4).

Siegel, S., and Tukey, J.W. (1960). A nonparametric sum of ranks procedure for relative spread in unpaired samples. *Journal of the American Statistical Association, 55,* 429–444 (corrections appear in Vol. 56, p. 1005) (5.1, 5.3).

Simon, G. (1974). Alternative analyses for the singly-ordered contingency table. *Journal of the American Statistical Association, 69,* 971–976 (4.2).

Simon, G. (1977a). A nonparametric test of total independence based on Kendall's tau. *Biometrika, 64,* 277–282 (5.4).

Simon, G. (1977b). Multivariate generalization of Kendall's tau with application to data reduction. *Journal of the American Statistical Association, 72,* 367–376 (5.4).

Slakter, M.J. (1965). A comparison of the Pearson chi-square and Kolmogorov goodness-of-fit tests with respect to validity. *Journal of the American Statistical Association, 60,* 854–858 (6.1).

Slakter, M.J. (1966). Comparative validity of the chi-square and two modified chi-square goodness-of-fit tests for small but equal expected frequencies. *Biometrika, 53,* 619–623 (4.5).

Slakter, M.J. (1968). Accuracy of an approximation to the power of the chi-square goodness of fit test with small but equal expected frequencies. *Journal of the American Statistical Association, 63,* 912–918 (4.5).

Slakter, M.J. (1973). Large values for the number of groups with the Pearson chi-squared goodness-of-fit test. *Biometrika, 60,* 420–421 (4.5).

Smirnov, N.V. (1936). Sui la distribution de w^2 (Criterium de M.R.v. Mises). *Comptes Rendus* (Paris), *202,* 449–452 (6.1).

Smirnov, N.V. (1939). Estimate of deviation between empirical distribution functions in two independent samples. (Russian) *Bulletin Moscow University, 2*(2), 3–16 (6.1, 6.3).

Smirnov, N.V. (1948). Table for estimating goodness of fit of empirical distributions. *The Annals of Mathematical Statistics, 19,* 279–281 (6.1).

Smith, K. (1953). Distribution-free statistical methods and the concept of power efficiency. In L. Festinger and D. Katz (eds.) *Research Methods in the Behavioral Sciences.* Dryden, New York, 536–577 (5.11).

Sobel, M. (1967). Nonparametric procedures for selecting the t populations with the largest α-quantile, *The Annals of Mathematical Statistics, 38,* 1804–1816 (5.2).

Soms, A.P. (1977). An algorithm for the discrete Fisher's permutation test. *Journal of the American Statistical Association, 72,* 662–664 (5.11).

Spearman, C. (1904). The proof and measurement of association between two things. *American Journal of Psychology, 15,* 72–101 (5,4).

Spurrier, J.D., and Hewett, J.E. (1976). Two-stage Wilcoxon tests of hypotheses. *Journal of the American Statistical Association, 71,* 982–987 (5.1, 5.7).

Srinivasan, R., and Godio, L.B. (1974). A Cramér-von Mises type statistic for testing symmetry. *Biometrika, 61,* 196–198 (6.1).

Srinivasan, R., and Wharton, R.M. (1973). The limit distribution of a random variable used in the construction of confidence bands. *Biometrika, 60,* 431–433 (6.2).

Srivastava, M.S. and Sen, A.K. (1973). On sequential confidence intervals based on Wilcoxon type estimates. *The Annals of Statistics, 1,* 1200–1202 (5.7).

Steck, G.P. (1968). A note on contingency-type bivariate distributions. *Biometrika, 55,* 262–264 (4.2).

Steck, G.P. (1969). The Smirnov two sample tests as rank tests. *The Annals of Mathematical Statistics, 40,* 1449–1466 (6.3).

Steel, R.G.D. (1960). A rank sum test for comparing all pairs of treatments. *Technometrics, 2,* 197–207 (5.2).

Stephens, M.A. (1964). The distribution of the goodness-of-fit statistic, $U_N{}^2$. II. *Biometrika, 51,* 393–398 (6.1).

Stephens, M.A. (1965a). The goodness-of-fit statistic, V_N: Distribution and significant points. *Biometrika, 52,* 309–322 (6.1).

Stephens, M.A. (1965b). Significance points for the two-sample statistic $U_{M,N}^2$. *Biometrika, 52,* 661–663 (6.3).

Stephens, M.A. (1969). A goodness-of-fit statistic for the circle, with some comparisons. *Biometrika, 56,* 161–168 (6.1).

Stephens, M.A. (1974). EDF statistics for goodness-of-fit and some comparisons. *Journal of the American Statistical Association, 69,* 730–73? (6.1).

Stephens, M.A. (1975). Asymptotic properties for covariance matrices of order statistics. *Biometrika, 62,* 23–28 (6.2).

Stephens, M.A., and Maag, U.R. (1968). Further percentage points for $W_N{}^2$. *Biometrika, 55,* 428–430 (6.1).

Stevens, S.S. (1946). On the theory of scales of measurement. *Science, 103,* 677–680 (2.1).

Stone, C.J. (1977). Consistent nonparametric regression. *The Annals of Statistics, 5,* 595–645 (5.5).

Stone, M. (1967). Extreme tail probabilities for the null distribution of the two-sample Wilcoxon statistic. *Biometrika, 54,* 629–640 (5.1).

Stone, M. (1968). Extreme tail probabilities for sampling without replacement and exact Bahadur efficiency of the two-sample normal scores test. *Biometrika, 55,* 371–376 (5.10).

Stuart, A. (1953). The estimation and comparisons of strenghts of association in contingency tables. *Biometrika, 40,* 105–110 (4.4).

Stuart, A. (1954). Asymptotic relative efficiency of tests and the derivatives of their power functions. *Skandinavisk Aktaurietidskrift,* Parts 3–4, pp. 163–169 (2.4, 5.4, 5.5).

Stuart, A. (1956). The efficiencies of test of randomness against normal regression. *Journal of the American Statistical Association, 51,* 285–287 (3.5, 5.4, 5.5).

Stuart, A. (1963). Calculation of Spearman's rho for ordered two-way classifications. *The American Statistician, 17*(4), 23–24 (5.4).

Sugiura, N., and Otake, M. (1968). Numerical comparison of improved methods of testing in contingency tables with small frequencies. *Annals of the Institute of Statistical Mathematics, 20,* 505–517 (4.2).

Susarla, V., and Van Ryzin, J. (1976). Nonparametric bayesian estimation of survival curves from incomplete observations. *Journal of the American Statistical Association, 71,* 897–902 (2.5).

Suzuki, G. (1967). On exact probabilities of some generalized Kolmogorov's D-statistics. *Annals of the Institute of Statistical Mathematics, 19*, 373–388 (6.1).

Suzuki, G. (1968). Kolmogorov-Smirnov tests of fit based on some general bounds. *Journal of the American Statistical Association, 63*, 919–924 (6.1).

Switzer, P. (1976). Confidence procedures for two-sample problems. *Biometrika, 63*, 13–25 (5.1).

Takacs, L. (1971). On the comparison of two empirical distribution functions. *The Annals of Mathematical Statistics, 42*, 1157–1166 (6.3).

Talwar, P.P., and Gentle, J.E. (1977). A robust test for the homogeneity of scales. *Communications in Statistics—Theory and Methods, A6*, 363–369 (5.3).

Tamura, R. (1963). On a modification of certain rank tests. *The Annals of Mathematical Statistics, 34*, 1101–1103 (5.1).

Tarone, R.E., and Ware, J. (1977). On distribution-free tests for equality of survival distributions. *Biometrika, 64*, 156–160 (2.5).

Tate, M.W., and Brown, S.M. (1970). Note on the Cochran Q test. *Journal of the American Statistical Association, 65*, 155–160 (4.6).

Teichroew, D. (1965). A history of distribution sampling prior to the era of the computer and its relevance to simulation. *Journal of the American Statistical Association, 60*, 27–49 (6.2).

Theil, H. (1950). A rank-invariant method of linear and polynomial regression analysis. *Indagationes Mathematicae, 12*, 85–91 (5.5).

Thompson, R. (1966). Bias of the one-sample Cramér-von Mises test. *Journal of the American Statistical Association, 61*, 246–247 (6.1).

Thompson, R., Govindarajulu, Z., and Doksum, K.A. (1967). Distribution and power of the absolute normal scores test. *Journal of the American Statistical Association, 62*, 966–975 (5.10).

Tiku, M.L. (1965). Chi-square approximations for the distribution of goodness-of-fit statistics, U_N^2 and W_N^2. *Biometrika, 52*, 630–633 (6.1).

Tobach, E., Smith, M., Rose, G., and Richter, D. (1967). A table for rank sum multiple paired comparisons. *Technometrics, 9*, 561–568 (5.2).

Tsai, W.S., Duran, B.S., and Lewis, T.O. (1975). Small sample behavior of some multisample nonparametric tests for scale. *Journal of the American Statistical Association, 70*, 791–796 (5.3).

Tsao, C.K. (1954). An extension of Massey's distribution of the maximum deviation between two sample cumulative step functions. *The Annals of Mathematical Statistics, 25*, 587–592 (6.3).

Tsutakawa, R.K., and Yang, S.L. (1974). Permutation tests applied to antibiotic drug resistance. *Journal of the American Statistical Association, 69*, 87–92 (5.11).

Tukey, J.W. (1949). The simplest signed-rank tests. Mimeographed Report No. 17, Statistical Research Group, Princeton University (5.7).

Tukey, J.W. (1977). *Exploratory Data Analysis*. Addison-Wesley, Reading, Mass. (5.0).

Upton, G.J.G., and Lee, R.D. (1976). The importance of the patient horizon in the sequential analysis of binomial clinical trials. *Biometrika, 63*, 335–342 (4.1).

Ury, H.K. (1966). Large-sample sign tests for trend in dispersion. *Biometrika, 53*, 289–291 (3.5).

Ury, H.K. (1972). On distribution-free confidence bounds for $Pr(Y < X)$. *Technometrics, 14*, 577–582 (5.1).

Ury, H.K. (1975). Efficiency of case-control studies with multiple controls per case: Continuous or dichotomous data. *Biometrics, 3*, 643–650 (3.5).

van der Parren, J.L. (1970). Tables for distribution-free confidence limits for the median. *Biometrika, 57*, 613–618 (3.2).

van der Parren, J.L. (1973). Tables of distribution-free one-sided and two-sided confidence limits for quantiles. *Biometrika, 73*, 433–434 (3.2).

van der Reyden, D. (1952). A simple statistical significance test. *Rhodesia Agricultural Journal, 49*, 96–104 (5.1).

van der Waerden, B.L. (1952/1953). Order tests for the two-sample problem and their power. *Proceedings Koninklijke Nederlandse Akademie van Wetenschappen (A), 55* (Indagationes Mathematical 14), 453–458, and *56* (Indagationes Mathematicae 15), 303–316 (corrections appear in Vol. 56, p. 80) (5.10).

van der Waerden, B.L. (1953). Testing a distribution function. *Proceedings Koninklijke Nederlandse Akademie van Wetenschappen (A), 56* (Indagationes Mathematicae 15), 201–207 (6.1).

van Eeden, C. (1963). The relation between Pitman's asymptotic relative efficiency of two tests and the correlation coefficient between their test statistics. *Annals of Mathematical Statistics, 34*, 1442–1451 (5.10).

van Eeden, C. (1964). Note on the consistency of some distribution-free tests for dispersion. *Journal of the American Statistical Association, 59*, 105–119 (5.3).

Verdooren, L.R. (1963). Extended tables of critical values for Wilcoxon's test statistic. *Biometrika, 50*, 177–186 (5.1).

von Mises, R. (1931). *Wahrscheinlichkeitsrechnung und Ihre Anwendung in der Statistik und Theoretischen Physik.* F. Deuticke, Leipzig (6.1).

Vorlickova, D. (1970). Asymptotic properties of rank tests under discrete distributions. *Zeitschrift fur Wahrscheinlichkeitstheorie, 14*, 275–289 (5.7).

Wagner, S.S. (1970). The maximum likelihood estimate for contingency tables with zero diagonal. *Journal of the American Statistical Association, 65*, 1362–1383 (4.7).

Wald, A., and Wolfowitz, J. (1939). Confidence limits for continuous distribution functions. *The Annals of Mathematical Statistics, 10*, 105–118 (6.1).

Walker, H.M., and Lev, J. (1953). *Statistical Inference.* Holt, New York (5.1, 5.7).

Wallis, W.A. (1939). The correlation ratio for ranked data. *Journal of the American Statistical Association, 34*, 533–538 (5.8).

Walsh, J.E. (1949). Some significance tests for the median which are valid under very general conditions. *The Annals of Mathematical Statistics, 20*, 64–81 (5.7).

Walsh, J.E. (1951). Some bounded significance level properties of the equal-tail sign test. *Annals of Mathematical Statistics, 22*, 408–417 (3.4).

Walsh, J.E. (1960). Probabilities for Cramér–von Mises–Smirnov test using grouped data. *Annals of the Institute of Statistical Mathematics, 12*, 143–145 (6.1).

Walsh, J.E. (1962). *Handbook of Nonparametric Statistics.* Van Nostrand, Princeton, N.J. (2.5).

Walsh, J.E. (1963). Bounded probability properties of Kolmogorov–Smirnov and similar statistics for discrete data. *Annals of the Institute of Statistical Mathematics, 15,* 153–158 (6.1).

Weed, H.D., Jr., Bradley, Ralph A., and Govindarajulu, Z. (1974). Stopping times of some one-sample sequential rank tests. *The Annals of Statistics, 2,* 1314–1322 (5.6).

Welch, B.L. (1937). On the z-test in randomized blocks and Latin squares. *Biometrika, 29,* 21–52 (5.11).

Wheeler, S., and Watson, G.S. (1964). A distribution-free two-sample test on a circle. *Biometrika, 51,* 256–257 (5.1).

White, C. (1952). The use of ranks in a test of significance for comparing two treatments. *Biometrics, 8,* 33–41 (5.1).

Wilcoxon, F. (1945). Individual comparisons by ranking methods. *Biometrics, 1,* 80–83 (5.1, 5.7).

Wilcoxon, F. (1949). *Some Rapid Approximate Statistical Procedures.* American Cyanamid Co., Stamford Research Laboratories (5.7).

Wilks, S.S. (1962). *Mathematical Statistics.* Wiley, New York (3.3).

Williams, O.S. and Grizzle, J.E. (1972). Analysis of contingency tables having ordered response categories. *Journal of the American Statistical Association, 67,* 55–63 (4.2).

Woinsky, M.N., and Kurz, L. (1969). Sequential nonparametric two-way classification with prescribed maximum asymptotic error probability. *The Annals of Mathematical Statistics, 40,* 445–455 (5.1).

Wolf, E.H., and Naus, J.I. (1973). Tables of critical values for a k-sample Kolmogorov–Smirnov test statistic. *Journal of the American Statistical Association, 68,* 994–997 (6.4).

Wolfe, D.A. (1977a). Two-stage two-sample median test. *Technometrics, 19,* 495–501 (4.3).

Wolfe, D.A. (1977b). A distribution-free test for related correlation coefficients. *Technometrics, 19,* 507–509 (5.4).

Wood, C.L., and Altavela, M.M. (1978). Large sample results for Kolmogorov–Smirnov statistics for discrete distributions. *Biometrika, 65*(1), 235–239 (6.1).

Woodbury, M.A., Manton, K.G., and Woodbury, L.A. (1977). An extension of the sign test for replicated measurements. *Biometrics, 33,* 453–461 (3.5).

Woodroofe, M. (1967). On the maximum deviation of the sample density. *The Annals of Mathematical Statistics, 38,* 475–481 (6.1).

Worsley, K.J. (1977). A nonparametric extension of a cluster analysis method by Scott and Knott. *Biometrics, 33,* 532–535 (5.12).

Yarnold, J.K. (1970). The minimum expectations in χ^2 goodness to fit tests and the accuracy of approximations for the null distribution. *Journal of the American Statistical Association, 65,* 864–886 (4.5).

Yates, F. (1934). Contingency tables involving small numbers and the χ^2 test. *Journal of the Royal Statistical Society, 1,* 217–235 (4.1).

Yule, G.U., and Kendall, M.G. (1950). *An Introduction to the Theory of Statistics,* 14th ed. Hafner, New York (4.4, 4.5).

Zahn, D.A., and Roberts, G.C. (1971). Exact χ^2 criterion tables with cell expectations one: An application to Coleman's measure of consensus. *Journal of the American Statistical Association, 66,* 145–148 (4.5).

Zar, J.H. (1972). Significance testing of the Spearman rank correlation coefficient. *Journal of the American Statistical Association, 67,* 578–580 (5.4).

Zaremba, S.K. (1965). Note on the Wilcoxon–Mann–Whitney statistic. *The Annals of Mathematical Statistics, 36,* 1058–1060 (5.1).

Zelen, M. (1971). The analysis of several 2×2 contingency tables. *Biometrika, 58,* 129–137 (4.1).

Appendix

TABLES

Table A1 Normal Distribution[a]

Selected values: $w_{0.0001} = -3.7190$, $w_{0.0005} = -3.2905$, $w_{0.025} = -1.9600$, $w_{0.05} = -1.6449$
$w_{0.9999} = 3.7190$, $w_{0.9995} = 3.2905$, $w_{0.975} = 1.9600$, $w_{0.95} = 1.6449$

p	0.000	0.001	0.002	0.003	0.004	0.005	0.006	0.007	0.008	0.009
.00		-3.0902	-2.8782	-2.7478	-2.6521	-2.5758	-2.5121	-2.4573	-2.4089	-2.3656
.01	-2.3263	-2.2904	-2.2571	-2.2262	-2.1973	-2.1701	-2.1444	-2.1201	-2.0969	-2.0749
.02	-2.0537	-2.0335	-2.0141	-1.9954	-1.9774	-1.9600	-1.9431	-1.9268	-1.9110	-1.8957
.03	-1.8808	-1.8663	-1.8522	-1.8384	-1.8250	-1.8119	-1.7991	-1.7866	-1.7744	-1.7624
.04	-1.7507	-1.7392	-1.7279	-1.7169	-1.7060	-1.6954	-1.6849	-1.6747	-1.6646	-1.6546
.05	-1.6449	-1.6352	-1.6258	-1.6164	-1.6072	-1.5982	-1.5893	-1.5805	-1.5718	-1.5632
.06	-1.5548	-1.5464	-1.5382	-1.5301	-1.5220	-1.5141	-1.5063	-1.4985	-1.4909	-1.4833
.07	-1.4758	-1.4684	-1.4611	-1.4538	-1.4466	-1.4395	-1.4325	-1.4255	-1.4187	-1.4118
.08	-1.4051	-1.3984	-1.3917	-1.3852	-1.3787	-1.3722	-1.3658	-1.3595	-1.3532	-1.3469
.09	-1.3408	-1.3346	-1.3285	-1.3225	-1.3165	-1.3106	-1.3047	-1.2988	-1.2930	-1.2873
.10	-1.2816	-1.2759	-1.2702	-1.2646	-1.2591	-1.2536	-1.2481	-1.2426	-1.2372	-1.2319
.11	-1.2265	-1.2212	-1.2160	-1.2107	-1.2055	-1.2004	-1.1952	-1.1901	-1.1850	-1.1800
.12	-1.1750	-1.1700	-1.1650	-1.1601	-1.1552	-1.1503	-1.1455	-1.1407	-1.1359	-1.1311
.13	-1.1264	-1.1217	-1.1170	-1.1123	-1.1077	-1.1031	-1.0985	-1.0939	-1.0893	-1.0848
.14	-1.0803	-1.0758	-1.0714	-1.0669	-1.0625	-1.0581	-1.0537	-1.0494	-1.0450	-1.0407
.15	-1.0364	-1.0322	-1.0279	-1.0237	-1.0194	-1.0152	-1.0110	-1.0069	-1.0027	-0.9986
.16	-0.9945	-0.9904	-0.9863	-0.9822	-0.9782	-0.9741	-0.9701	-0.9661	-0.9621	-0.9581
.17	-0.9542	-0.9502	-0.9463	-0.9424	-0.9385	-0.9346	-0.9307	-0.9269	-0.9230	-0.9192
.18	-0.9154	-0.9116	-0.9078	-0.9040	-0.9002	-0.8965	-0.8927	-0.8890	-0.8853	-0.8816
.19	-0.8779	-0.8742	-0.8705	-0.8669	-0.8633	-0.8596	-0.8560	-0.8524	-0.8488	-0.8452
.20	-0.8416	-0.8381	-0.8345	-0.8310	-0.8274	-0.8239	-0.8204	-0.8169	-0.8134	-0.8099
.21	-0.8064	-0.8030	-0.7995	-0.7961	-0.7926	-0.7892	-0.7858	-0.7824	-0.7790	-0.7756
.22	-0.7722	-0.7688	-0.7655	-0.7621	-0.7588	-0.7554	-0.7521	-0.7488	-0.7454	-0.7421
.23	-0.7388	-0.7356	-0.7323	-0.7290	-0.7257	-0.7225	-0.7192	-0.7160	-0.7128	-0.7095
.24	-0.7063	-0.7031	-0.6999	-0.6967	-0.6935	-0.6903	-0.6871	-0.6840	-0.6808	-0.6776

p	0.000	0.001	0.002	0.003	0.004	0.005	0.006	0.007	0.008	0.009
.25	-0.6745	-0.6713	-0.6682	-0.6651	-0.6620	-0.6588	-0.6557	-0.6526	-0.6495	-0.6464
.26	-0.6433	-0.6403	-0.6372	-0.6341	-0.6311	-0.6280	-0.6250	-0.6219	-0.6189	-0.6158
.27	-0.6128	-0.6098	-0.6068	-0.6038	-0.6008	-0.5978	-0.5948	-0.5918	-0.5888	-0.5858
.28	-0.5828	-0.5799	-0.5769	-0.5740	-0.5710	-0.5681	-0.5651	-0.5622	-0.5592	-0.5563
.29	-0.5534	-0.5505	-0.5476	-0.5446	-0.5417	-0.5388	-0.5359	-0.5330	-0.5302	-0.5273
.30	-0.5244	-0.5215	-0.5187	-0.5158	-0.5129	-0.5101	-0.5072	-0.5044	-0.5015	-0.4987
.31	-0.4959	-0.4930	-0.4902	-0.4874	-0.4845	-0.4817	-0.4789	-0.4761	-0.4733	-0.4705
.32	-0.4677	-0.4649	-0.4621	-0.4593	-0.4565	-0.4538	-0.4510	-0.4482	-0.4454	-0.4427
.33	-0.4399	-0.4372	-0.4344	-0.4316	-0.4289	-0.4261	-0.4234	-0.4207	-0.4179	-0.4152
.34	-0.4125	-0.4097	-0.4070	-0.4043	-0.4016	-0.3989	-0.3961	-0.3934	-0.3907	-0.3880
.35	-0.3853	-0.3826	-0.3799	-0.3772	-0.3745	-0.3719	-0.3692	-0.3665	-0.3638	-0.3611
.36	-0.3585	-0.3558	-0.3531	-0.3505	-0.3478	-0.3451	-0.3425	-0.3398	-0.3372	-0.3345
.37	-0.3319	-0.3292	-0.3266	-0.3239	-0.3213	-0.3186	-0.3160	-0.3134	-0.3107	-0.3081
.38	-0.3055	-0.3029	-0.3002	-0.2976	-0.2950	-0.2924	-0.2898	-0.2871	-0.2845	-0.2819
.39	-0.2793	-0.2767	-0.2741	-0.2715	-0.2689	-0.2663	-0.2637	-0.2611	-0.2585	-0.2559
.40	-0.2533	-0.2508	-0.2482	-0.2456	-0.2430	-0.2404	-0.2378	-0.2353	-0.2327	-0.2301
.41	-0.2275	-0.2250	-0.2224	-0.2198	-0.2173	-0.2147	-0.2121	-0.2096	-0.2070	-0.2045
.42	-0.2019	-0.1993	-0.1968	-0.1942	-0.1917	-0.1891	-0.1866	-0.1840	-0.1814	-0.1789
.43	-0.1764	-0.1738	-0.1713	-0.1687	-0.1662	-0.1637	-0.1611	-0.1586	-0.1560	-0.1535
.44	-0.1510	-0.1484	-0.1459	-0.1434	-0.1408	-0.1383	-0.1358	-0.1332	-0.1307	-0.1282
.45	-0.1257	-0.1231	-0.1206	-0.1181	-0.1156	-0.1130	-0.1105	-0.1080	-0.1055	-0.1030
.46	-0.1004	-0.0979	-0.0954	-0.0929	-0.0904	-0.0878	-0.0853	-0.0828	-0.0803	-0.0778
.47	-0.0753	-0.0728	-0.0702	-0.0677	-0.0652	-0.0627	-0.0602	-0.0577	-0.0552	-0.0527
.48	-0.0502	-0.0476	-0.0451	-0.0426	-0.0401	-0.0376	-0.0351	-0.0326	-0.0301	-0.0276
.49	-0.0251	-0.0226	-0.0201	-0.0175	-0.0150	-0.0125	-0.0100	-0.0075	-0.0050	-0.0025
.50	0.0000	0.0025	0.0050	0.0075	0.0100	0.0125	0.0150	0.0175	0.0201	0.0226
.51	0.0251	0.0276	0.0301	0.0326	0.0351	0.0376	0.0401	0.0426	0.0451	0.0476
.52	0.0502	0.0527	0.0552	0.0577	0.0602	0.0627	0.0652	0.0677	0.0702	0.0728
.53	0.0753	0.0778	0.0803	0.0828	0.0853	0.0878	0.0904	0.0929	0.0954	0.0979
.54	0.1004	0.1030	0.1055	0.1080	0.1105	0.1130	0.1156	0.1181	0.1206	0.1231

Table A1 (Continued)

p	0.000	0.001	0.002	0.003	0.004	0.005	0.006	0.007	0.008	0.009
.55	0.1257	0.1282	0.1307	0.1332	0.1358	0.1383	0.1408	0.1434	0.1459	0.1484
.56	0.1510	0.1535	0.1560	0.1586	0.1611	0.1637	0.1662	0.1687	0.1713	0.1738
.57	0.1764	0.1789	0.1815	0.1840	0.1866	0.1891	0.1917	0.1942	0.1968	0.1993
.58	0.2019	0.2045	0.2070	0.2096	0.2121	0.2147	0.2173	0.2198	0.2224	0.2250
.59	0.2275	0.2301	0.2327	0.2353	0.2378	0.2404	0.2430	0.2456	0.2482	0.2508
.60	0.2533	0.2559	0.2585	0.2611	0.2637	0.2663	0.2689	0.2715	0.2741	0.2767
.61	0.2793	0.2819	0.2845	0.2871	0.2898	0.2924	0.2950	0.2976	0.3002	0.3029
.62	0.3055	0.3081	0.3107	0.3134	0.3160	0.3186	0.3213	0.3239	0.3266	0.3292
.63	0.3319	0.3345	0.3372	0.3398	0.3425	0.3451	0.3478	0.3505	0.3531	0.3558
.64	0.3585	0.3611	0.3638	0.3665	0.3692	0.3719	0.3745	0.3772	0.3799	0.3826
.65	0.3853	0.3880	0.3907	0.3934	0.3961	0.3989	0.4016	0.4043	0.4070	0.4097
.66	0.4125	0.4152	0.4179	0.4207	0.4234	0.4261	0.4289	0.4316	0.4344	0.4372
.67	0.4399	0.4427	0.4454	0.4482	0.4510	0.4538	0.4565	0.4593	0.4621	0.4649
.68	0.4677	0.4705	0.4733	0.4761	0.4789	0.4817	0.4845	0.4874	0.4902	0.4930
.69	0.4959	0.4987	0.5015	0.5044	0.5072	0.5101	0.5129	0.5158	0.5187	0.5215
.70	0.5244	0.5273	0.5302	0.5330	0.5359	0.5388	0.5417	0.5446	0.5476	0.5505
.71	0.5534	0.5563	0.5592	0.5622	0.5651	0.5681	0.5710	0.5740	0.5769	0.5799
.72	0.5828	0.5858	0.5888	0.5918	0.5948	0.5978	0.6008	0.6038	0.6068	0.6098
.73	0.6128	0.6158	0.6189	0.6219	0.6250	0.6280	0.6311	0.6341	0.6372	0.6403
.74	0.6433	0.6464	0.6495	0.6526	0.6557	0.6588	0.6620	0.6651	0.6682	0.6713
.75	0.6745	0.6776	0.6808	0.6840	0.6871	0.6903	0.6935	0.6967	0.6999	0.7031
.76	0.7063	0.7095	0.7128	0.7160	0.7192	0.7225	0.7257	0.7290	0.7323	0.7356
.77	0.7388	0.7421	0.7454	0.7488	0.7521	0.7554	0.7588	0.7621	0.7655	0.7688
.78	0.7722	0.7756	0.7790	0.7824	0.7858	0.7892	0.7926	0.7961	0.7995	0.8030
.79	0.8064	0.8099	0.8134	0.8169	0.8204	0.8239	0.8274	0.8310	0.8345	0.8381
.80	0.8416	0.8452	0.8488	0.8524	0.8560	0.8596	0.8633	0.8669	0.8705	0.8742
.81	0.8779	0.8816	0.8853	0.8890	0.8927	0.8965	0.9002	0.9040	0.9078	0.9116
.82	0.9154	0.9192	0.9230	0.9269	0.9307	0.9346	0.9385	0.9424	0.9463	0.9502

p	0.000	0.001	0.002	0.003	0.004	0.005	0.006	0.007	0.008	0.009
.83	0.9542	0.9581	0.9621	0.9661	0.9701	0.9741	0.9782	0.9822	0.9863	0.9904
.84	0.9945	0.9986	1.0027	1.0069	1.0110	1.0152	1.0194	1.0237	1.0279	1.0322
.85	1.0364	1.0407	1.0450	1.0494	1.0537	1.0581	1.0625	1.0669	1.0714	1.0758
.86	1.0803	1.0848	1.0893	1.0939	1.0985	1.1077	1.1077	1.1123	1.1170	1.1217
.87	1.1264	1.1311	1.1359	1.1407	1.1455	1.1503	1.1552	1.1601	1.1650	1.1700
.88	1.1750	1.1800	1.1850	1.1901	1.1952	1.2004	1.2055	1.2107	1.2160	1.2212
.89	1.2265	1.2319	1.2372	1.2426	1.2481	1.2536	1.2591	1.2646	1.2702	1.2759
.90	1.2816	1.2873	1.2930	1.2988	1.3047	1.3106	1.3165	1.3225	1.3285	1.3346
.91	1.3408	1.3469	1.3532	1.3595	1.3658	1.3722	1.3787	1.3852	1.3917	1.3984
.92	1.4051	1.4118	1.4187	1.4255	1.4325	1.4395	1.4466	1.4538	1.4611	1.4684
.93	1.4758	1.4833	1.4909	1.4985	1.5063	1.5141	1.5220	1.5301	1.5382	1.5464
.94	1.5548	1.5632	1.5718	1.5805	1.5893	1.5982	1.6072	1.6164	1.6258	1.6352
.95	1.6449	1.6546	1.6646	1.6747	1.6849	1.6954	1.7060	1.7169	1.7279	1.7392
.96	1.7507	1.7624	1.7744	1.7866	1.7991	1.8119	1.8250	1.8384	1.8522	1.8663
.97	1.8808	1.8957	1.9110	1.9268	1.9431	1.9600	1.9774	1.9954	2.0141	2.0335
.98	2.0537	2.0749	2.0969	2.1201	2.1444	2.1701	2.1973	2.2262	2.2571	2.2904
.99	2.3263	2.3656	2.4089	2.4573	2.5121	2.5758	2.6521	2.7478	2.8782	3.0902

SOURCE.   Adapted from Tables 3 and 4, Pearson and Hartley (1970), with permission from the *Biometrika* Trustees.

[a] The entries in this table are quantiles w_p of the standard normal random variable W, selected so $P(W \leq w_p) = p$ and $P(W > w_p) = 1 - p$. Note that the value of p to two decimal places determines which row to use; the third decimal place of p determines which column to use to find w_p.

Table A2 CHI-SQUARE DISTRIBUTION[a]

	p = .750	.900	.950	.975	.990	.995	.999
k = 1	1.323	2.706	3.841	5.024	6.635	7.879	10.83
2	2.773	4.605	5.991	7.378	9.210	10.60	13.82
3	4.108	6.251	7.815	9.348	11.34	12.84	16.27
4	5.385	7.779	9.488	11.14	13.28	14.86	18.47
5	6.626	9.236	11.07	12.83	15.09	16.75	20.51
6	7.841	10.64	12.59	14.45	16.81	18.55	22.46
7	9.037	12.02	14.07	16.01	18.48	20.28	24.32
8	10.22	13.36	15.51	17.53	20.09	21.96	26.13
9	11.39	14.68	16.92	19.02	21.67	23.59	27.88
10	12.55	15.99	18.31	20.48	23.21	25.19	29.59
11	13.70	17.28	19.68	21.92	24.73	26.76	31.26
12	14.85	18.55	21.03	23.34	26.22	28.30	32.91
13	15.98	19.81	22.36	24.74	27.69	29.82	34.53
14	17.12	21.06	23.68	26.12	29.14	31.32	36.12
15	18.25	22.31	25.00	27.49	30.58	32.80	37.70
16	19.37	23.54	26.30	28.85	32.00	34.27	39.25
17	20.49	24.77	27.59	30.19	33.41	35.72	40.79
18	21.60	25.99	28.87	31.53	34.81	37.16	42.31
19	22.72	27.20	30.14	32.85	36.19	38.58	43.82
20	23.83	28.41	31.41	34.17	37.57	40.00	45.32
21	24.93	29.62	32.67	35.48	38.93	41.40	46.80
22	26.04	30.81	33.92	36.78	40.29	42.80	48.27
23	27.14	32.01	35.17	38.08	41.64	44.18	49.73
24	28.24	33.20	36.42	39.37	42.98	45.56	51.18
25	29.34	34.38	37.65	40.65	44.31	46.93	52.62
26	30.43	35.56	38.89	41.92	45.64	48.29	54.05
27	31.53	36.74	40.11	43.19	46.96	49.64	55.48
28	32.62	37.92	41.34	44.46	48.28	50.99	56.89
29	33.71	39.09	42.56	45.72	49.59	52.34	58.30
30	34.80	40.26	43.77	46.98	50.89	53.67	59.70
40	45.62	51.81	55.76	59.34	63.69	66.77	73.40
50	56.33	63.17	67.50	71.42	76.15	79.49	86.66
60	66.98	74.40	79.08	83.30	88.38	91.95	99.61
70	77.58	85.53	90.53	95.02	100.4	104.2	112.3
80	88.13	96.58	101.9	106.6	112.3	116.3	124.8
90	98.65	107.6	113.1	118.1	124.1	128.3	137.2
100	109.1	118.5	124.3	129.6	135.8	140.2	149.4
x_p	.675	1.282	1.645	1.960	2.326	2.576	3.090

For $k > 100$ use the approximation $w_p = (1/2)(x_p + \sqrt{2k-1})^2$, or the more accurate

$$w_p = k\left(1 - \frac{2}{9k} + r_p\sqrt{\frac{2}{9k}}\right)^3,$$ where r_p is the value from the standardized normal distribution shown in the bottom of the table.

SOURCE. Abridged from Table 8, Pearson and Hartley (1970), with permission from the *Biometrika*, Trustees.

[a] The entries in this table are quantiles w_p of a chi-square random variable W with k degrees of freedom, selected so $P(W \le w_p) = p$ and $P(W > w_p) = 1 - p$.

Table A3 Binomial Distribution[a]

n	y	p = .05	.10	.15	.20	.25	.30	.35	.40	.45
1	0	.9500	.9000	.8500	.8000	.7500	.7000	.6500	.6000	.5500
	1	1.0000	1.0000	1.0000	1.0000	1.0000	1.0000	1.0000	1.0000	1.0000
2	0	.9025	.8100	.7225	.6400	.5625	.4900	.4225	.3600	.3025
	1	.9975	.9900	.9775	.9600	.9375	.9100	.8775	.8400	.7975
	2	1.0000	1.0000	1.0000	1.0000	1.0000	1.0000	1.0000	1.0000	1.0000
3	0	.8574	.7290	.6141	.5120	.4219	.3430	.2746	.2160	.1664
	1	.9928	.9720	.9392	.8960	.8438	.7840	.7182	.6480	.5748
	2	.9999	.9990	.9966	.9920	.9844	.9730	.9571	.9360	.9089
	3	1.0000	1.0000	1.0000	1.0000	1.0000	1.0000	1.0000	1.0000	1.0000
4	0	.8145	.6561	.5220	.4096	.3164	.2401	.1785	.1296	.0915
	1	.9860	.9477	.8905	.8192	.7383	.6517	.5630	.4752	.3910
	2	.9995	.9963	.9880	.9728	.9492	.9163	.8735	.8208	.7585
	3	1.0000	.9999	.9995	.9984	.9961	.9919	.9850	.9743	.9590
	4	1.0000	1.0000	1.0000	1.0000	1.0000	1.0000	1.0000	1.0000	1.0000
5	0	.7738	.5905	.4437	.3277	.2373	.1681	.1160	.0778	.0503
	1	.9774	.9185	.8352	.7373	.6328	.5282	.4284	.3370	.2562
	2	.9988	.9914	.9734	.9421	.8965	.8369	.7648	.6826	.5931
	3	1.0000	.9995	.9978	.9933	.9844	.9692	.9460	.9130	.8688
	4	1.0000	1.0000	.9999	.9997	.9990	.9976	.9947	.9898	.9815
	5	1.0000	1.0000	1.0000	1.0000	1.0000	1.0000	1.0000	1.0000	1.0000
6	0	.7351	.5314	.3771	.2621	.1780	.1176	.0754	.0467	.0277
	1	.9672	.8857	.7765	.6554	.5339	.4202	.3191	.2333	.1636
	2	.9978	.9842	.9527	.9011	.8306	.7443	.6471	.5443	.4415
	3	.9999	.9987	.9941	.9830	.9624	.9295	.8826	.8208	.7447
	4	1.0000	.9999	.9996	.9984	.9954	.9891	.9777	.9590	.9308
	5	1.0000	1.0000	1.0000	.9999	.9998	.9993	.9982	.9959	.9917
	6	1.0000	1.0000	1.0000	1.0000	1.0000	1.0000	1.0000	1.0000	1.0000
7	0	.6983	.4783	.3206	.2097	.1335	.0824	.0490	.0280	.0152
	1	.9556	.8503	.7166	.5767	.4449	.3294	.2338	.1586	.1024
	2	.9962	.9743	.9262	.8520	.7564	.6471	.5323	.4199	.3164
	3	.9998	.9973	.9879	.9667	.9294	.8740	.8002	.7102	.6083
	4	1.0000	.9998	.9988	.9953	.9871	.9712	.9444	.9037	.8471
	5	1.0000	1.0000	.9999	.9996	.9987	.9962	.9910	.9812	.9643
	6	1.0000	1.0000	1.0000	1.0000	.9999	.9998	.9994	.9984	.9963
	7	1.0000	1.0000	1.0000	1.0000	1.0000	1.0000	1.0000	1.0000	1.0000

[a] Y has the binomial distribution with parameters n and p. The entries are the values of $P(Y \le y) = \sum_{i=0}^{y} \binom{n}{i} p^i (1-p)^{n-i}$, for p ranging from .05 to .95.

Table A3 (Continued)

n	y	p = .50	.55	.60	.65	.70	.75	.80	.85	.90	.95
1	0	.5000	.4500	.4000	.3500	.3000	.2500	.2000	.1500	.1000	.0500
	1	1.0000	1.0000	1.0000	1.0000	1.0000	1.0000	1.0000	1.0000	1.0000	1.0000
2	0	.2500	.2025	.1600	.1225	.0900	.0625	.0400	.0225	.0100	.0025
	1	.7500	.6975	.6400	.5775	.5100	.4375	.3600	.2775	.1900	.0975
	2	1.0000	1.0000	1.0000	1.0000	1.0000	1.0000	1.0000	1.0000	1.0000	1.0000
3	0	.1250	.0911	.0640	.0429	.0270	.0156	.0080	.0034	.0010	.0001
	1	.5000	.4252	.3520	.2818	.2160	.1562	.1040	.0608	.0280	.0072
	2	.8750	.8336	.7840	.7254	.6570	.5781	.4880	.3859	.2710	.1426
	3	1.0000	1.0000	1.0000	1.0000	1.0000	1.0000	1.0000	1.0000	1.0000	1.0000
4	0	.0625	.0410	.0256	.0150	.0081	.0039	.0016	.0005	.0001	.0000
	1	.3125	.2415	.1792	.1265	.0837	.0508	.0272	.0120	.0037	.0005
	2	.6875	.6090	.5248	.4370	.3483	.2617	.1808	.1095	.0523	.0140
	3	.9375	.9085	.8704	.8215	.7599	.6836	.5904	.4780	.3439	.1855
	4	1.0000	1.0000	1.0000	1.0000	1.0000	1.0000	1.0000	1.0000	1.0000	1.0000
5	0	.0312	.0185	.0102	.0053	.0024	.0010	.0003	.0001	.0000	.0000
	1	.1875	.1312	.0870	.0540	.0308	.0156	.0067	.0022	.0005	.0000
	2	.5000	.4069	.3174	.2352	.1631	.1035	.0579	.0266	.0086	.0012
	3	.8125	.7438	.6630	.5716	.4718	.3672	.2627	.1648	.0815	.0226
	4	.9688	.9497	.9222	.8840	.8319	.7627	.6723	.5563	.4095	.2262
	5	1.0000	1.0000	1.0000	1.0000	1.0000	1.0000	1.0000	1.0000	1.0000	1.0000
6	0	.0156	.0083	.0041	.0018	.0007	.0002	.0001	.0000	.0000	.0000
	1	.1094	.0692	.0410	.0223	.0109	.0046	.0016	.0004	.0001	.0000
	2	.3438	.2553	.1792	.1174	.0705	.0376	.0170	.0059	.0013	.0001
	3	.6562	.5585	.4557	.3529	.2557	.1694	.0989	.0473	.0158	.0022
	4	.8906	.8364	.7667	.6809	.5798	.4661	.3446	.2235	.1143	.0328
	5	.9844	.9723	.9533	.9246	.8824	.8220	.7379	.6229	.4686	.2649
	6	1.0000	1.0000	1.0000	1.0000	1.0000	1.0000	1.0000	1.0000	1.0000	1.0000
7	0	.0078	.0037	.0016	.0006	.0002	.0001	.0000	.0000	.0000	.0000
	1	.0625	.0357	.0188	.0090	.0038	.0013	.0004	.0001	.0000	.0000
	2	.2266	.1529	.0963	.0556	.0288	.0129	.0047	.0012	.0002	.0000
	3	.5000	.3917	.2898	.1998	.1260	.0706	.0333	.0121	.0027	.0002
	4	.7734	.6836	.5801	.4677	.3529	.2436	.1480	.0738	.0257	.0038
	5	.9375	.8976	.8414	.7662	.6706	.5551	.4233	.2834	.1497	.0444
	6	.9922	.9848	.9720	.9510	.9176	.8665	.7903	.6794	.5217	.3017
	7	1.0000	1.0000	1.0000	1.0000	1.0000	1.0000	1.0000	1.0000	1.0000	1.0000

n	y	p = .05	.10	.15	.20	.25	.30	.35	.40	.45
8	0	.6634	.4305	.2725	.1678	.1001	.0576	.0319	.0168	.0084
	1	.9428	.8131	.6572	.5033	.3671	.2553	.1691	.1064	.0632
	2	.9942	.9619	.8948	.7969	.6785	.5518	.4278	.3154	.2201
	3	.9996	.9950	.9786	.9437	.8862	.8059	.7064	.5941	.4770
	4	1.0000	.9996	.9971	.9896	.9727	.9420	.8939	.8263	.7396
	5	1.0000	1.0000	.9998	.9988	.9958	.9887	.9747	.9502	.9115
	6	1.0000	1.0000	1.0000	.9999	.9996	.9987	.9964	.9915	.9819
	7	1.0000	1.0000	1.0000	1.0000	1.0000	.9999	.9988	.9993	.9983
	8	1.0000	1.0000	1.0000	1.0000	1.0000	1.0000	1.0000	1.0000	1.0000
9	0	.6302	.3874	.2316	.1342	.0751	.0404	.0207	.0101	.0046
	1	.9288	.7748	.5995	.4362	.3003	.1960	.1211	.0705	.0385
	2	.9916	.9470	.8591	.7382	.6007	.4628	.3373	.2318	.1495
	3	.9994	.9917	.9661	.9144	.8343	.7297	.6089	.4826	.3614
	4	1.0000	.9991	.9944	.9804	.9511	.9012	.8283	.7334	.6214
	5	1.0000	.9999	.9994	.9969	.9900	.9747	.9464	.9006	.8342
	6	1.0000	1.0000	1.0000	.9997	.9987	.9957	.9888	.9750	.9502
	7	1.0000	1.0000	1.0000	1.0000	.9999	.9996	.9986	.9962	.9909
	8	1.0000	1.0000	1.0000	1.0000	1.0000	1.0000	.9999	.9997	.9992
	9	1.0000	1.0000	1.0000	1.0000	1.0000	1.0000	1.0000	1.0000	1.0000
10	0	.5987	.3487	.1969	.1074	.0563	.0282	.0135	.0060	.0025
	1	.9139	.7361	.5443	.3758	.2440	.1493	.0860	.0464	.0233
	2	.9885	.9298	.8202	.6778	.5256	.3828	.2616	.1673	.0996
	3	.9990	.9872	.9500	.8791	.7759	.6496	.5138	.3823	.2660
	4	.9999	.9984	.9901	.9672	.9219	.8497	.7515	.6331	.5044
	5	1.0000	.9999	.9986	.9936	.9803	.9527	.9051	.8338	.7384
	6	1.0000	1.0000	.9999	.9991	.9965	.9894	.9740	.9452	.8980
	7	1.0000	1.0000	1.0000	.9999	.9996	.9984	.9952	.9877	.9726
	8	1.0000	1.0000	1.0000	1.0000	1.0000	.9999	.9995	.9983	.9955
	9	1.0000	1.0000	1.0000	1.0000	1.0000	1.0000	1.0000	.9999	.9997
	10	1.0000	1.0000	1.0000	1.0000	1.0000	1.0000	1.0000	1.0000	1.0000
11	0	.5688	.3138	.1673	.0859	.0422	.0198	.0088	.0036	.0014
	1	.8981	.6974	.4922	.3221	.1971	.1130	.0606	.0302	.0139
	2	.9848	.9104	.7788	.6174	.4552	.3127	.2001	.1189	.0652
	3	.9984	.9815	.9306	.8389	.7133	.5696	.4256	.2963	.1911
	4	.9999	.9972	.9841	.9496	.8854	.7897	.6683	.5328	.3971
	5	1.0000	.9997	.9973	.9883	.9657	.9218	.8513	.7535	.6331
	6	1.0000	1.0000	.9997	.9980	.9924	.9784	.9499	.9006	.8262
	7	1.0000	1.0000	1.0000	.9998	.9988	.9957	.9878	.9707	.9390
	8	1.0000	1.0000	1.0000	1.0000	.9999	.9994	.9980	.9941	.9852
	9	1.0000	1.0000	1.0000	1.0000	1.0000	1.0000	.9998	.9993	.9978
	10	1.0000	1.0000	1.0000	1.0000	1.0000	1.0000	1.0000	1.0000	.9998
	11	1.0000	1.0000	1.0000	1.0000	1.0000	1.0000	1.0000	1.0000	1.0000

n	y	p = .50	.55	.60	.65	.70	.75	.80	.85	.90	.95
8	0	.0039	.0017	.0007	.0002	.0001	.0000	.0000	.0000	.0000	.0000
	1	.0352	.0181	.0085	.0036	.0013	.0004	.0001	.0000	.0000	.0000
	2	.1445	.0885	.0498	.0253	.0113	.0042	.0012	.0002	.0000	.0000
	3	.3633	.2604	.1737	.1061	.0580	.0273	.0104	.0029	.0004	.0000
	4	.6367	.5230	.4059	.2936	.1941	.1138	.0563	.0214	.0050	.0004
	5	.8555	.7799	.6846	.5722	.4482	.3215	.2031	.1052	.0381	.0058
	6	.9648	.9368	.8936	.8309	.7447	.6329	.4967	.3428	.1869	.0572
	7	.9961	.9916	.9832	.9681	.9424	.8999	.8322	.7275	.5695	.3366
	8	1.0000	1.0000	1.0000	1.0000	1.0000	1.0000	1.0000	1.0000	1.0000	1.0000
9	0	.0020	.0008	.0003	.0001	.0000	.0000	.0000	.0000	.0000	.0000
	1	.0195	.0091	.0038	.0014	.0004	.0001	.0000	.0000	.0000	.0000
	2	.0898	.0498	.0250	.0112	.0043	.0013	.0003	.0000	.0000	.0000
	3	.2539	.1658	.0994	.0536	.0253	.0100	.0031	.0006	.0001	.0000
	4	.5000	.3786	.2666	.1717	.0988	.0489	.0196	.0056	.0009	.0000
	5	.7461	.6386	.5174	.3911	.2703	.1657	.0856	.0339	.0083	.0006
	6	.9102	.8505	.7682	.6627	.5372	.3993	.2618	.1409	.0530	.0084
	7	.9805	.9615	.9295	.8789	.8040	.6997	.5638	.4005	.2252	.0712
	8	.9980	.9954	.9899	.9793	.9596	.9249	.8658	.7684	.6126	.3698
	9	1.0000	1.0000	1.0000	1.0000	1.0000	1.0000	1.0000	1.0000	1.0000	1.0000
10	0	.0010	.0003	.0001	.0000	.0000	.0000	.0000	.0000	.0000	.0000
	1	.0107	.0045	.0017	.0005	.0001	.0000	.0000	.0000	.0000	.0000
	2	.0547	.0274	.0123	.0048	.0016	.0004	.0001	.0000	.0000	.0000
	3	.1719	.1020	.0548	.0260	.0106	.0035	.0009	.0001	.0000	.0000
	4	.3770	.2616	.1662	.0949	.0473	.0197	.0064	.0014	.0001	.0000
	5	.6230	.4956	.3669	.2485	.1503	.0781	.0328	.0099	.0016	.0001
	6	.8281	.7340	.6177	.4862	.3504	.2241	.1209	.0500	.0128	.0010
	7	.9453	.9004	.8327	.7384	.6172	.4744	.3222	.1798	.0702	.0115
	8	.9893	.9767	.9536	.9140	.8507	.7560	.6242	.4557	.2639	.0861
	9	.9990	.9975	.9940	.9865	.9718	.9437	.8926	.8031	.6513	.4013
	10	1.0000	1.0000	1.0000	1.0000	1.0000	1.0000	1.0000	1.0000	1.0000	1.0000
11	0	.0005	.0002	.0000	.0000	.0000	.0000	.0000	.0000	.0000	.0000
	1	.0059	.0022	.0007	.0002	.0000	.0000	.0000	.0000	.0000	.0000
	2	.0327	.0148	.0059	.0020	.0006	.0001	.0000	.0000	.0000	.0000
	3	.1133	.0610	.0293	.0122	.0043	.0012	.0002	.0000	.0000	.0000
	4	.2744	.1738	.0994	.0501	.0216	.0076	.0020	.0003	.0000	.0000
	5	.5000	.3669	.2465	.1487	.0782	.0343	.0117	.0027	.0003	.0000
	6	.7256	.6029	.4672	.3317	.2103	.1146	.0504	.0159	.0028	.0001
	7	.8867	.8089	.7037	.5744	.4304	.2867	.1611	.0694	.0185	.0016
	8	.9673	.9348	.8811	.7999	.6873	.5448	.3826	.2212	.0896	.0152
	9	.9941	.9861	.9698	.9394	.8870	.8029	.6779	.5078	.3026	.1019
	10	.9995	.9986	.9964	.9912	.9802	.9578	.9141	.8327	.6862	.4312
	11	1.0000	1.0000	1.0000	1.0000	1.0000	1.0000	1.0000	1.0000	1.0000	1.0000

Table A3 (Continued)

n	y	p = .05	.10	.15	.20	.25	.30	.35	.40	.45
12	0	.5404	.2824	.1422	.0687	.0317	.0138	.0057	.0022	.0008
	1	.8816	.6590	.4435	.2749	.1584	.0850	.0424	.0196	.0083
	2	.9804	.8891	.7358	.5583	.3907	.2528	.1513	.0834	.0421
	3	.9978	.9744	.9078	.7946	.6488	.4925	.3467	.2253	.1345
	4	.9998	.9957	.9761	.9274	.8424	.7237	.5833	.4382	.3044
	5	1.0000	.9995	.9954	.9806	.9456	.8822	.7873	.6652	.5269
	6	1.0000	.9999	.9993	.9961	.9857	.9614	.9154	.8418	.7393
	7	1.0000	1.0000	.9999	.9994	.9972	.9905	.9745	.9427	.8883
	8	1.0000	1.0000	1.0000	.9999	.9996	.9983	.9944	.9847	.9644
	9	1.0000	1.0000	1.0000	1.0000	1.0000	.9998	.9992	.9972	.9921
	10	1.0000	1.0000	1.0000	1.0000	1.0000	1.0000	.9999	.9997	.9989
	11	1.0000	1.0000	1.0000	1.0000	1.0000	1.0000	1.0000	1.0000	.9999
	12	1.0000	1.0000	1.0000	1.0000	1.0000	1.0000	1.0000	1.0000	1.0000
13	0	.5133	.2542	.1209	.0550	.0238	.0097	.0037	.0013	.0004
	1	.8646	.6213	.3983	.2336	.1267	.0637	.0296	.0126	.0049
	2	.9755	.8661	.7296	.5017	.3326	.2025	.1132	.0579	.0269
	3	.9969	.9658	.9033	.7473	.5843	.4206	.2783	.1686	.0929
	4	.9997	.9935	.9740	.9009	.7940	.6543	.5005	.3530	.2279
	5	1.0000	.9991	.9947	.9700	.9198	.8346	.7159	.5744	.4268
	6	1.0000	.9999	.9987	.9930	.9757	.9376	.8705	.7712	.6437
	7	1.0000	1.0000	.9998	.9988	.9944	.9818	.9538	.9023	.8212
	8	1.0000	1.0000	1.0000	.9998	.9990	.9960	.9874	.9679	.9302
	9	1.0000	1.0000	1.0000	1.0000	.9999	.9993	.9975	.9922	.9797
	10	1.0000	1.0000	1.0000	1.0000	1.0000	.9999	.9997	.9987	.9959
	11	1.0000	1.0000	1.0000	1.0000	1.0000	1.0000	1.0000	.9999	.9995
	12	1.0000	1.0000	1.0000	1.0000	1.0000	1.0000	1.0000	1.0000	1.0000
	13	1.0000	1.0000	1.0000	1.0000	1.0000	1.0000	1.0000	1.0000	1.0000
14	0	.4877	.2288	.1028	.0440	.0178	.0068	.0024	.0008	.0002
	1	.8470	.5846	.3567	.1979	.1010	.0475	.0205	.0081	.0029
	2	.9699	.8416	.6479	.4481	.2811	.1608	.0839	.0398	.0170
	3	.9958	.9559	.8535	.6982	.5213	.3552	.2205	.1243	.0632
	4	.9996	.9908	.9533	.8702	.7415	.5842	.4227	.2793	.1672
	5	1.0000	.9985	.9885	.9561	.8883	.7805	.6405	.4859	.3373
	6	1.0000	.9998	.9978	.9884	.9617	.9067	.8164	.6925	.5461
	7	1.0000	1.0000	.9997	.9976	.9897	.9685	.9247	.8499	.7414
	8	1.0000	1.0000	1.0000	.9996	.9978	.9917	.9757	.9417	.8811
	9	1.0000	1.0000	1.0000	1.0000	.9997	.9983	.9940	.9825	.9574
	10	1.0000	1.0000	1.0000	1.0000	1.0000	.9998	.9989	.9961	.9886
	11	1.0000	1.0000	1.0000	1.0000	1.0000	1.0000	.9999	.9994	.9978
	12	1.0000	1.0000	1.0000	1.0000	1.0000	1.0000	1.0000	.9999	.9997
	13	1.0000	1.0000	1.0000	1.0000	1.0000	1.0000	1.0000	1.0000	1.0000
	14	1.0000	1.0000	1.0000	1.0000	1.0000	1.0000	1.0000	1.0000	1.0000

n	y	p = .50	.55	.60	.65	.70	.75	.80	.85	.90	.95
12	0	.0002	.0001	.0000	.0000	.0000	.0000	.0000	.0000	.0000	.0000
	1	.0032	.0011	.0003	.0001	.0000	.0000	.0000	.0000	.0000	.0000
	2	.0193	.0079	.0028	.0008	.0002	.0000	.0000	.0000	.0000	.0000
	3	.0730	.0356	.0153	.0056	.0017	.0004	.0001	.0000	.0000	.0000
	4	.1938	.1117	.0573	.0255	.0095	.0028	.0006	.0001	.0000	.0000
	5	.3872	.2607	.1582	.0846	.0386	.0143	.0039	.0007	.0001	.0000
	6	.6128	.4731	.3348	.2127	.1178	.0544	.0194	.0046	.0005	.0000
	7	.8062	.6956	.5618	.4167	.2763	.1576	.0726	.0239	.0043	.0002
	8	.9270	.8655	.7747	.6533	.5075	.3512	.2054	.0922	.0256	.0022
	9	.9807	.9579	.9166	.8487	.7472	.6093	.4417	.2642	.1109	.0196
	10	.9968	.9917	.9804	.9576	.9150	.8416	.7251	.5565	.3410	.1184
	11	.9998	.9992	.9978	.9943	.9862	.9683	.9313	.8578	.7176	.4596
	12	1.0000	1.0000	1.0000	1.0000	1.0000	1.0000	1.0000	1.0000	1.0000	1.0000
13	0	.0001	.0000	.0000	.0000	.0000	.0000	.0000	.0000	.0000	.0000
	1	.0017	.0005	.0001	.0000	.0000	.0000	.0000	.0000	.0000	.0000
	2	.0112	.0041	.0013	.0003	.0001	.0000	.0000	.0000	.0000	.0000
	3	.0461	.0203	.0078	.0025	.0007	.0001	.0000	.0000	.0000	.0000
	4	.1334	.0698	.0321	.0126	.0040	.0010	.0002	.0000	.0000	.0000
	5	.2905	.1788	.0977	.0462	.0182	.0056	.0012	.0002	.0000	.0000
	6	.5000	.3563	.2288	.1295	.0624	.0243	.0070	.0013	.0001	.0000
	7	.7095	.5732	.4256	.2841	.1654	.0802	.0300	.0053	.0009	.0000
	8	.8666	.7721	.6470	.4995	.3457	.2060	.0991	.0260	.0065	.0003
	9	.9539	.9071	.8314	.7217	.5794	.4157	.2527	.0967	.0342	.0031
	10	.9888	.9731	.9421	.8868	.7975	.6674	.4983	.2704	.1339	.0245
	11	.9983	.9951	.9874	.9704	.9363	.8733	.7664	.6017	.3787	.1354
	12	.9999	.9996	.9987	.9963	.9903	.9762	.9450	.8791	.7458	.4867
	13	1.0000	1.0000	1.0000	1.0000	1.0000	1.0000	1.0000	1.0000	1.0000	1.0000
14	0	.0000	.0000	.0000	.0000	.0000	.0000	.0000	.0000	.0000	.0000
	1	.0009	.0003	.0001	.0000	.0000	.0000	.0000	.0000	.0000	.0000
	2	.0065	.0022	.0006	.0001	.0000	.0000	.0000	.0000	.0000	.0000
	3	.0287	.0114	.0039	.0011	.0002	.0000	.0000	.0000	.0000	.0000
	4	.0898	.0462	.0175	.0060	.0017	.0003	.0000	.0000	.0000	.0000
	5	.2120	.1189	.0583	.0243	.0083	.0022	.0004	.0000	.0000	.0000
	6	.3953	.2586	.1501	.0753	.0315	.0103	.0024	.0003	.0000	.0000
	7	.6047	.4539	.3075	.1836	.0933	.0383	.0116	.0022	.0002	.0000
	8	.7880	.6627	.5141	.3595	.2195	.1117	.0439	.0115	.0015	.0000
	9	.9102	.8328	.7207	.5773	.4158	.2585	.1298	.0467	.0092	.0004
	10	.9713	.9368	.8757	.7795	.6448	.4787	.3018	.1465	.0441	.0042
	11	.9935	.9830	.9602	.9161	.8392	.7189	.5519	.3521	.1584	.0301
	12	.9991	.9971	.9919	.9795	.9525	.8990	.8021	.6433	.4154	.1530
	13	.9999	.9998	.9992	.9976	.9932	.9822	.9560	.8972	.7712	.5123
	14	1.0000	1.0000	1.0000	1.0000	1.0000	1.0000	1.0000	1.0000	1.0000	1.0000

n	y	p = .05	.10	.15	.20	.25	.30	.35	.40	.45
15	0	.4633	.2059	.0874	.0352	.0134	.0047	.0016	.0005	.0001
	1	.8290	.5490	.3186	.1671	.0802	.0353	.0142	.0052	.0017
	2	.9638	.8159	.6042	.3980	.2361	.1268	.0617	.0271	.0107
	3	.9945	.9444	.8227	.6482	.4613	.2969	.1727	.0905	.0424
	4	.9994	.9873	.9383	.8358	.6865	.5155	.3519	.2173	.1204
	5	.9999	.9978	.9832	.9389	.8516	.7216	.5643	.4032	.2608
	6	1.0000	.9997	.9964	.9819	.9434	.8689	.7548	.6098	.4522
	7	1.0000	1.0000	.9994	.9958	.9827	.9500	.8868	.7869	.6535
	8	1.0000	1.0000	.9999	.9992	.9958	.9848	.9578	.9050	.8182
	9	1.0000	1.0000	1.0000	.9999	.9992	.9963	.9876	.9662	.9231
	10	1.0000	1.0000	1.0000	1.0000	.9999	.9993	.9972	.9907	.9745
	11	1.0000	1.0000	1.0000	1.0000	1.0000	.9999	.9995	.9981	.9937
	12	1.0000	1.0000	1.0000	1.0000	1.0000	1.0000	.9999	.9997	.9989
	13	1.0000	1.0000	1.0000	1.0000	1.0000	1.0000	1.0000	1.0000	.9999
	14	1.0000	1.0000	1.0000	1.0000	1.0000	1.0000	1.0000	1.0000	1.0000
	15	1.0000	1.0000	1.0000	1.0000	1.0000	1.0000	1.0000	1.0000	1.0000
16	0	.4401	.1853	.0743	.0281	.0100	.0033	.0010	.0003	.0001
	1	.8108	.5147	.2839	.1407	.0635	.0261	.0098	.0033	.0010
	2	.9571	.7892	.5614	.3518	.1971	.0994	.0451	.0183	.0066
	3	.9930	.9316	.7899	.5981	.4050	.2459	.1339	.0651	.0281
	4	.9991	.9830	.9209	.7982	.6302	.4499	.2892	.1666	.0853
	5	.9999	.9967	.9765	.9183	.8103	.6598	.4900	.3288	.1976
	6	1.0000	.9995	.9944	.9733	.9204	.8247	.6881	.5272	.3660
	7	1.0000	.9999	.9989	.9930	.9729	.9256	.8406	.7161	.5629
	8	1.0000	1.0000	.9998	.9985	.9925	.9743	.9329	.8577	.7441
	9	1.0000	1.0000	1.0000	.9998	.9984	.9929	.9771	.9417	.8759
	10	1.0000	1.0000	1.0000	1.0000	.9997	.9984	.9938	.9809	.9514
	11	1.0000	1.0000	1.0000	1.0000	1.0000	.9997	.9987	.9951	.9851
	12	1.0000	1.0000	1.0000	1.0000	1.0000	1.0000	.9998	.9991	.9965
	13	1.0000	1.0000	1.0000	1.0000	1.0000	1.0000	1.0000	.9999	.9994
	14	1.0000	1.0000	1.0000	1.0000	1.0000	1.0000	1.0000	1.0000	.9999
	15	1.0000	1.0000	1.0000	1.0000	1.0000	1.0000	1.0000	1.0000	1.0000
	16	1.0000	1.0000	1.0000	1.0000	1.0000	1.0000	1.0000	1.0000	1.0000

n	y	p = .50	.55	.60	.65	.70	.75	.80	.85	.90	.95
15	0	.0000	.0000	.0000	.0000	.0000	.0000	.0000	.0000	.0000	.0000
	1	.0005	.0001	.0000	.0000	.0000	.0000	.0000	.0000	.0000	.0000
	2	.0037	.0011	.0003	.0001	.0000	.0000	.0000	.0000	.0000	.0000
	3	.0176	.0063	.0019	.0005	.0001	.0000	.0000	.0000	.0000	.0000
	4	.0592	.0255	.0093	.0028	.0007	.0001	.0000	.0000	.0000	.0000
	5	.1509	.0769	.0338	.0124	.0037	.0008	.0001	.0000	.0000	.0000
	6	.3036	.1818	.0950	.0422	.0152	.0042	.0008	.0001	.0000	.0000
	7	.5000	.3465	.2131	.1132	.0500	.0173	.0042	.0006	.0000	.0000
	8	.6964	.5478	.3902	.2452	.1311	.0566	.0181	.0036	.0003	.0000
	9	.8491	.7392	.5968	.4357	.2784	.1484	.0611	.0168	.0022	.0001
	10	.9408	.8796	.7827	.6481	.4845	.3135	.1642	.0617	.0127	.0006
	11	.9824	.9576	.9095	.8273	.7031	.5387	.3518	.1773	.0556	.0055
	12	.9963	.9893	.9729	.9383	.8732	.7639	.6020	.3958	.1841	.0362
	13	.9995	.9983	.9948	.9858	.9647	.9198	.8329	.6814	.4510	.1710
	14	1.0000	.9999	.9995	.9984	.9953	.9866	.9648	.9126	.7941	.5367
	15	1.0000	1.0000	1.0000	1.0000	1.0000	1.0000	1.0000	1.0000	1.0000	1.0000
16	0	.0000	.0000	.0000	.0000	.0000	.0000	.0000	.0000	.0000	.0000
	1	.0003	.0001	.0000	.0000	.0000	.0000	.0000	.0000	.0000	.0000
	2	.0021	.0006	.0001	.0000	.0000	.0000	.0000	.0000	.0000	.0000
	3	.0106	.0035	.0009	.0002	.0000	.0000	.0000	.0000	.0000	.0000
	4	.0384	.0149	.0049	.0013	.0003	.0000	.0000	.0000	.0000	.0000
	5	.1051	.0486	.0191	.0062	.0016	.0003	.0000	.0000	.0000	.0000
	6	.2272	.1241	.0583	.0229	.0071	.0016	.0002	.0000	.0000	.0000
	7	.4018	.2559	.1423	.0671	.0257	.0075	.0015	.0002	.0000	.0000
	8	.5982	.4371	.2839	.1594	.0744	.0271	.0070	.0011	.0001	.0000
	9	.7728	.6340	.4728	.3119	.1753	.0796	.0267	.0056	.0005	.0000
	10	.8949	.8024	.6712	.5100	.3402	.1897	.0817	.0235	.0033	.0001
	11	.9616	.9147	.8334	.7108	.5501	.3698	.2018	.0791	.0170	.0009
	12	.9894	.9719	.9349	.8661	.7541	.5950	.4019	.2101	.0684	.0070
	13	.9979	.9934	.9817	.9549	.9006	.8729	.6482	.4386	.2108	.0429
	14	.9997	.9990	.9967	.9902	.9739	.9365	.8593	.7161	.4853	.1892
	15	1.0000	.9999	.9997	.9990	.9967	.9900	.9719	.9257	.8147	.5599
	16	1.0000	1.0000	1.0000	1.0000	1.0000	1.0000	1.0000	1.0000	1.0000	1.0000

n	y	p = .05	.10	.15	.20	.25	.30	.35	.40	.45
17	0	.4181	.1668	.0631	.0225	.0075	.0023	.0007	.0002	.0000
	1	.7922	.4818	.2525	.1182	.0501	.0193	.0067	.0021	.0006
	2	.9497	.7618	.5198	.3096	.1637	.0774	.0327	.0123	.0041
	3	.9912	.9174	.7556	.5489	.3530	.2019	.1028	.0464	.0184
	4	.9988	.9779	.9013	.7582	.5739	.3887	.2348	.1260	.0596
	5	.9999	.9953	.9681	.8943	.7653	.5968	.4197	.2639	.1471
	6	1.0000	.9992	.9917	.9623	.8929	.7752	.6188	.4478	.2902
	7	1.0000	.9999	.9983	.9891	.9598	.8954	.7872	.6405	.4743
	8	1.0000	1.0000	.9997	.9974	.9876	.9597	.9006	.8011	.6626
	9	1.0000	1.0000	1.0000	.9995	.9969	.9873	.9617	.9081	.8166
	10	1.0000	1.0000	1.0000	.9999	.9994	.9968	.9880	.9652	.9174
	11	1.0000	1.0000	1.0000	1.0000	.9999	.9993	.9970	.9894	.9699
	12	1.0000	1.0000	1.0000	1.0000	1.0000	.9999	.9994	.9975	.9914
	13	1.0000	1.0000	1.0000	1.0000	1.0000	1.0000	.9999	.9995	.9981
	14	1.0000	1.0000	1.0000	1.0000	1.0000	1.0000	1.0000	.9999	.9997
	15	1.0000	1.0000	1.0000	1.0000	1.0000	1.0000	1.0000	1.0000	1.0000
	16	1.0000	1.0000	1.0000	1.0000	1.0000	1.0000	1.0000	1.0000	1.0000
	17	1.0000	1.0000	1.0000	1.0000	1.0000	1.0000	1.0000	1.0000	1.0000
18	0	.3972	.1501	.0536	.0180	.0056	.0016	.0004	.0001	.0000
	1	.7735	.4503	.2241	.0991	.0395	.0142	.0046	.0013	.0003
	2	.9419	.7338	.4797	.2713	.1353	.0600	.0236	.0082	.0025
	3	.9891	.9018	.7202	.5010	.3057	.1646	.0783	.0328	.0120
	4	.9985	.9718	.8794	.7164	.5187	.3327	.1886	.0942	.0411
	5	.9998	.9936	.9581	.8671	.7175	.5344	.3550	.2088	.1077
	6	1.0000	.9988	.9882	.9487	.8610	.7217	.5491	.3743	.2258
	7	1.0000	.9998	.9973	.9837	.9431	.8593	.7283	.5634	.3915
	8	1.0000	1.0000	.9995	.9957	.9807	.9404	.8609	.7368	.5778
	9	1.0000	1.0000	.9999	.9991	.9946	.9790	.9403	.8653	.7473
	10	1.0000	1.0000	1.0000	.9998	.9988	.9939	.9788	.9424	.8720
	11	1.0000	1.0000	1.0000	1.0000	.9998	.9986	.9938	.9797	.9463
	12	1.0000	1.0000	1.0000	1.0000	1.0000	.9997	.9986	.9942	.9817
	13	1.0000	1.0000	1.0000	1.0000	1.0000	1.0000	.9997	.9987	.9951
	14	1.0000	1.0000	1.0000	1.0000	1.0000	1.0000	1.0000	.9998	.9990
	15	1.0000	1.0000	1.0000	1.0000	1.0000	1.0000	1.0000	1.0000	.9999
	16	1.0000	1.0000	1.0000	1.0000	1.0000	1.0000	1.0000	1.0000	1.0000
	17	1.0000	1.0000	1.0000	1.0000	1.0000	1.0000	1.0000	1.0000	1.0000
	18	1.0000	1.0000	1.0000	1.0000	1.0000	1.0000	1.0000	1.0000	1.0000

n	y	p = .50	.55	.60	.65	.70	.75	.80	.85	.90	.95
17	0	.0000	.0000	.0000	.0000	.0000	.0000	.0000	.0000	.0000	.0000
	1	.0001	.0000	.0000	.0000	.0000	.0000	.0000	.0000	.0000	.0000
	2	.0012	.0003	.0001	.0000	.0000	.0000	.0000	.0000	.0000	.0000
	3	.0064	.0019	.0005	.0001	.0000	.0000	.0000	.0000	.0000	.0000
	4	.0245	.0086	.0025	.0006	.0001	.0000	.0000	.0000	.0000	.0000
	5	.0717	.0301	.0106	.0030	.0007	.0001	.0000	.0000	.0000	.0000
	6	.1662	.0826	.0348	.0120	.0032	.0006	.0001	.0000	.0000	.0000
	7	.3145	.1834	.0383	.0383	.0127	.0031	.0005	.0000	.0000	.0000
	8	.5000	.3374	.1989	.0994	.0403	.0124	.0026	.0003	.0000	.0000
	9	.6855	.5257	.3595	.2128	.1046	.0402	.0109	.0017	.0001	.0000
	10	.8338	.7098	.5522	.3812	.2248	.1071	.0377	.0083	.0008	.0000
	11	.9283	.8529	.7361	.5803	.4032	.2347	.1057	.0319	.0047	.0001
	12	.9755	.9404	.8740	.7652	.6113	.4261	.2418	.0987	.0221	.0012
	13	.9936	.9816	.9536	.8972	.7981	.6470	.4511	.2444	.0826	.0088
	14	.9988	.9959	.9877	.9673	.9226	.8363	.6904	.4802	.2382	.0503
	15	.9999	.9994	.9979	.9933	.9807	.9499	.8818	.7475	.5182	.2078
	16	1.0000	1.0000	.9998	.9993	.9977	.9925	.9775	.9369	.8332	.5819
	17	1.0000	1.0000	1.0000	1.0000	1.0000	1.0000	1.0000	1.0000	1.0000	1.0000
18	0	.0000	.0000	.0000	.0000	.0000	.0000	.0000	.0000	.0000	.0000
	1	.0001	.0000	.0000	.0000	.0000	.0000	.0000	.0000	.0000	.0000
	2	.0007	.0001	.0000	.0000	.0000	.0000	.0000	.0000	.0000	.0000
	3	.0038	.0010	.0002	.0000	.0000	.0000	.0000	.0000	.0000	.0000
	4	.0154	.0049	.0013	.0003	.0000	.0000	.0000	.0000	.0000	.0000
	5	.0481	.0183	.0058	.0014	.0003	.0000	.0000	.0000	.0000	.0000
	6	.1189	.0537	.0203	.0062	.0014	.0002	.0000	.0000	.0000	.0000
	7	.2403	.1280	.0576	.0212	.0061	.0012	.0002	.0000	.0000	.0000
	8	.4073	.2527	.1347	.0597	.0210	.0054	.0009	.0001	.0000	.0000
	9	.5927	.4222	.2632	.1391	.0596	.0193	.0043	.0005	.0000	.0000
	10	.7597	.6085	.4366	.2717	.1407	.0569	.0163	.0027	.0002	.0000
	11	.8811	.7742	.6257	.4509	.2783	.1390	.0513	.0118	.0012	.0000
	12	.9519	.8923	.7912	.6450	.4656	.2825	.1329	.0419	.0064	.0002
	13	.9846	.9589	.9058	.8114	.6673	.4813	.2836	.1206	.0282	.0015
	14	.9962	.9880	.9672	.9217	.8354	.6943	.4990	.2798	.0982	.0109
	15	.9993	.9975	.9918	.9764	.9400	.8647	.7287	.5203	.2662	.0581
	16	.9999	.9997	.9987	.9954	.9858	.9605	.9009	.7759	.5497	.2265
	17	1.0000	1.0000	.9999	.9996	.9984	.9944	.9820	.9464	.8499	.6028
	18	1.0000	1.0000	1.0000	1.0000	1.0000	1.0000	1.0000	1.0000	1.0000	1.0000

Table A3 (Continued)

n	y	p = .05	.10	.15	.20	.25	.30	.35	.40	.45
19	0	.3774	.1351	.0456	.0144	.0042	.0011	.0003	.0001	.0000
	1	.7547	.4203	.1985	.0829	.0310	.0104	.0031	.0008	.0002
	2	.9335	.7054	.4413	.2369	.1113	.0462	.0170	.0055	.0015
	3	.9869	.8850	.6841	.4551	.2631	.1332	.0591	.0230	.0077
	4	.9980	.9648	.8556	.6733	.4654	.2822	.1500	.0696	.0280
	5	.9998	.9914	.9463	.8369	.6678	.4739	.2968	.1629	.0777
	6	1.0000	.9983	.9837	.9324	.8251	.6655	.4812	.3081	.1727
	7	1.0000	.9997	.9959	.9767	.9225	.8180	.6656	.4878	.3169
	8	1.0000	1.0000	.9992	.9933	.9713	.9161	.8145	.6675	.4940
	9	1.0000	1.0000	.9999	.9984	.9911	.9674	.9125	.8139	.6710
	10	1.0000	1.0000	1.0000	.9997	.9977	.9895	.9653	.9115	.8159
	11	1.0000	1.0000	1.0000	1.0000	.9995	.9972	.9886	.9648	.9129
	12	1.0000	1.0000	1.0000	1.0000	.9999	.9994	.9969	.9884	.9658
	13	1.0000	1.0000	1.0000	1.0000	1.0000	.9999	.9993	.9969	.9891
	14	1.0000	1.0000	1.0000	1.0000	1.0000	1.0000	.9999	.9994	.9972
	15	1.0000	1.0000	1.0000	1.0000	1.0000	1.0000	1.0000	.9999	.9995
	16	1.0000	1.0000	1.0000	1.0000	1.0000	1.0000	1.0000	1.0000	.9999
	17	1.0000	1.0000	1.0000	1.0000	1.0000	1.0000	1.0000	1.0000	1.0000
	18	1.0000	1.0000	1.0000	1.0000	1.0000	1.0000	1.0000	1.0000	1.0000
	19	1.0000	1.0000	1.0000	1.0000	1.0000	1.0000	1.0000	1.0000	1.0000
20	0	.3585	.1216	.0388	.0115	.0032	.0008	.0002	.0000	.0000
	1	.7358	.3917	.1756	.0692	.0243	.0076	.0021	.0005	.0001
	2	.9245	.6769	.4049	.2061	.0913	.0355	.0121	.0036	.0009
	3	.9841	.8670	.6477	.4114	.2252	.1071	.0444	.0160	.0049
	4	.9974	.9568	.8298	.6296	.4148	.2375	.1182	.0510	.0189
	5	.9997	.9887	.9327	.8042	.6172	.4164	.2454	.1256	.0553
	6	1.0000	.9976	.9781	.9133	.7858	.6080	.4166	.2500	.1299
	7	1.0000	.9996	.9941	.9679	.8982	.7723	.6010	.4159	.2520
	8	1.0000	.9999	.9987	.9900	.9591	.8867	.7624	.5956	.4143
	9	1.0000	1.0000	.9998	.9974	.9861	.9520	.8782	.7553	.5914
	10	1.0000	1.0000	1.0000	.9994	.9961	.9829	.9468	.8725	.7507
	11	1.0000	1.0000	1.0000	.9999	.9991	.9949	.9804	.9435	.8692
	12	1.0000	1.0000	1.0000	1.0000	.9998	.9987	.9940	.9790	.9420
	13	1.0000	1.0000	1.0000	1.0000	1.0000	.9997	.9985	.9935	.9786
	14	1.0000	1.0000	1.0000	1.0000	1.0000	1.0000	.9997	.9984	.9936
	15	1.0000	1.0000	1.0000	1.0000	1.0000	1.0000	1.0000	.9997	.9985
	16	1.0000	1.0000	1.0000	1.0000	1.0000	1.0000	1.0000	1.0000	.9997
	17	1.0000	1.0000	1.0000	1.0000	1.0000	1.0000	1.0000	1.0000	1.0000
	18	1.0000	1.0000	1.0000	1.0000	1.0000	1.0000	1.0000	1.0000	1.0000
	19	1.0000	1.0000	1.0000	1.0000	1.0000	1.0000	1.0000	1.0000	1.0000
	20	1.0000	1.0000	1.0000	1.0000	1.0000	1.0000	1.0000	1.0000	1.0000

Table A3 (Continued)

n	y	p = .50	.55	.60	.65	.70	.75	.80	.85	.90	.95
19	0	.0000	.0000	.0000	.0000	.0000	.0000	.0000	.0000	.0000	.0000
	1	.0000	.0000	.0000	.0000	.0000	.0000	.0000	.0000	.0000	.0000
	2	.0004	.0001	.0000	.0000	.0000	.0000	.0000	.0000	.0000	.0000
	3	.0022	.0005	.0001	.0000	.0000	.0000	.0000	.0000	.0000	.0000
	4	.0096	.0028	.0006	.0001	.0000	.0000	.0000	.0000	.0000	.0000
	5	.0318	.0109	.0031	.0007	.0001	.0000	.0000	.0000	.0000	.0000
	6	.0835	.0342	.0116	.0031	.0006	.0001	.0000	.0000	.0000	.0000
	7	.1796	.0871	.0352	.0114	.0028	.0005	.0000	.0000	.0000	.0000
	8	.3238	.1841	.0885	.0347	.0105	.0023	.0003	.0000	.0000	.0000
	9	.5000	.3290	.1861	.0875	.0326	.0089	.0016	.0001	.0000	.0000
	10	.6762	.5060	.3325	.1855	.0839	.0287	.0067	.0008	.0000	.0000
	11	.8204	.6831	.5122	.3344	.1820	.0775	.0233	.0041	.0003	.0000
	12	.9165	.8273	.6919	.5188	.3345	.1749	.0676	.0163	.0017	.0000
	13	.9682	.9223	.8371	.7032	.5261	.3322	.1631	.0537	.0086	.0002
	14	.9904	.9720	.9304	.8500	.7178	.5346	.3267	.1444	.0352	.0020
	15	.9978	.9923	.9770	.9409	.8668	.7369	.5449	.3159	.1150	.0132
	16	.9996	.9985	.9945	.9830	.9538	.8887	.7631	.5587	.2946	.0665
	17	1.0000	.9998	.9992	.9969	.9896	.9690	.9171	.8015	.5797	.2453
	18	1.0000	1.0000	.9999	.9997	.9989	.9958	.9856	.9544	.8649	.6226
	19	1.0000	1.0000	1.0000	1.0000	1.0000	1.0000	1.0000	1.0000	1.0000	1.0000
20	0	.0000	.0000	.0000	.0000	.0000	.0000	.0000	.0000	.0000	.0000
	1	.0000	.0000	.0000	.0000	.0000	.0000	.0000	.0000	.0000	.0000
	2	.0002	.0000	.0000	.0000	.0000	.0000	.0000	.0000	.0000	.0000
	3	.0013	.0003	.0000	.0000	.0000	.0000	.0000	.0000	.0000	.0000
	4	.0059	.0015	.0003	.0000	.0000	.0000	.0000	.0000	.0000	.0000
	5	.0207	.0064	.0016	.0003	.0000	.0000	.0000	.0000	.0000	.0000
	6	.0577	.0214	.0065	.0015	.0003	.0000	.0000	.0000	.0000	.0000
	7	.1316	.0580	.0210	.0060	.0013	.0002	.0000	.0000	.0000	.0000
	8	.2517	.1308	.0565	.0196	.0051	.0009	.0001	.0000	.0000	.0000
	9	.4119	.2493	.1275	.0532	.0171	.0039	.0006	.0000	.0000	.0000
	10	.5881	.4086	.2447	.1218	.0480	.0139	.0026	.0002	.0000	.0000
	11	.7483	.5857	.4044	.2376	.1133	.0409	.0100	.0013	.0001	.0000
	12	.8684	.7480	.5841	.3990	.2277	.1018	.0321	.0059	.0004	.0000
	13	.9423	.8701	.7500	.5834	.3920	.2142	.0867	.0219	.0024	.0000
	14	.9793	.9447	.8744	.7546	.5836	.3828	.1958	.0673	.0113	.0003
	15	.9941	.9811	.9490	.8818	.7625	.5852	.3704	.1702	.0432	.0026
	16	.9987	.9951	.9840	.9556	.8929	.7748	.5886	.3523	.1330	.0159
	17	.9998	.9991	.9964	.9879	.9645	.9087	.7939	.5951	.3231	.0755
	18	1.0000	.9999	.9995	.9979	.9924	.9757	.9308	.8244	.6083	.2642
	19	1.0000	1.0000	1.0000	.9998	.9992	.9968	.9885	.9612	.8784	.6415
	20	1.0000	1.0000	1.0000	1.0000	1.0000	1.0000	1.0000	1.0000	1.0000	1.0000

For n larger than 20, the rth quantile y_r of a binomial random variable may be approximated using $y_r = np + w_r \sqrt{np(1-p)}$, where w_r is the rth quantile of a standard normal random variable, obtained from Table A1.

Table A4 CHART PROVIDING CONFIDENCE LIMITS FOR p IN BINOMIAL SAMPLING, GIVEN A SAMPLE FRACTION Y/n. CONFIDENCE COEFFICIENT, $1 - 2\alpha = .95$

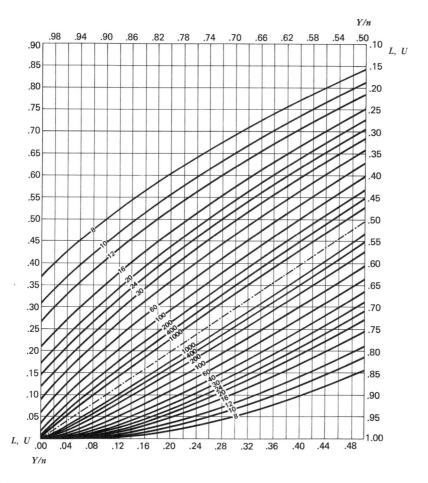

The numbers printed along the curves indicate the sample size n. If for a given value of the abscissa Y/n, L, and U are the ordinates read from (or interpolated between) the appropriate lower and upper curves, then

$$P(L \leq p \leq U) \geq 1 - 2\alpha$$

Table A4 (Continued) Confidence Coefficient, $1 - 2\alpha = .99$

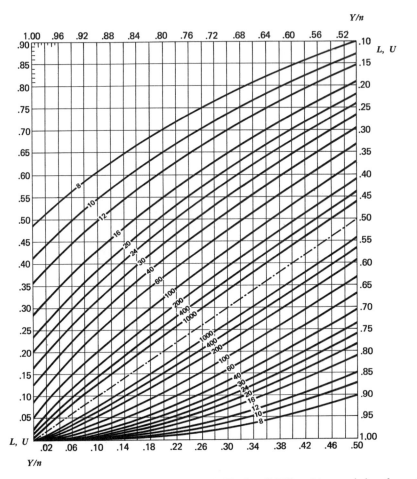

SOURCE. Adapted from Table 41, Pearson and Hartley (1970), with permission from the *Biometrika* Trustees.

NOTE. The process of reading from the curves can be simplified with the help of the right-angled corner of a loose sheet of paper or thin card, along the edges of which are marked off the scales shown in the top left-hand corner of each chart.

Table A5 Sample Sizes for One-Sided Nonparametric Tolerance Limits[a]

1 − α	q = .500	.700	.750	.800	.850	.900	.950	.975	.980	.990
.500	1	2	3	4	5	7	14	28	35	69
.700	2	4	5	6	8	12	24	48	60	120
.750	2	4	5	7	9	14	28	55	69	138
.800	3	5	6	8	10	16	32	64	80	161
.850	3	6	7	9	12	19	37	75	94	189
.900	4	7	9	11	15	22	45	91	144	230
.950	5	9	11	14	19	29	59	119	149	299
.975	6	11	13	17	23	36	72	146	183	368
.980	6	11	14	18	25	38	77	155	194	390
.990	7	13	17	21	29	44	90	182	228	459
.995	8	15	19	24	33	51	104	210	263	528
.999	10	20	25	31	43	66	135	273	342	688

[a] The quantity tabled is the sample size n such that $q^n \le \alpha$, for use in finding the tolerance limits

$$P(X^{(1)} \le p \text{ of the population}) \ge 1 - a$$

or

$$P(q \text{ of the population} \le X^{(n)}) \ge 1 - a$$

as described in Section 3.3.

Table A6 Sample Sizes for Nonparametric Tolerance Limits When $r + m = 2$[a]

1 − α	q = .500	.700	.750	.800	.850	.900	.950	.975	.980	.990
.500	3	6	7	9	11	17	34	67	84	168
.700	5	8	10	12	16	24	49	97	122	244
.750	5	9	10	13	18	27	53	107	134	269
.800	5	9	11	14	19	29	59	119	149	299
.850	6	10	13	16	22	33	67	134	168	337
.900	7	12	15	18	25	38	77	155	194	388
.950	8	14	18	22	30	46	93	188	236	473
.975	9	17	20	26	35	54	110	221	277	555
.980	9	17	21	27	37	56	115	231	290	581
.990	11	20	24	31	42	64	130	263	330	662
.995	12	22	27	34	47	72	146	294	369	740
.999	14	27	33	42	58	89	181	366	458	920

[a] The quantity tabled is that sample size n such that $q^n + nq^{n-1}(1-q) \le \alpha$ for use in finding the tolerance limits

$$P(X^{(r)} \le q \text{ of the population} \le X^{(n+1-m)}) \ge 1 - \alpha$$

when $r + m = 2$.

Table A7 Quantiles of the Mann–Whitney Test Statistic[a]

n	p	m = 2	3	4	5	6	7	8	9	10	11	12	13	14	15	16	17	18	19	20
2	.001	3	3	3	3	3	3	3	3	3	3	3	3	3	3	3	3	3	3	3
	.005	3	3	3	3	3	3	3	3	3	3	3	3	3	3	3	3	3	4	4
	.01	3	3	3	3	3	3	3	3	3	3	3	4	4	4	4	4	4	5	5
	.025	3	3	3	3	3	3	4	4	4	5	5	5	5	5	5	6	6	6	6
	.05	3	3	3	4	4	4	5	5	5	5	6	6	7	7	7	7	8	8	8
	.10	3	4	4	5	5	5	6	6	7	7	8	8	8	9	9	10	10	11	11
3	.001	6	6	6	6	6	6	6	6	6	6	6	6	6	6	6	7	7	7	7
	.005	6	6	6	6	6	6	6	7	7	7	8	8	8	9	9	9	9	10	10
	.01	6	6	6	6	6	7	7	8	8	8	9	9	9	10	10	11	11	11	12
	.025	6	6	6	7	8	8	9	9	10	10	11	11	12	12	13	13	14	14	15
	.05	6	7	7	8	9	9	10	11	11	12	12	13	14	14	15	16	16	17	18
	.10	7	8	8	9	10	11	12	12	13	14	15	16	17	17	18	19	20	21	22
4	.001	10	10	10	10	10	10	10	10	11	11	11	12	12	12	13	13	14	14	14
	.005	10	10	10	10	11	11	12	12	13	13	14	14	15	16	16	17	17	18	19
	.01	10	10	10	11	12	12	13	14	14	15	16	16	17	18	18	19	20	20	21
	.025	10	10	11	12	13	14	15	15	16	17	18	19	20	21	22	22	23	24	25
	.05	10	11	12	13	14	15	16	17	18	19	20	21	22	23	25	26	27	28	29
	.10	11	12	14	15	16	17	18	20	21	22	23	24	26	27	28	29	31	32	33
5	.001	15	15	15	15	15	15	16	17	17	18	18	19	19	20	21	21	22	23	23
	.005	15	15	15	16	17	17	18	19	20	21	22	23	23	24	25	26	27	28	29
	.01	15	15	16	17	18	19	20	21	22	23	24	25	26	27	28	29	30	31	32
	.025	15	16	17	18	19	21	22	23	24	25	27	28	29	30	31	33	34	35	36
	.05	16	17	18	20	21	22	24	25	27	28	29	31	32	34	35	36	38	39	41
	.10	17	18	20	21	23	24	26	28	29	31	33	34	36	38	39	41	43	44	46

6	.001	21	21	21	21	21	21	23	24	25	26	26	27	28	29	30	31	32	33	34
	.005	21	21	22	23	24	25	26	27	28	29	31	32	33	34	35	37	38	39	40
	.01	21	21	23	24	25	26	28	29	30	31	33	34	35	37	38	40	41	42	44
	.025	21	23	24	25	27	28	30	32	33	35	36	38	39	41	43	44	46	47	49
	.05	22	24	25	27	29	30	32	34	36	38	39	41	43	45	47	48	50	52	54
	.10	23	25	27	29	31	33	35	37	39	41	43	45	47	49	51	53	56	58	60
7	.001	28	28	28	28	29	30	31	32	34	35	36	37	38	39	40	42	43	44	45
	.005	28	28	29	30	32	33	35	36	38	39	41	42	44	45	47	48	50	51	53
	.01	28	29	30	32	33	35	36	38	40	41	43	45	46	48	50	52	53	55	57
	.025	28	30	32	34	35	37	39	41	43	45	47	49	51	53	55	57	59	61	63
	.05	29	31	33	35	37	40	42	44	46	48	50	53	55	57	59	62	64	66	68
	.10	30	33	35	37	40	42	45	47	50	52	55	57	60	62	65	67	70	72	75
8	.001	36	36	36	37	38	39	41	42	43	45	46	48	49	51	52	54	55	57	58
	.005	36	36	37	39	41	43	44	46	48	50	52	54	55	57	59	61	63	65	67
	.01	36	37	39	41	43	44	46	48	50	52	54	56	59	61	63	65	67	69	71
	.025	37	39	41	43	45	47	50	52	54	56	59	61	63	66	68	71	73	75	78
	.05	38	40	42	45	47	50	52	55	57	60	63	65	68	70	73	76	78	81	84
	.10	39	42	44	47	50	53	56	59	61	64	67	70	73	76	79	82	85	88	91
9	.001	45	45	45	47	48	49	51	53	54	56	58	60	61	63	65	67	69	71	72
	.005	45	46	47	49	51	53	55	57	59	62	64	66	68	70	73	75	77	79	82
	.01	45	47	49	51	53	55	57	60	62	64	67	69	72	74	77	79	82	84	86
	.025	46	48	50	53	56	58	61	63	66	69	72	74	77	80	83	85	88	91	94
	.05	47	50	52	55	58	61	64	67	70	73	76	79	82	85	88	91	94	97	100
	.10	48	51	55	58	61	64	68	71	74	77	81	84	87	91	94	98	101	104	108
10	.001	55	55	56	57	59	61	62	64	66	68	70	73	75	77	79	81	83	85	88
	.005	55	56	58	60	62	65	67	69	72	74	77	80	82	85	87	90	93	95	98
	.01	55	57	59	62	64	67	69	72	75	78	80	83	86	89	92	94	97	100	103
	.025	56	59	61	64	67	70	73	76	79	82	85	89	92	95	98	101	104	108	111
	.05	57	60	63	67	70	73	76	80	83	87	90	93	97	100	104	107	111	114	118
	.10	59	62	66	69	73	77	80	84	88	92	95	99	103	107	110	114	118	122	126

Table A7 (CONTINUED)

n	p	m = 2	3	4	5	6	7	8	9	10	11	12	13	14	15	16	17	18	19	20
	.001	66	66	67	69	71	73	75	77	79	82	84	87	89	91	94	96	99	101	104
	.005	66	67	69	72	74	77	80	83	85	88	91	94	97	100	103	106	109	112	115
11	.01	66	68	71	74	76	79	82	85	89	92	95	98	101	104	108	111	114	117	120
	.025	67	70	73	76	80	83	86	90	93	97	100	104	107	111	114	118	122	125	129
	.05	68	72	75	79	83	86	90	94	98	101	105	109	113	117	121	124	128	132	136
	.10	70	74	78	82	86	90	94	98	103	107	111	115	119	124	128	132	136	140	145
	.001	78	78	79	81	83	86	88	91	93	96	98	102	104	106	110	113	116	118	121
	.005	78	80	82	85	88	91	94	97	100	103	106	110	113	116	120	123	126	130	133
12	.01	78	81	84	87	90	93	96	100	103	107	110	114	117	121	125	128	132	135	139
	.025	80	83	86	90	93	97	101	105	108	112	116	120	124	128	132	136	140	144	148
	.05	81	84	88	92	96	100	105	109	111	117	121	126	130	134	139	143	147	151	156
	.10	83	87	91	96	100	105	109	114	118	123	128	132	137	142	146	151	156	160	165
	.001	91	91	93	95	97	100	103	106	109	112	115	118	121	124	127	130	134	137	140
	.005	91	93	95	99	102	105	109	112	116	119	123	126	130	134	137	141	145	149	152
13	.01	92	94	97	101	104	108	112	115	119	123	127	131	135	139	143	147	151	155	159
	.025	93	96	100	104	108	112	116	120	125	129	133	137	142	146	151	155	159	164	168
	.05	94	98	102	107	111	116	120	125	129	134	139	143	148	153	157	162	167	172	176
	.10	96	101	105	110	115	120	125	130	135	140	145	150	155	160	166	171	176	181	186
	.001	105	105	107	109	112	115	118	121	125	128	131	135	138	142	145	149	152	156	160
	.005	105	107	110	113	117	121	124	128	132	136	140	144	148	152	156	160	164	169	173
14	.01	106	108	112	116	119	123	128	132	136	140	144	149	153	157	162	166	171	175	179
	.025	107	111	115	119	123	128	132	137	142	146	151	156	161	165	170	175	180	184	189
	.05	109	113	117	122	127	132	137	142	147	152	157	162	167	172	177	183	188	193	198
	.10	110	116	121	126	131	137	142	147	153	158	164	169	175	180	186	191	197	203	208

n	α																			
15	.001	120	120	122	125	128	133	135	138	142	145	149	153	157	161	164	168	172	176	180
	.005	120	123	126	129	133	137	141	145	150	154	158	163	167	172	176	181	185	190	194
	.01	121	124	128	132	136	140	145	149	154	158	163	168	172	177	182	187	191	196	201
	.025	122	126	131	135	140	145	150	155	160	165	170	175	180	185	191	196	201	206	211
	.05	124	128	133	139	144	149	154	160	165	171	176	182	187	193	198	204	209	215	221
	.10	126	131	137	143	148	154	160	166	172	178	184	189	195	201	207	213	219	225	231
16	.001	136	136	139	142	145	148	152	156	160	164	168	172	176	180	185	189	193	197	202
	.005	136	139	142	146	150	155	159	164	168	173	178	182	187	192	197	202	207	211	216
	.01	137	140	144	149	153	158	163	168	173	178	183	188	193	198	203	208	213	219	224
	.025	138	143	148	152	158	163	168	174	179	184	190	196	201	207	212	218	223	229	235
	.05	140	145	151	156	162	167	173	179	185	191	197	202	208	214	220	226	232	238	244
	.10	142	148	154	160	166	173	179	185	191	198	204	211	217	223	230	236	243	249	256
17	.001	153	154	156	159	163	167	171	175	179	183	188	192	197	201	206	211	215	220	224
	.005	153	156	160	164	169	173	178	183	188	193	198	203	208	214	219	224	229	235	240
	.01	154	158	162	167	172	177	182	187	192	198	203	209	214	220	225	231	236	242	247
	.025	156	160	165	171	176	182	188	193	199	205	211	217	223	229	235	241	247	253	259
	.05	157	163	169	174	180	187	193	199	205	211	218	224	231	237	243	250	256	263	269
	.10	160	166	172	179	185	192	199	206	212	219	226	233	239	246	253	260	267	274	281
18	.001	171	172	175	178	182	186	190	195	199	204	209	214	218	223	228	233	238	243	248
	.005	171	174	178	183	188	193	198	203	209	214	219	225	230	236	242	247	253	259	264
	.01	172	176	181	186	191	196	202	208	213	219	225	231	237	242	248	254	260	266	272
	.025	174	179	184	190	196	202	208	214	220	227	233	239	246	252	258	265	271	278	284
	.05	176	181	188	194	200	207	213	220	227	233	240	247	254	260	267	274	281	288	295
	.10	178	185	192	199	206	213	220	227	234	241	249	256	263	270	278	285	292	300	307
19	.001	190	191	194	198	202	206	211	216	220	225	231	236	241	246	251	257	262	268	273
	.005	191	194	198	203	208	213	219	224	230	236	242	248	254	260	265	272	278	284	290
	.01	192	195	200	206	211	217	223	229	235	241	247	254	260	266	273	279	285	292	298
	.025	193	198	204	210	216	223	229	236	243	249	256	263	269	276	283	290	297	304	310
	.05	195	201	208	214	221	228	235	242	249	256	263	271	278	285	292	300	307	314	321
	.10	198	205	212	219	227	234	242	249	257	264	272	280	288	295	303	311	319	326	334

Table A7 (CONTINUED)

n	p	m = 2	3	4	5	6	7	8	9	10	11	12	13	14	15	16	17	18	19	20
	.001	210	211	214	218	223	227	232	237	243	248	253	259	265	270	276	281	287	293	299
	.005	211	214	219	224	229	235	241	247	253	259	265	271	278	284	290	297	303	310	316
	.01	212	216	221	227	233	239	245	251	258	264	271	278	284	291	298	304	311	318	325
20	.025	213	219	225	231	238	245	251	259	266	273	280	287	294	301	309	316	323	330	338
	.05	215	222	229	236	243	250	258	265	273	280	288	295	303	311	318	326	334	341	349
	.10	218	226	233	241	249	257	265	273	281	289	297	305	313	321	330	338	346	354	362

For n or m greater than 20, the pth quantile w_p of the Mann–Whitney test statistic may be approximated by

$$w_p = n(N+1)/2 + x_p \sqrt{nm(N+1)/12}$$

where x_p is the pth quantile of a standard normal random variable, obtained from Table A1, and where $N = m + n$.

[a] The entries in this table are quantiles w_p of the Mann–Whitney test statistic T, given by Equation 5.1.1, for selected values of p. Note that $P(T < w_p) \leq p$. Upper quantiles may be found from the equation

$$w_p = n(n + m + 1) - w_{1-p}$$

Critical regions correspond to values of T less than (or greater than) but not equal to the appropriate quantile.

Table A8 QUANTILES OF THE KRUSKAL-WALLIS TEST STATISTIC
FOR SMALL SAMPLE SIZES[a]

Sample Sizes	$w_{0.90}$	$w_{0.95}$	$w_{0.99}$
2, 2, 2	3.7143	4.5714	4.5714
3, 2, 1	3.8571	4.2857	4.2857
3, 2, 2	4.4643	4.5000	5.3571
3, 3, 1	4.0000	4.5714	5.1429
3, 3, 2	4.2500	5.1389	6.2500
3, 3, 3	4.6000	5.0667	6.4889
4, 2, 1	4.0179	4.8214	4.8214
4, 2, 2	4.1667	5.1250	6.0000
4, 3, 1	3.8889	5.0000	5.8333
4, 3, 2	4.4444	5.4000	6.3000
4, 3, 3	4.7000	5.7273	6.7091
4, 4, 1	4.0667	4.8667	6.1667
4, 4, 2	4.4455	5.2364	6.8727
4, 4, 3	4.773	5.5758	7.1364
4, 4, 4	4.5000	5.6538	7.5385
5, 2, 1	4.0500	4.4500	5.2500
5, 2, 2	4.2933	5.0400	6.1333
5, 3, 1	3.8400	4.8711	6.4000
5, 3, 2	4.4946	5.1055	6.8218
5, 3, 3	4.4121	5.5152	6.9818
5, 4, 1	3.9600	4.8600	6.8400
5, 4, 2	4.5182	5.2682	7.1182
5, 4, 3	4.5231	5.6308	7.3949
5, 4, 4	4.6187	5.6176	7.7440
5, 5, 1	4.0364	4.9091	6.8364
5, 5, 2	4.5077	5.2462	7.2692
5, 5, 3	4.5363	5.6264	7.5429
5, 5, 4	4.5200	5.6429	7.7914
5, 5, 5	4.5000	5.6600	7.9800

SOURCE. Adapted from Iman, Quade, and Alexander (1975), with permission from the American Mathematical Society.

[a] The null hypothesis may be rejected at the level α if the Kruskal-Wallis test statistic, given by Equation 5.2.5, exceeds the $1 - \alpha$ quantile given in the table.

Table A9 QUANTILES OF THE SQUARED RANKS TEST STATISTIC[a]

n	p	m = 3	4	5	6	7	8	9	10
	.005	14	14	14	14	14	14	21	21
	.01	14	14	14	14	21	21	26	26
	.025	14	14	21	26	29	30	35	41
	.05	21	21	26	30	38	42	49	54
3	.10	26	29	35	42	50	59	69	77
	.90	65	90	117	149	182	221	260	305
	.95	70	101	129	161	197	238	285	333
	.975	77	110	138	170	213	257	308	362
	.99	77	110	149	194	230	285	329	394
	.995	77	110	149	194	245	302	346	413
	.005	30	30	30	39	39	46	50	54
	.01	30	30	39	46	50	51	62	66
	.025	30	39	50	54	63	71	78	90
	.05	39	50	57	66	78	90	102	114
4	.10	50	62	71	85	99	114	130	149
	.90	111	142	182	222	270	321	375	435
	.95	119	154	197	246	294	350	413	476
	.975	126	165	206	255	311	374	439	510
	.99	126	174	219	270	334	401	470	545
	.995	126	174	230	281	351	414	494	567
	.005	55	55	66	75	79	88	99	110
	.01	55	66	75	82	90	103	115	127
	.025	66	79	88	100	114	130	145	162
	.05	75	88	103	120	135	155	175	195
5	.10	87	103	121	142	163	187	212	239
	.90	169	214	264	319	379	445	514	591
	.95	178	228	282	342	410	479	558	639
	.975	183	235	297	363	433	508	592	680
	.99	190	246	310	382	459	543	631	727
	.995	190	255	319	391	478	559	654	754
	.005	91	104	115	124	136	152	167	182
	.01	91	115	124	139	155	175	191	210
	.025	115	130	143	164	184	208	231	255
	.05	124	139	164	187	211	239	268	299
6	.10	136	163	187	215	247	280	315	352
	.90	243	300	364	435	511	592	679	772
	.95	255	319	386	463	545	634	730	831
	.975	259	331	406	486	574	670	771	880
	.99	271	339	424	511	607	706	817	935
	.995	271	346	431	526	624	731	847	970

SOURCE. Adapted from tables generated by R.L. Iman.

[a] The entries in this table are selected quantiles w_p of the squared ranks test statistic T, given by Equation 5.3.3. Note that $P(T < w_p) \leq p$ and $P(T > w_p) \leq 1 - p$. Critical regions correspond to values less than (or greater than) but not including the appropriate quantile.

n	p	m = 3	4	5	6	7	8	9	10
	.005	140	155	172	195	212	235	257	280
	.01	155	172	191	212	236	260	287	315
	.025	172	195	217	245	274	305	338	372
	.05	188	212	240	274	308	344	384	425
	.10	203	236	271	308	350	394	440	489
7	.90	335	407	487	572	665	764	871	984
	.95	347	428	515	608	707	814	929	1051
	.975	356	443	536	635	741	856	979	1108
	.99	364	456	560	664	779	900	1032	1172
	.995	371	467	571	683	803	929	1067	1212
	.005	204	236	260	284	311	340	368	401
	.01	221	249	276	309	340	372	408	445
	.025	249	276	311	345	384	425	468	513
	.05	268	300	340	381	426	473	524	576
	.10	285	329	374	423	476	531	590	652
8	.90	447	536	632	735	846	965	1091	1224
	.95	464	560	664	776	896	1023	1159	1303
	.975	476	579	689	807	935	1071	1215	1368
	.99	485	599	716	840	980	1124	1277	1442
	.995	492	604	731	863	1005	1156	1319	1489
	.005	304	325	361	393	429	466	508	549
	.01	321	349	384	423	464	508	553	601
	.025	342	380	423	469	517	570	624	682
	.05	365	406	457	510	567	626	689	755
	.10	390	444	501	561	625	694	766	843
9	.90	581	689	803	925	1056	1195	1343	1498
	.95	601	717	840	972	1112	1261	1420	1587
	.975	615	741	870	1009	1158	1317	1485	1662
	.99	624	757	900	1049	1209	1377	1556	1745
	.995	629	769	916	1073	1239	1417	1601	1798
	.005	406	448	486	526	573	620	672	725
	.01	425	470	513	561	613	667	725	785
	.025	457	505	560	616	677	741	808	879
	.05	486	539	601	665	734	806	883	963
	.10	514	580	649	724	801	885	972	1064
10	.90	742	866	1001	1144	1296	1457	1627	1806
	.95	765	901	1045	1197	1360	1533	1715	1907
	.975	778	925	1078	1241	1413	1596	1788	1991
	.99	793	949	1113	1286	1470	1664	1869	2085
	.995	798	961	1130	1314	1505	1708	1921	2145

For n or m greater than 10, the pth quantile w_p of the squared ranks test statistic may be approximated by

$$w_p = \frac{n(N+1)(2N+1)}{6} + x_p \sqrt{\frac{mn(N+1)(2N+1)(8N+11)}{180}}$$

where $N = n + m$, and where x_p is the pth quantile of a standard normal random variable, obtained from Table A1.

Table A10 QUANTILES OF THE SPEARMAN TEST STATISTIC[a]

n	p = .900	.950	.975	.990	.995	.999
4	.8000	.8000				
5	.7000	.8000	.9000	.9000		
6	.6000	.7714	.8286	.8857	.9429	
7	.5357	.6786	.7450	.8571	.8929	.9643
8	.5000	.6190	.7143	.8095	.8571	.9286
9	.4667	.5833	.6833	.7667	.8167	.9000
10	.4424	.5515	.6364	.7333	.7818	.8667
11	.4182	.5273	.6091	.7000	.7455	.8364
12	.3986	.4965	.5804	.6713	.7273	.8182
13	.3791	.4780	.5549	.6429	.6978	.7912
14	.3626	.4593	.5341	.6220	.6747	.7670
15	.3500	.4429	.5179	.6000	.6536	.7464
16	.3382	.4265	.5000	.5824	.6324	.7265
17	.3260	.4118	.4853	.5637	.6152	.7083
18	.3148	.3994	.4716	.5480	.5975	.6904
19	.3070	.3895	.4579	.5333	.5825	.6737
20	.2977	.3789	.4451	.5203	.5684	.6586
21	.2909	.3688	.4351	.5078	.5545	.6455
22	.2829	.3597	.4241	.4963	.5426	.6318
23	.2767	.3518	.4150	.4852	.5306	.6186
24	.2704	.3435	.4061	.4748	.5200	.6070
25	.2646	.3362	.3977	.4654	.5100	.5962
26	.2588	.3299	.3894	.4564	.5002	.5856
27	.2540	.3236	.3822	.4481	.4915	.5757
28	.2490	.3175	.3749	.4401	.4828	.5660
29	.2443	.3113	.3685	.4320	.4744	.5567
30	.2400	.3059	.3620	.4251	.4665	.5479

For n greater than 30 the approximate quantiles of ρ may be obtained from

$$w_p \cong \frac{x_p}{\sqrt{n-1}}$$

where x_p is the p quantile of a standard normal random variable obtained from Table 1.

SOURCE. Adapted from Glasser and Winter (1961), with corrections, with permission from the *Biometrika* Trustees.

[a] The entries in this table are selected quantiles w_p of the Spearman rank correlation coefficient ρ when used as a test statistic. The lower quantiles may be obtained from the equation

$$w_p = -w_{1-p}$$

The critical region corresponds to values of ρ smaller than (or greater than) but not including the appropriate quantile. Note that the median of ρ is 0.

Table A11 QUANTILES OF THE HOTELLING-PABST TEST STATISTIC[a]

n	$p = .001$	.005	.010	.025	.050	.100	$\frac{1}{3}n(n^2-1)$
4					2	2	20
5			2	2	4	6	40
6		2	4	6	8	14	70
7	2	6	8	14	18	26	112
8	6	12	16	24	32	42	168
9	12	22	28	38	50	64	240
10	22	36	44	60	74	92	330
11	36	56	66	86	104	128	440
12	52	78	94	120	144	172	572
13	76	110	130	162	190	226	728
14	106	148	172	212	246	290	910
15	142	194	224	270	312	364	1120
16	186	250	284	340	390	450	1360
17	238	314	356	420	480	550	1632
18	300	390	438	512	582	664	1938
19	372	476	532	618	696	790	2280
20	454	574	638	738	826	934	2660
21	546	686	758	870	972	1092	3080
22	652	810	892	1020	1134	1270	3542
23	772	950	1042	1184	1312	1464	4048
24	904	1104	1208	1366	1510	1678	4600
25	1050	1274	1390	1566	1726	1912	5200
26	1212	1462	1590	1786	1960	2168	5850
27	1390	1666	1808	2024	2216	2444	6552
28	1586	1890	2046	2284	2494	2744	7308
29	1800	2134	2306	2564	2796	3068	8120
30	2032	2398	2584	2868	3120	3416	8990

For n greater than 30, the quantiles of T may be approximated by

$$w_p \cong \tfrac{1}{6}n(n^2-1) + x_p \frac{1}{6}\frac{n(n^2-1)}{\sqrt{n-1}}$$

where x_p is the pth quantile of a standard normal random variable given in Table 1.

SOURCE. Adapted from Glasser and Winter (1961), with corrections, with permission from the *Biometrika* Trustees.

[a] The entries in this table are the quantiles w_p of the Hotelling-Pabst test statistic T, defined by Equation 5.4.11, for selected values of p. Note that $P(T < w_p) \leq p$. Upper quantiles may be found from the equation

$$w_{1-p} = \tfrac{1}{3}n(n^2-1) - w_p$$

Critical regions correspond to values of T less than (or greater than) but not including the appropriate quantiles. Note that the median of T is given by

$$w_{.50} = \tfrac{1}{6}n(n^2-1)$$

Table A12 Quantiles of the Kendall Test Statistic[a]

n	p = .900	.950	.975	.990	.995
4	4	4	6	6	6
5	6	6	8	8	10
6	7	9	11	11	13
7	9	11	13	15	17
8	10	14	16	18	20
9	12	16	18	22	24
10	15	19	21	25	27
11	17	21	25	29	31
12	18	24	28	34	36
13	22	26	32	38	42
14	23	31	35	41	45
15	27	33	39	47	51
16	28	36	44	50	56
17	32	40	48	56	62
18	35	43	51	61	67
19	37	47	55	65	73
20	40	50	60	70	78
21	42	54	64	76	84
22	45	59	69	81	89
23	49	63	73	87	97
24	52	66	78	92	102
25	56	70	84	98	108
26	59	75	89	105	115
27	61	79	93	111	123
28	66	84	98	116	128
29	68	88	104	124	136
30	73	93	109	129	143
31	75	97	115	135	149
32	80	102	120	142	158
33	84	106	126	150	164
34	87	111	131	155	173
35	91	115	137	163	179
36	94	120	144	170	188
37	98	126	150	176	196
38	103	131	155	183	203
39	107	137	161	191	211
40	110	142	168	198	220

n	p = .900	.950	.975	.990	.995
41	114	146	174	206	228
42	119	151	181	213	235
43	123	157	187	221	245
44	128	162	194	228	252
45	132	168	200	236	262
46	135	173	207	245	271
47	141	179	213	253	279
48	144	186	220	260	288
49	150	190	228	268	296
50	153	197	233	277	305
51	159	203	241	285	315
52	162	208	248	294	324
53	168	214	256	302	334
54	173	221	263	311	343
55	177	227	269	319	353
56	182	232	276	328	362
57	186	240	284	336	372
58	191	245	291	345	381
59	197	251	299	355	391
60	202	258	306	364	402

For n greater than 60, approximate quantiles of T may be obtained from

$$w_p \cong x_p \sqrt{\frac{n(n-1)(2n+5)}{18}}$$

where x_p is from the standard normal distribution given by Table A1.

[a] The entries in this table are selected quantiles w_p of the Kendall test statistic T, defined by Equation 5.4.13, for selected values of p. Only upper quantiles are given here, but lower quantiles may be obtained from the relationship

$$w_p = -w_{1-p}$$

Critical regions correspond to values of T greater than (or less than) but not including the appropriate quantile. Note that the median of T is 0.

SOURCE. Adapted from Table 1, Best (1974), with permission from the author.

Table A13 Quantiles of the Wilcoxon Signed Ranks Test Statistic

	$w_{0.005}$	$w_{0.01}$	$w_{0.025}$	$w_{0.05}$	$w_{0.10}$	$w_{0.20}$	$w_{0.30}$	$w_{0.40}$	$w_{0.50}$	$\dfrac{n(n+1)}{2}$
$n = 4$	0	0	0	0	1	3	3	4	5	10
5	0	0	0	1	3	4	5	6	7.5	15
6	0	0	1	3	4	6	8	9	10.5	21
7	0	1	3	4	6	9	11	12	14	28
8	1	2	4	6	9	12	14	16	18	36
9	2	4	6	9	11	15	18	20	22.5	45
10	4	6	9	11	15	19	22	25	27.5	55
11	6	8	11	14	18	23	27	30	33	66
12	8	10	14	18	22	28	32	36	39	78
13	10	13	18	22	27	33	38	42	45.5	91
14	13	16	22	26	32	39	44	48	52.5	105
15	16	20	26	31	37	45	51	55	60	120
16	20	24	30	36	43	51	58	63	68	136
17	24	28	35	42	49	58	65	71	76.5	153
18	28	33	41	48	56	66	73	80	85.5	171
19	33	38	47	54	63	74	82	89	95	190
20	38	44	53	61	70	83	91	98	105	210
21	44	50	59	68	78	91	100	108	115.5	131
22	49	56	67	76	87	100	110	119	126.5	153
23	55	63	74	84	95	110	120	130	138	176
24	62	70	82	92	105	120	131	141	150	300
25	69	77	90	101	114	131	143	153	162.5	325
26	76	85	99	111	125	142	155	165	175.5	351
27	84	94	108	120	135	154	167	178	189	378
28	92	102	117	131	146	166	180	192	203	406
29	101	111	127	141	158	178	193	206	217.5	435
30	110	121	138	152	170	191	207	220	232.5	465
31	119	131	148	164	182	205	221	235	248	496
32	129	141	160	176	195	219	236	250	264	528
33	139	152	171	188	208	233	251	266	280.5	561
34	149	163	183	201	222	248	266	282	297.5	595
35	160	175	196	214	236	263	283	299	315	630
36	172	187	209	228	251	279	299	317	333	666
37	184	199	222	242	266	295	316	335	351.5	703
38	196	212	236	257	282	312	334	353	370.5	741
39	208	225	250	272	298	329	352	372	390	780
40	221	239	265	287	314	347	371	391	410	820
41	235	253	280	303	331	365	390	411	430.5	861
42	248	267	295	320	349	384	409	431	451.5	903
43	263	282	311	337	366	403	429	452	473	946
44	277	297	328	354	385	422	450	473	495	990
45	292	313	344	372	403	442	471	495	517.5	1035
46	308	329	362	390	423	463	492	517	540.5	1081
47	324	346	379	408	442	484	514	540	564	1128

	$w_{0.005}$	$w_{0.01}$	$w_{0.025}$	$w_{0.05}$	$w_{0.10}$	$w_{0.20}$	$w_{0.30}$	$w_{0.40}$	$w_{0.50}$	$\dfrac{n(n+1)}{2}$
48	340	363	397	428	463	505	536	563	588	1176
49	357	381	416	447	483	527	559	587	612.5	1225
50	374	398	435	467	504	550	583	611	637.5	1275

For n larger than 50, the pth quantile w_p of the Wilcoxon signed ranks test statistic may be approximated by $w_p = [n(n+1)/4] + x_p\sqrt{n(n+1)(2n+1)/24}$, where x_p is the pth quantile of a standard normal random variable, obtained from Table A1.

SOURCE. Adapted from Harter and Owen (1970), with permission from the Institute of Mathematical Statistics.

^{a} The entries in this table are quantiles w_p of the Wilcoxon signed ranks test statistic T, given by Equation 5.7.6, for selected values of $p \leq .50$. Quantiles w_p for $p > .50$ may be computed from the equation

$$w_p = n(n+1)/2 - w_{1-p}$$

where $n(n+1)/2$ is given in the right hand column in the table. Note that $P(T < w_p) \leq p$ and $P(T > w_p) \leq 1 - p$ if H_0 is true. Critical regions correspond to values of T less than (or greater than) but not including the appropriate quantile.

One-Sided Test $p = .90$	.95	.975	.99	.995		One-Sided Test $p = .90$	.95	.975	.99	.995
Two-Sided Test $p = .80$	.90	.95	.98	.99		Two-Sided Test $p = .80$	.90	.95	.98	.99
$n = 1$ .900	.950	.975	.990	.995	$n = 21$ .226	.259	.287	.321	.344
2 .684	.776	.842	.900	.929	22 .221	.253	.281	.314	.337
3 .565	.636	.708	.785	.829	23 .216	.247	.275	.307	.330
4 .493	.565	.624	.689	.734	24 .212	.242	.269	.301	.323
5 .447	.509	.563	.627	.669	25 .208	.238	.264	.295	.317
6 .410	.468	.519	.577	.617	26 .204	.233	.259	.290	.311
7 .381	.436	.483	.538	.576	27 .200	.229	.254	.284	.305
8 .358	.410	.454	.507	.542	28 .197	.225	.250	.279	.300
9 .339	.387	.430	.480	.513	29 .193	.221	.246	.275	.295
10 .323	.369	.409	.457	.489	30 .190	.218	.242	.270	.290
11 .308	.352	.391	.437	.468	31 .187	.214	.238	.266	.285
12 .296	.338	.375	.419	.449	32 .184	.211	.234	.262	.281
13 .285	.325	.361	.404	.432	33 .182	.208	.231	.258	.277
14 .275	.314	.349	.390	.418	34 .179	.205	.227	.254	.273
15 .266	.304	.338	.377	.404	35 .177	.202	.224	.251	.269
16 .258	.295	.327	.366	.392	36 .174	.199	.221	.247	.265
17 .250	.286	.318	.355	.381	37 .172	.196	.218	.244	.262
18 .244	.279	.309	.346	.371	38 .170	.194	.215	.241	.258
19 .237	.271	.301	.337	.361	39 .168	.191	.213	.238	.255
20 .232	.265	.294	.329	.352	40 .165	.189	.210	.235	.252
				Approximation for $n > 40$		$\dfrac{1.07}{\sqrt{n}}$	$\dfrac{1.22}{\sqrt{n}}$	$\dfrac{1.36}{\sqrt{n}}$	$\dfrac{1.52}{\sqrt{n}}$	$\dfrac{1.63}{\sqrt{n}}$

SOURCE. Adapted from Table of Miller (1956).

[a] The entries in this table are selected quantiles w_p of the Kolmogorov test statistics T, T^+, and T^- as defined by Equation 6.1.1 for two-sided tests and by Equations 6.1.2 and 6.1.3 for one-sided tests. Reject H_0 at the level α if T exceeds the $1 - \alpha$ quantile given in this table. These quantiles are exact for $n \leq 40$ in the two-tailed test. The other quantiles are approximations that are equal to the exact quantiles in most cases. A better approximation for $n > 40$ results if $(n + \sqrt{n/10})^{1/2}$ is used instead of $\sqrt{n}$ in the denominator.

Table A15 QUANTILES OF THE LILLIEFORS TEST STATISTIC FOR NORMALITY[a]

	$p = .80$	.85	.90	.95	.99
Sample size n = 4	.300	.319	.352	.381	.417
5	.285	.299	.315	.337	.405
6	.265	.277	.294	.319	.364
7	.247	.258	.276	.300	.348
8	.233	.244	.261	.285	.331
9	.223	.233	.249	.271	.311
10	.215	.224	.239	.258	.294
11	.206	.217	.230	.249	.284
12	.199	.212	.223	.242	.275
13	.190	.202	.214	.234	.268
14	.183	.194	.207	.227	.261
15	.177	.187	.201	.220	.257
16	.173	.182	.195	.213	.250
17	.169	.177	.189	.206	.245
18	.166	.173	.184	.200	.239
19	.163	.169	.179	.195	.235
20	.160	.166	.174	.190	.231
25	.142	.147	.158	.173	.200
30	.131	.136	.144	.161	.187
Over 30	$\dfrac{.736}{\sqrt{n}}$	$\dfrac{.768}{\sqrt{n}}$	$\dfrac{.805}{\sqrt{n}}$	$\dfrac{.886}{\sqrt{n}}$	$\dfrac{1.031}{\sqrt{n}}$

SOURCE. Adapted from Table 1 of Lilliefors (1967), with corrections.

[a] The entries in this table are the approximate quantiles w_p of the Lilliefors test statistic T_1 as defined by Equation 6.2.4. Reject H_0 at the level α if T_1 exceeds $w_{1-\alpha}$ for the particular sample size n.

Table A16 QUANTILES OF THE LILLIEFORS TEST STATISTIC FOR THE EXPONENTIAL DISTRIBUTION[a]

	p = .05	.10	.20	.30	.50	.70	.80	.90	.95	.99	.999
n = 2	.3127	.3200	.3337	.3617	.4337	.5034	.5507	.5934	.6133	.6284	.6317
3	.2299	.2544	.2899	.3166	.3645	.4122	.4508	.5111	.5508	.6003	.6296
4	.2072	.2281	.2545	.2766	.3163	.3685	.4007	.4442	.4844	.5574	.6215
5	.1884	.2052	.2290	.2483	.2877	.3317	.3603	.4045	.4420	.5127	.5814
6	.1726	.1882	.2102	.2290	.2645	.3045	.3320	.3732	.4085	.4748	.5497
7	.1604	.1750	.1961	.2136	.2458	.2838	.3098	.3481	.3811	.4459	.5181
8	.1506	.1646	.1845	.2006	.2309	.2671	.2914	.3274	.3590	.4208	.4913
9	.1426	.1561	.1746	.1897	.2186	.2529	.2758	.3101	.3404	.3995	.4679
10	.1359	.1486	.1661	.1805	.2082	.2407	.2626	.2955	.3244	.3813	.4473
12	.1249	.1364	.1524	.1657	.1912	.2209	.2411	.2714	.2981	.3511	.4132
14	.1162	.1268	.1418	.1542	.1778	.2054	.2242	.2525	.2774	.3272	.3858
16	.1091	.1191	.1332	.1448	.1669	.1929	.2105	.2371	.2606	.3076	.3632
18	.1032	.1127	.1260	.1369	.1578	.1824	.1990	.2242	.2465	.2911	.3441
20	.0982	.1073	.1199	.1303	.1501	.1735	.1893	.2132	.2345	.2771	.3277
22	.0939	.1025	.1146	.1245	.1434	.1657	.1809	.2038	.2241	.2649	.3135
24	.0901	.0984	.1099	.1195	.1376	.1590	.1735	.1954	.2150	.2542	.3010
26	.0868	.0947	.1058	.1150	.1324	.1530	.1670	.1881	.2069	.2447	.2899
28	.0838	.0914	.1021	.1110	.1278	.1477	.1611	.1815	.1997	.2362	.2799
30	.0811	.0885	.0988	.1074	.1236	.1428	.1559	.1756	.1932	.2286	.2709
35	.0754	.0822	.0918	.0997	.1148	.1326	.1447	.1630	.1793	.2123	.2517
40	.0707	.0771	.0861	.0935	.1077	.1243	.1356	.1528	.1681	.1990	.2361
45	.0668	.0729	.0814	.0884	.1017	.1174	.1281	.1443	.1588	.1880	.2231
50	.0636	.0693	.0774	.0840	.0966	.1116	.1217	.1371	.1509	.1787	.2121
60	.0582	.0635	.0708	.0769	.0885	.1021	.1114	.1255	.1381	.1635	.1943

70	.0541	.0589	.0658	.0714	.0821	.0946	.1033	.1164	.1281	.1517 [b]
80	.0507	.0553	.0616	.0669	.0769	.0887	.0968	.1090	.1200	.1421 [b]
90	.0479	.0522	.0582	.0632	.0726	.0838	.0914	.1029	.1132	.1341 [b]
$n = 100$	.0455	.0496	.0553	.0600	.0690	.0796	.0868	.0977	.1075	.1274 [b]
Approximation for $n > 100$	$\dfrac{.4550}{\sqrt{n}}$	$\dfrac{.4959}{\sqrt{n}}$	$\dfrac{.5530}{\sqrt{n}}$	$\dfrac{.6000}{\sqrt{n}}$	$\dfrac{.6898}{\sqrt{n}}$	$\dfrac{.7957}{\sqrt{n}}$	$\dfrac{.8678}{\sqrt{n}}$	$\dfrac{.9773}{\sqrt{n}}$	$\dfrac{1.0753}{\sqrt{n}}$	$\dfrac{1.2743}{\sqrt{n}}$ [b]

SOURCE. Adapted from Durbin (1975), with permission from the *Biometrika* Trustees.

[a] The entries in this table are selected quantiles w_p of the Lilliefors test statistic T_2 as given by Equation 6.2.7. Reject at the level of significance α if T_2 is greater than the $1 - \alpha$ quantile given in the table. The approximation for $n > 100$ is merely the exact value for $n = 100$. More accurate approximations for $n > 100$ may be obtained from Table 54 of Pearson and Hartley (1972).

[b] These quantiles are not presently available.

n / i	2	3	4	5	6	7	8	9	10
1	0.7071	0.7071	0.6872	0.6646	0.6431	0.6233	0.6052	0.5888	0.5739
2	—	0.0000	0.1667	0.2413	0.2806	0.3031	0.3164	0.3244	0.3291
3	—	—	—	0.0000	0.0875	0.1401	0.1743	0.1976	0.2141
4	—	—	—	—	—	0.0000	0.0561	0.0947	0.1224
5	—	—	—	—	—	—	—	0.0000	0.0399

n / i	11	12	13	14	15	16	17	18	19	20
1	0.5601	0.5475	0.5359	0.5251	0.5150	0.5056	0.4968	0.4886	0.4808	0.4734
2	0.3315	0.3325	0.3325	0.3318	0.3306	0.3290	0.3273	0.3253	0.3232	0.3211
3	0.2260	0.2347	0.2412	0.2460	0.2495	0.2521	0.2540	0.2553	0.2561	0.2565
4	0.1429	0.1586	0.1707	0.1802	0.1878	0.1939	0.1988	0.2027	0.2059	0.2085
5	0.0695	0.0922	0.1099	0.1240	0.1353	0.1447	0.1524	0.1587	0.1641	0.1686
6	0.0000	0.0303	0.0539	0.0727	0.0880	0.1005	0.1109	0.1197	0.1271	0.1334
7	—	—	0.0000	0.0240	0.0433	0.0593	0.0725	0.0837	0.0932	0.1013
8	—	—	—	—	0.0000	0.0196	0.0359	0.0496	0.0612	0.0711
9	—	—	—	—	—	—	0.0000	0.0163	0.0303	0.0422
10	—	—	—	—	—	—	—	—	0.0000	0.0140

n / i	21	22	23	24	25	26	27	28	29	30
1	0.4643	0.4590	0.4542	0.4493	0.4450	0.4407	0.4366	0.4328	0.4291	0.4254
2	0.3185	0.3156	0.3126	0.3098	0.3069	0.3043	0.3018	0.2992	0.2968	0.2944
3	0.2578	0.2571.	0.2563	0.2554	0.2543	0.2533	0.2522	0.2510	0.2499	0.2487
4	0.2119	0.2131	0.2139	0.2145	0.2148	0.2151	0.2152	0.2151	0.2150	0.2148
5	0.1736	0.1764	0.1787	0.1807	0.1822	0.1836	0.1848	0.1857	0.1864	0.1870
6	0.1399	0.1443	0.1480	0.1512	0.1539	0.1563	0.1584	0.1601	0.1616	0.1630
7	0.1092	0.1150	0.1201	0.1245	0.1283	0.1316	0.1346	0.1372	0.1395	0.1415
8	0.0804	0.0878	0.0941	0.0997	0.1046	0.1089	0.1128	0.1162	0.1192	0.1219
9	0.0530	0.0618	0.0696	0.0764	0.0823	0.0876	0.0923	0.0965	0.1002	0.1036
10	0.0263	0.0368	0.0459	0.0539	0.0610	0.0672	0.0728	0.0778	0.0822	0.0862
11	0.0000	0.0122	0.0228	0.0321	0.0403	0.0476	0.0540	0.0598	0.0650	0.0697
12	—	—	0.0000	0.0107	0.0200	0.0284	0.0358	0.0424	0.0483	0.0537
13	—	—	—	—	0.0000	0.0094	0.0178	0.0253	0.0320	0.0381
14	—	—	—	—	—	—	0.0000	0.0084	0.0159	0.0227
15	—	—	—	—	—	—	—	—	0.0000	0.0076

SOURCE. Reprinted from Pearson and Hartley (1972), with permission from the *Biometrika* Trustees.

[a] The entries in this table are the coefficients a_i for use in the Shapiro–Wilk test statistic for normality given by Equation 6.2.9.

i \ n	31	32	33	34	35	36	37	38	39	40
1	0.4220	0.4188	0.4156	0.4127	0.4096	0.4068	0.4040	0.4015	0.3989	0.3964
2	0.2921	0.2898	0.2876	0.2854	0.2834	0.2813	0.2794	0.2774	0.2755	0.2737
3	0.2475	0.2462	0.2451	0.2439	0.2427	0.2415	0.2403	0.2391	0.2380	0.2368
4	0.2145	0.2141	0.2137	0.2132	0.2127	0.2121	0.2116	0.2110	0.2104	0.2098
5	0.1874	0.1878	0.1880	0.1882	0.1883	0.1883	0.1883	0.1881	0.1880	0.1878
6	0.1641	0.1651	0.1660	0.1667	0.1673	0.1678	0.1683	0.1686	0.1689	0.1691
7	0.1433	0.1449	0.1463	0.1475	0.1487	0.1496	0.1505	0.1513	0.1520	0.1526
8	0.1243	0.1265	0.1284	0.1301	0.1317	0.1331	0.1344	0.1356	0.1366	0.1376
9	0.1066	0.1093	0.1118	0.1140	0.1160	0.1179	0.1196	0.1211	0.1225	0.1237
10	0.0899	0.0931	0.0961	0.0988	0.1013	0.1036	0.1056	0.1075	0.1092	0.1108
11	0.0739	0.0777	0.0812	0.0844	0.0873	0.0900	0.0924	0.0947	0.0967	0.0986
12	0.0585	0.0629	0.0669	0.0706	0.0739	0.0770	0.0798	0.0824	0.0848	0.0870
13	0.0435	0.0485	0.0530	0.0572	0.0610	0.0645	0.0677	0.0706	0.0733	0.0759
14	0.0289	0.0344	0.0395	0.0441	0.0484	0.0523	0.0559	0.0592	0.0622	0.0651
15	0.0144	0.0206	0.0262	0.0314	0.0361	0.0404	0.0444	0.0481	0.0515	0.0546
16	0.0000	0.0068	0.0131	0.0187	0.0239	0.0287	0.0331	0.0372	0.0409	0.0444
17	—	—	0.0000	0.0062	0.0119	0.0172	0.0220	0.0264	0.0305	0.0343
18	—	—	—	—	0.0000	0.0057	0.0110	0.0158	0.0203	0.0244
19	—	—	—	—	—	—	0.0000	0.0053	0.0101	0.0146
20	—	—	—	—	—	—	—	—	0.0000	0.0049

i \ n	41	42	43	44	45	46	47	48	49	50
1	0.3940	0.3917	0.3894	0.3872	0.3850	0.3830	0.3808	0.3789	0.3770	0.3751
2	0.2719	0.2701	0.2684	0.2667	0.2651	0.2635	0.2620	0.2604	0.2589	0.2574
3	0.2357	0.2345	0.2334	0.2323	0.2313	0.2302	0.2291	0.2281	0.2271	0.2260
4	0.2091	0.2085	0.2078	0.2072	0.2065	0.2058	0.2052	0.2045	0.2038	0.2032
5	0.1876	0.1874	0.1871	0.1868	0.1865	0.1862	0.1859	0.1855	0.1851	0.1847
6	0.1693	0.1694	0.1695	0.1695	0.1695	0.1695	0.1695	0.1693	0.1692	0.1691
7	0.1531	0.1535	0.1539	0.1542	0.1545	0.1548	0.1550	0.1551	0.1553	0.1554
8	0.1384	0.1392	0.1398	0.1405	0.1410	0.1415	0.1420	0.1423	0.1427	0.1430
9	0.1249	0.1259	0.1269	0.1278	0.1286	0.1293	0.1300	0.1306	0.1312	0.1317
10	0.1123	0.1136	0.1149	0.1160	0.1170	0.1180	0.1189	0.1197	0.1205	0.1212
11	0.1004	0.1020	0.1035	0.1049	0.1062	0.1073	0.1085	0.1095	0.1105	0.1113
12	0.0891	0.0909	0.0927	0.0943	0.0959	0.0972	0.0986	0.0998	0.1010	0.1020
13	0.0782	0.0804	0.0824	0.0842	0.0860	0.0876	0.0892	0.0906	0.0919	0.0932
14	0.0677	0.0701	0.0724	0.0745	0.0765	0.0783	0.0801	0.0817	0.0832	0.0846
15	0.0575	0.0602	0.0628	0.0651	0.0673	0.0694	0.0713	0.0731	0.0748	0.0764
16	0.0476	0.0506	0.0534	0.0560	0.0584	0.0607	0.0628	0.0648	0.0667	0.0685
17	0.0379	0.0411	0.0442	0.0471	0.0497	0.0522	0.0546	0.0568	0.0588	0.0608
18	0.0283	0.0318	0.0352	0.0383	0.0412	0.0439	0.0465	0.0489	0.0511	0.0532
19	0.0188	0.0227	0.0263	0.0296	0.0328	0.0357	0.0385	0.0411	0.0436	0.0459
20	0.0094	0.0136	0.0175	0.0211	0.0245	0.0277	0.0307	0.0335	0.0361	0.0386
21	0.0000	0.0045	0.0087	0.0126	0.0163	0.0197	0.0229	0.0259	0.0288	0.0314
22		—	0.0000	0.0042	0.0081	0.0118	0.0153	0.0185	0.0215	0.0244
23	—	—	—	—	0.0000	0.0039	0.0076	0.0111	0.0143	0.0174
24	—	—	—	—	—	—	0.0000	0.0037	0.0071	0.0104
25	—	—	—	—	—	—	—	—	0.0000	0.0035

Table A18 Quantiles of the Shapiro–Wilk Test Statistic[a]

n	0.01	0.02	0.05	0.10	0.50	0.90	0.95	0.98	0.99
3	0.753	0.756	0.767	0.789	0.959	0.998	0.999	1.000	1.000
4	0.687	0.707	0.748	0.792	0.935	0.987	0.992	0.996	0.997
5	0.686	0.715	0.762	0.806	0.927	0.979	0.986	0.991	0.993
6	0.713	0.743	0.788	0.826	0.927	0.974	0.981	0.986	0.989
7	0.730	0.760	0.803	0.838	0.928	0.972	0.979	0.985	0.988
8	0.749	0.778	0.818	0.851	0.932	0.972	0.978	0.984	0.987
9	0.764	0.791	0.829	0.859	0.935	0.972	0.978	0.984	0.986
10	0.781	0.806	0.842	0.869	0.938	0.972	0.978	0.983	0.986
11	0.792	0.817	0.850	0.876	0.940	0.973	0.979	0.984	0.986
12	0.805	0.828	0.859	0.883	0.943	0.973	0.979	0.984	0.986
13	0.814	0.837	0.866	0.889	0.945	0.974	0.979	0.984	0.986
14	0.825	0.846	0.874	0.895	0.947	0.975	0.980	0.984	0.986
15	0.835	0.855	0.881	0.901	0.950	0.975	0.980	0.984	0.987
16	0.844	0.863	0.887	0.906	0.952	0.976	0.981	0.985	0.987
17	0.851	0.869	0.892	0.910	0.954	0.977	0.981	0.985	0.987
18	0.858	0.874	0.897	0.914	0.956	0.978	0.982	0.986	0.988
19	0.863	0.879	0.901	0.917	0.957	0.978	0.982	0.986	0.988
20	0.868	0.884	0.905	0.920	0.959	0.979	0.983	0.986	0.988
21	0.873	0.888	0.908	0.923	0.960	0.980	0.983	0.987	0.989
22	0.878	0.892	0.911	0.926	0.961	0.980	0.984	0.987	0.989
23	0.881	0.895	0.914	0.928	0.962	0.981	0.984	0.987	0.989
24	0.884	0.898	0.916	0.930	0.963	0.981	0.984	0.987	0.989
25	0.888	0.901	0.918	0.931	0.964	0.981	0.985	0.988	0.989
26	0.891	0.904	0.920	0.933	0.965	0.982	0.985	0.988	0.989
27	0.894	0.906	0.923	0.935	0.965	0.982	0.985	0.988	0.990
28	0.896	0.908	0.924	0.936	0.966	0.982	0.985	0.988	0.990
29	0.898	0.910	0.926	0.937	0.966	0.982	0.985	0.988	0.990
30	0.900	0.912	0.927	0.939	0.967	0.983	0.985	0.988	0.990
31	0.902	0.914	0.929	0.940	0.967	0.983	0.986	0.988	0.990
32	0.904	0.915	0.930	0.941	0.968	0.983	0.986	0.988	0.990
33	0.906	0.917	0.931	0.942	0.968	0.983	0.986	0.989	0.990
34	0.908	0.919	0.933	0.943	0.969	0.983	0.986	0.989	0.990
35	0.910	0.920	0.934	0.944	0.969	0.984	0.986	0.989	0.990
36	0.912	0.922	0.935	0.945	0.970	0.984	0.986	0.989	0.990
37	0.914	0.924	0.936	0.946	0.970	0.984	0.987	0.989	0.990
38	0.916	0.925	0.938	0.947	0.971	0.984	0.987	0.989	0.990
39	0.917	0.927	0.939	0.948	0.971	0.984	0.987	0.989	0.991
40	0.919	0.928	0.940	0.949	0.972	0.985	0.987	0.989	0.991
41	0.920	0.929	0.941	0.950	0.972	0.985	0.987	0.989	0.991
42	0.922	0.930	0.942	0.951	0.972	0.985	0.987	0.989	0.991
43	0.923	0.932	0.943	0.951	0.973	0.985	0.987	0.990	0.991
44	0.924	0.933	0.944	0.952	0.973	0.985	0.987	0.990	0.991
45	0.926	0.934	0.945	0.953	0.973	0.985	0.988	0.990	0.991
46	0.927	0.935	0.945	0.953	0.974	0.985	0.988	0.990	0.991
47	0.928	0.936	0.946	0.954	0.974	0.985	0.988	0.990	0.991

Table A18 (Continued)

n	0.01	0.02	0.05	0.10	0.50	0.90	0.95	0.98	0.99
48	0.929	0.937	0.947	0.954	0.974	0.985	0.988	0.990	0.991
49	0.929	0.937	0.947	0.955	0.974	0.985	0.988	0.990	0.991
50	0.930	0.938	0.947	0.955	0.974	0.985	0.988	0.990	0.991

SOURCE. Reprinted from Pearson and Hartley (1972), with permission from the *Biometrika* Trustees.

[a] The entries in this table are quantiles w_p of the Shapiro–Wilk test statistic given by Equation 6.2.9. Reject H_0 at the level p if $T_3 < w_p$.

Table A19 A Method for Converting the Shapiro-Wilk Statistic to Approximate Normality

v \ n (d_n)	3 (0.7500)	4 (0.6297)	5 (0.5521)	6 (0.4963)	v n (d_n)	3 (0.7500)	4 (0.6297)	5 (0.5521)	6 (0.4963)
−7.0	−3.29	—	—	—	**2.2**	0.52	0.74	0.75	0.64
−5.4	−2.81	—	—	—	**2.6**	0.67	1.00	1.09	1.06
−5.0	−2.68	—	—	—	**3.0**	0.81	1.23	1.40	1.45
−4.6	−2.54	—	—	—	**3.4**	0.95	1.44	1.67	1.83
−4.2	−2.40	—	—	—	**3.8**	1.07	1.65	1.91	2.17
−3.8	−2.25	−3.50	—	—	**4.2**	1.19	1.85	2.15	2.50
−3.4	−2.10	−3.27	—	—	**4.6**	1.31	2.03	2.47	2.77
−3.0	−1.94	−3.05	−4.01	—	**5.0**	1.42	2.19	2.85	3.09
−2.6	−1.77	−2.84	−3.70	—	**5.4**	1.52	2.34	3.24	3.54
−2.2	−1.59	−2.64	−3.38	—	**5.8**	1.62	2.48	3.64	—
−1.8	−1.40	−2.44	−3.11	—	**6.2**	1.72	2.62	—	—
−1.4	−1.21	−2.22	−2.87	—	**6.6**	1.81	2.75	—	—
−1.0	−1.01	−1.96	−2.56	−3.72	**7.0**	1.90	2.87	—	—
−0.6	−0.80	−1.66	−2.20	−2.88	**7.4**	1.98	2.97	—	—
−0.2	−0.60	−1.31	−1.81	−2.27	**7.8**	2.07	3.08	—	—
0.2	−0.39	−0.94	−1.41	−1.85	**8.2**	2.15	3.22	—	—
0.6	−0.19	−0.57	−0.97	−1.38	**8.6**	2.23	3.36	—	—
1.0	−0.00	−0.19	−0.51	−0.84	**9.0**	2.31	—	—	—
1.4	0.18	0.15	−0.06	−0.33	**9.4**	2.38	—	—	—
1.8	0.35	0.45	0.37	0.18	**9.8**	2.45	—	—	—

SOURCE. Reprinted from Pearson and Hartley (1972), with permission from the *Biometrika* Trustees.

For $3 \le n \le 6$, first compute $v = \ln[(T - d_n)/(1 - T)]$ where d_n is given at the top of the table and T is the Shapiro-Wilk statistic. Then enter the table with v and n to find G, which is approximately normal.

For $7 \le n \le 50$, enter the table with n to find the coefficients b_n, c_n, and d_n. Then compute

$$G = b_n + c_n \ln\{(T - d_n)/(1 - T)\}$$

which is approximately standard normal.

n	b_n	c_n	d_n	n	b_n	c_n	d_n
7	−2.356	1.245	0.4533	29	−6.074	1.934	0.1907
8	−2.696	1.333	0.4186	30	−6.150	1.949	0.1872
9	−2.968	1.400	0.3900				
10	−3.262	1.471	0.3600	31	−6.248	1.965	0.1840
				32	−6.324	1.976	0.1811
11	−3.485	1.515	0.3451	33	−6.402	1.988	0.1781
12	−3.731	1.571	0.3270	34	−6.480	2.000	0.1755
13	−3.936	1.613	0.3111	35	−6.559	2.012	0.1727
14	−4.155	1.655	0.2969				
15	−4.373	1.695	0.2842	36	−6.640	2.024	0.1702
				37	−6.721	2.037	0.1677
16	−4.567	1.724	0.2727	38	−6.803	2.049	0.1656
17	−4.713	1.739	0.2622	39	−6.887	2.062	0.1633
18	−4.885	1.770	0.2528	40	−6.961	2.075	0.1612
19	−5.018	1.786	0.2440				
20	−5.153	1.802	0.2359	41	−7.035	2.088	0.1591
				42	−7.111	2.101	0.1572
21	−5.291	1.818	0.2264	43	−7.188	2.114	0.1552
22	−5.413	1.835	0.2207	44	−7.266	2.128	0.1534
23	−5.508	1.848	0.2157	45	−7.345	2.141	0.1516
24	−5.605	1.862	0.2106				
25	−5.704	1.876	0.2063	46	−7.414	2.155	0.1499
				47	−7.484	2.169	0.1482
26	−5.803	1.890	0.2020	48	−7.555	2.183	0.1466
27	−5.905	1.905	0.1980	49	−7.615	2.198	0.1451
28	−5.988	1.919	0.1943	50	−7.677	2.212	0.1436

Table A20 QUANTILES OF THE SMIRNOV TEST STATISTIC FOR TWO SAMPLES OF EQUAL SIZE n [a]

One-Sided Test: p = .90	.95	.975	.99	.995		One-Sided Test: p = .90	.95	.975	.99	.995
Two-Sided Test: p = .80	.90	.95	.98	.99		Two-sided Test: p = .80	.90	.95	.98	.99
$n = 3$ 2/3	2/3				$n = 20$ 6/20	7/20	8/20	9/20	10/20
4 3/4	3/4	3/4			21 6/21	7/21	8/21	9/21	10/21
5 3/5	3/5	4/5	4/5	4/5	22 7/22	8/22	8/22	10/22	10/22
6 3/6	4/6	4/6	5/6	5/6	23 7/23	8/23	9/23	10/23	10/23
7 4/7	4/7	5/7	5/7	5/7	24 7/24	8/24	9/24	10/24	11/24
8 4/8	4/8	5/8	5/8	6/8	25 7/25	8/25	9/25	10/25	11/25
9 4/9	5/9	5/9	6/9	6/9	26 7/26	8/26	9/26	10/26	11/26
10 4/10	5/10	6/10	6/10	7/10	27 7/27	8/27	9/27	11/27	11/27
11 5/11	5/11	6/11	7/11	7/11	28 8/28	9/28	10/28	11/28	12/28
12 5/12	5/12	6/12	7/12	7/12	29 8/29	9/29	10/29	11/29	12/29
13 5/13	6/13	6/13	7/13	8/13	30 8/30	9/30	10/30	11/30	12/30
14 5/14	6/14	7/14	7/14	8/14	31 8/31	9/31	10/31	11/31	12/31
15 5/15	6/15	7/15	8/15	8/15	32 8/32	9/32	10/32	12/32	12/32
16 6/16	6/16	7/16	8/16	9/16	34 8/34	10/34	11/34	12/34	13/34
17 6/17	7/17	7/17	8/17	9/17	36 9/36	10/36	11/36	12/36	13/36
18 6/18	7/18	8/18	9/18	9/19	38 9/38	10/38	11/38	13/38	14/38
19 6/19	7/19	8/19	9/19	9/19	40 9/40	10/40	12/40	13/40	14/40
					Approximation for $n > 40$: $\dfrac{1.52}{\sqrt{n}}$	$\dfrac{1.73}{\sqrt{n}}$	$\dfrac{1.92}{\sqrt{n}}$	$\dfrac{2.15}{\sqrt{n}}$	$\dfrac{2.30}{\sqrt{n}}$

SOURCE. Adapted from Birnbaum and Hall (1960), with permission from the Institute of Mathematical Statistics.

[a] The entries in this table are selected quantiles w_p of the Smirnov two-sample test statistic T defined by Equations 6.3.2 and 6.3.3 for the one-tailed test and defined by Equation 6.3.1 for the two-tailed test. Reject H_0 at the level α if T exceeds the $1 - \alpha$ quantile of T as given in this table. The test statistic is a discrete random variable, so the exact level of significance may be less than the apparent α used in this table.

Table A21 QUANTILES OF THE SMIRNOV TEST STATISTIC FOR TWO SAMPLES OF DIFFERENT SIZE n AND m[a]

One-Sided Test: Two-Sided Test:	$p = .90$ $p = .80$	.95 .90	.975 .95	.99 .98	.995 .99
$N_1 = 1$ $N_2 = 9$	17/18				
10	9/10				
$N_1 = 2$ $N_2 = 3$	5/6				
4	3/4				
5	4/5	4/5			
6	5/6	5/6			
7	5/7	6/7			
8	3/4	7/8	7/8		
9	7/9	8/9	8/9		
10	7/10	4/5	9/10		
$N_1 = 3$ $N_2 = 4$	3/4	3/4			
5	2/3	4/5	4/5		
6	2/3	2/3	5/6		
7	2/3	5/7	6/7	6/7	
8	5/8	3/4	3/4	7/8	
9	2/3	2/3	7/9	8/9	8/9
10	3/5	7/10	4/5	9/10	9/10
12	7/12	2/3	3/4	5/6	11/12
$N_1 = 4$ $N_2 = 5$	3/5	3/4	4/5	4/5	
6	7/12	2/3	3/4	5/6	5/6
7	17/28	5/7	3/4	6/7	6/7
8	5/8	5/8	3/4	7/8	7/8
9	5/9	2/3	3/4	7/9	8/9
10	11/20	13/20	7/10	4/5	4/5
12	7/12	2/3	2/3	3/4	5/6
16	9/16	5/8	11/16	3/4	13/16
$N_1 = 5$ $N_2 = 6$	3/5	2/3	2/3	5/6	5/6
7	4/7	23/35	5/7	29/35	6/7
8	11/20	5/8	27/40	4/5	4/5
9	5/9	3/5	31/45	7/9	4/5
10	1/2	3/5	7/10	7/10	4/5
15	8/15	3/5	2/3	11/15	11/15
20	1/2	11/20	3/5	7/10	3/4
$N_1 = 6.$ $N_2 = 7$	23/42	4/7	29/42	5/7	5/6
8	1/2	7/12	2/3	3/4	3/4
9	1/2	5/9	2/3	13/18	7/9
10	1/2	17/30	19/30	7/10	11/15
12	1/2	7/12	7/12	2/3	3/4
18	4/9	5/9	11/18	2/3	13/18
24	11/24	1/2	7/12	5/8	2/3

One-Sided Test:			$p = .90$	.95	.975	.99	.995
Two-Sided Test:			$p = .80$	.90	.95	.98	.99
$N_1 = 7$	$N_2 =$	8	27/56	33/56	5/8	41/56	3/4
		9	31/63	5/9	40/63	5/7	47/63
		10	33/70	39/70	43/70	7/10	5/7
		14	3/7	1/2	4/7	9/14	5/7
		28	3/7	13/28	15/28	17/28	9/14
$N_1 = 8$	$N_2 =$	9	4/9	13/24	5/8	2/3	3/4
		10	19/40	21/40	23/40	27/40	7/10
		12	11/24	1/2	7/12	5/8	2/3
		16	7/16	1/2	9/16	5/8	5/8
		32	13/32	7/16	1/2	9/16	19/32
$N_1 = 9$	$N_2 =$	10	7/15	1/2	26/45	2/3	31/45
		12	4/9	1/2	5/9	11/18	2/3
		15	19/45	22/45	8/15	3/5	29/45
		18	7/18	4/9	1/2	5/9	11/18
		36	13/36	5/12	17/36	19/36	5/9
$N_1 = 10$	$N_2 =$	15	2/5	7/15	1/2	17/30	19/30
		20	2/5	9/20	1/2	11/20	3/5
		40	7/20	2/5	9/20	1/2	
$N_1 = 12$	$N_2 =$	15	23/60	9/20	1/2	11/20	7/12
		16	3/8	7/16	23/48	13/24	7/12
		18	13/36	5/12	17/36	19/36	5/9
		20	11/30	5/12	7/15	31/60	17/30
$N_1 = 15$	$N_2 = 20$		7/20	2/5	13/30	29/60	31/60
$N_1 = 16$	$N_2 = 20$		27/80	31/80	17/40	19/40	41/80
Large-sample approximation			$1.07\sqrt{\dfrac{m+n}{mn}}$	$1.22\sqrt{\dfrac{m+n}{mn}}$	$1.36\sqrt{\dfrac{m+n}{mn}}$	$1.52\sqrt{\dfrac{m+n}{mn}}$	$1.63\sqrt{\dfrac{m+n}{mn}}$

SOURCE. Adapted from Massey (1952), with permission from the Institute of Mathematical Statistics.

[a] The entries in this table are selected quantities w_p of the Smirnov test statistic T for two samples, defined by Equations 6.3.1, 6.3.2, and 6.3.3. To enter the table let N_1 be the smaller sample size and let N_2 be the larger sample size. Reject H_0 at the level α if T exceeds $w_{1-\alpha}$ as given in the table. If n and m are not covered by this table, use the large sample approximation given at the end of the table, or consult exact tables by Kim and Jennrich, which appear in Harter and Owen (1970) for $n, m \le 100$.

Table A22 QUANTILES OF THE BIRNBAUM-HALL, OR THREE-SAMPLE SMIRNOV, STATISTIC[a]

	$p = .80$	.90	.95	.98	.99
$n = 4$	3/4	3/4			
5	3/5	4/5	4/5		
6	4/6	4/6	5/6	5/6	5/6
7	4/7	5/7	5/7	6/7	6/7
8	5/8	5/8	5/8	6/8	6/8
9	5/9	5/9	6/9	6/9	7/9
10	5/10	6/10	6/10	7/10	7/10
11	5/11	6/11	7/11	7/11	8/11
12	6/12	6/12	7/12	8/12	8/12
13	6/13	7/13	7/13	8/13	8/13
14	6/14	7/14	8/14	8/14	9/14
15	6/15	7/15	8/15	9/15	9/15
16	7/16	7/16	8/16	9/16	9/16
17	7/17	8/17	8/17	9/17	10/17
18	7/18	8/18	9/18	9/18	10/18
19	7/19	8/19	9/19	10/19	10/19
20	7/20	8/20	9/20	10/20	11/20
22	8/22	9/22	10/22	11/22	11/22
24	8/24	9/24	10/24	11/24	12/24
26	9/26	10/26	10/26	11/26	12/26
28	9/28	10/28	11/28	12/28	13/28
30	9/30	10/30	11/30	12/30	13/30
32	10/32	11/32	12/32	13/32	14/32
34	10/34	11/34	12/34	13/34	14/34
36	10/36	11/36	12/36	14/36	14/36
38	10/38	12/38	13/38	14/38	15/38
40	11/40	12/40	13/40	14/40	15/40
Approximation for $n > 40$	$2.02/\sqrt{n}$	$2.18/\sqrt{n}$	$2.34/\sqrt{n}$	$2.53/\sqrt{n}$	$2.66/\sqrt{n}$

SOURCE. The entries for $n \le 40$ were obtained from Table 1 of Birnbaum and Hall (1960), with permission from the Institute of Mathematical Statistics. The large sample approximations result from a conjecture by the author.

[a] The entries in this table are selected quantiles w_p of the three-sample Smirnov statistic T_1 given by Equation 6.4.1. All three samples are of size n. Reject H_0 at the level α if T_1 exceeds $w_{1-\alpha}$ as given in this table.

Table A23 Quantiles of the One-Sided k-Sample Smirnov Statistic[a]

	$k = 2$					$k = 3$					$k = 4$				
$p =$	.90	.95	.975	.99	.995	.90	.95	.975	.99	.995	.90	.95	.975	.99	.995
$n = 2$	2	2													
3	3	3	3			2									
4	3	4	4	4	4	3	3	4	4		3	3			
5	3	4	5	5	5	3	4	5	5	5	4	4	4	5	5
6	3	4	5	5	6	4	4	5	5	6	4	4	5	6	6
7	4	4	5	6	6	4	5	5	6	6	4	5	5	6	6
8	4	5	5	6	7	4	5	6	6	7	5	5	6	7	7
9	4	5	6	6	7	5	5	6	7	7	5	6	6	7	7
10	4	5	6	7	8	5	6	7	7	8	5	6	7	8	8
12	5	6	7	7	8	5	6	7	8	8	6	6	8	8	9
14	5	6	7	8	9	6	7	8	9	9	6	7	8	9	9
16	6	7	8	9	9	6	7	8	9	10	7	8	9	10	10
18	6	7	8	9	10	7	8	9	10	10	7	8	9	10	11
20	6	8	9	10	10	7	8	10	11	11	8	9	10	11	12
25	7	9	10	11	11	8	9	11	12	12	9	10	11	12	13
30	8	9	10	12	12	9	10	12	13	13	10	10	12	14	14
35	8	10	11	13	13	10	11	13	14	14	10	11	13	15	15
40	9	10	12	13	14	10	12	14	15	15	11	12	14	15	16
45	10	11	12	14	15	11	12	14	16	16	12	13	15	15	16
50	10	12	13	15	16	12	13	15	16	17	13	14	15	16	17
Approximation for $n > 50$	$\dfrac{1.52}{\sqrt{n}}$	$\dfrac{1.73}{\sqrt{n}}$	$\dfrac{1.92}{\sqrt{n}}$	$\dfrac{2.15}{\sqrt{n}}$	$\dfrac{2.30}{\sqrt{n}}$	$\dfrac{1.73}{\sqrt{n}}$	$\dfrac{1.92}{\sqrt{n}}$	$\dfrac{2.09}{\sqrt{n}}$	$\dfrac{2.30}{\sqrt{n}}$	$\dfrac{2.45}{\sqrt{n}}$	$\dfrac{1.85}{\sqrt{n}}$	$\dfrac{2.02}{\sqrt{n}}$	$\dfrac{2.19}{\sqrt{n}}$	$\dfrac{2.39}{\sqrt{n}}$	$\dfrac{2.53}{\sqrt{n}}$

[a] entries in this table, divided by n are selected quantiles of the one-sided k-sample Smirnov statistic T_2 defined by Equation 6.4.2. To obtain the p quantile, enter the table with the number of samples k and the sample size n. The value obtained from the table is then divided by n to become the p quantile. Reject H_0 if T_2 exceeds the q quantile. The approximate quantiles given at the bottom of the table require only division by $\sqrt{n}$, as indicated.

Table A23 (Continued)

		k = 5					k = 6					k = 7		
p = .90	.95	.975	.99	.995	.90	.95	.975	.99	.995	.90	.95	.975	.99	.995
n = 2														
3														
4	3					3					3				
5	4	4	4			4	4	4			4	4	4		
6	4	5	5	5	5	4	5	5	5		4	5	5		
7	5	5	5	6	6	5	5	5	6	6	5	5	5	5	6
8	5	5	6	6	6	5	5	6	6	7	5	6	6	6	7
9	5	6	6	7	7	5	6	6	7	7	5	6	6	7	7
10	6	6	6	7	7	6	6	7	7	8	6	6	7	7	8
12	6	7	7	8	8	6	7	7	8	8	6	7	7	8	8
14	7	7	8	8	9	7	7	8	9	9	7	8	8	9	9
16	7	8	8	9	10	7	8	9	9	10	8	8	9	9	10
18	8	8	9	10	10	8	9	9	10	10	8	9	9	10	11
20	8	9	9	10	11	8	9	10	10	11	8	9	10	10	11
25	9	10	11	12	12	9	10	11	12	12	9	10	11	12	13
30	10	11	12	13	14	10	11	12	13	14	10	11	12	13	14
35	11	12	13	14	15	11	12	13	14	15	11	12	13	14	15
40	12	13	14	15	16	12	13	14	15	16	12	13	14	15	16
45	12	13	15	16	17	13	14	15	16	17	13	14	15	16	17
50	13	14	15	17	18	13	15	16	17	18	14	15	16	17	18
Approximation for n > 50	$\frac{1.92}{\sqrt{n}}$	$\frac{2.09}{\sqrt{n}}$	$\frac{2.25}{\sqrt{n}}$	$\frac{2.45}{\sqrt{n}}$	$\frac{2.59}{\sqrt{n}}$	$\frac{1.97}{\sqrt{n}}$	$\frac{2.14}{\sqrt{n}}$	$\frac{2.30}{\sqrt{n}}$	$\frac{2.49}{\sqrt{n}}$	$\frac{2.63}{\sqrt{n}}$	$\frac{2.02}{\sqrt{n}}$	$\frac{2.18}{\sqrt{n}}$	$\frac{2.34}{\sqrt{n}}$	$\frac{2.53}{\sqrt{n}}$	$\frac{2.66}{\sqrt{n}}$

Table A23 (CONTINUED)

n	k = 8 .90	.95	.975	.99	.995	k = 9 .90	.95	.975	.99	.995	k = 10 .90	.95	.975	.99	.995
n = 2															
3	3														
4	4					4									
5	4	5				5	5								
6	5	5	5			5	5	5			4	4			
7	5	6	6	5	6	5	6	6	5	6	5	5	5	5	6
8	5	6	6	6	7	6	6	6	6	7	5	5	6	6	7
9	6	6	7	7	7	6	6	7	7	7	5	6	6	7	7
10	6	7	8	7	8	7	7	8	7	8	6	6	7	7	8
12	7	8	8	8	9	7	8	8	8	9	6	7	7	8	9
14	7	8	9	9	9	8	8	9	9	9	7	7	8	9	9
16	8	8	9	10	10	8	8	9	10	10	7	8	8	10	10
18	8	9	10	10	11	9	9	10	10	11	8	8	9	10	11
20	9	9	10	11	11	9	9	10	11	11	8	9	10	11	12
25	10	11	11	12	13	10	11	11	12	13	9	10	10	12	13
30	11	12	12	13	14	11	12	13	14	14	10	11	12	14	14
35	12	13	13	15	15	12	13	14	15	15	11	12	13	15	16
40	12	13	14	16	16	13	14	15	16	17	12	13	14	16	17
45	13	14	15	17	17	13	15	16	17	18	13	14	15	17	18
50	14	15	16	17	18	14	15	16	18	19	14	15	16	18	19
Approximation for n > 50	$\frac{2.05}{\sqrt{n}}$	$\frac{2.22}{\sqrt{n}}$	$\frac{2.37}{\sqrt{n}}$	$\frac{2.55}{\sqrt{n}}$	$\frac{2.69}{\sqrt{n}}$	$\frac{2.09}{\sqrt{n}}$	$\frac{2.25}{\sqrt{n}}$	$\frac{2.40}{\sqrt{n}}$	$\frac{2.58}{\sqrt{n}}$	$\frac{2.72}{\sqrt{n}}$	$\frac{2.11}{\sqrt{n}}$	$\frac{2.27}{\sqrt{n}}$	$\frac{2.42}{\sqrt{n}}$	$\frac{2.61}{\sqrt{n}}$	$\frac{2.74}{\sqrt{n}}$

Table A24 Quantiles of the Two-Sided k-Sided Smirnov Statistic[a]

	$p = .90$	$p = .95$	$p = .975$	$p = .99$	$p = .995$
$n = 3$	2 ($k = 2$)				
$n = 4$	3 ($2 \leq k \leq 6$)	3 ($k = 2$)			
$n = 5$	3 ($k = 2$); 4 ($3 \leq k \leq 10$)	4 ($2 \leq k \leq 10$)	4 ($2 \leq k \leq 4$)		
$n = 6$	4 ($2 \leq k \leq 8$); 5 ($k = 9, 10$)	4 ($k = 2, 3$); 5 ($4 \leq k \leq 10$)	4 ($k = 2$); 5 ($3 \leq k \leq 5$)	4 ($k = 2$)	5 ($k = 2, 3$)
$n = 7$	4 ($2 \leq k \leq 4$); 5 ($5 \leq k \leq 10$)	4 ($k = 2$); 5 ($3 \leq k \leq 10$)	5 ($2 \leq k \leq 5$); 6 ($6 \leq k \leq 10$)	5 ($2 \leq k \leq 6$)	6 ($2 \leq k \leq 10$)
$n = 8$	4 ($k = 2$); 5 ($3 \leq k \leq 10$)	5 ($2 \leq k \leq 6$); 6 ($7 \leq k \leq 10$)	5 ($k = 2$); 6 ($3 \leq k \leq 10$)	5 ($k = 2$); 6 ($3 \leq k \leq 10$)	6 ($k = 2, 3$); 7 ($4 \leq k \leq 10$)
$n = 9$	4 ($k = 2$); 5 ($3 \leq k \leq 10$)	5 ($k = 2, 3$); 6 ($4 \leq k \leq 10$)	6 ($2 \leq k \leq 9$); 7 ($k = 10$)	6 ($2 \leq k \leq 7$); 7 ($8 \leq k \leq 10$)	7 ($2 \leq k \leq 10$)
$n = 10$	5 ($2 \leq k \leq 6$); 6 ($7 \leq k \leq 10$)	5 ($k = 2$); 6 ($3 \leq k \leq 10$)	6 ($2 \leq k \leq 5$); 7 ($6 \leq k \leq 10$)	6 ($k = 2, 3$); 7 ($4 \leq k \leq 10$)	7 ($2 \leq k \leq 4$); 8 ($5 \leq k \leq 10$)
$n = 12$	5 ($k = 2, 3$); 6 ($4 \leq k \leq 10$)	6 ($2 \leq k \leq 4$); 7 ($5 \leq k \leq 10$)	6 ($k = 2$); 7 ($3 \leq k \leq 10$)	7 ($k = 2, 3$); 8 ($4 \leq k \leq 10$)	8 ($2 \leq k \leq 7$); 9 ($8 \leq k \leq 10$)
$n = 14$	6 ($2 \leq k \leq 7$); 7 ($8 \leq k \leq 10$)	6 ($k = 2$); 7 ($3 \leq k \leq 10$)	7 ($k = 2, 3$); 8 ($4 \leq k \leq 10$)	8 ($2 \leq k \leq 5$); 9 ($6 \leq k \leq 10$)	8 ($k = 2$); 9 ($3 \leq k \leq 10$)
$n = 16$	6 ($k = 2, 3$); 7 ($4 \leq k \leq 10$)	7 ($2 \leq k \leq 5$); 8 ($6 \leq k \leq 10$)	8 ($2 \leq k \leq 8$); 9 ($k = 9, 10$)	8 ($k = 2$); 9 ($3 \leq k \leq 10$)	9 ($2 \leq k \leq 4$); 10 ($5 \leq k \leq 10$)

[a] The entries in this table, divided by n, are selected quantiles of the two-sided k-sample Smirnov statistic T_3 defined by Equation 6.4.3. To obtain the p quantile of T_3 enter the table with n, the number of observations in each of the k samples, and obtain the entry listed with the number of samples k and under the desired p. The value obtained from the table is then divided by n to become the p quantile. Reject H_0 if T_3 exceeds the p quantile. The approximate quantiles given at the end of the table require only division by $\sqrt{n}$, as indicated, and are valid for all values of k.

Table A24 (Continued)

	$p = .90$	$p = .95$	$p = .975$	$p = .99$	$p = .995$
$n = 18$	6 ($k = 2$) 7 ($3 \le k \le 10$)	7 ($k = 2$) 8 ($3 \le k \le 10$)	8 ($2 \le k \le 4$) 9 ($5 \le k \le 10$)	9 ($2 \le k \le 4$) 10 ($5 \le k \le 10$)	10 ($2 \le k \le 9$) 11 ($k = 10$)
$n = 20$	7 ($2 \le k \le 6$) 8 ($7 \le k \le 10$)	8 ($2 \le k \le 7$) 9 ($8 \le k \le 10$)	8 ($k = 2$) 9 ($3 \le k \le 10$)	9 ($k = 2$) 10 ($3 \le k \le 10$)	10 ($k = 2, 3$) 11 ($4 \le k \le 10$)
$n = 25$	8 ($2 \le k \le 8$)	9 ($2 \le k \le 8$)	9 ($k = 2$) 10 ($3 \le k \le 9$)	11 ($2 \le k \le 8$)	11 ($k = 2$)
$n = 30$	9 ($k = 9, 10$) 8 ($k = 2$)	10 ($k = 9, 10$) 9 ($k = 2$)	11 ($k = 10$) 10 ($k = 2$)	12 ($k = 9, 10$) 12 ($2 \le k \le 8$)	12 ($3 \le k \le 10$) 12 ($k = 2$)
$n = 35$	9 ($3 \le k \le 10$) 9 ($2 \le k \le 4$)	10 ($3 \le k \le 10$) 10 ($k = 2, 3$)	11 ($3 \le k \le 10$) 11 ($k = 2$)	13 ($k = 9, 10$) 13 ($2 \le k \le 8$)	13 ($3 \le k \le 10$) 13 ($k = 2$)
$n = 40$	10 ($5 \le k \le 10$) 10 ($2 \le k \le 8$)	11 ($4 \le k \le 10$) 11 ($2 \le k \le 5$)	12 ($3 \le k \le 10$) 12 ($k = 2, 3$)	14 ($k = 9, 10$) 13 ($k = 2$)	14 ($3 \le k \le 10$) 14 ($k = 2$)
$n = 45$	11 ($k = 9, 10$) 10 ($k = 2, 3$)	12 ($6 \le k \le 10$) 12 ($2 \le k \le 8$)	13 ($4 \le k \le 10$) 13 ($2 \le k \le 5$)	14 ($3 \le k \le 10$) 14 ($k = 2$)	15 ($3 \le k \le 10$) 15 ($k = 2$)
$n = 50$	11 ($4 \le k \le 10$) 11 ($2 \le k \le 6$) 12 ($7 \le k \le 10$)	13 ($k = 9, 10$) 12 ($k = 2, 3$) 13 ($4 \le k \le 10$)	14 ($6 \le k \le 10$) 14 ($2 \le k \le 9$) 15 ($k = 10$)	15 ($3 \le k \le 10$) 15 ($k = 2, 3$) 16 ($4 \le k \le 10$)	16 ($3 \le k \le 10$) 16 ($k = 2, 3$) 17 ($4 \le k \le 10$)
Approxi-mation for $n > 50$	$1.52/\sqrt{n}$	$1.73/\sqrt{n}$	$1.92/\sqrt{n}$	$2.15/\sqrt{n}$	$2.30/\sqrt{n}$

Table A25 The t Distribution[a]

Degrees of Freedom p =	.6	.75	.9	.95	.975	.99	.995	.9975	.999	.9995
1	0.325	1.000	3.078	6.314	12.706	31.821	63.657	127.32	318.31	636.62
2	0.289	0.816	1.886	2.920	4.303	6.965	9.925	14.089	22.327	31.598
3	0.277	0.765	1.638	2.353	3.182	4.541	5.841	7.453	10.214	12.924
4	0.271	0.741	1.533	2.132	2.776	3.747	4.604	5.598	7.173	8.610
5	0.267	0.727	1.476	2.015	2.571	3.365	4.032	4.773	5.893	6.869
6	0.265	0.718	1.440	1.943	2.447	3.143	3.707	4.317	5.208	5.959
7	0.263	0.711	1.415	1.895	2.365	2.998	3.499	4.029	4.785	5.408
8	0.262	0.706	1.397	1.860	2.306	2.896	3.355	3.833	4.501	5.041
9	0.261	0.703	1.383	1.833	2.262	2.821	3.250	3.690	4.297	4.781
10	0.260	0.700	1.372	1.812	2.228	2.764	3.169	3.581	4.144	4.587
11	0.260	0.697	1.363	1.796	2.201	2.718	3.106	3.497	4.025	4.437
12	0.259	0.695	1.356	1.782	2.179	2.681	3.055	3.428	3.930	4.318
13	0.259	0.694	1.350	1.771	2.160	2.650	3.012	3.372	3.852	4.221
14	0.258	0.692	1.345	1.761	2.145	2.624	2.977	3.326	3.787	4.140
15	0.258	0.691	1.341	1.753	2.131	2.602	2.947	3.286	3.733	4.073
16	0.258	0.690	1.337	1.746	2.120	2.583	2.921	3.252	3.686	4.015
17	0.257	0.689	1.333	1.740	2.110	2.567	2.898	3.222	3.646	3.965
18	0.257	0.688	1.330	1.734	2.101	2.552	2.878	3.197	3.610	
19	0.257	0.688	1.328	1.729	2.093	2.539	2.861	3.174	3.579	3.883
20	0.257	0.687	1.325	1.725	2.086	2.528	2.845	3.153	3.552	3.850
21	0.257	0.686	1.323	1.721	2.080	2.518	2.831	3.135	3.527	3.819
22	0.256	0.686	1.321	1.717	2.074	2.508	2.819	3.119	3.505	3.792
23	0.256	0.685	1.319	1.714	2.069	2.500	2.807	3.104	3.485	3.767
24	0.256	0.685	1.318	1.711	2.064	2.492	2.797	3.091	3.467	3.745
25	0.256	0.684	1.316	1.708	2.060	2.485	2.787	3.078	3.450	3.725
26	0.256	0.684	1.315	1.706	2.056	2.479	2.779	3.067	3.435	3.707
27	0.256	0.684	1.314	1.703	2.052	2.473	2.771	3.057	3.421	3.690
28	0.256	0.683	1.313	1.701	2.048	2.467	2.763	3.047	3.408	3.674
29	0.256	0.683	1.311	1.699	2.045	2.462	2.756	3.038	3.396	3.659
30	0.256	0.683	1.310	1.697	2.042	2.457	2.750	3.030	3.385	3.646
40	0.255	0.681	1.303	1.684	2.021	2.423	2.704	2.971	3.307	3.551
60	0.254	0.679	1.296	1.671	2.000	2.390	2.660	2.915	3.232	3.460
120	0.254	0.677	1.289	1.658	1.980	2.358	2.617	2.860	3.160	3.373
∞	0.253	0.674	1.282	1.645	1.960	2.326	2.576	2.807	3.090	3.291

SOURCE. Reprinted from Pearson and Hartley (1970), with permission from the *Biometrika* Trustees.

[a] The entries in this table are quantiles w_p of the t distribution for various degrees of freedom. Quantiles w_p for $p < .5$ may be computed from the equation

$$w_p = -w_{1-p}$$

Note that $w_{0.50} = 0$ for all degrees of freedom.

Table A26 THE F DISTRIBUTION WITH k_1 AND k_2 DEGREES OF FREEDOM (0.75 QUANTILES)

k_2 \ k_1	1	2	3	4	5	6	7	8	9	10	12	15	20	24	30	40	60	120	∞
1	5.83	7.50	8.20	8.58	8.82	8.98	9.10	9.19	9.26	9.32	9.41	9.49	9.58	9.63	9.67	9.71	9.76	9.80	9.85
2	2.57	3.00	3.15	3.23	3.28	3.31	3.34	3.35	3.37	3.38	3.39	3.41	3.43	3.43	3.44	3.45	3.46	3.47	3.48
3	2.02	2.28	2.36	2.39	2.41	2.42	2.43	2.44	2.44	2.44	2.45	2.46	2.46	2.46	2.47	2.47	2.47	2.47	2.47
4	1.81	2.00	2.05	2.06	2.07	2.08	2.08	2.08	2.08	2.08	2.08	2.08	2.08	2.08	2.08	2.08	2.08	2.08	2.08
5	1.69	1.85	1.88	1.89	1.89	1.89	1.89	1.89	1.89	1.89	1.89	1.89	1.88	1.88	1.88	1.88	1.87	1.87	1.87
6	1.62	1.76	1.78	1.79	1.79	1.78	1.78	1.78	1.77	1.77	1.77	1.76	1.76	1.75	1.75	1.75	1.74	1.74	1.74
7	1.57	1.70	1.72	1.72	1.71	1.71	1.70	1.70	1.69	1.69	1.68	1.68	1.67	1.67	1.66	1.66	1.65	1.65	1.65
8	1.54	1.66	1.67	1.66	1.66	1.65	1.64	1.64	1.63	1.63	1.62	1.62	1.61	1.60	1.60	1.59	1.59	1.58	1.58
9	1.51	1.62	1.63	1.63	1.62	1.61	1.60	1.60	1.59	1.59	1.58	1.57	1.56	1.56	1.55	1.54	1.54	1.53	1.53
10	1.49	1.60	1.60	1.59	1.59	1.58	1.57	1.56	1.56	1.55	1.54	1.53	1.52	1.52	1.51	1.51	1.50	1.49	1.48
11	1.47	1.58	1.58	1.57	1.56	1.55	1.54	1.53	1.53	1.52	1.51	1.50	1.49	1.49	1.48	1.47	1.47	1.46	1.45
12	1.46	1.56	1.56	1.55	1.54	1.53	1.52	1.51	1.51	1.50	1.49	1.48	1.47	1.46	1.45	1.45	1.44	1.43	1.42
13	1.45	1.55	1.55	1.53	1.52	1.51	1.50	1.49	1.49	1.48	1.47	1.46	1.45	1.44	1.43	1.42	1.42	1.41	1.40
14	1.44	1.53	1.53	1.52	1.51	1.50	1.49	1.48	1.47	1.46	1.45	1.44	1.43	1.42	1.41	1.41	1.40	1.39	1.38
15	1.43	1.52	1.52	1.51	1.49	1.48	1.47	1.46	1.46	1.45	1.44	1.43	1.41	1.41	1.40	1.39	1.38	1.37	1.36
16	1.42	1.51	1.51	1.50	1.48	1.47	1.46	1.45	1.44	1.44	1.43	1.41	1.40	1.39	1.38	1.37	1.36	1.35	1.34
17	1.42	1.51	1.50	1.49	1.47	1.46	1.45	1.44	1.43	1.43	1.41	1.40	1.39	1.38	1.37	1.36	1.35	1.34	1.33
18	1.41	1.50	1.49	1.48	1.46	1.45	1.44	1.43	1.42	1.42	1.40	1.39	1.38	1.37	1.36	1.35	1.34	1.33	1.32
19	1.41	1.49	1.49	1.47	1.46	1.44	1.43	1.42	1.41	1.41	1.40	1.38	1.37	1.36	1.35	1.34	1.33	1.32	1.30
20	1.40	1.49	1.48	1.47	1.45	1.44	1.43	1.42	1.41	1.40	1.39	1.37	1.36	1.35	1.34	1.33	1.32	1.31	1.29
21	1.40	1.48	1.48	1.46	1.44	1.43	1.42	1.41	1.40	1.39	1.38	1.37	1.35	1.34	1.33	1.32	1.31	1.30	1.28
22	1.40	1.48	1.47	1.45	1.44	1.42	1.41	1.40	1.39	1.39	1.37	1.36	1.34	1.33	1.32	1.31	1.30	1.29	1.28
23	1.39	1.47	1.47	1.45	1.43	1.42	1.41	1.40	1.39	1.38	1.37	1.35	1.34	1.33	1.32	1.31	1.30	1.28	1.27
24	1.39	1.47	1.46	1.44	1.43	1.41	1.40	1.39	1.38	1.38	1.36	1.35	1.33	1.32	1.31	1.30	1.29	1.28	1.26
25	1.39	1.47	1.46	1.44	1.42	1.41	1.40	1.39	1.38	1.37	1.36	1.34	1.33	1.32	1.31	1.29	1.28	1.27	1.25
26	1.38	1.46	1.45	1.44	1.42	1.41	1.39	1.38	1.37	1.37	1.35	1.34	1.32	1.31	1.30	1.29	1.28	1.26	1.25
27	1.38	1.46	1.45	1.43	1.42	1.40	1.39	1.38	1.37	1.36	1.35	1.33	1.32	1.31	1.30	1.28	1.27	1.26	1.24
28	1.38	1.46	1.45	1.43	1.41	1.40	1.39	1.38	1.37	1.36	1.34	1.33	1.31	1.30	1.29	1.28	1.27	1.25	1.24
29	1.38	1.45	1.45	1.43	1.41	1.40	1.38	1.37	1.36	1.35	1.34	1.32	1.31	1.30	1.29	1.27	1.26	1.25	1.23
30	1.38	1.45	1.44	1.42	1.41	1.39	1.38	1.37	1.36	1.35	1.34	1.32	1.30	1.29	1.28	1.27	1.26	1.24	1.23
40	1.36	1.44	1.42	1.40	1.39	1.37	1.36	1.35	1.34	1.33	1.31	1.30	1.28	1.26	1.25	1.24	1.22	1.21	1.19
60	1.35	1.42	1.41	1.38	1.37	1.35	1.33	1.32	1.31	1.30	1.29	1.27	1.25	1.24	1.22	1.21	1.19	1.17	1.15
120	1.34	1.40	1.39	1.37	1.35	1.33	1.31	1.30	1.29	1.28	1.26	1.24	1.22	1.21	1.19	1.18	1.16	1.13	1.10
∞	1.32	1.39	1.37	1.35	1.33	1.31	1.29	1.28	1.27	1.25	1.24	1.22	1.19	1.18	1.16	1.14	1.12	1.08	1.00

SOURCE. Reprinted from Pearson and Hartley (1970), with permission from the *Biometrika* Trustees.

Table A26 (Continued) (0.90 Quantiles)

k_2 \ k_1	1	2	3	4	5	6	7	8	9	10	12	15	20	24	30	40	60	120	∞
1	39.86	49.50	53.59	55.83	57.24	58.20	58.91	59.44	59.86	60.19	60.71	61.22	61.74	62.00	62.26	62.53	62.79	63.06	63.33
2	8.53	9.00	9.16	9.24	9.29	9.33	9.35	9.37	9.38	9.39	9.41	9.42	9.44	9.45	9.46	9.47	9.47	9.48	9.49
3	5.54	5.46	5.39	5.34	5.31	5.28	5.27	5.25	5.24	5.23	5.22	5.20	5.18	5.18	5.17	5.16	5.15	5.14	5.13
4	4.54	4.32	4.19	4.11	4.05	4.01	3.98	3.95	3.94	3.92	3.90	3.87	3.84	3.83	3.82	3.80	3.79	3.78	3.76
5	4.06	3.78	3.62	3.52	3.45	3.40	3.37	3.34	3.32	3.30	3.27	3.24	3.21	3.19	3.17	3.16	3.14	3.12	3.10
6	3.78	3.46	3.29	3.18	3.11	3.05	3.01	2.98	2.96	2.94	2.90	2.87	2.84	2.82	2.80	2.78	2.76	2.74	2.72
7	3.59	3.26	3.07	2.96	2.88	2.83	2.78	2.75	2.72	2.70	2.67	2.63	2.59	2.58	2.56	2.54	2.51	2.49	2.47
8	3.46	3.11	2.92	2.81	2.73	2.67	2.62	2.59	2.56	2.54	2.50	2.46	2.42	2.40	2.38	2.36	2.34	2.32	2.29
9	3.36	3.01	2.81	2.69	2.61	2.55	2.51	2.47	2.44	2.42	2.38	2.34	2.30	2.28	2.25	2.23	2.21	2.18	2.16
10	3.29	2.92	2.73	2.61	2.52	2.46	2.41	2.38	2.35	2.32	2.28	2.24	2.20	2.18	2.16	2.13	2.11	2.08	2.06
11	3.23	2.86	2.66	2.54	2.45	2.39	2.34	2.30	2.27	2.25	2.21	2.17	2.12	2.10	2.08	2.05	2.03	2.00	1.97
12	3.18	2.81	2.61	2.48	2.39	2.33	2.28	2.24	2.21	2.19	2.15	2.10	2.06	2.04	2.01	1.99	1.96	1.93	1.90
13	3.14	2.76	2.56	2.43	2.35	2.28	2.23	2.20	2.16	2.14	2.10	2.05	2.01	1.98	1.96	1.93	1.90	1.88	1.85
14	3.10	2.73	2.52	2.39	2.31	2.24	2.19	2.15	2.12	2.10	2.05	2.01	1.96	1.94	1.91	1.89	1.86	1.83	1.80
15	3.07	2.70	2.49	2.36	2.27	2.21	2.16	2.12	2.09	2.06	2.02	1.97	1.92	1.90	1.87	1.85	1.82	1.79	1.76
16	3.05	2.67	2.46	2.33	2.24	2.18	2.13	2.09	2.06	2.03	1.99	1.94	1.89	1.87	1.84	1.81	1.78	1.75	1.72
17	3.03	2.64	2.44	2.31	2.22	2.15	2.10	2.06	2.03	2.00	1.96	1.91	1.86	1.84	1.81	1.78	1.75	1.72	1.69
18	3.01	2.62	2.42	2.29	2.20	2.13	2.08	2.04	2.00	1.98	1.93	1.89	1.84	1.81	1.78	1.75	1.72	1.69	1.68
19	2.99	2.61	2.40	2.27	2.18	2.11	2.06	2.02	1.98	1.96	1.91	1.86	1.81	1.79	1.76	1.73	1.70	1.67	1.63
20	2.97	2.59	2.38	2.25	2.16	2.09	2.04	2.00	1.96	1.94	1.89	1.84	1.79	1.77	1.74	1.71	1.68	1.64	1.61
21	2.96	2.57	2.36	2.23	2.14	2.08	2.02	1.98	1.95	1.92	1.87	1.83	1.78	1.75	1.72	1.69	1.66	1.62	1.59
22	2.95	2.56	2.35	2.22	2.13	2.06	2.01	1.97	1.93	1.90	1.86	1.81	1.76	1.73	1.70	1.67	1.64	1.60	1.57
23	2.94	2.55	2.34	2.21	2.11	2.05	1.99	1.95	1.92	1.89	1.84	1.80	1.74	1.72	1.69	1.66	1.62	1.59	1.55
24	2.93	2.54	2.33	2.19	2.10	2.04	1.98	1.94	1.91	1.88	1.83	1.78	1.73	1.70	1.67	1.64	1.61	1.57	1.53
25	2.92	2.53	2.32	2.18	2.09	2.02	1.97	1.93	1.89	1.87	1.82	1.77	1.72	1.69	1.66	1.63	1.59	1.56	1.52
26	2.91	2.52	2.31	2.17	2.08	2.01	1.96	1.92	1.88	1.86	1.81	1.76	1.71	1.68	1.65	1.61	1.58	1.54	1.50
27	2.90	2.51	2.30	2.17	2.07	2.00	1.95	1.91	1.87	1.85	1.80	1.75	1.70	1.67	1.64	1.60	1.57	1.53	1.49
28	2.89	2.50	2.29	2.16	2.06	2.00	1.94	1.90	1.87	1.84	1.79	1.74	1.69	1.66	1.63	1.59	1.56	1.52	1.48
29	2.89	2.50	2.28	2.15	2.06	1.99	1.93	1.89	1.86	1.83	1.78	1.73	1.68	1.65	1.62	1.58	1.55	1.51	1.47
30	2.88	2.49	2.28	2.14	2.05	1.98	1.93	1.88	1.85	1.82	1.77	1.72	1.67	1.64	1.61	1.57	1.54	1.50	1.46
40	2.84	2.44	2.23	2.09	2.00	1.93	1.87	1.83	1.79	1.76	1.71	1.66	1.61	1.57	1.54	1.51	1.47	1.42	1.38
60	2.79	2.39	2.18	2.04	1.95	1.87	1.82	1.77	1.74	1.71	1.66	1.60	1.54	1.51	1.48	1.44	1.40	1.35	1.29
120	2.75	2.35	2.13	1.99	1.90	1.82	1.77	1.72	1.68	1.65	1.60	1.55	1.48	1.45	1.41	1.37	1.32	1.26	1.19
∞	2.71	2.30	2.08	1.94	1.85	1.77	1.72	1.67	1.63	1.60	1.55	1.49	1.42	1.38	1.34	1.30	1.24	1.17	1.00

Table A26 (CONTINUED) (0.95 QUANTILES)

k_1 / k_2	1	2	3	4	5	6	7	8	9	10	12	15	20	24	30	40	60	120	∞
1	161.4	199.5	215.7	224.6	230.2	234.0	236.8	238.9	240.5	241.9	243.9	245.9	248.0	249.1	250.1	251.1	252.2	253.3	254.3
2	18.51	19.00	19.16	19.25	19.30	19.33	19.35	19.37	19.38	19.40	19.41	19.43	19.45	19.45	19.46	19.47	19.48	19.49	19.50
3	10.13	9.55	9.28	9.12	9.01	8.94	8.89	8.85	8.81	8.79	8.74	8.70	8.66	8.64	8.62	8.59	8.57	8.55	8.53
4	7.71	6.94	6.59	6.39	6.26	6.16	6.09	6.04	6.00	5.96	5.91	5.86	5.80	5.77	5.75	5.72	5.69	5.66	5.63
5	6.61	5.79	5.41	5.19	5.05	4.95	4.88	4.82	4.77	4.74	4.68	4.62	4.56	4.53	4.50	4.46	4.43	4.40	4.36
6	5.99	5.14	4.76	4.53	4.39	4.28	4.21	4.15	4.10	4.06	4.00	3.94	3.87	3.84	3.81	3.77	3.74	3.70	3.67
7	5.59	4.74	4.35	4.12	3.97	3.87	3.79	3.73	3.68	3.64	3.57	3.51	3.44	3.41	3.38	3.34	3.30	3.27	3.23
8	5.32	4.46	4.07	3.84	3.69	3.58	3.50	3.44	3.39	3.35	3.28	3.22	3.15	3.12	3.08	3.04	3.01	2.97	2.93
9	5.12	4.26	3.86	3.63	3.48	3.37	3.29	3.23	3.18	3.14	3.07	3.01	2.94	2.90	2.86	2.83	2.79	2.75	2.71
10	4.96	4.10	3.71	3.48	3.33	3.22	3.14	3.07	3.02	2.98	2.91	2.85	2.77	2.74	2.70	2.66	2.62	2.58	2.54
11	4.84	3.98	3.59	3.36	3.20	3.09	3.01	2.95	2.90	2.85	2.79	2.72	2.65	2.61	2.57	2.53	2.49	2.45	2.40
12	4.75	3.89	3.49	3.26	3.11	3.00	2.91	2.85	2.80	2.75	2.69	2.62	2.54	2.51	2.47	2.43	2.38	2.34	2.30
13	4.67	3.81	3.41	3.18	3.03	2.92	2.83	2.77	2.71	2.67	2.60	2.53	2.46	2.42	2.38	2.34	2.30	2.25	2.21
14	4.60	3.74	3.34	3.11	2.96	2.85	2.76	2.70	2.65	2.60	2.53	2.46	2.39	2.35	2.31	2.27	2.22	2.18	2.13
15	4.54	3.68	3.29	3.06	2.90	2.79	2.71	2.64	2.59	2.54	2.48	2.40	2.33	2.29	2.25	2.20	2.16	2.11	2.07
16	4.49	3.63	3.24	3.01	2.85	2.74	2.66	2.59	2.54	2.49	2.42	2.35	2.28	2.24	2.19	2.15	2.11	2.06	2.01
17	4.45	3.59	3.20	2.96	2.81	2.70	2.61	2.55	2.49	2.45	2.38	2.31	2.23	2.19	2.15	2.10	2.06	2.01	1.96
18	4.41	3.55	3.16	2.93	2.77	2.66	2.58	2.51	2.46	2.41	2.34	2.27	2.19	2.15	2.11	2.06	2.02	1.97	1.92
19	4.38	3.52	3.13	2.90	2.74	2.63	2.54	2.48	2.42	2.38	2.31	2.23	2.16	2.11	2.07	2.03	1.98	1.93	1.88
20	4.35	3.49	3.10	2.87	2.71	2.60	2.51	2.45	2.39	2.35	2.28	2.20	2.12	2.08	2.04	1.99	1.95	1.90	1.84
21	4.32	3.47	3.07	2.84	2.68	2.57	2.49	2.42	2.37	2.32	2.25	2.18	2.10	2.05	2.01	1.96	1.92	1.87	1.81
22	4.30	3.44	3.05	2.82	2.66	2.55	2.46	2.40	2.34	2.30	2.23	2.15	2.07	2.03	1.98	1.94	1.89	1.84	1.78
23	4.28	3.42	3.03	2.80	2.64	2.53	2.44	2.37	2.32	2.27	2.20	2.13	2.05	2.01	1.96	1.91	1.86	1.81	1.76
24	4.26	3.40	3.01	2.78	2.62	2.51	2.42	2.36	2.30	2.25	2.18	2.11	2.03	1.98	1.94	1.89	1.84	1.79	1.73
25	4.24	3.39	2.99	2.76	2.60	2.49	2.40	2.34	2.28	2.24	2.16	2.09	2.01	1.96	1.92	1.87	1.82	1.77	1.71
26	4.23	3.37	2.98	2.74	2.59	2.47	2.39	2.32	2.27	2.22	2.15	2.07	1.99	1.95	1.90	1.85	1.80	1.75	1.69
27	4.21	3.35	2.96	2.73	2.57	2.46	2.37	2.31	2.25	2.20	2.13	2.06	1.97	1.93	1.88	1.84	1.79	1.73	1.67
28	4.20	3.34	2.95	2.71	2.56	2.45	2.36	2.29	2.24	2.19	2.12	2.04	1.96	1.91	1.87	1.82	1.77	1.71	1.65
29	4.18	3.33	2.93	2.70	2.55	2.43	2.35	2.28	2.22	2.18	2.10	2.03	1.94	1.90	1.85	1.81	1.75	1.70	1.64
30	4.17	3.32	2.92	2.69	2.53	2.42	2.33	2.27	2.21	2.16	2.09	2.01	1.93	1.89	1.84	1.79	1.74	1.68	1.62
40	4.08	3.23	2.84	2.61	2.45	2.34	2.25	2.18	2.12	2.08	2.00	1.92	1.84	1.79	1.74	1.69	1.64	1.58	1.51
60	4.00	3.15	2.76	2.53	2.37	2.25	2.17	2.10	2.04	1.99	1.92	1.84	1.75	1.70	1.65	1.59	1.53	1.47	1.39
120	3.92	3.07	2.68	2.45	2.29	2.17	2.09	2.02	1.96	1.91	1.83	1.75	1.66	1.61	1.55	1.50	1.43	1.35	1.25
∞	3.84	3.00	2.60	2.37	2.21	2.10	2.01	1.94	1.88	1.83	1.75	1.67	1.57	1.52	1.46	1.39	1.32	1.22	1.00

Table A26 (CONTINUED) (0.975 QUANTILES)

k_2 \ k_1	1	2	3	4	5	6	7	8	9	10	12	15	20	24	30	40	60	120	∞
1	647.8	799.5	864.2	899.6	921.8	937.1	948.2	956.7	963.3	963.3	976.7	984.9	993.1	997.2	1001	1006	1010	1014	1018
2	38.51	39.00	39.17	39.25	39.30	39.33	39.36	39.37	39.39	39.40	39.41	39.43	39.45	39.46	39.46	39.46	39.48	39.49	39.50
3	17.44	16.04	15.44	15.10	14.88	14.73	14.62	14.54	14.47	14.42	14.34	14.25	14.17	14.12	14.08	14.04	13.99	13.95	13.90
4	12.22	10.65	9.98	9.60	9.36	9.20	9.07	8.98	8.90	8.84	8.75	8.66	8.56	8.51	8.46	8.41	8.36	8.31	8.26
5	10.01	8.43	7.76	7.39	7.15	6.98	6.85	6.76	6.68	6.62	6.52	6.43	6.33	6.28	6.23	6.18	6.12	6.07	6.02
6	8.81	7.26	6.60	6.23	5.99	5.82	5.70	5.60	5.52	5.46	5.37	5.27	5.17	5.12	5.07	5.01	4.96	4.90	4.85
7	8.07	6.54	5.89	5.52	5.29	5.12	4.99	4.90	4.82	4.76	4.67	4.57	4.47	4.42	4.36	4.31	4.25	4.20	4.14
8	7.57	6.06	5.42	5.05	4.82	4.65	4.53	4.43	4.36	4.30	4.20	4.10	4.00	3.95	3.89	3.84	3.78	3.73	3.67
9	7.21	5.71	5.08	4.72	4.48	4.32	4.20	4.10	4.03	3.96	3.87	3.77	3.67	3.61	3.56	3.51	3.45	3.39	3.33
10	6.94	5.46	4.83	4.47	4.24	4.07	3.95	3.85	3.78	3.72	3.62	3.52	3.42	3.37	3.31	3.26	3.20	3.14	3.08
11	6.72	5.26	4.63	4.28	4.04	3.88	3.76	3.66	3.59	3.53	3.43	3.33	3.23	3.17	3.12	3.06	3.00	2.94	2.88
12	6.55	5.10	4.47	4.12	3.89	3.73	3.61	3.51	3.44	3.37	3.28	3.18	3.07	3.02	2.96	2.91	2.85	2.79	2.72
13	6.41	4.97	4.35	4.00	3.77	3.60	3.48	3.39	3.31	3.25	3.15	3.05	2.95	2.89	2.84	2.78	2.72	2.66	2.60
14	6.30	4.86	4.24	3.89	3.66	3.50	3.38	3.29	3.21	3.15	3.05	2.95	2.84	2.79	2.73	2.67	2.61	2.55	2.49
15	6.20	4.77	4.15	3.80	3.58	3.41	3.29	3.20	3.12	3.06	2.96	2.86	2.76	2.70	2.64	2.59	2.52	2.46	2.40
16	6.12	4.69	4.08	3.73	3.50	3.34	3.22	3.12	3.05	2.99	2.89	2.79	2.68	2.63	2.57	2.51	2.45	2.38	2.32
17	6.04	4.62	4.01	3.66	3.44	3.28	3.16	3.06	2.98	2.92	2.82	2.72	2.62	2.56	2.50	2.44	2.38	2.32	2.25
18	5.98	4.56	3.95	3.61	3.38	3.22	3.10	3.01	2.93	2.87	2.77	2.67	2.56	2.50	2.44	2.38	2.32	2.26	2.19
19	5.92	4.51	3.90	3.56	3.33	3.17	3.05	2.96	2.88	2.82	2.72	2.62	2.51	2.45	2.39	2.33	2.27	2.20	2.13
20	5.87	4.46	3.86	3.51	3.29	3.13	3.01	2.91	2.84	2.77	2.68	2.57	2.46	2.41	2.35	2.29	2.22	2.16	2.09
21	5.83	4.42	3.82	3.48	3.25	3.09	2.97	2.87	2.80	2.73	2.64	2.53	2.42	2.37	2.31	2.25	2.18	2.11	2.04
22	5.79	4.38	3.78	3.44	3.22	3.05	2.93	2.84	2.76	2.70	2.60	2.50	2.39	2.33	2.27	2.21	2.14	2.08	2.00
23	5.75	4.35	3.75	3.41	3.18	3.02	2.90	2.81	2.73	2.67	2.57	2.47	2.36	2.30	2.24	2.18	2.11	2.04	1.97
24	5.72	4.32	3.72	3.38	3.15	2.99	2.87	2.78	2.70	2.64	2.54	2.44	2.33	2.27	2.21	2.15	2.08	2.01	1.94
25	5.69	4.29	3.69	3.35	3.13	2.97	2.85	2.75	2.68	2.61	2.51	2.41	2.30	2.24	2.18	2.12	2.05	1.98	1.91
26	5.66	4.27	3.67	3.33	3.10	2.94	2.82	2.73	2.65	2.59	2.49	2.39	2.28	2.22	2.16	2.09	2.03	1.95	1.88
27	5.63	4.24	3.65	3.31	3.08	2.92	2.80	2.71	2.63	2.57	2.47	2.36	2.25	2.19	2.13	2.07	2.00	1.93	1.85
28	5.61	4.22	3.63	3.29	3.06	2.90	2.78	2.69	2.61	2.55	2.45	2.34	2.23	2.17	2.11	2.05	1.98	1.91	1.83
29	5.59	4.20	3.61	3.27	3.04	2.88	2.76	2.67	2.59	2.53	2.43	2.32	2.21	2.15	2.09	2.03	1.96	1.89	1.81
30	5.57	4.18	3.59	3.25	3.03	2.87	2.75	2.65	2.57	2.51	2.41	2.31	2.20	2.14	2.07	2.01	1.94	1.87	1.79
40	5.42	4.05	3.46	3.13	2.90	2.74	2.62	2.53	2.45	2.39	2.29	2.18	2.07	2.01	1.94	1.88	1.80	1.72	1.64
60	5.29	3.93	3.34	3.01	2.79	2.63	2.51	2.41	2.33	2.27	2.17	2.06	1.94	1.88	1.82	1.74	1.67	1.58	1.48
120	5.15	3.80	3.23	2.89	2.67	2.52	2.39	2.30	2.22	2.16	2.05	1.94	1.82	1.76	1.69	1.61	1.53	1.43	1.31
∞	5.02	3.69	3.12	2.79	2.57	2.41	2.29	2.19	2.11	2.05	1.94	1.83	1.71	1.64	1.57	1.48	1.39	1.27	1.00

Table A26 (Continued) (0.99 Quantiles)

k_2 \ k_1	1	2	3	4	5	6	7	8	9	10	12	15	20	24	30	40	60	120	∞
1	4052	4999.5	5403	5625	5764	5859	5928	5981	6022	6056	6106	6157	6209	6235	6261	6287	6313	6339	6366
2	98.50	99.00	99.17	99.25	99.30	99.33	99.36	99.37	99.39	99.40	99.42	99.43	99.45	99.46	99.47	99.47	99.48	99.49	99.50
3	34.12	30.82	29.46	28.71	28.24	27.91	27.67	27.49	27.35	27.23	27.05	26.87	26.69	26.60	26.50	26.41	26.32	26.22	26.13
4	21.20	18.00	16.69	15.98	15.52	15.21	14.98	14.80	14.66	14.55	14.37	14.20	14.02	13.93	13.84	13.75	13.65	13.56	13.46
5	16.26	13.27	12.06	11.39	10.97	10.67	10.46	10.29	10.16	10.05	9.89	9.72	9.55	9.47	9.38	9.29	9.20	9.11	9.02
6	13.75	10.92	9.78	9.15	8.75	8.47	8.26	8.10	7.98	7.87	7.72	7.56	7.40	7.31	7.23	7.14	7.06	6.97	6.88
7	12.25	9.55	8.45	7.85	7.46	7.19	6.99	6.84	6.72	6.62	6.47	6.31	6.16	6.07	5.99	5.91	5.82	5.74	5.65
8	11.26	8.65	7.59	7.01	6.63	6.37	6.18	6.03	5.91	5.81	5.67	5.52	5.36	5.28	5.20	5.12	5.03	4.95	4.86
9	10.56	8.02	6.99	6.42	6.06	5.80	5.61	5.47	5.35	5.26	5.11	4.96	4.81	4.73	4.65	4.57	4.48	4.40	4.31
10	10.04	7.56	6.55	5.99	5.64	5.39	5.20	5.06	4.94	4.85	4.71	4.56	4.41	4.33	4.25	4.17	4.08	4.00	3.91
11	9.65	7.21	6.22	5.67	5.32	5.07	4.89	4.74	4.63	4.54	4.40	4.25	4.10	4.02	3.94	3.86	3.78	3.69	3.60
12	9.33	6.93	5.95	5.41	5.06	4.82	4.64	4.50	4.39	4.30	4.16	4.01	3.86	3.78	3.70	3.62	3.54	3.45	3.36
13	9.07	6.70	5.74	5.21	4.86	4.62	4.44	4.30	4.19	4.10	3.96	3.82	3.66	3.59	3.51	3.43	3.34	3.25	3.17
14	8.86	6.51	5.56	5.04	4.69	4.46	4.28	4.14	4.03	3.94	3.80	3.66	3.51	3.43	3.35	3.27	3.18	3.09	3.00
15	8.68	6.36	5.42	4.89	4.56	4.32	4.14	4.00	3.89	3.80	3.67	3.52	3.37	3.29	3.21	3.13	3.05	2.96	2.87
16	8.53	6.23	5.29	4.77	4.44	4.20	4.03	3.89	3.78	3.69	3.55	3.41	3.26	3.18	3.10	3.02	2.93	2.84	2.75
17	8.40	6.11	5.18	4.67	4.34	4.10	3.93	3.79	3.68	3.59	3.46	3.31	3.16	3.08	3.00	2.92	2.83	2.75	2.65
18	8.29	6.01	5.09	4.58	4.25	4.01	3.84	3.71	3.60	3.51	3.37	3.23	3.08	3.00	2.92	2.84	2.75	2.66	2.57
19	8.18	5.93	5.01	4.50	4.17	3.94	3.77	3.63	3.52	3.43	3.30	3.15	3.00	2.92	2.84	2.76	2.67	2.58	2.49
20	8.10	5.85	4.94	4.43	4.10	3.87	3.70	3.56	3.46	3.37	3.23	3.09	2.94	2.86	2.78	2.69	2.61	2.52	2.42
21	8.02	5.78	4.87	4.37	4.04	3.81	3.64	3.51	3.40	3.31	3.17	3.03	2.88	2.80	2.72	2.64	2.55	2.46	2.36
22	7.95	5.72	4.82	4.31	3.99	3.76	3.59	3.45	3.35	3.26	3.12	2.98	2.83	2.75	2.67	2.58	2.50	2.40	2.31
23	7.88	5.66	4.76	4.26	3.94	3.71	3.54	3.41	3.30	3.21	3.07	2.93	2.78	2.70	2.62	2.54	2.45	2.35	2.26
24	7.82	5.61	4.72	4.22	3.90	3.67	3.50	3.36	3.26	3.17	3.03	2.89	2.74	2.66	2.58	2.49	2.40	2.31	2.21
25	7.77	5.57	4.68	4.18	3.85	3.63	3.46	3.32	3.22	3.13	2.99	2.85	2.70	2.62	2.54	2.45	2.36	2.27	2.17
26	7.72	5.53	4.64	4.14	3.82	3.59	3.42	3.29	3.18	3.09	2.96	2.81	2.66	2.58	2.50	2.42	2.33	2.23	2.13
27	7.68	5.49	4.60	4.11	3.78	3.56	3.39	3.26	3.15	3.06	2.93	2.78	2.63	2.55	2.47	2.38	2.29	2.20	2.10
28	7.64	5.45	4.57	4.07	3.75	3.53	3.36	3.23	3.12	3.03	2.90	2.75	2.60	2.52	2.44	2.35	2.26	2.17	2.06
29	7.60	5.42	4.54	4.04	3.73	3.50	3.33	3.20	3.09	3.00	2.87	2.73	2.57	2.49	2.41	2.33	2.23	2.14	2.03
30	7.56	5.39	4.51	4.02	3.70	3.47	3.30	3.17	3.07	2.98	2.84	2.70	2.55	2.47	2.39	2.30	2.21	2.11	2.01
40	7.31	5.18	4.31	3.83	3.51	3.29	3.12	2.99	2.89	2.80	2.66	2.52	2.37	2.29	2.20	2.11	2.02	1.92	1.80
60	7.08	4.98	4.13	3.65	3.34	3.12	2.95	2.82	2.72	2.63	2.50	2.35	2.20	2.12	2.03	1.94	1.84	1.73	1.60
120	6.85	4.79	3.95	3.48	3.17	2.96	2.79	2.66	2.56	2.47	2.34	2.19	2.03	1.95	1.86	1.76	1.66	1.53	1.38
∞	6.63	4.61	3.78	3.32	3.02	2.80	2.64	2.51	2.41	2.32	2.18	2.04	1.88	1.79	1.70	1.59	1.47	1.32	1.00

Answers to Odd-Numbered Exercises

CHAPTER 1

Sec. 1.1: 1. 10,000; 3. 24; 5. 220; 7. 105; 9. 14; 11. 232/729.

Sec. 1.2: 1. HHH, HHT, HTH, THH, TTH, THT, HTT, TTT; 3. .85; 5. $9/2^8$; 7. .008; 9. 1/8; 11. 1/7.

Sec. 1.3: 1.(a) 1/729, (b) 64/729, (c) 0, (d) 496/729, (e) 0, (f) 1; 3.(a)1/128, (b) 4/128, (c) 0, (d) 4/128, (e) 0, (f) 1/128, (g) 8/128, (h) 99/128; 5.(a) A, B, C, D, E, F, (b) $P(A) = 1/6$, $P(B) = 1/6$, ..., $P(F) = 1/6$, (c) $X(A) = 1$, $X(B) = 2$, etc.

Sec. 1.4: 1.(a) 7/6, (b) 41/36, (c) 29/6, (d) 1, (e) 0.5, (f) 1; 3.(a) 1/2, (b) 1/2, (c) 1/4, (d) 1, (e) 0, (f) 1/2, (g) 1/2, (h) yes; 5. 2211; 7.(a) 7/2, (b) 35/12, (c) 56/3.

Sec. 1.5: 1.(a) .5, (b) .975, (c) .159, (d) .682, (e) .5, (f) .6745; 3. .001; 5. 207; 7.(a) 9.488, (b) 15.51, (c) 233.7.

CHAPTER 2

Sec. 2.1: 1.(a) all high schools in the U.S.A., (b) all high schools in the Washington, D.C., area, (c) $\binom{100}{5}$, (d) $1 \Big/ \binom{100}{5}$; 3.(a) ordinal, (b) the sum of the points, for awarding the trophy; 5. all nominal; 7. no.

Sec. 2.2: 1.(b) $4300, (c) $8600, (d) $97,296,000, (e) $9863.87; 3. (number of sample values $\leq c)/n$, 1/5.

Sec. 2.3: 1.(a) H_0: the new method is no better than the existing one, H_1: the new method is better than the existing one, (b) the probability of deciding the new method is better, when it is not, (c) the probability of deciding the new method is better when it is better; 3.(a) Fertilizer B is not as good as

Fertilizer A, (b) my opponent is cheating, (c) the occurrence of sun spots does affect the economic cycle; 5. $\alpha = .1875$, power $= .33696$.

Sec 2.4: 1.(a) 1/32, (b) p^5, $p > .5$, (d) yes; 3. yes; 5.(a) 0.57, (b) 1.75

CHAPTER 3

Sec. 3.1: 1. No, $T = 0$, $\hat{\alpha} = .0343$; 3.(a) yes, $\hat{\alpha} = .11$, (b) (0.15, 0.28), (c) (0.15, 0.27); 5. (0.06, 0.44).

Sec. 3.2: 1. $T_1 = 6$, $T_2 = 4$, $\hat{\alpha} = .1154$; 3. $T_1 = 4$, $\hat{\alpha} = .2375$; 5. No, $T_1 = 90$, $T_2 = 90$, $\hat{\alpha} \doteq .02$.

Sec. 3.3: 1.(a) 77, (b) 76.35, so use 77; 3.(a) 15, (b) 14.198, so use 15; 5. 26.

Sec. 3.4: 1. $T = 1$, $\hat{\alpha} = .1094$; 3. $T = 23$, $\hat{\alpha} < .0001$; 5. $T = 60$, $T' = 42$, $\hat{\alpha} = .037$.

Sec. 3.5: 1. $T_1 = 0.5455$, $\hat{\alpha} > .25$; 3. $T = 7$, $T' = 0$, $\hat{\alpha} = 0.0078$; 5. $T = 4$, $\hat{\alpha} = .003$; 7. $T = 0$, $\hat{\alpha} = .0312$.

CHAPTER 4

Sec. 4.1: 1. $T = 0.6395$, $\hat{\alpha} = .427$; 3. $T = 8742$, $\hat{\alpha} = .355$.

Sec. 4.2: 1. 2×3 ($R_1 = 8$, $R_2 = 10$), $T = 11.06$, $\hat{\alpha} = .005$; 3. $T = 6.81$, $\hat{\alpha} = .08$, fixed marginal totals.

Sec. 4.3: 1. Exact $T = 1.10$, approximately $T = 1.20$, $\hat{\alpha} > .25$; 3. $T = 34.875$, $\hat{\alpha} < .001$.

Sec. 4.4: 1.(a) 6.34, (b) .18, (c) 0.0317, (d) 0.244, (e) 0.0634, (f) 0.1780; 3.(a) $R_5 = 0.149$, $\hat{\alpha} = .12$, (b) 0.3636, (c) 0.

Sec. 4.5: The answers depend on the choice of intervals.

Sec. 4.6: 1.(a) $T = 4$, $\hat{\alpha} = .05$, (b) $T = 4$, $\hat{\alpha} = .05$, (c) $T = 2.67$, $\hat{\alpha} = .11$; 3. $T = 10.4$, $\hat{\alpha} = .005$.

CHAPTER 5

Sec. 5.1: 1. $T = 117$, $\hat{\alpha} < .001$; 3. (0, 14).

Sec. 5.2: 1. $T = 8.40$, $\hat{\alpha} < .025$, difference between A and B, A and C at $\alpha = .05$; 3. $T = 40.83$, $\hat{\alpha} < .001$ differences between minimum tillage and terrace, minimum tillage and other, contour and terrace, contour and other at $\alpha = .05$.

Sec. 5.3: 1. $T = 1107.5$, $\hat{\alpha} = .088$; 3. $T = 5.15$, $\hat{\alpha} = .08$.

Sec. 5.4: 1.(a) -0.603, (b) -0.511, (c) $\hat{\alpha} = .07$, (d) $\hat{\alpha} = .037$; 3.(a) $\rho = -0.9025$, $\hat{\alpha} < .001$, (b) $T = -57$, $\hat{\alpha} < .005$.

Sec. 5.5: 1.(b) $A = 8.69$, $B = 13.6$, (d) $\rho = -0.6606$, $\hat{\alpha} = .04$, (e) $(S^{(12)}, S^{(33)}) = (10.7, 16.8)$.

Sec. 5.6: 1.(a) it might be linear in the middle; it certainly looks monotonic, (b) 46.14, (c) 53.51, (d) line segments connecting $(X, Y) = (0.50, 0)$, (0.85, 0), (1.94, 20), (2.75, 40), (3.56, 60), (4.11, 80), (4.92, 100), (5.00, 100).

Sec. 5.7:   1. $T = 3.2640$, $\hat{\alpha} < .002$; 3. $T = 39$, $\hat{\alpha} \doteq .50$; 5. 5.5 to 7.45, 6. 5.

Sec. 5.8:   1.(a) $T_1 = 4.43$, $\hat{\alpha} = .019$, differences between spring and the other three seasons, (b) $T_2 = 2.946$, $\hat{\alpha} = .06$, (c) the larger hospitals received more weight in the first test.

Sec. 5.9:   1. $T = 15.8$, $\hat{\alpha} = .016$, tires 3 and 6 are better than tires 1, 2, or 4.

Sec. 5.10: 1. $T_1 = 8.73$, $\hat{\alpha} = .014$, same differences as with $K - W$ test; 3. $T_2 = 3.31023$, $\hat{\alpha} < .002$, same as before; 5. $T_3 = 0.3530$, $\hat{\alpha} = .362$, the results are weaker; 7. $\rho = -0.7153$, larger in absolute value, $\hat{\alpha} = .032$, slightly smaller.

Sec. 5.11: 1. $T_1 = 6$, $\hat{\alpha} = 6/56$; 3. $T_2 = 20$, $\hat{\alpha} = 1/8$.

CHAPTER 6

Sec. 6.1:   1. $S_n(x) \pm 0.509$; 3. $T = 0.425$, $\hat{\alpha} = .20$.

Sec. 6.2:   1. $T_1 = 0.103$, $\hat{\alpha} > .20$; 3. $T_3 = 0.9569$, $\hat{\alpha} > .50$; 5. $T_2 = 0.2155$, $\hat{\alpha} \doteq .20$; 7. $Z = -1.9083$, $\hat{\alpha} > -.028$.

Sec. 6.3:   1. $T_1^+ = 0.8$, $\hat{\alpha} = .05$; 3. $T_2 = 0.556$, $\hat{\alpha} = .04$.

Sec. 6.4:   1. $T_1 = 5/6$, $\hat{\alpha} = .10$, $T_3 = 5/6$, $\hat{\alpha} = .05$; 3. $T_3 = 0.9$, $\hat{\alpha} = .005$.

Index